Advances in Coal Processing, Utilization, and Process Safety

Advances in Coal Processing, Utilization, and Process Safety

Guest Editors

Feng Du
Leilei Si
Chaojun Fan
Qiming Zhuo

MDPI

Basel • Beijing • Wuhan • Barcelona • Belgrade • Novi Sad • Cluj • Manchester

Guest Editors
Feng Du
School of Emergency Management and Safety Engineering
China University of Mining and Technology-Beijing
Beijing
China

Leilei Si
State Key Laboratory Cultivation Base for Gas Geology and Gas Control
Henan Polytechnic University
Jiaozuo
China

Chaojun Fan
College of Mining
Liaoning Technical University
Fuxin
China

Qiming Zhuo
School of Chemical and Environmental Engineering
China University of Mining and Technology-Beijing
Beijing
China

Editorial Office
MDPI AG
Grosspeteranlage 5
4052 Basel, Switzerland

This is a reprint of the Special Issue, published open access by the journal *Processes* (ISSN 2227-9717), freely accessible at: https://www.mdpi.com/journal/processes/special_issues/01A0J40BNO.

For citation purposes, cite each article independently as indicated on the article page online and as indicated below:

Lastname, A.A.; Lastname, B.B. Article Title. *Journal Name* **Year**, *Volume Number*, Page Range.

ISBN 978-3-7258-8383-7 (Hbk)
ISBN 978-3-7258-8384-4 (PDF)
https://doi.org/10.3390/books978-3-7258-8384-4

Contents

About the Editors

Feng Du

Feng Du is with the School of Emergency Management and Safety Engineering at China University of Mining and Technology—Beijing. His main research interests include mine gas prevention and control, and prevention and control of coal–rock–gas compound dynamic disasters. He has been selected for the Young Talent Support Project of the China Association for Science and Technology, the Young Talent Support Project of the Beijing Association for Science and Technology, and the National Conditions Research Workshop for Young Scientific and Technological Leading Talents under the Pilot Program of the China Association for Science and Technology. In recent years, he has published 50 academic papers as the first or corresponding author in leading journals in the field and has authored 2 monographs. He is currently the principal investigator of multiple national research projects, including the National Natural Science Foundation of China (General Program, Youth Program, and subproject of Key Program), subtasks of the National Science and Technology Major Project and the National Key R&D Program of China, and the China Postdoctoral Science Foundation. He serves as a young Editorial Board Member, Guest Editor of Special Issues, and regular reviewer for several well-known journals. He also holds the position of Deputy Secretary-General of the Beijing Energy and Environment Society. His honors include the Excellent Doctoral Dissertation Award from the China Occupational Safety and Health Association, the Outstanding Undergraduate Graduation Thesis Advisor Award of Beijing Municipal Colleges and Universities, and the Outstanding Advisor Award of the National College Students' Practice and Innovation Competition in Safety Science and Engineering.

Leilei Si

Leilei Si is with the State Key Laboratory Cultivation Base for Gas Geology and Gas Control at Henan Polytechnic University. His main research interests include theory and technology of mine gas prevention and control, occupational health and dust prevention and control, and carbon dioxide geological storage. He has presided over multiple scientific research projects, including the General Program and Youth Program of the National Natural Science Foundation of China, the subproject of the National Key R&D Program of China, the National Postdoctoral Program for Innovative Talents, the Excellent Youth Science Foundation of Henan Province, the Science and Technology Research Project of Henan Province, and the subproject of the Key Consulting Project of the Chinese Academy of Engineering. He has independently published 1 academic monograph, more than 20 SCI/EI-indexed papers as the first or corresponding author, including 2 ESI highly cited papers, and has been granted more than 20 international and domestic invention patents.

Chaojun Fan

Chaojun Fan is with the College of Mining at Liaoning Technical University. His main research interests include mine pressure and strata control, mine safety, and disaster prevention and control, with a focus on mine gas disaster prevention, coal-based carbon dioxide sequestration, and abandoned mine resource development. He has proposed the geological dynamic catastrophe theory of coal and gas outbursts and developed the key "unloading–pressure reduction–impedance" coordinated outburst prevention technology.

He has presided over 15 national and provincial research projects including grants from the National Natural Science Foundation of China and China Postdoctoral Science Foundation. He has published 103 academic papers in journals such as *Energy* and *Journal of China Coal Society*, with 58 as first/corresponding author (2 ESI hot papers, 7 ESI highly cited papers). His works have received 2660 total citations on Google Scholar as of December 2025. He has been granted 21 invention patents and authored 3 monographs. His research achievements have been appraised at international leading and advanced levels, winning more than 10 provincial and ministerial science and technology awards. He has been selected for the Young Talent Support Project of CAST, Liaoning "Xingliao Talents" Young Top Talents, and the "Global Top 2% Scientists" list. He is also a recipient of the National Coal Youth May Fourth Medal and the China Coal Society Youth Science and Technology Award.

Qiming Zhuo

Qiming Zhuo is with the School of Chemical and Environmental Engineering at China University of Mining and Technology—Beijing. His main research interests include deep ash reduction, quality improvement and fine separation of coal, comprehensive utilization of mineral resources, and resource utilization of urban solid waste.

He has presided over and participated in more than 10 research projects, covering the National Natural Science Foundation of China (Youth and General Programs), National Key R&D Program of China, State Key Laboratory Open Fund, Excellent Youth Team Program of Central Universities, Chinese Academy of Engineering Strategic Research Project, national mandatory standard compilation, and enterprise technology development projects.

He has published more than 30 academic papers in leading journals such as *Fuel*, *Applied Surface Science*, *Minerals Engineering*, *Journal of China Coal Society*, and *Coal Science and Technology*, and has been granted more than 10 patents. He serves as an expert for the National Postgraduate Education Evaluation and Monitoring of the Ministry of Education and a regular reviewer for several well-known international journals.

processes

Editorial

Special Issue "Advances in Coal Processing, Utilization, and Process Safety"

Feng Du [1,*], Leilei Si [2], Chaojun Fan [3] and Qiming Zhuo [4]

[1] School of Emergency Management and Safety Engineering, China University of Mining and Technology-Beijing, Beijing 100083, China
[2] State Key Laboratory Cultivation Base for Gas Geology and Gas Control, Henan Polytechnic University, Jiaozuo 454000, China
[3] College of Mining, Liaoning Technical University, Fuxin 123000, China
[4] School of Chemical and Environmental Engineering, China University of Mining and Technology-Beijing, Beijing 100083, China
* Correspondence: fdu@cumtb.edu.cn

Safety in production has always been the top priority in the coal industry. Although renewable energy is growing rapidly, coal resources still occupy an important place in the energy mix of many countries [1]. In recent years, as coal mining has continued to extend into deeper formations, mining conditions have become extremely harsh under the complex deep geological environment characterized by "three highs and one low". The risk of dynamic coal-rock disasters has increased significantly, making it particularly important to ensure the safe extraction of coal [2–4]. Meanwhile, the subsequent processing and utilization of coal are equally important. This concerns not only the efficient conversion and clean use of coal resources, but also serves as an important support for ensuring the safe and sustainable operation of the entire energy industry chain [5].

In response to the needs for safety, efficiency, and green development in the process of coal development and utilization, scholars at home and abroad have carried out extensive research in recent years on mineral processing and the efficient utilization of resources, process safety control, and mine disaster prevention and control. Relevant studies focus not only on new technologies for mineral processing and pathways to improve efficiency [6–8], but also on issues such as dust explosions [9–11], spontaneous coal combustion [12,13], and safety in coal-related processes [14].

Meanwhile, in the field of mine safety engineering and disaster prevention, research on the mechanisms of coal-rock and gas dynamic disasters [15,16]—as well as coalbed methane migration [17]—has shown a development trend toward multi-scale and multi-method interdisciplinary integration. In particular, with the continuous advancement of intelligent prediction and early-warning technologies [18,19] as well as numerical simulation [20], research on risk identification, mechanism analysis, and prevention-and-control optimization throughout the entire process of coal processing, mining, gas drainage, and utilization is being continuously deepened, providing important theoretical foundations and technical support for achieving the safe extraction of coal resources, their clean utilization, and the coordinated development of the industrial chain.

The Special Issue "Advances in Coal Processing, Utilization, and Process Safety" in the journal *Processes* brought together research papers presented at the forum on safe coal mining, mine disaster prevention and control, and coal utilization. Its content covers topics such as mine pressure and surrounding rock control, coalbed methane migration, and the monitoring and early warning of dynamic disasters in coal mines. Many of the studies are interdisciplinary in nature, but their overarching theme is coal mine production safety.

In the area of mine pressure and surrounding rock control, different scholars have revealed the evolution patterns of disasters and the corresponding control mechanisms under deep mining conditions from various engineering perspectives. Liu [21] conducted triaxial permeability tests on three types of coal-rock samples by combining physical simulation with FLAC3D numerical simulation, revealing the coupled evolution mechanism of the stress–fracture–seepage field in the overburden, and providing a basis for the identification and prevention of water inrush risks. Li [22] used PFC2D to analyze the effects of floor stiffness, strength, and joint distribution on the heave mechanism during mining, providing a theoretical basis for the optimal groove-cutting design. Liu [23] used PFC2D to investigate the corresponding evolution of overburden fractures during continuous multi-seam mining, providing a basis for determining the optimal location for roadway construction in coal seams. Shi [24] theoretical analysis and FLAC3D numerical simulation combined with field conditions, proposed a method to address the hazards caused by roof hanging. Wang [25] used FLAC3D to simulate the stress field distribution during coal seam extraction under folded geological conditions and proposed a coordinated control strategy to ensure the stability of the surrounding rock. Zhang [26] systematically studied the coordinated variation mechanism of the stress distribution and plastic zone morphology of the surrounding rock in a roadway on one side of a goaf under mining disturbance by comprehensively applying theoretical analysis, numerical simulation, and case studies. A quantitative evaluation method for assessing the stability of roadway surrounding rock was established. Zhang [27] investigated the mechanical parameters, failure characteristics, acoustic emission characteristic parameters, and energy evolution of specimens through laboratory tests and numerical simulations under different drilling parameters, providing a useful reference for the selection and optimization of pressure-relief drilling parameters in rock burst mines.

In terms of coalbed methane migration, Cheng [28] developed a permeability evolution model and a long short-term memory (LSTM) prediction model, formulated a dynamic regulation strategy for intelligent gas drainage, and carried out related applications. To address the unique CBM reservoir problems associated with superimposed pressure systems, Ren [29] independently developed an experimental apparatus for the joint production of CBM under multiple pressure systems. Chen [30] characterized the pore structures of lignite and anthracite using the liquid nitrogen adsorption method, and investigated the ultramicroscopic structural characteristics and adsorption properties of coal pores. In addition, Zhang [31] investigated the deformation behavior and pore-structure characteristics of lignite from room temperature to 650 °C through high-temperature triaxial experiments, nuclear magnetic resonance, and X-ray computed tomography, providing a theoretical basis for coalbed methane extraction and displacement. Tang [32] conducted triaxial loading and seepage tests to analyze the mechanical failure characteristics and permeability variation in coal–grout consolidation bodies under loading after grouting-based borehole sealing, providing guidance for evaluating the leakage-prevention performance of sealing materials and optimizing the borehole-sealing effectiveness of grouting materials in field engineering. Shen [33] used FLAC3D numerical simulation to investigate the effects of three protective coal-pillar heights and widths on stress evolution and the diffusion process of leaked methane, revealing the spatiotemporal migration patterns of leaked methane under multi-field coupling. In addition, displacement technology plays a key role in improving the extraction efficiency of coalbed methane (CBM) and mine gas. Yang [34] systematically reviewed advances in physical field modeling, such as flow and diffusion, with a particular focus on the evolution of multiphysics coupling mechanisms and permeability models.

In the field of monitoring and early warning for dynamic disasters in coal mines, Zhang [35] investigated the relationship between the spectral characteristics of artificial

https://doi.org/10.3390/10.3390/pr14091427

acoustic signals and gas pressure as well as stress state. Based on a modular design concept, a testing device and system capable of generating acoustic signals through mechanically induced vibrations was developed, which can provide a reference for disaster identification and prediction based on artificial acoustic signals. Liu [36] conducted an in-depth analysis of data cleaning, feature extraction, and model optimization in gas emission prediction, and proposed a high-accuracy prediction model based on KPCA-ICSA-SVR, providing important technical support for mine gas management. Wang [37] successfully synthesized $La_{0.7}Gd_{0.3}Fe_{0.9}Co_{0.1}O_3$ (S-2) perovskite microspheres via an efficient ultrasound-assisted hydrothermal method, demonstrating strong potential for use as a high-performance low-temperature methane sensor.

In terms of coal utilization and its environmental governance, Bao [38] proposed the use of a thermoelectric power generation device to convert the heat energy produced by coal fires into electrical energy, and established an experimental testing system for evaluation. Coal is an important energy source in industrial processes and is commonly used in power plant boilers. Flue gas desulfurization is essential for reducing pollutant emissions, but the desulfurization process generates a large amount of desulfurization wastewater that pollutes the environment. Liao [39] reviewed the main zero-liquid-discharge technologies used for treating desulfurization wastewater (DWW) in coal-fired power plants, providing ideas for technological improvement.

In other aspects of coal mine safety, He [40] used theoretical derivation and PyroSim numerical simulation to analyze tunnel fire spread under different wind-speed conditions, and established the relationship between wind speed and water-mist spray trajectory, providing theoretical support for the study of tunnel fire patterns and emergency response measures. Yuan [41] analyzed the causes of airflow oscillation under parallel upward ventilation and its influencing factors through theoretical analysis and COMSOL Multiphysics 5.5 numerical simulation, and found that the initial airflow velocity and branch length were the key factors affecting flow reversal. Rybár R [42] analyzed the representation of geological, mining, processing, and environmental processes in Minecraft, and conducted a methodological comparison between the mechanisms within the platform and real-world technological and geoscientific processes, evaluating the accuracy, degree of simplification, and educational applicability of various processes related to the extraction and use of mineral resources.

The above studies address many issues in coal mine safety from multiple perspectives. In the face of the many challenges confronting safe coal production, processing, and utilization, there remain many areas worthy of further exploration and in-depth discussion. The journal will continue to solicit high-quality submissions in these fields.

We would like to thank all contributors and editors for their enthusiastic support of this Special Issue, and we also appreciate the efforts of the editorial team of *Processes*.

Conflicts of Interest: The authors declare no conflicts of interest.

References

1. Du, P.; Ren, Y.G.; Liu, Z.F.; He, J.J.; Wang, L.T. From waste to resources: Coal gangue utilization-A comprehensive analysis. *Process Saf. Environ.* **2025**, *201*, 107558. [CrossRef]
2. Wang, K.; Sun, J.Z.; Du, F.; Zhang, X.; Li, K.N.; Zuo, X.H.; Wang, D.X. Mechanisms of Energy Release in Fracture and Gas Expansion-Driven Instabilities of Coal-Rock Composite Structure: Theoretical Modeling and Experimental Validation. *Rock Mech. Rock Eng.* **2025**, *58*, 6359–6379. [CrossRef]
3. Wang, K.; Sun, J.; Du, F.; Zuo, X.H.; Zhao, Y.; Feng, C.T.; Wang, C.L. Energy mutual feedback mechanism between surrounding rock and coal seam in deep coal-gas compound dynamic disasters. *J. China Univ. Min. Technol.* **2025**, *54*, 1242–1260. (In Chinese)

https://doi.org/10.3390/10.3390/pr14091427

4. Du, F.; Wang, K.; Sun, J.; Feng, C.T.; Ju, Y.; Xu, C.; Wang, L.; Wang, C.J. Research on Macroscopic and Microscopic Failure Characteristics and Energy Mechanism of Deep Stress-Dominated Coal and Gas Outburst. *J. China Coal Soc.* **2026**, *51*, 1357–1376. (In Chinese)
5. Shang, K.; Ma, X.; Dong, X.; Sun, W.; Fan, Y.P.; Fu, Y.P.; Zhang, C.Y.; Guo, K.K. Research progress on supercritical fluid extraction in the field of clean coal processing and utilization. *J. Cent. South Univ.* **2025**, *56*, 4034–4049. (In Chinese)
6. Sriramoju, S.K.; Babu, V.; Dash, P.S.; Majumdar, S.; Shee, D. Effective Utilization of Coal Processing Waste: Separation of Low Ash Clean Coal from Washery Rejects by Hydrothermal Treatment. *Miner. Process. Extr. Metall. Rev.* **2022**, *43*, 165–181. [CrossRef]
7. Dmitrienko, M.A.; Strizhak, P.A. Environmentally and economically efficient utilization of coal processing waste. *Sci. Total Environ.* **2017**, *598*, 21–27. [CrossRef]
8. Li, W.Y.; Xu, X.T.; Bei, P.Z. Coal-based solar thermal fuels: A novel perspective on clean utilization of coal tar. *Chem. Eng. Process. Process Intensif.* **2024**, *205*, 109976. [CrossRef]
9. Song, J.H.; Gao, K.; Liu, Y.J.; Li, S.N.; Wang, Y.H.; Yang, Z.M.; Liu, Z.Y. Study on the microscopic and remnants characteristics of coal dust explosion in a closed space. *Powder Technol.* **2025**, *465*, 121359. [CrossRef]
10. Ren, X.F.; Zhang, J.S.; Jia, H.F. Influence of sodium hydrogen carbonate, ammonium dihydrogen phosphate, and lignin-based explosion inhibitors on the sensitivity of coal dust explosion. *Process Saf. Environ.* **2024**, *189*, 1193–1206. [CrossRef]
11. Guo, R.; Xu, C.; Li, N.; Huang, Y.X.; Zhang, Y.S.; Zhang, X.Y. Suppression effect of nanocomposite inhibitor on Methane-Coal dust explosion flame propagation. *Adv. Powder Technol.* **2024**, *35*, 104314. [CrossRef]
12. Pan, H.G.; Fan, Y.B.; Deng, J.; Shi, K.K.; Wang, C.P.; Lei, X.Y.; Wei, Z.C.; Bai, J.M. GCN-based prediction method for coal spontaneous combustion temperature. *Process Saf. Environ.* **2025**, *196*, 106855. [CrossRef]
13. Liu, B.C.; Wen, H.; Mi, W.S.; Fan, S.X.; Xu, Y.; Li, R.J.; Chen, Y.; Han, L.T. Inhibition of Coal Spontaneous Combustion During Dynamic Mining in Adjacent Gob with Narrow Coal Pillars. *Combust. Sci. Technol.* **2025**, *197*, 6831–6853. [CrossRef]
14. Wang, Y.R.; Li, W.X.; Wang, L.; Li, R.; Wang, H.N. Research on optimization method of coal mine safety investment driven by digital twin-enabled INFO-SVR. *Process Saf. Environ.* **2025**, *200*, 107345. [CrossRef]
15. Zhang, J.W.; Bai, X.Y.; Song, Z.X.; Zhang, Y.; Dong, X.K.; Wu, S.K.; Xing, C.R.; Li, X.; Xu, W.Z.; Zhang, S.L. Research Progress and Perspectives on Prevention and Control Technologies for Coal–Rock–Gas Composite Dynamic Disasters: New Types of Induced Classifications, Discriminant Criteria, and Structural Control Schemes. *Rock Mech. Rock Eng.* **2025**, *58*, 10143–10181. [CrossRef]
16. Wang, K.; Du, F. Coal-gas compound dynamic disasters in China: A review. *Process Saf. Environ.* **2020**, *133*, 1–17. [CrossRef]
17. Chu, P.; Xie, H.P.; Gao, M.Z.; Li, C.B.; Shang, D.L.; Liu, Q.Q.; Wang, L. Influence of desorption hysteresis effects on coalbed methane migration and production based on dual-porosity medium model incorporating hysteresis pressure. *Comput. Geotech.* **2024**, *165*, 105893. [CrossRef]
18. Wang, K.; Li, K.; Du, F.; Zhao, W.; Zhao, Y.; Zhang, J.W.; Zhao, M.H. Risk level early warning for coal gas compound dynamic disasters viamulti-source information fusion and deep learning. *J. China Coal Soc.* **2026**, *51*, 461–479. (In Chinese)
19. Du, F.; Li, K.; Wang, K.; Dai, L.C.; Zhao, M.H.; Wang, C.J.; Jiang, L.X.; Wang, L. Coal and gas outburst risk prediction based onimproved DBO optimized CNN. *J. Min. Sci. Technol.* **2025**, *10*, 912–922+1029. (In Chinese)
20. Tan, P.; Xiao, Y.J.; Wang, M.; Yang, T.; Zhang, C.C.; Li, W.Q. Insights into particle breakage induced macroscopic and microscopic behavior of railway ballast via DEM. *Comput. Geotech.* **2025**, *182*, 107135. [CrossRef]
21. Liu, Y.; Fang, S.X.; Hu, T.F.; Zhang, C.; Guo, Y.; Li, F.Z.; Huang, J.W. Study on Coupled Evolution Mechanisms of Stress–Fracture–Seepage Fields in Overburden Strata During Fully Mechanized Coal Mining. *Processes* **2025**, *13*, 1753. [CrossRef]
22. Li, T.X.; Hao, B.Y.; Shi, C.Y. Study on the Mechanisms of Floor Heave in Roadways and Control Strategies via Grooving and Pressure Relief. *Processes* **2025**, *13*, 1642. [CrossRef]
23. Liu, Y. Study on Rational Roadway Layout and Air Leakage Prevention in Shallow Close-Distance Coal Seam Mining. *Processes* **2025**, *13*, 1641. [CrossRef]
24. Shi, H.; Hao, B.Y.; Ren, X.Y.; Zhang, J. Mechanism and Application of Static Stress Intervention for Controlled Directional Roof Caving in Fully Mechanized Mining Faces. *Processes* **2025**, *13*, 1552. [CrossRef]
25. Wang, J.Y.; Li, G.; Wang, W.C.; Liu, H.; Wang, R.; Zhang, H.; Yuan, S.X. Distribution Characteristics of Mining-Induced Stress Fields and Surrounding Rock Control Technology in Adjacent Working Faces Within Fold Structure Zones. *Processes* **2025**, *13*, 1534. [CrossRef]
26. Zhang, H.K.; Guo, X.F.; Yang, K.L. Stress Zoning Characteristics of Goaf Side and Quantitative Evaluation of Surrounding Rock Stability. *Processes* **2024**, *12*, 2490. [CrossRef]
27. Zhang, Y.J.; Chen, Y.X.; Zhang, S.; Feng, G.R.; Wang, Y.G.; Li, S.L.; Wang, Q.; Wang, B.; Zhao, L. Study on the Influence of Drilling Parameters on the Mechanical Properties and Pressure Relief Effect of Coal Rock. *Processes* **2025**, *13*, 993. [CrossRef]
28. Cheng, X.Y.; Cheng, C.; Wang, H.; Xiao, L.; Ma, X.Y. Research and Application of Gas Drainage Negative Pressure Regulation Method Considering Permeability Differences. *Processes* **2025**, *13*, 1236. [CrossRef]
29. Ren, J.W.; Li, Q.X.; Zhang, M.C.; Xu, J.; Li, Y.; Yang, P.B. On the Fluid Behavior Response Characteristics During Early Stage of CBM Co-Production in Superimposed Pressure Systems: Insights from Experimental Analysis. *Processes* **2025**, *13*, 1095. [CrossRef]

https://doi.org/10.3390/10.3390/pr14091427

30. Chen, P.; Wang, Y.P.; Zhao, Y.X.; Wang, Q.; Wen, Z.H.; Tang, L.G. Molecular Simulation of Ultra-Microstructural Characteristics of Adsorption Pores in Terms of Coal and Gas Adsorption Properties. *Processes* **2025**, *13*, 771. [CrossRef]
31. Zhang, F.; Niu, S.W.; He, J.W.; Zhang, K.; Qin, Z.C. Deformation and Pore Structure Characteristics of Lignite Pyrolysis with Temperature Under Triaxial Stress. *Processes* **2025**, *13*, 1444. [CrossRef]
32. Tang, Y.C.; Lu, P.; Zhang, J.X.; Wang, J. Characterization of Damage and Infiltration Modeling of Coal-Slurry Consolidation Mechanics Under Loaded Conditions. *Processes* **2025**, *13*, 400. [CrossRef]
33. Shen, J.H.; Liang, S.; Hao, Y.S.; Ma, Z.; He, W.S.; Liang, X.; Xu, S.Y.; Luo, C.H. Stress Zoning Characteristics and Migration of Leaked Methane from Gas Wells Penetrating Protective Coal Pillars in Longwall Mining Areas. *Processes* **2025**, *13*, 47. [CrossRef]
34. Yang, X.; Du, F.; Zhang, Q.C.; Zuo, Y.F.; Tan, F.Y.; Zhang, Y.Y.; Xu, Y.Y. Numerical Simulation of Gas Injection Displacement in Coal Seams: A Mini-Review. *Processes* **2025**, *13*, 3463. [CrossRef]
35. Zhang, J.G.; Fu, W.L.; Cui, W.L.; Ma, J.; Du, F. Experimental Study on Spectral Response Characteristics and Mechanical State of Coal Based on Artificial Acoustic Signals. *Processes* **2024**, *12*, 2752. [CrossRef]
36. Liu, L.; Dai, L.C.; Mao, X.Y.; Chen, Y.T.; Jing, Y.H. Research on Gas Emission Prediction Based on KPCA-ICSA-SVR. *Processes* **2024**, *12*, 2655. [CrossRef]
37. Wang, J.W. Nanostructure-Governed Methane Sensing and Response Mechanism in Engineered $LaFeO_3$. *Processes* **2026**, *14*, 836. [CrossRef]
38. Bao, Q.F.; Guo, X.T.; Li, B.; Chen, W.Y.; Wang, Z.P.; Xiao, Y. Performance and Reliability of Thermoelectric Conversion Using a Crooked Thermosyphon to Enhance Heat Transfer from Coal Fires. *Processes* **2024**, *12*, 2692. [CrossRef]
39. Liao, B.S.; Zeng, X.Y.; Ling, Z.Q.; Zhao, S.M.; Li, B.; Han, X.L. Recent Advances in Zero Discharge Treatment Technologies for Desulfurization Wastewater in Coal-Fired Power Plants: A Mini-Review. *Processes* **2025**, *13*, 982. [CrossRef]
40. He, W.J.; Fu, M.H.; Xiong, L.; Zheng, S.Q. Fire Characteristics and Water Mist Cooling Measures in the Coal Transportation Process of a Heavy-Haul Railway Tunnel in Shanxi Province. *Processes* **2025**, *13*, 1789. [CrossRef]
41. Yuan, X.P.; Zhang, Q.H.; Wang, Z.J. Impact of Gas Accumulation on the Stability of Parallel Upward Ventilation in High-Temperature Sloped Shafts of Deep Wells. *Processes* **2024**, *12*, 2530. [CrossRef]
42. Rybár, R.; Beer, M.; Bednárová, L. Mining and Processing of Mineral Resources: A Comparative Study of Simulated and Operational Processes. *Processes* **2025**, *13*, 2823. [CrossRef]

https://doi.org/10.3390/10.3390/pr14091427

processes

MDPI

Article

Nanostructure-Governed Methane Sensing and Response Mechanism in Engineered $LaFeO_3$

Jianwei Wang [1,2]

1 State Key Laboratory of Coal Mine Disaster Prevention and Control, Chongqing 400037, China; wjwmky@163.com

2 CCTEG Chongqing Research Institute, Chongqing 400037, China

Abstract

Methane, a highly hazardous gas mixture when exposed to open flames, is commonly encountered in coal mines. Its primary component is CH_4, making the detection of its concentration, especially under diverse environmental conditions, highly significant. In this study, $La_{0.7}Gd_{0.3}Fe_{0.9}Co_{0.1}O_3$ nanomaterials were prepared using an ultrasound-assisted hydrothermal method. Through dual-site synergistic regulation involving Gd doping at the A-site and Co doping at the B-site, rapid detection of CH_4 at low temperatures was achieved. At 150 °C, the sensor demonstrated a significantly enhanced response to 100 ppm CH_4, with a sensitivity of 10.22. This value represents an approximately tenfold improvement over that achieved with pure $LaFeO_3$. In addition, the sensor responded rapidly to the gas exposure within 6.3 s and recovered within 5.4 s, respectively, at the same gas concentration. Such swift recovery capabilities enable reliable detection across multiple environmental conditions. Moreover, the sensor not only shows excellent repeatability but also maintains a high response value of 9 even under highly humid conditions (95% RH). The performance enhancement is attributed mainly to lattice distortion induced by A-site doping and the increased active sites provided by B-site doping. The development of this sensor lays a foundation for future CH_4 detection and industrial safety applications.

Keywords: methane detection; coal mine safety; low-temperature sensing; fast response and recovery; high humidity stability

1. Introduction

Given the frequent occurrence of gas disasters, which pose a serious threat to coal mine safety production, the development of CH_4 sensors that meet practical needs has become increasingly important. During coal mining, flammable, explosive, and harmful gases such as methane (CH_4), carbon monoxide (CO), hydrogen (H_2) and nitrogen dioxide are generated or released. When these gases reach certain concentration levels, they can harm human health and even lead to major mine accidents such as fires and explosions, endangering workers and the entire mining operation [1,2]. For instance, when the volume fraction of H_2S reaches 0.1%, it can cause fatal poisoning within seconds. In recent years, extensive field practices have shown that abnormal CO concentrations can also appear in coal mining spaces. CO has been detected in mine atmospheres without any signs of spontaneous combustion of coal, and even a small amount of CO can be lethal. Moreover, certain concentrations of CH_4 and H_2 can trigger severe explosions upon contact with sparks, resulting in catastrophic accidents. Therefore, the detection, monitoring, and early warning of toxic and harmful gases such as CH_4, CO, and H_2S in underground coal mines are crucial links and prerequisites for safe coal production. CH_4, a colorless, odorless, flammable gas with a typical tetrahedral molecular structure, is non-toxic in itself. However, in confined environments, it can reduce oxygen levels, leading to asphyxiation and serious harm to humans. Furthermore, when CH_4 mixes with air within a concentration range of 5% to 16%, it can ignite upon contact with an open flame, producing a blue flame and potentially causing an instantaneous explosion [3,4]. The rapid temperature rise from such an explosion can bring devastating disasters to coal mine safety. Thus, research on new nanostructured gas-sensitive materials for detecting CH_4 and other hazardous gases in coal mines is of great scientific importance and will offer vital technical support for safe mining operations, greatly reducing accidents and casualties. The research and development of oxide semiconductor materials for detecting harmful gases in coal mines, the analysis of the relationship between their composition, structure, and sensing performance, the exploration of mechanisms behind the selectivity, sensitivity, and stability of gas-sensitive materials, the reduction in power consumption in semiconductor gas sensors, and the improvement of detection efficiency are all essential scientific and technological needs for the sustainable development of coal mine safety. Hence, conducting research on novel nanostructured gas-sensitive materials for detecting flammable, explosive, and harmful gases in underground coal mines holds profound scientific significance. Such work will provide important technical support for coal mine safety production and significantly reduce casualties caused by gases such as methane, carbon monoxide, hydrogen sulfide, and other hazardous substances [5,6].

Currently, various technologies such as optical trapping, mass spectrometry, and chemical catalysis have been applied to detect CH_4 gas. However, these methods suffer from significant drawbacks, including high cost, large size, and cumbersome testing procedures. Owing to their low cost, ease of integration, and compact size, metal oxide semiconductor sensors have found extensive application in diverse scenarios aimed at enhancing quality of life. Some two-dimensional materials, such as ZnO, MnO_2, and CuO, were among the earliest used in gas sensors. These materials not only exhibit ultra-wide band gaps and high carrier mobility but are also easy to synthesize, making them a focus of recent gas sensor research. However, as demand grows, two-dimensional materials have revealed certain limitations, such as long response and recovery times, high operating temperatures, and poor stability. Many researchers have attempted to modify these materials to improve their performance, but due to structural constraints, the conditions for modification remain limited. In light of these shortcomings, ABO_3-type perovskite materials with p-type semiconductor characteristics have attracted research attention. These materials not only allow flexible structural tuning at both A and B sites but are also simple to prepare [7,8]. Among these perovskite materials, $LaFeO_3$ has emerged as a promising candidate. Its stable perovskite structure ensures good chemical and thermal stability, enabling long-term operation in complex or high-temperature environments. Its surface, rich in oxygen vacancies and active sites, provides an ideal platform for gas adsorption and reaction, demonstrating high sensitivity and fast response-recovery characteristics to various redox gases such as ethanol, acetone, and formaldehyde. Moreover, by doping with ions (e.g., Sr^{2+}, Co^{3+}) or tuning the nanostructure, its conductivity, band structure, and selectivity can be effectively optimized, enabling selective detection of specific gases [9,10]. Thus, $LaFeO_3$ is a sensing material with tunable performance, good stability, broad application prospects. In recent years, many researchers have investigated dual-site doping in $LaFeO_3$ for detecting various gases. For example, Wang et al. used electrospinning to modify $LaFeO_3$ with Keggin-type phosphotungstic acid as an acceptor, achieving high-performance detection of acetone [11]. Sun et al. applied Au surface modification to $LaFeO_3$, forming a porous nanofiber structure that enabled rapid detection of formaldehyde at a low temperature of 120 °C with a high re-

https://doi.org/10.3390/pr14050836

sponse of 52.26 [12]. Fu et al. employed a unique ultrasound-assisted hydrothermal method to create porous microspheres, realizing fast detection of ethanol with impressive response and recovery times [13]. Meng et al. used a hydrothermal method to compound In_2O_3 with $LaFeO_3$, which showed excellent response characteristics to 2-butanone [14]. While these studies have achieved outstanding gas detection performance with modified $LaFeO_3$ for various gases, reports on $LaFeO_3$-based sensors for CH_4 detection remain scarce.

In this study, we employed dual-site modulation at both A and B sites to synthesize a high-performance gas-sensitive material, $La_{0.7}Gd_{0.3}Fe_{0.9}Co_{0.1}O_3$, for methane detection. The material was physically characterized using XRD, SEM, and ICP-MS, confirming the formation of the perovskite structure. A series of gas-sensing tests were conducted to evaluate the sensor's performance. At 150 °C, the device revealed a response of 10.22 to CH_4, along with response and recovery times of 6.3 s and 5.4 s, respectively. It also exhibited excellent humidity resistance, stability, and remarkable repeatability. Finally, this paper provides an in-depth explanation of the gas-sensing mechanism based on various characterization results, offering insights into the enhanced performance of the doped perovskite material in detecting methane under conditions relevant to coal mine safety applications. The development of such sensors have a good performance in creating reliable, efficient, and durable detection systems for hazardous gases in mining environments, ultimately contributing to the reduction in gas-related accidents and the promotion of safer working conditions in the coal mining industry.

2. Experimental Section

2.1. Materials

All starting reagents were procured from the Aladdin website, of analytical-grade purity, and used without further purification, specifically including lanthanum(III) nitrate hexahydrate ($La(NO_3)_3 \cdot 6H_2O$, 99.9%), iron(III) nitrate nonahydrate ($Fe(NO_3)_3 \cdot 9H_2O$, 98.5%), gadolinium(III) nitrate hexahydrate ($Gd(NO_3)_3 \cdot 6H_2O$, 99.99%), cobalt(II) nitrate hexahydrate ($Co(NO_3)_2 \cdot 6H_2O$, 99.9%), and anhydrous citric acid ($C_6H_8O_7$, 99.9%) as the chelating agent.

2.2. Synthesis of $LaFeO_3$

The synthesis began with the preparation of two separate precursor solutions. First, 1.08 g of lanthanum(III) nitrate hexahydrate was dissolved in 35 mL of deionized water. This solution was stirred continuously on a magnetic stirrer hotplate at a constant temperature of 60 °C for 20 min to ensure complete dissolution and homogeneity. Concurrently, a second sample was obtained by dissolving nitrate nonahydrate and 2.00 g of anhydrous citric acid in DI water. This mixture was also stirred magnetically at 60 °C, but for a duration of 30 min, allowing for thorough mixing and the initial formation of metal-citrate complexes. Subsequently, both clear solutions were carefully combined and transferred into the polytetrafluoroethylene (PTFE) liner. The sealed container was then ultrasonicated for 1 h at a power of 350 W. This ultrasonication step promotes intense mixing at the molecular level and further facilitates the complexation between the metal cations and the citrate ions. Following sonication, the autoclave was placed in a preheated oven for a hydrothermal treatment. Heating at 200 °C was continued for 15 h to facilitate the reaction. Under these elevated temperature and pressure conditions, the precursor complexes undergo decomposition and rearrangement, initiating the formation of the desired perovskite oxide framework. After the reactor cooled naturally to room temperature, a brownish solid–liquid mixture was obtained. The product was collected and purified via centrifugation. Sequential washing was performed with absolute ethanol (×3) and deionized water (×1). This process removed residual solvents, unreacted precursors, and reaction by-products.

https://doi.org/10.3390/pr14050836

After each washing cycle, the solid was re-dispersed and isolated by centrifugation. The purified wet solid was then transferred to an oven and dried at a constant temperature for 12 h, yielding a brown-colored precursor powder. Finally, this precursor was placed in a crucible, then calcined. A heat treatment of 10 h at 1000 °C was performed under static air conditions. The slow heating and extended holding time at high temperature are crucial for achieving complete crystallinity, removing organic residues, and forming the phase-pure, final $LaFeO_3$ product.

2.3. *Synthesis of* $La_{0.7}Gd_{0.3}Fe_{0.9}Co_{0.1}O_3$

The same high-purity reagents from Aladdin, as listed for the $LaFeO_3$ synthesis, were used for this doped variant. The synthesis for the gadolinium- and cobalt-substituted perovskite followed a similar hydrothermal route but with adjusted stoichiometries. The process commenced with the preparation of the A-site and B-site cation solutions separately. For the A-site solution, 0.707 g of lanthanum(III) nitrate hexahydrate and gadolinium(III) nitrate hexahydrate were co-dissolved in 30 mL of DI water. To this mixture, 0.51 g of anhydrous citric acid was added. The sample was subjected to magnetic stirring at 60 °C for 20 min to obtain a clear solution, ensuring the lanthanum and gadolinium ions were uniformly mixed and chelated. Simultaneously, the B-site solution was obtained by dissolving 0.909 g of iron(III) nitrate nonahydrate, 0.0727 g of cobalt(II) nitrate hexahydrate, and 0.50 g of anhydrous citric acid in 15 mL of deionized water. This solution was also heated and stirred at 60 °C for 30 min, promoting the formation of a homogeneous solution containing the transition metal ions complexed with citrate. The two precursor solutions were then combined quantitatively in the PTFE liner of the autoclave. The mixture underwent the same intensive ultrasonic processing step for 1 h at 350 W to ensure a perfectly homogeneous mixture at the atomic level, which is critical for achieving a uniform doped perovskite phase. The subsequent hydrothermal reaction was identical: the autoclave was maintained at 200 °C for 15 h. This step drives the co-precipitation and condensation reactions necessary to form the mixed-metal oxide framework incorporating all four cationic species (La^{3+}, Gd^{3+}, Fe^{3+}, Co^{3+}) [15]. After the hydrothermal treatment, the resulting yellowish-brown mixture was processed identically to the undoped material. It was subjected to three centrifugal washes with ethanol followed by one wash (DI water). The product was then oven-dried for 12 h, yielding a yellowish-brown precursor powder. The final and critical crystallization step involved calcining this precursor powder at 1000 °C for 10 h in air. This high-temperature annealing ensures the complete decomposition of any residual nitrates and organic material, drives the solid-state diffusion required for homogeneous cation distribution, and results in the crystallographically defined, single-phase multiferroic perovskite material $La_{0.7}Gd_{0.3}Fe_{0.9}Co_{0.1}O_3$, the flowchart is shown in Figure 1.

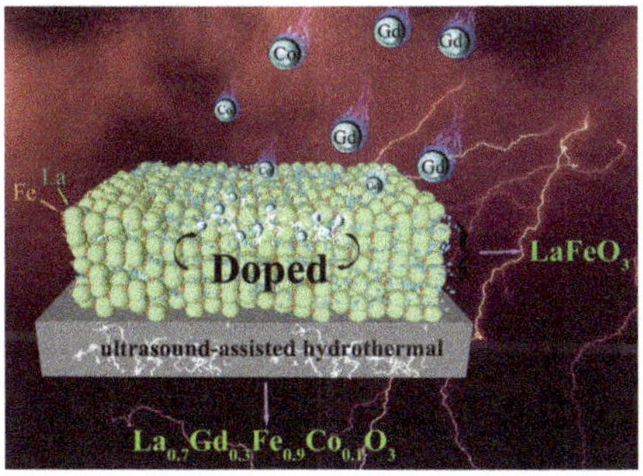

Figure 1. Schematic of the synthesis of $La_{0.7}Gd_{0.3}Fe_{0.9}Co_{0.1}O_3$ (Illustration created by the author).

https://doi.org/10.3390/pr14050836

2.4. Material Characterization

X-ray diffraction (XRD) was initially employed to determine the crystal structure and assess the phase purity of the powder samples (Horiba, Ltd., Kyoto, Japan). XRD patterns were collected over a defined 2θ range (e.g., 10° to 80°) with a slow scan rate to ensure high-resolution data. The obtained patterns were compared with standard reference patterns (JCPDS cards) to identify the crystalline phases present. To evaluate the morphology, particle size, and microstructure, scanning electron microscopy (SEM) was employed. SEM micrographs were taken at various magnifications to provide a detailed view of the sample's topography (Hitachi High-Tech Corporation, Tokyo, Japan). The particle size distribution was also estimated from these images. The morphology is critically important as it directly influences the material's surface area and, consequently, its performance in applications like gas sensing or catalysis. BET (Brunauer–Emmett–Teller) analysis (Micromeritics Instrument Corporation, Norcross, GA, USA) was performed by measuring nitrogen adsorption–desorption isotherms at 77 K (liquid nitrogen temperature). To obtain precise quantitative information on the elemental composition, inductively coupled plasma mass spectrometry was conducted (Shimadzu Corporation, Kyoto, Japan). The electrical transport properties and interfacial characteristics of the materials were probed using electrochemical impedance spectroscopy, performed with an electrochemical workstation (CH Instruments, Inc., Shanghai, China). Finally, the gas sensing performance, a key functional property for these perovskite oxides, was systematically evaluated using a commercial CGS-8 (Computerized Gas Sensing) analysis station.

2.5. Sensing Measurement of Gas Sensors

The gas sensors utilized in this study primarily consist of a sensor base and a ceramic tube equipped with gold-platinum electrodes. The fabrication process begins by weighing 0.15 g of the sensing material, which is then thoroughly ground. Subsequently, a mixture of 0.005 mL of absolute ethanol and 0.01 mL of deionized water is added to the powder. This combination is further mixed and ground to form a slurry. The slurry is moved into a custom-made spray nozzle, which operates at an air pressure of 2 N. The slurry is then uniformly sprayed onto the surface of the ceramic tube, specifically covering the gold-platinum electrodes. Following the coating process, the ceramic tube is placed in an oven for heat treatment. It is heated at 60 °C for 12 h. This step is crucial for evaporating the solvent and firmly fixing the sensing material onto the substrate surface, thereby forming a stable and adherent sensitive layer. Once dried, the ceramic tube is carefully welded onto the sensor base. Next, a heating filament is threaded through the interior of the ceramic tube and also welded onto the sensor base. This filament serves the essential function of controlling and maintaining the optimal operating temperature of the sensor during gas detection.

Finally, the fully assembled gas sensor is placed into a gas sensor aging platform for a stabilization or aging process. This aging procedure is conducted to remove any residual impurities or volatile compounds from the surface of the sensing material, thereby enhancing the sensor's baseline stability, reproducibility, and long-term performance.

The test gases employed in this study, all sourced from pure, certified gas cylinders with a guaranteed purity of 99.99%.

For performance evaluation, the response of the sensor is defined differently for reducing and oxidizing gases, reflecting the fundamental difference in their interaction mechanisms with the semiconductor sensing material.

For reducing gases (such as CH_4, H_2S, NH_3), the sensor response is defined as [16]:

$$\text{Response} = R_g/R_a \tag{1}$$

https://doi.org/10.3390/pr14050836

At the optimal operating temperature, the sensor resistance in the target gas and in air are designated as R_g and R_a, respectively.

For oxidizing gases (such as NO_2, NO), the response is defined as [17]:

$$\text{Response} = R_a/R_g \tag{2}$$

All other measurement standards and conditions remain consistent with those used for reducing gases.

Furthermore, the response time and recovery time are critical dynamic parameters. In this study, the response time is defined as the time required for the sensor's resistance to change by 90% of the total difference between its baseline resistance (in air) and its steady-state resistance in the target gas atmosphere upon gas exposure.

3. Results and Discussion

3.1. Characterizations

The XRD patterns for both S-1 and S-2 are presented in Figure 2a. A preliminary visual inspection confirms that all the characteristic diffraction peaks for each sample align well with the reference pattern for orthorhombic $LaFeO_3$ (PDF#37-1493, space group Pnma) [18]. No detectable peaks corresponding to secondary or impurity phases are observed, indicating the high phase purity of the synthesized powders. The prominent diffraction peaks located at 2θ values of approximately 22.6°, 32.9°, 40.0°, 46.1°, 52.0°, 53.2°, 57.4°, 63.8°, 67.1°, 71.9°, and 76.5° are readily indexed to the (101), (121), (220), (202), (141), (311), (240), (331), (242), (143), and (204) crystallographic planes of the perovskite structure, respectively. This excellent correspondence confirms that both the parent and the co-doped materials crystallize in the typical orthorhombic perovskite structure.

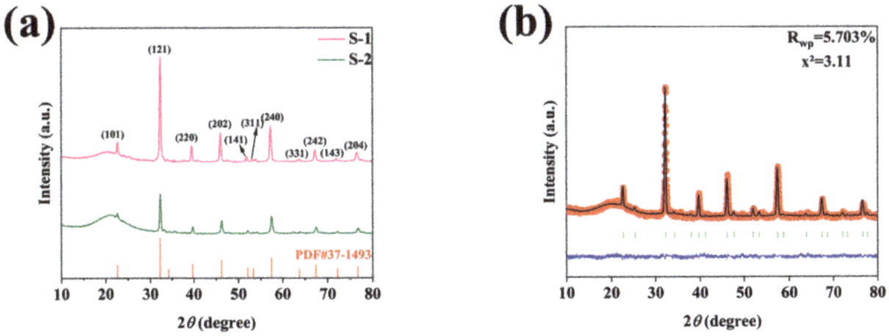

Figure 2. Wide-angle XRD patterns of (**a**) S-1 and S-2; (**b**) XRD refinement results of S-2.

To gain deeper insights into the structural modifications induced by doping, a comprehensive Rietveld refinement analysis was conducted on the XRD pattern of sample S-2 using the GSAS-II (version 5455) software package. The refinement was initiated using the structural model of $LaFeO_3$ (Pnma space group) as the starting point. As clearly demonstrated in the Rietveld plot (Figure 2b), an excellent agreement exists between the observed diffraction profile (red circles) and the calculated pattern (black solid line). The difference curve (blue line at the bottom) is flat and featureless, confirming the quality of the fit. All diffraction peaks were successfully accounted for by the Pnma structural model, validating that the Gd and Co co-doping does not induce a phase transformation but rather leads to a homogeneous solid solution. The reliability factors of the refinement converged to $\chi^2 = 5.703$ and $R_{wp} = 3.11\%$, which are within the statistically acceptable range for such materials, further verifying the accuracy of the refined structural parameters and the phase purity of S-2.

https://doi.org/10.3390/pr14050836

The refined lattice parameters and unit cell volumes for both S-1 and S-2 are compared in Table 1. The data reveal a clear and consistent trend: the introduction of smaller cations at both the A- and B-sites of the perovskite lattice induces a significant contraction of the unit cell. In sample S-2, the larger La^{3+} ion (ionic radius = 1.36 Å for 12-coordination) on the A-site is partially substituted by the smaller Gd^{3+} ion (1.27 Å, CN = 12). Concurrently, on the B-site, a fraction of the Fe^{3+} ion (0.645 Å for high-spin, CN = 6) is replaced by Co ions, which, depending on their oxidation state (Co^{3+}: 0.61 Å for high-spin, CN = 6; Co^{2+}: 0.745 Å for high-spin, CN = 6, though typically less stable in this structure) [19,20], generally have an effective ionic radius that is smaller or comparable. This dual-site substitution strategy creates a cooperative chemical pressure, leading to an isotropic lattice contraction. As a direct consequence, all three lattice parameters (a, b, c) and, most notably, the total unit cell volume of S-2 show a measurable decrease compared to the undoped S-1 sample. This observation is in perfect agreement with Vegard's law and established crystal chemical principles, where the unit cell dimensions scale with the average ionic radii of the constituent ions. The successful refinement and the resultant lattice parameter data provide direct structural evidence for the incorporation of Gd and Co into the perovskite lattice, confirming the formation of the targeted $La_{0.7}Gd_{0.3}Fe_{0.9}Co_{0.1}O_3$ compound.

Table 1. Comparison of Lattice Parameters between S-1 and S-2.

Different Samples	Crystallite Size (nm)	Lattice Parameters (Å)			Lattice Volume (Å^3)
		a	b	c	
S-1	29.7	5.551	7.852	5.551	242.1
S-2	28.93	5.531	7.843	5.545	240.5

Figure 3 presents the scanning electron microscopy (SEM) images of S-1 and S-2. As clearly observed, the undoped sample S-1 exhibits an irregular block-like morphology with extensive agglomeration, leading to a relatively dense and non-uniform microstructure. In contrast, the dual-site doped sample S-2 shows a distinct morphological transformation: the particle size is significantly reduced, the distribution becomes more homogeneous, and the overall structure evolves into well-defined microspheres. Additionally, the surface roughness of S-2 is noticeably increased, which can be attributed to the lattice strain and altered surface energy induced by the doping process. The introduction of dopant atoms disrupts the regular crystal growth, promotes the formation of finer grains, and encourages the assembly of particles into spherical aggregates with higher surface-to-volume ratios. This refined morphology, characterized by smaller particle size, improved dispersion, spherical shape, and enhanced surface roughness, collectively contributes to a larger effective surface area and more exposed active sites, thereby facilitating gas adsorption and surface reactions [21,22].

To verify the successful incorporation of the target elements into the crystal lattice, we further performed elemental content analysis on the samples. First, both materials were completely digested using a microwave digestion system. Each sample was dissolved in 35 mL of aqua regia to convert the solid material into a clear solution. Subsequently, standard solutions of each target element were prepared, and the instrument was calibrated using blank samples and repeated measurements of the standards to eliminate matrix interference. Table 2 compares the actually measured elemental contents of the two samples with their theoretically calculated values. As shown in the table, the measured values agree well with the theoretical ones, with relative errors all below 90% [23,24], confirming that the target elements have been successfully doped into the material according to the designed stoichiometric ratio. Moreover, we observed that the total elemental content does not sum to 100%. This indicates a deficiency of oxygen, which indirectly proves the formation of

https://doi.org/10.3390/pr14050836

oxygen vacancies. Such oxygen vacancies are known to enhance gas-sensing performance by increasing active sites on the material surface and promoting charge transfer during gas adsorption, thereby improving both sensitivity and response kinetics. In summary, the elemental analysis not only validates the effective doping of the intended elements but also provides indirect evidence for the creation of oxygen vacancies. These vacancies play a crucial role in boosting the sensor's gas-detection capability, further supporting the superior performance demonstrated in the earlier sensing tests [25,26].

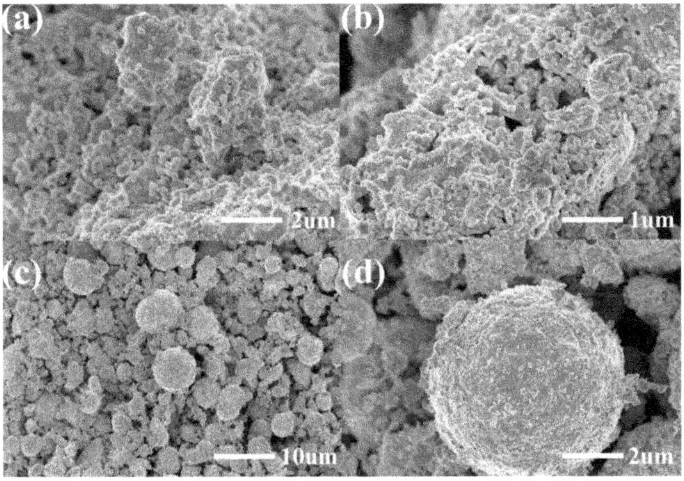

Figure 3. SEM morphological features of (**a**,**b**) S-1, (**c**,**d**) S-2.

Table 2. Quantitative analysis of S-1 and S-2 elements.

Elements	S-1		Elements	S-2	
	mg/L	At %		mg/L	At%
La	27.91	44.29	La	20.5	29.2
Fe	24.55	39.77	Fe	22.1	37.54
			Gd	12.8	12.51
Total		84.06	Co	2.7	4.17
	Fe:La =			La:Fe = 77.8%	
	87.97%			Co:Gd = 33.3%	
			Total		83.42

Figure 4 displays the BET surface area analysis curves of the two samples. As shown in the figure, the specific surface area of S-1 is 5.22 m^2/g, whereas that of the dual-doped material S-2 reaches 29.11 m^2/g, approximately four times larger than that of S-1. This remarkable increase is primarily attributed to the introduction of dopant elements, which effectively inhibit grain growth, induce lattice distortion, and likely alter the nucleation and aggregation behavior during synthesis. Consequently, the doped material forms smaller particles and a more porous microstructure. The significantly enlarged specific surface area provides a greater number of surface-active sites and adsorption points, thereby substantially enhancing gas-sensing efficiency. A high surface area facilitates more intimate contact between the target gas molecules and the sensing material, promotes efficient gas diffusion within the porous network, and ultimately leads to improved sensor response and faster reaction kinetics. In summary, the combination of a substantially increased specific surface area and improved electronic transport properties, both resulting from the

https://doi.org/10.3390/pr14050836

strategic dual-site doping, provides a solid physicochemical foundation for the outstanding gas-sensing behavior of the S-2 sensor. These material-level advantages align well with the superior functional performance demonstrated in previous gas-sensing tests, underscoring the effectiveness of the doping strategy in advancing the development of high-performance gas sensors [27].

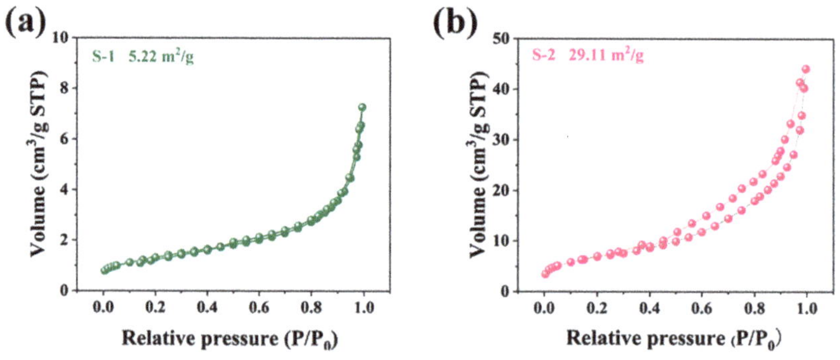

Figure 4. N_2 adsorption–desorption isotherms: (**a**) S-1; (**b**) S-2.

To gain deeper insight into the electrical transport properties of the materials, we performed alternating current (AC) impedance spectroscopy on the samples. This technique allows us to extract key electrical parameters such as bulk resistance, grain-boundary resistance, and related electrical characteristics, thereby elucidating the influence of doping on the electron/ion mobility mechanism. The corresponding Nyquist plots are presented in Figure 5, where Figures 5a and 5b correspond to the spectra of S-1 and S-2, respectively. It can be clearly observed that the diameter of the semicircular arc in the impedance spectra gradually decreases with increasing temperature. At the same temperature, the semicircle diameter of S-2 is notably smaller than that of S-1, indicating a lower overall resistance and, more importantly, higher electron mobility in the doped material. The enhanced electron mobility is a direct consequence of the modified electronic structure and reduced energy barriers for charge transport introduced by dual-site doping. This leads to an increase in charge-carrier concentration and a more efficient charge-transfer process across the material. Such improved electrical characteristics directly influence the efficiency of gas-surface interactions, as faster charge exchange between the adsorbed gas molecules and the sensing layer accelerates the response and recovery dynamics. Consequently, the superior electrical properties of S-2 contribute significantly to its enhanced gas-sensing performance, including higher sensitivity, faster response, and better overall stability [28–30].

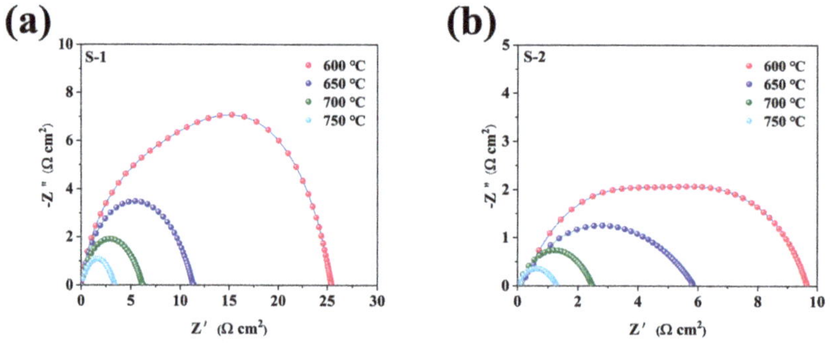

Figure 5. EIS curves of (**a**) S-1 and (**b**) S-2 at different temperatures.

https://doi.org/10.3390/pr14050836

In the temperature range of 100 °C to 300 °C, the S-1 sensor begins to show a detectable response, as the thermal energy at these levels finally enables gas molecules to reach and interact with the material surface, as shown in Figure 6. Notably, at their respective optimal operating temperatures, the response of the S-2 sensor is approximately three times higher than that of the S-1 sensor. Furthermore, the optimal working temperature of the S-2 sensor is about 100 °C lower than that of S-1. This significant enhancement in both response magnitude and reduction in operating temperature vividly demonstrates the advantageous effects of the dual-site doping strategy, which effectively modulates the surface chemistry and electronic structure of the sensing material, thereby improving its reactivity and lowering the energy required for gas detection. As the temperature further increases from 150 °C to 300 °C, the response of the S-2 sensor gradually declines. This decrease occurs because, at excessively high temperatures, the desorption rate of gas molecules from the material surface exceeds the adsorption rate. Consequently, gas molecules cannot reside on the active sites long enough to sustain a high response, leading to a reduction in the overall sensor signal. This typical volcano-shaped trend, increase followed by decrease, highlights the existence of an optimal temperature window that balances adsorption and desorption processes for maximum sensor performance. In summary, the temperature-dependent studies clearly reveal the superior sensing characteristics of the dual-doped S-2 sensor, including a significantly lower optimal operating temperature, a substantially higher response amplitude, and a well-defined activation behavior at lower temperatures. These improvements underscore the effectiveness of dual-site doping in tailoring the sensing properties, making the S-2 sensor a more energy-efficient and sensitive candidate for practical gas-sensing applications.

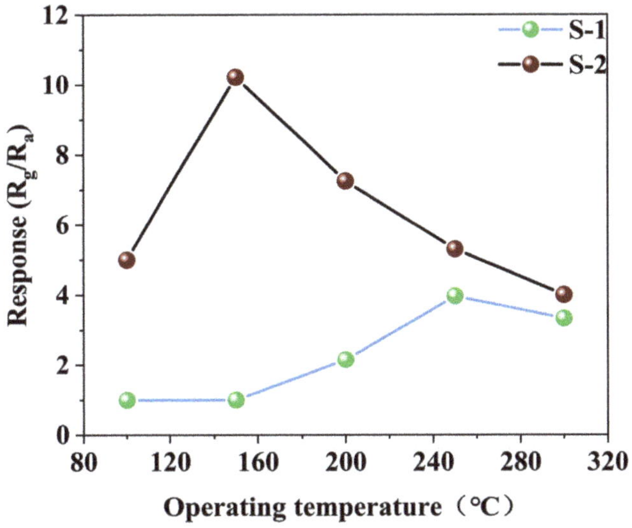

Figure 6. Different response values of two sensors at different working temperatures.

As shown in Figure 7a, the sensor's response and recovery curves were recorded at varying gas concentrations. The test was conducted across seven concentration levels: 100, 75, 50, 25, 10, 5, and 1 ppm. As shown in the figure, the response values of both sensors increase with rising gas concentration, exhibiting well-defined peak profiles. Taking the S-2 sensor as an example, at its optimal operating temperature of 150 °C, the response values from low to high concentrations are 1.63, 2.33, 3.09, 4.11, 6.01, 8.68, and 10.22, respectively. Response and recovery times are also critical performance indicators for gas sensors. The

https://doi.org/10.3390/pr14050836

dynamic response and recovery profiles of the two sensors are displayed in Figure 7b,c. It can be observed that the S-1 sensor exhibits response and recovery times of 11.7 s and 8.1 s, respectively. In contrast, the dual-doped S-2 sensor demonstrates significantly faster response and recovery times of only 6.3 s and 5.4 s. This improvement indicates that the S-2 sensor possesses enhanced gas-molecule transport capability, allowing quicker adsorption and desorption processes, which in turn supports higher detection standards. Furthermore, Figure 7d illustrates the fitted curve of the S-2 sensor's response as a function of gas concentration. Clearly, in the low-concentration range (0–10 ppm), the relationship can be described by the linear equation y = 0.1618x + 1.48705, with a coefficient of determination (R^2) of 0.996, indicating an excellent fit. In the high-concentration range (25–100 ppm), the fitted curve follows y = 0.084x + 2.005, with an R^2 (0.986). This indicates that sensor maintains a strong linear correlation across both concentration ranges, with particularly sensitive detection performance in the low-concentration region, making it well-suited for trace-level methane monitoring.

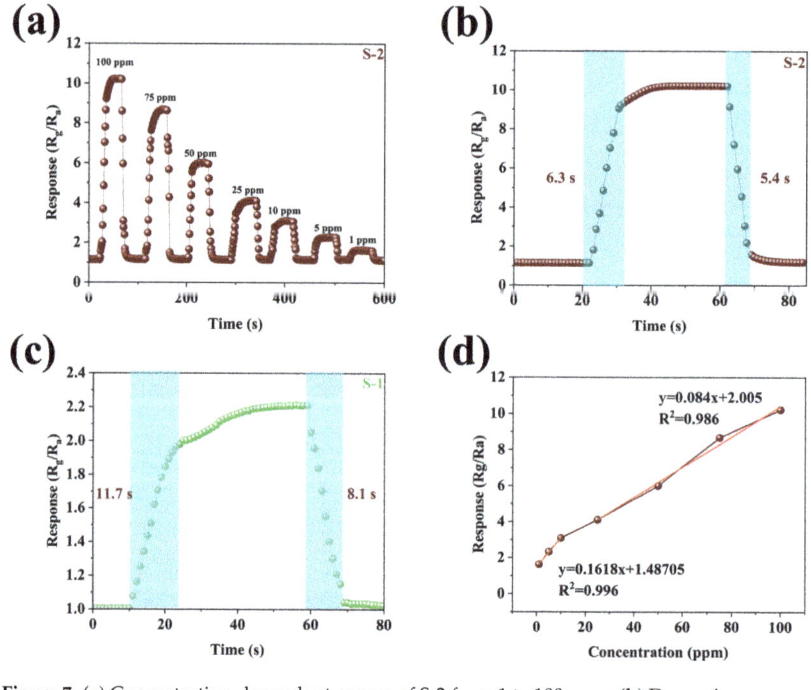

Figure 7. (**a**) Concentration-dependent curves of S-2 from 1 to 100 ppm; (**b**) Dynamic response and recovery curve of S-2; (**c**) Dynamic response and recovery curve of S-1; (**d**) Fitted response values of S-2 as a function of concentration from 1 to 100 ppm.

Subsequently, to further evaluate the performance of the sensor, we conducted additional gas-sensing tests on the S-2 sensor. Figure 8a shows the response values of the S-2 sensor at its optimal operating temperature toward various reducing and oxidizing gases, all at a concentration of 100 ppm. The reducing gases included CH_4, ethanol, and methanol, while the oxidizing gases comprised CO_2, NO_2, and NH_3. As can be seen from the figure, the sensor exhibits the highest response toward CH_4, with a response value approximately eight times that toward CO_2, highlighting the excellent selectivity of this sensor. Figure 8b presents the fitting curve of the sensor's response as a function of humidity under different relative humidity (RH) conditions. The humidity environments were generated using pre-prepared saturated salt solutions, covering relative humidity

https://doi.org/10.3390/pr14050836

levels of 35%, 44%, 54%, 75%, 85%, and 95%. The salts employed were saturated calcium carbonate, saturated iron(II) chloride, saturated sodium chloride, saturated potassium chloride, and saturated potassium sulfate, respectively. It can be clearly observed that the response gradually decreases as the humidity rises. This decline is attributed to the adsorption of a large number of water molecules onto the material surface in moist air; these molecules hinder the adsorption of target gases, thereby leading to a reduction in the response. Nevertheless, even at 95% RH, the sensor retains about 90% of its maximum response, demonstrating its outstanding moisture resistance. In the low-humidity range, corresponding to typical indoor conditions up to 54% RH, the fitted curve for the response is $y = -0.01103x + 10.64274$. In the high-humidity range (75–95% RH), the fitted relationship follows $y = -0.01912x + 11.13233$. Both fitting results indicate a good correlation between the experimental data and the linear models, confirming that the sensor performs consistently across a wide spectrum of humidity levels. Figure 8c displays the repeatability profile of the sensor, obtained over six consecutive response–recovery cycles. Each cycle shows complete and stable response and recovery characteristics, underscoring the excellent repeatability imparted by the dual-site doping strategy adopted in the material design. Finally, Figure 8d illustrates the long-term stability test conducted over a period of 60 days. The initial resistance and response values exhibit minor fluctuations over time, which can be ascribed to day-to-day variations in the ambient laboratory conditions. Importantly, the amplitude of these fluctuations does not exceed $\pm$ 0.5, confirming the robust long-term stability of the sensor. Overall, the comprehensive evaluation reveals that the S-2 sensor possesses remarkable gas-sensing properties: high selectivity toward methane, strong tolerance to moisture, reliable repeatability, and excellent operational stability, making it a promising sensing device for practical applications in complex environmental conditions.

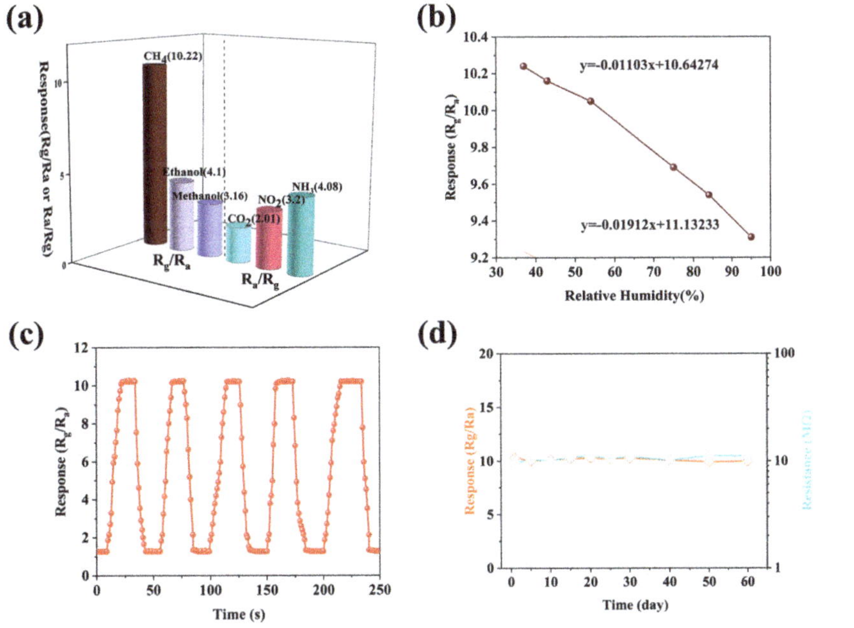

Figure 8. (**a**) Response of S-2 to different gases; (**b**) Fitted curve of the response of S-2 under different humidity levels; (**c**) Repeatability curve of S-2; (**d**) Long-term stability test of S-2.

https://doi.org/10.3390/pr14050836

3.2. Gas Sensing Mechanism

The gas-sensing mechanism observed in this study is interpreted using the widely accepted oxygen ionosorption model. Methane (CH_4) serves as a typical reducing gas. When it interacts with different types of chemisorbed oxygen ions on the material surface, a series of redox reactions occur, leading to measurable changes in the electrical properties of the sensor. A schematic diagram illustrating this mechanism is presented in Figure 9. Based on comprehensive material characterization and gas-sensing tests, it has been firmly established that the dual-site doped material behaves as a p-type semiconductor, where the majority charge carriers are holes. The sensing mechanism is explained in two distinct stages: the first describes the material's behavior in air, and the second explains its response in the presence of the target gas. The process begins with the adsorption of atmospheric oxygen molecules onto the sensor surface when in air. These oxygen molecules capture free electrons from the conduction band of the p-type semiconductor, becoming chemisorbed oxygen ions (O_{2-}, O^-, or O^{2-}) [31]. The specific type of ion formed depends significantly on the operating temperature. At lower temperatures (typically below 150 °C), O_{2-} is predominant, while at higher temperatures (like the optimal 150 °C used in this study), O^- becomes the more stable and reactive species. This extraction of electrons leads to the widening of the hole-accumulation layer (for a p-type semiconductor) or, equivalently, the formation and thickening of an electron depletion layer at the surface. The increased hole concentration in the near-surface region caused small resistance in air, which serves as the baseline resistance (Ra). Upon exposure to methane (CH_4), the pre-adsorbed reactive oxygen ions (O^-) on the surface react with the reducing gas molecules. Methane is oxidized, typically forming carbon dioxide (CO_2) and water vapor (H_2O), while the oxygen ions are consumed. The freed electrons subsequently re-populate the semiconductor's conduction band. The surface reaction can be summarized as [32,33]:

$$O_2 \text{ (gas)} \rightarrow O_2 \text{ (ads)} \quad (3)$$

$$O_2 \text{ (ads)} + 2e^- = 2O^- \text{ (150 °C)} \quad (4)$$

$$CH_4 + O^- = CO_2 + H_2O \quad (5)$$

For a p-type semiconductor, the injection of these electrons leads to the recombination of excess holes at the surface. This recombination reduces the hole concentration, causing the hole-accumulation layer to shrink and the electron depletion layer to become even thicker. Consequently, the sensor's electrical resistance increases significantly in the presence of the target gas (R_g). The sensor response (S) is defined as R_g/R_a for a p-type material, and the magnitude of this resistance change is directly correlated with the concentration of the target gas. The exceptional performance of the S-2 sensor can be directly linked to this mechanism and the effects of dual-site doping. The increased specific surface area provides more sites for oxygen adsorption. The higher carrier concentration and improved electron mobility, confirmed by impedance spectroscopy, facilitate faster electron exchange during both the initial oxygen ionosorption and the subsequent redox reaction with methane. Furthermore, the presence of oxygen vacancies, indicated by elemental analysis, acts as active centers for oxygen dissociation and enhances the surface reactivity [34,35]. This synergistic effect, more adsorption sites, faster charge transfer, and higher surface activity, explains the significantly enhanced response, faster kinetics, and superior selectivity of the dual-doped S-2 sensor compared to the undoped S-1, as comprehensively demonstrated in the experimental results.

https://doi.org/10.3390/pr14050836

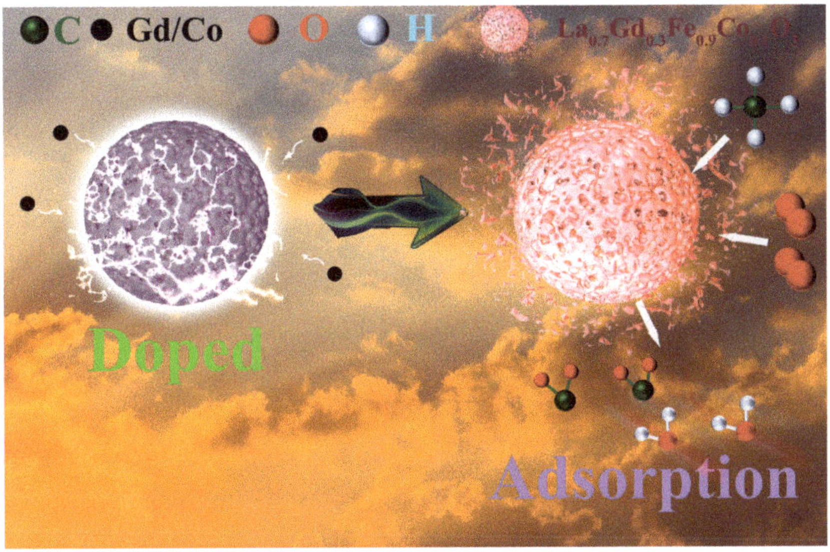

Figure 9. Schematic Diagram of the Sensing Mechanism (of S-2) (Illustration created by the author).

Through the introduction of dual-site doping, the material not only generates a high density of surface adsorption sites but also undergoes a significant microstructural transformation. The successful incorporation of dopant elements and the formation of the designed material have been verified by ICP-MS analysis and detailed XRD refinement. Furthermore, impedance spectroscopy measurements confirm that the dual-site doped material possesses a higher charge-carrier concentration during electrical transport. The remarkable enhancement in electron transport capability, which is a key reason for the improved gas-sensing performance, stems from a synergistic enhancement effect triggered by the co-modification at the A-site (La/Gd) and the B-site (Fe/Co). The underlying mechanism can be elaborated as follows. First, the core driving force originates from the co-doping of Fe and Co at the B-site. These transition metal ions possess variable valence states. Their coexistence directly introduces abundant multi-valent pairs, such as Fe^{3+}/Fe^{4+} and Co^{3+}/Co^{4+} [36]. This creates a robust foundation for electron hopping via a small-polaron mechanism between different ions, effectively establishing microscopic pathways for charge transfer within the lattice. Second, the partial substitution of La^{3+} with the smaller Gd^{3+} at the A-site induces lattice contraction and internal strain. More importantly, this strain precisely modulates the bond lengths and angles of the Fe/Co–O bonds. Such fine-tuning optimizes the geometric configuration of the electron-hopping paths mentioned above, effectively reducing the activation energy required for electron migration. Concurrently, this lattice adjustment works in concert with the B-site doping to collectively influence the overall charge balance of the system. This coordinated effect regulates the formation and distribution of oxygen vacancies, which in turn modulates the concentration of electronic charge carriers, creating a more favorable local chemical environment for efficient electron transport. Finally, the strategy of simultaneously introducing heteroatoms at both A and B sites effectively disrupts any potential long-range ordered arrangement of elements in the lattice. It suppresses the formation of local charge ordering or clusters, promoting a uniform atomic-scale distribution of ions with different valence states [37–40]. Consequently, a continuous, interconnected, and homogeneous electron transport network is formed throughout the entire material. This percolating network significantly reduces scattering and barriers encountered by electrons during long-range migration, leading to drastically improved bulk conductivity. In summary, the dual-site doping creates a synergistic effect: B-site doping establishes the fundamental charge-hopping centers, while A-site doping fine-tunes the lattice to lower the energy barriers for hopping and optimizes the carrier concentration. Together, they construct a highly efficient and uniform conductive network [41–47]. This significantly enhanced charge transport capability facilitates faster charge exchange between the sensing material and adsorbed gas molecules, which is a fundamental reason for the superior response speed, higher sensitivity, and improved overall gas-sensing performance observed in the S-2 sensor.

4. Conclusions

This study successfully synthesized $La_{0.7}Gd_{0.3}Fe_{0.9}Co_{0.1}O_3$ (S-2) perovskite microspheres via an efficient ultrasound-assisted hydrothermal method. The characterization results (XRD, SEM, ICP) confirmed high purity, homogeneous doping, and well-defined spherical morphology. The fabricated sensor exhibited excellent low-temperature CH_4 sensing performance: at 150 °C, the response reached 10.22 to 100 ppm CH_4 with fast response/recovery times of 6.3 s and 5.4 s. The device also showed outstanding repeatability, stability, humidity resistance, and maintained performance over 60 days. These results demonstrate the material's strong potential as a high-performance low-temperature CH_4 sensor.

Supplementary Materials: The following supporting information can be downloaded at: https://www.mdpi.com/article/10.3390/pr14050836/s1, Figure S1: S-2 response to high-concentration CH_4 and the corresponding fitting curve: (a) Gas response curves to 1000 ppm, 3000 ppm, and 5000 ppm CH_4; (b) Fitting curve within the high-concentration range; Figure S2: Response of S-2 to H_2S, CO, and H_2; Table S1: Comparison of CH_4-sensing performance of gas sensors. Refs. [48–53] are cited in Table S1.

Funding: Key Science and Technology Project of Ministry of Emergency Management of the People's Republic of China (2024EMST070703).

Data Availability Statement: The original contributions presented in this study are included in the article/Supplementary Materials. Further inquiries can be directed to the corresponding author.

Conflicts of Interest: The author declares no conflicts of interest.

References

1. Sun, H.; Qiao, S.; He, Y.; Liu, Y.; Ma, Y. Highly sensitive CH_4,C_2H_2 and CO simultaneous measurement LITES sensor based on multi-pass cell with overlapped spots pattern and QTFs with low resonant frequency. *Opt. Express* **2024**, *32*, 28183–28194. [CrossRef]
2. Zhang, M.; Wu, G.; Guan, Y.; Xing, J.; Gao, T.; Sun, J.; Gong, Z.; Peng, W. A miniaturized PET-silicon double cantilever-enhanced fiber-optic photoacoustic sensor for rapid and high sensitivity detection of C_2H_2 and CH_4. *Sens. Actuators B Chem.* **2026**, *451*, 139428. [CrossRef]
3. Lv, C.; Lv, X.L.; Wang, Z.; Zhao, T.; Tian, W.; Zhou, Q.; Zeng, L.; Wan, M.; Liu, C. A focal quotient gradient system method for deep neural network training. *Appl. Soft Comput.* **2025**, *184*, 113704. [CrossRef]
4. Bai, H.; Feng, C.; Guo, R.; Chen, Y.; Yuan, W.; Du, Y.; Feng, Y.; Liu, W.; Liu, K.; Guo, F.; et al. Wireless gas sensor based on monolayer $Ti_3C_2T_x$-modified GaN/CuO heterostructures for ultrafast and trace detection of NO_2 gas. *Chem. Eng. J.* **2025**, *524*, 169751. [CrossRef]
5. Lv, C.L.; Cao, Y.D.; Feng, Y.; Fan, L.L.; Zhang, J.Y.; Song, Y.H.; Liu, H.; Gao, G.G. Crosslinked Vinyl-Capped Polyoxometalates to Construct a Three-Dimensional Porous Inorganic–Organic Catalyst to Effectively Suppress Polysulfide Shuttle in Li–S Batteries. *Adv. Energy Mater.* **2025**, e05978. [CrossRef]
6. Guo, S.; Song, Y.; Wu, Y.; Hu, J.; Lu, J.; Wang, C.; Jia, Z.; Hua, L. Microscopic mechanism of enhanced strength-plasticity synergy in pre-damaged TC11 titanium alloys via novel electroshock treatment. *J. Alloy. Compd.* **2026**, *1052*, 186139. [CrossRef]
7. Zhang, S.; Lin, J.; Yu, C.; Guo, Z.; Tang, C.; Huang, Y. An ultra-high response humidity sensor based on porous boron nitride nanofibers: For respiratory monitoring and non-contact sensing. *Sens. Actuators B Chem.* **2025**, *437*, 137744. [CrossRef]

8. Mao, Z.; Qiu, D.; Xu, Z.; Hao, B.; Zhang, H.-Y.; Sun, H.; Pei, X.; Wang, M.; Li, Y.; Gu, Z.-B.; et al. A Two-Dimensional Superconducting Electron Gas at $LaFeO_3/SrTiO_3$ Interfaces. *Nano Lett.* **2024**, *25*, 586–592. [CrossRef] [PubMed]
9. Sun, H.; Yin, C.; Attfield, J.P.; Yang, M. Tunable Magnetoresistive Gas Sensing Enabled by La-Deficient $LaFeO_3$. *ACS Sens.* **2025**, *10*, 6800–6807. [CrossRef]
10. Al-Majali, M.R.; Zhang, M.; Al-Majali, Y.T.; Trembly, J.P. Impact of raw material on thermo-physical properties of carbon foam. *Can. J. Chem. Eng.* **2025**, *103*, 1309–1318. [CrossRef]
11. Wang, D.; Tan, R.; Yu, H.; Chen, H.; Ma, X.; Zhao, H.; Kang, X.; Fang, H. High response H_2S chemiresistive gas sensor based on multidimensional CuO/CeO_2: The application for real-time portable alarm device. *Sens. Actuators B Chem.* **2025**, *424*, 136884. [CrossRef]
12. Zhao, J.; Zhang, M.; Wang, C.; Yu, W.; Zhu, Y.; Zhu, P. First-Principles Study of CO, NH_3, HCN, CNCl, and Cl_2 Gas Adsorption Behaviors of Metal and Cyclic C–Metal B-and N-Site-Doped h-BNs. *Electron. Mater. Lett.* **2025**, *21*, 268–288. [CrossRef]
13. Fu, B.; Bi, L.; Lin, J.; Fu, J.; Wen, J.; Zou, B.; Wang, C.; Wang, Y. Low-temperature detection and excellent selectivity of ethanol using $LaFeO_3$ gas sensors with dual regulation of doping and non-stoichiometry. *J. Alloys Compd.* **2025**, *1013*, 178590. [CrossRef]
14. Meng, F.; Qin, L.; Gao, H.; Zhu, H.; Yuan, Z. Perovskite-structured $LaFeO_3$ modified In_2O_3 gas sensor with high selectivity and ultra-low detection limit for 2-butanone. *J. Alloys Compd.* **2024**, *970*, 172464. [CrossRef]
15. Zhang, G.H.; Chen, Q.; Deng, X.Y.; Jiao, H.Y.; Wang, P.Y.; Gengzang, D.J. Synthesis and characterization of In-doped $LaFeO_3$ hollow nanofibers with enhanced formaldehyde sensing properties. *Mater. Lett.* **2019**, *236*, 229–232. [CrossRef]
16. Li, C.; Li, C.; Tian, S.; Zhou, X.; Zhou, B.; Qian, L.; Wang, L.; Li, D.; Li, H.; Li, M. Surface acoustic wave humidity sensors based on Al-doped ZnO ultrathin nanosheets architecture and their applications for skin moisture detection. *Sens. Actuators B Chem.* **2025**, *443*, 138222. [CrossRef]
17. Wang, Z.; Yuan, J.; Zhao, L.; Ren, J. Significant improvement of the thermal conductivity and dielectric properties of cyanoethyl cellulose films by introducing barium titanate decorated boron nitride nanosheet. *Mater. Lett.* **2022**, *323*, 132588. [CrossRef]
18. Hu, J.; Duan, Z.; Tan, C.; Wang, Y.; Yuan, Z.; Jiang, Y.; Tai, H. Self-powered electrochemical humidity sensor based on Cu/(cellulose nanocrystals/NaCl)/Zn structure for respiratory rate detection. *Inorg. Chem. Commun.* **2025**, *180*, 114994. [CrossRef]
19. Shakya, P.; Khan, M.S.; Kumar, S.; Jhankal, D.; Verma, N.; Sachdev, K. Understanding the gas sensing mechanism of MoSe2 nanoroses and rGO composite-based chemiresistive sensors for room temperature NO_2 detection. *Chem. Eng. J.* **2025**, *503*, 158178. [CrossRef]
20. Wan, Y.; Liu, Y.; Huang, Y.; Zhao, L.; Lu, H.-L.; Gao, L. Realizing fast response speed and enhanced selectivity for ethanol gas sensors with Au modified NiO/ZnO heterostructures. *Sens. Actuators B Chem.* **2025**, *444*, 138521. [CrossRef]
21. Wang, Y.; Li, J.; Xiao, D.; Zhang, D.; Liu, Y.; Sun, M.; Chen, S.; Sun, M. Progress in functionalized WO_3-based gas sensors for selective H_2S and NH_3: A review. *Ceram. Int.* **2024**, *50*, 40631–40665. [CrossRef]
22. Malik, P.; Duhan, S.; Malik, R. A high-performance humidity sensor based on 3D porous SnO_2-encapsulated MCM-48 for real-time breath monitoring and contactless gesture detection. *Mater. Adv.* **2024**, *5*, 2510–2525. [CrossRef]
23. Lee, J.-S.; Lee, S.; Seo, J.-W.; Choi, S.-H.; Kim, S.-J.; Choi, S.-J. Porous one-dimensional $CuWO_4$-WO_3 nanofibers with enhanced gas accessibility and catalytic sensitization for selective H2S sensors. *Sens. Actuators B Chem.* **2025**, *444*, 138350. [CrossRef]
24. Davydova, M.; Ashcheulov, P.; Palla Papavlu, A.; Laposa, A.; Taylor, A.; Kroutil, J.; Lippert, T. Boron-doped diamond as a functional semiconductive layer in chemiresistive sensors for the enhanced gas sensing of NO_2 at room temperature. *Sens. Actuators A Phys.* **2025**, *389*, 116525. [CrossRef]
25. Zhu, S.-G.; Cui, Y.; Zhang, L.-X.; Shao, H.; Yin, J.; Bie, L.-J. Cation-oxygen dual-defective $ACu_3Ti_4O_{12}$ (A = Sr, Ba) perovskites enable high-performance humidity sensors for human-body related moisture monitoring. *Ceram. Int.* **2024**, *50*, 43182–43191. [CrossRef]
26. Zhang, Z.-A.; Nie, L.-F.; Zhou, Q.; Song, Z.-D.; Pan, G.-B. Chemiresistive H_2S gas sensors based on composites of ZnO nanocrystals and foam-like GaN fabricated by photoelectrochemical etching and a sol-gel method. *Sens. Actuators B Chem.* **2023**, *393*, 134148. [CrossRef]
27. Yue, C.; Zhang, Z.; Liu, Z.; Mu, Y.; Yang, Z.; Dastan, D.; Yin, X.-T.; Ma, X. High-performance ethanol gas sensor with fast response/recovery rate based on the construction of SnO_2-CdS heterojunction. *J. Alloys Compd.* **2024**, *981*, 173742. [CrossRef]
28. Li, Q.; Zhu, G. Controlling negative permittivity and permeability behavior in LaFeO3 through sintering temperature. *Ceram. Int.* **2021**, *47*, 5244–5248. [CrossRef]
29. Kim, S.; Kim, I.H.; Fattahimoghaddam, H.; Cho, M.-K.; Jeong, Y.J.; Lee, W. Copper-doped tellurium nanowire-based gas sensor for chemiresistive detection of nitrogen dioxide. *Sens. Actuators B Chem.* **2025**, *433*, 137587. [CrossRef]
30. Luo, Q.; Chen, Y.; Feng, C.; Gao, S.; Lan, T.; Fu, Q.; Teng, R. Efficient detection of n-butanol vapor using Er^{3+} cations doped $LaFeO_3$ nanofibers. *Ceram. Int.* **2025**, *51*, 35529–35540. [CrossRef]
31. Zhu, L.; Wang, J.; Liu, J.; Chen, X.; Xu, Z.; Ma, Q.; Wang, Z.; Liang, J.; Li, S.; Yan, W. Designing highly sensitive formaldehyde sensors via A-site cation deficiency in $LaFeO_3$ hollow nanofibers. *Appl. Surf. Sci.* **2022**, *590*, 153085. [CrossRef]

32. Gao, Y.; Jing, P.; Yan, N.; Hilbers, M.; Zhang, H.; Rothenberg, G.; Tanase, S. Dual-mode humidity detection using a lanthanide-based metal–organic framework: Towards multifunctional humidity sensors. *Chem. Commun.* **2017**, *53*, 4465–4468. [CrossRef] [PubMed]
33. Liu, J.; Du, S.; Huang, Z.; Liu, N.; Shao, Z.; Qin, N.; Wang, Y.; Wang, H.; Ni, Z.; Yang, L. Enhanced Reduction of Nitrate to Ammonia at the Co-N Heteroatomic Interface in MOF-Derived Porous Carbon. *Materials* **2025**, *18*, 2976. [CrossRef]
34. Su, X.; Shan, H.; Tian, Y.; Guo, W.; Zhao, P.; Xue, L.; Zhang, Y. Enhancing the sensitivity of Co-doped $LaFeO_3$ thick-film sensors for low-concentration CO_2 detection: Investigating the synergistic effect of complex defects and adsorbed oxygen. *J. Environ. Chem. Eng.* **2025**, *13*, 116517. [CrossRef]
35. Zhang, J.; Yue, Y.; Chen, Y.; Xu, H.; Xie, J.; Zhang, G. Enhanced n-butanol sensing performance of $Co_3O_4/LaFeO_3$ nanocomposites. *Mater. Lett.* **2024**, *377*, 137574. [CrossRef]
36. Deng, C.; Li, Y.; Xiong, F.; Liu, H.; Li, X.; Zeng, Y. Detection of Rupture Damage Degree in Laminated Rubber Bearings Using a Piezoelectric-Based Active Sensing Method and Hybrid Machine Learning Algorithms. *Struct. Control Health Monit.* **2025**, *1*, 6694610. [CrossRef]
37. Malik, P.; Duhan, S.; Sheoran, G.; Sehrawat, S.; Boora, A.; Rohilla, B.; Kumar, P. High Performance Humidity Sensor based on Fe_2O_3/MCM-48 as Wearable Multifunctional Device. *Mater. Res. Bull.* **2025**, *192*, 113626. [CrossRef]
38. Tseng, S.-F.; Lin, C.-I.; Chang, T.-L.; Huang, C.-Y.; Chiu, P.-S. High performance humidity sensors based on electrospinning CuO nanofibers on GZO/glass substrates. *Ceram. Int.* **2025**, *51*, 41821–41830. [CrossRef]
39. Xiao, C.; Tang, Z.; Ma, Z.; Gao, X.; Wang, H.; Jia, L. High performance porous $LaFeO_3$ gas sensor with embedded p-n junctions enabling ppb-level formaldehyde detection. *Sens. Actuators B Chem.* **2023**, *397*, 134670. [CrossRef]
40. Huang, J.; He, X.; Zou, S.; Ling, K.; Zhu, H.; Jiang, Q.; Zhang, Y.; Feng, Z.; Wang, P.; Duan, X.; et al. A Flexible Electrochemical Sensor Based on Porous Ceria Hollow Microspheres Nanozyme for Sensitive Detection of H_2O_2. *Biosensors* **2025**, *15*, 664. [CrossRef]
41. Li, Y.; Han, T.; Zhao, M.; Han, J.; Zhao, R.; Xu, Z.; Cao, W.; Ma, M.; Zhang, C.; Luo, J. Modulating interfacial shear of nanoconfined hydration layer via surface charging. *Nano Energy* **2026**, *148*, 111692. [CrossRef]
42. Yang, B.; Li, Y.; Lin, H.; Li, W.; Liu, X.; Shen, J. Study on crack resistance and impermeability of functional material modified concrete. *Compos. Commun.* **2026**, *62*, 102703. [CrossRef]
43. Zhao, Z.; Meng, T.; Tang, R.; Cai, L.; He, Z.; Wen, W.; Cao, M. A Homo-Triatomic Molybdenum Cluster Catalyst Enables Synergistic Optimization of Multi-Step Sulfur Reduction Chemistry. *Adv. Mater.* **2026**, e72365. [CrossRef]
44. Wu, S.; Li, D.; Wei, M.; Li, B.; Wu, L. Polyoxometalate-mediated film bearing near-infrared photothermal self-healing property for seawater desalination. *Chem. Eng. J.* **2024**, *499*, 156030. [CrossRef]
45. Zhao, C.; Meng, Z.; Yi, J.; Chen, C.Q. Auxetic metamaterials with double re-entrant configuration. *Int. J. Mech. Sci.* **2025**, *301*, 110505. [CrossRef]
46. Zhang, D.; Li, R.; Luo, H.; Meng, Z.; Yao, J.; Liu, H.; Huang, Y.; Li, S.; Yu, P.; Yang, J.; et al. Break-through amplified spontaneous emission with ultra-low threshold in perovskite via synergetic moisture and BHT dual strategies. *Light Sci. Appl.* **2026**, *15*, 99. [CrossRef] [PubMed]
47. Liu, Y.; Zhou, L.; Wan, X.; Tang, Y.; Liu, Q.; Li, W.; Liao, J. Synthesis and Characterization of a Temperature-Sensitive Microcapsule Gelling Agent for High-Temperature Acid Release. *ACS Omega* **2024**, *9*, 20849–20858. [CrossRef] [PubMed]
48. Vatandoust, L.; Habibi, A.; Naghshara, H.; Aref, S.M. Fabrication and investigation of TiO1.5/ZnO nanocomposite nanosensor for detection of CO and CH_4 gases. *Surf. Interfaces* **2022**, *31*, 102001. [CrossRef]
49. Pholoana, M.G.; Malgas, G.F.; Morulane, K.L.; Tshilongo, J.; Swart, H.C.; Motaung, D.E. Dual selective sensing of CH_4 and ultra-low NO gas utilizing Ag-decorated CeO_2-CuO nanorods: Role of humidity in p-n conductivity transition. *Inorg. Chem. Commun.* **2026**, *185*, 116098. [CrossRef]
50. Bunpang, K.; Wisitsoraat, A.; Tuantranont, A.; Singkammo, S.; Phanichphant, S.; Liewhiran, C. Highly selective and sensitive CH_4 gas sensors based on flame-spray-made Cr-doped SnO_2 particulate films. *Sens. Actuators B Chem.* **2019**, *291*, 177–191. [CrossRef]
51. Wang, Y.; Cui, Y.; Meng, X.; Zhang, Z.; Cao, J. A gas sensor based on Ag-modified ZnO flower-like microspheres: Temperature-modulated dual selectivity to CO and CH_4. *Surf. Interfaces* **2021**, *24*, 101110. [CrossRef]
52. Yin, X.-T.; Zhou, W.-D.; Dastan, D.; Li, J.; Tan, X.-M.; Liu, Y.; Gao, X.-C.; Ma, X.-G. Selectivity sensing response of ZnO-xCo_3O_4 based sensor to CO against CH_4. *Mater. Sci. Semicond. Process.* **2022**, *149*, 106883. [CrossRef]
53. Sun, X.; Tang, M.; Yu, M.; Fan, Y.; Qin, C.; Cao, J.; Wang, Y. UV-activated CH_4 gas sensor based on Pd@Ni/ZnO microspheres. *Mater. Today Commun.* **2024**, *40*, 109551. [CrossRef]

https://doi.org/10.3390/pr14050836

processes

MDPI

Article

Mining and Processing of Mineral Resources: A Comparative Study of Simulated and Operational Processes

Radim Rybár *, Martin Beer and Lucia Bednárová

Institute of Earth Sources, Faculty of Mining, Ecology, Process Technologies and Geotechnology, Technical University of Košice, Letná 9, 042 00 Košice, Slovakia; lucia.bednarova@tuke.sk (L.B.)
* Correspondence: radim.rybar@tuke.sk

Abstract

The aim of this study is to analyze the representation of geological, mining, processing, and environmental processes in platform Minecraft. Based on a methodological comparison of in-platform mechanics with technological and geoscientific procedures, the article assesses the degree of accuracy, simplification, and didactic applicability of individual processes related to the extraction and use of mineral resources. The analysis is structured into seven main thematic areas covering the entire resource value chain—from geological exploration through mining, ore beneficiation and processing, to quantitative indicators (e.g., waste-to-ore ratio), fluid resources, and environmental impacts. Special attention is given to the potential of modifications that significantly enhance the complexity and accuracy of simulated processes. The results show that Minecraft, enriched with thematic mods, can serve as an accessible and flexible tool for the popularization and education of industrial and geoscientific processes, while engaging a wide range of audiences.

Keywords: Minecraft; mineral resources; mining; processing engineering; resource value chain; environmental impact

1. Introduction

Today's world faces challenges that are closely tied to the ways in which we extract and utilize mineral resources. As noted by [1], mineral resource extraction has been one of the key activities shaping the development of human society from prehistory to the present day. Climate change, geopolitical instability, and the growing demand for critical raw materials—such as lithium, copper, cobalt, and rare earth elements—highlight the urgent need for greater awareness of the processes involved in geological exploration, mining, and resource processing [2]. Nevertheless, this topic remains complex, poorly understood by the general public, and often controversial. The communication gap between the expert community and the wider society can result in misunderstandings, oversimplifications, or even rejection of the scientific and technological aspects of mineral resource utilization.

As noted in [3], there is a growing importance of informal education and popularization tools that can convey scientific knowledge in an engaging and accessible manner. One of the most prominent digital phenomena proving to be a suitable platform for developing scientific thinking and acquiring geoscientific concepts is the platform Minecraft. This platform, developed by Mojang Studios, is based on a system of interaction with an environment composed of unit-based blocks representing various materials and objects [4]. With more than 140 million active players per month and a vast community of users, educators, and content creators, Minecraft represents a unique space for bridging gameplay with scientific education [5–9].

One of the key mechanics in Minecraft is the extraction of mineral resources. The player explores a procedurally generated world, searches for and mines various types of ores—from coal and iron to copper, gold, diamonds, and emeralds. These ores appear in the game based on depth, rock type, and environment. In newer versions of Minecraft (particularly since version 1.18), the ore distribution system partially reflects geological principles such as stratigraphic layering, localized occurrences, and large ore veins. Examples of raw material blocks, the distribution of raw material blocks within the rock environment, and the approach to designing a mining excavation for surface mineral extraction in Minecraft are shown in Figure 1. While these processes are simplified in the base version of the Minecraft, various community-developed modifications (e.g., BetterGeo, Immersive Geology, Geoimmersion) greatly enhance the level of geological realism, allowing players to simulate complex processes related to exploration, mining, and mineral processing [2]. These mods introduce advanced features such as chemical separation, gravity-based sorting, oil extraction, fuel refining, and realistic mining technologies involving appropriate equipment and tools.

Gamified learning is an approach in which game elements—such as points, levels, challenges, or storylines—are integrated into educational activities to increase participant engagement and motivation. In the context of geoscience disciplines, this approach has proven particularly effective, as it enables the combination of visually appealing and interactive environments with solving real-world problems and modeling natural processes [10]. Studies confirm that gamification supports deeper cognitive engagement, experimentation, and collaboration, while enhancing the ability to transfer acquired knowledge into practical situations [11]. The latest systematic review of gamification in geoscience education highlights that this approach has high potential for developing spatial thinking, understanding complex processes, and ensuring long-term knowledge retention, especially when specialized digital tools and simulations are utilized [12].

Numerous scientific studies and projects over the past decade have confirmed that Minecraft holds significant potential for education and the popularization of scientific topics, including geoscientific disciplines [13–21]. A systematic review of research published in [22] demonstrated that Minecraft is utilized in educational contexts across a wide range of fields—from mathematics and physics to biology and technology—while emphasizing that the application of the game in geology and mineral resource sciences remains relatively uncommon. Similarly, the study in [13] identified three primary areas of educational use: integration into formal instruction, enhancement of student engagement, and development of practical skills. Although these studies highlight a high degree of motivation and interactivity, they rarely address content accuracy or analogies with real-world geological processes.

In the Minecraft environment, there is a concept of modifications (mods), which are external software extensions that alter or expand the original game content. These add-ons, most often created by the gaming community or independent developers, can change existing game mechanics, add new objects, blocks, tools, game modes, or entire technological systems. The scope of modifications ranges from simple add-ons (e.g., new decorative elements or changes to the user interface) to complex expansions that simulate realistic geological processes, mining techniques, raw material processing, and energy systems that are absent in the base game. For instance, the BetterGeoEdu project, supported by the EIT RawMaterials initiative, developed the BetterGeo extension, which enriches the world of Minecraft with realistic geological layers, new minerals, rocks, and processing techniques [23]. The aim of the project was to introduce students to basic geological principles, responsible resource management, and the environmental aspects of mining. Similarly, the British Geological Survey (BGS) used Minecraft to visualize the geological structure of the United Kingdom, allowing the public to perceive rock layers that would otherwise remain inaccessible. Several case studies also demonstrate that Minecraft can serve as an effective tool for conveying concepts such as stratigraphy, the rock cycle, or geochemical phenomena through simulations of layering, precipitation, deformation, and player interaction with the environment [24–26]. Research published in [27] shows how students modeled cave systems, planetary structures, or landscape evolution, highlighting the platform's strong potential for developing spatial imagination and fundamental geoscience concepts.

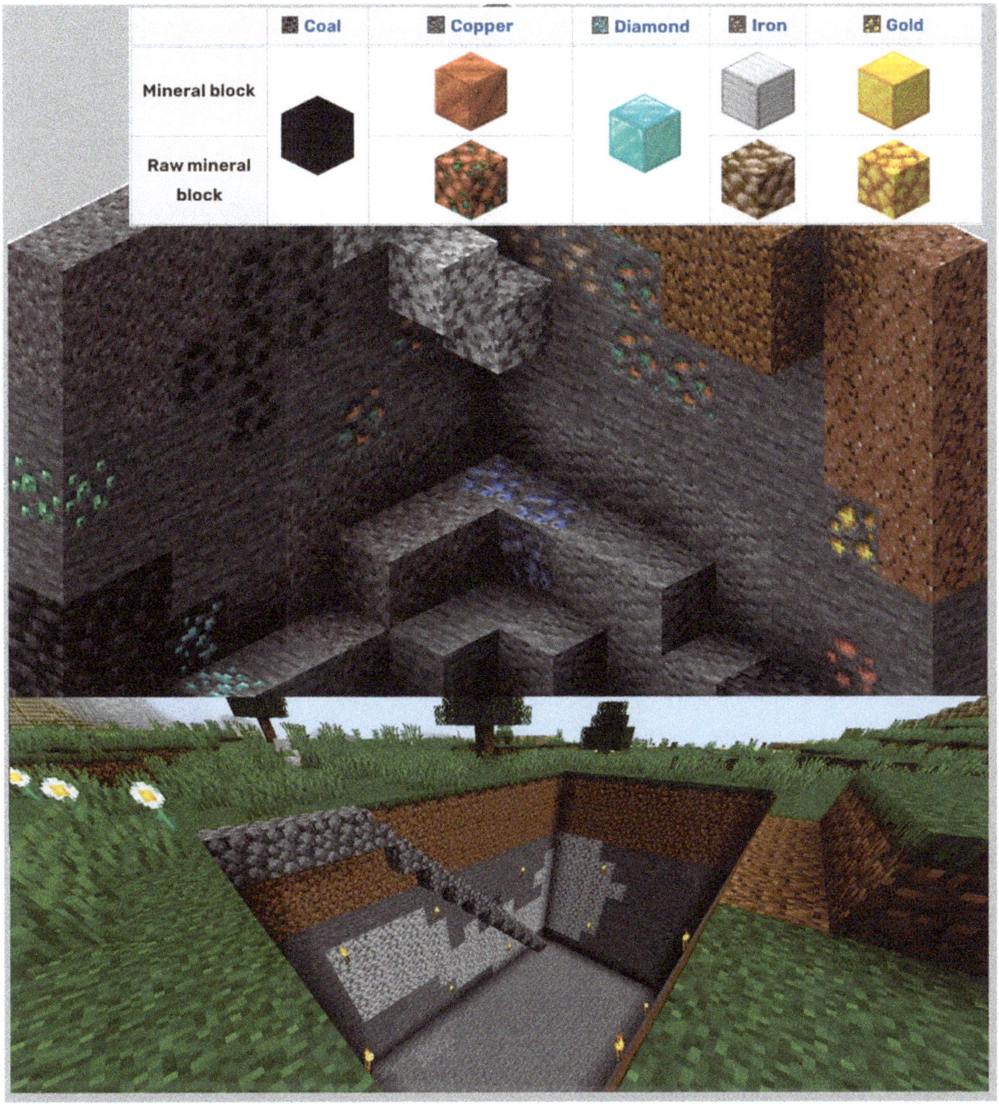

Figure 1. A composite image showing examples of raw material blocks, the distribution of raw material blocks within the rock environment, and the design of a mining excavation for surface mineral extraction in Minecraft.

Despite these positive examples, the academic literature has so far only marginally addressed deeper process-level analysis—that is, the extent to which Minecraft's mechanisms correspond to real-world processes of exploration, extraction, beneficiation, and processing of mineral resources. Most existing studies focus on qualitative aspects of learning, player motivation, or specific classroom applications. What remains largely absent is a systematic comparison of concrete process steps in the platform versus those in industry—such as deposit formation, geological prospecting, rock fragmentation, sorting, flotation, or chemical treatment of ores. This gap provides the impetus for our research, which focuses precisely on these aspects.

In this study, we concentrate on a detailed analysis of how specific processes are represented in Minecraft, how closely they align with actual geoscientific and engineering practices, and what their educational potential and limitations are. In doing so, we aim to expand current knowledge about the use of digital gaming platforms in education and scientific communication, with special attention given to the complex nature of the raw material cycle as a system of interconnected natural and technological processes.

This study aims to systematically examine how accurately Minecraft represents individual aspects of the real-world mineral resource acquisition processes. It focuses on the following key research questions:

(1) To what extent are the game mechanics of mining, ore distribution, and processing aligned with geological and technological realities?
(2) What analogies can be drawn between mining techniques in the game and real-world extraction methods (e.g., underground vs. surface mining)?
(3) What is the educational potential of the base Minecraft game and its modifications in the context of teaching geoscientific and engineering disciplines?
(4) Where are the limits of simplification in the game's model, and what risks may arise from its uncritical use in educational contexts?

A specific contribution of this work lies in the comparison of quantitative parameters—such as the waste-to-ore ratio, depth-based ore distribution, or separation processes—with real geological data and industrial practices. Such comparisons make it possible to identify both the strengths and limitations of the game model, while also creating opportunities to propose potential modifications or expansions aimed at increasing its scientific accuracy and didactic value.

2. Materials and Methods

This study focuses on the analysis of selected processes within the environment of Minecraft—including its educational and modified versions—that resemble real-world geological, mining, and mineral processing operations. The objective was to systematically examine the extent to which in-game mechanisms represent the fundamental stages of the real raw material cycle and to assess their didactic potential for use in education and the popularization of mineral resource acquisition processes. In this context, Minecraft was approached as a simplified, model-representational system that enables the simulation of complex natural and technological phenomena in an accessible and interactive digital form.

The methodological framework of the study was based on a multi-level comparison between virtual game processes and their real industrial or geoscientific counterparts. The analysis combined qualitative and comparative approaches, with the primary aim not to evaluate the completeness or accuracy of individual models, but rather their ability to faithfully reflect the structure and logic of actual raw material processes.

Data collection was conducted in parallel from multiple sources [28–32]. The first group consisted of official documentation and community wiki platforms focused on Minecraft (e.g., Minecraft Wiki, BetterGeo Wiki, and the Modrinth and CurseForge mod-ding sites), which describe game mechanics such as ore distribution, crafting recipes, generation of geological structures, and tool usage. These sources provided essential information on the functionality of individual game elements in the context of resource acquisition. The second group included monographs [33–35], as well as scientific and technical articles dealing with real-world processes of geological exploration, mining, separation, and mineral processing [36–48]. These materials were crucial for establishing a reference framework through which the degree of analogy and fidelity of in-game processes could be assessed.

An important component of the study involved practical observation of game mechanics. This included the use of gameplay videos—particularly "mod showcases" on the YouTube platform—as well as direct observation of the game environment in both the base version of Minecraft (version 1.18 and above) and modified versions incorporating mods such as BetterGeo, BuildCraft, Immersive Engineering, Immersive Petroleum, Geoimmersion, TerraFirmaCraft, and Rocks Done Right. These modifications were selected based on their thematic focus on geological and mining elements and analyzed in terms of implemented processes, levels of simulation, structure of input–output operations, and visualization capabilities [49]. The selection of mods was guided by two main criteria: (1) the mod had to expand or more realistically simulate a specific phase of the mineral resource value chain, and (2) it had to be technically stable and compatible with the current version of the game. Only publicly available and well-documented mods were included.

Supplementary sources included discussion forums and player communities on platforms such as Reddit and Discord, which provided feedback and insights on the practical use of mods, including their advantages and limitations from the players' perspective—an important consideration when evaluating their didactic application.

Numerical data on ore occurrence and generation parameters were obtained through a combination of direct observation in three independently generated worlds (default seed, ver. 1.18+) and verified by comparison with results from community measurements available on the Minecraft Wiki and player forums. Each world was explored for approximately 6–8 h, with vertical profiles, ore counts per chunk, and average W:O ratios being mapped.

For the purposes of analysis, in-game processes were classified into thematic categories reflecting the real-world raw material value chain:

(1) Geological formation and distribution of resources,
(2) Geological exploration processes,
(3) Mining techniques and resource extraction,
(4) Beneficiation and processing of raw materials,
(5) Waste-to-ore ratio,
(6) Fluid-phase resources,
(7) Environmental and systemic processes.

Within each category, specific game elements (blocks, structures, tools, machines) that represent a given process or operation were identified. These elements were subsequently compared with real-world industrial technologies and standards applied in mining and mineral processing practices.

The comparison was conducted according to several criteria. The first was technological accuracy, meaning the extent to which game mechanics correspond to real-world processes (e.g., deposit formation, stratigraphic structure, extraction techniques, separation principles). The second criterion was processual consistency, the coherence of individual operations and their logical arrangement into a functional sequence. The third was quantitative realism, which involved examining factors such as the waste-to-ore ratio, ore yield, vertical distribution, rarity of elements, and related parameters. Additional criteria included the degree of simplification (abstraction from real-world models) and didactic potential (accessibility, clarity, memorability, and the ability to convey core principles).

The analysis was carried out in iterative steps: identification of an in-game process → determination of its real-world equivalent → qualitative and quantitative comparison → evaluation of similarities, discrepancies, and educational applicability. This methodological approach enabled not only a descriptive but also an evaluative assessment of how effectively and realistically Minecraft, as a digital model, can convey the complex domain of geological and technological processes. From the perspective of result validation and reproducibility verification, it is essential to note that the primary information base relied on the most reputable platforms in the field, such as Reddit (e.g., subreddits r/Minecraft or r/feedthebeast) and Discord servers, which contain empirical insights not explicitly stated in official descriptions and obtained directly by users. From a formalized perspective, this study has several methodological limitations, as full data triangulation was not applied—the results are based primarily on observations and the analysis of available in-game and community sources, without direct verification through experimental testing or surveys among players and educators. The scope of examined game worlds, modifications, and versions was also limited, although this was intended to eliminate marginal forms. However, given the nature of the examined processes and environment, these factors should not affect the broader interpretation of the results or their application in educational or research contexts. The authors believe that implementing some of the steps listed here as methodological limitations could, in fact, lead to a narrowing of validity or possible distortion of results, due to the increased influence of subjective inputs from a limited group of users. The results of this analysis are presented in the following sections of the article.

3. Results and Discussion

The results of this study are based on a systematic analysis aimed at comparing the individual processes related to the exploration, extraction, beneficiation, and processing of mineral resources within the Minecraft game environment with their real-world industrial and geological counterparts. Drawing from observations and examinations of both the base game mechanics and selected technical modifications, the study explored the degree of process accuracy, complexity, and didactic potential of specific activities. Attention was given not only to technical aspects—such as mining techniques, raw material processing, or the treatment of fluid resources—but also to broader systemic connections, including the waste-to-ore ratio, environmental impacts, and the game's ability to convey complex technological chains.

Each of the following subsections focuses on a specific type of process and its representation within the game, with the objective of evaluating the extent to which Minecraft can be utilized as a tool for illustrative, publicly accessible, and scientifically informed education on the acquisition and processing of mineral resources.

3.1. Geological Formation and Resource Distribution

One of the key aspects of realistically simulating mining processes is how geological conditions—particularly the formation of the rock environment and the distribution of mineral resources—are represented in the model. In the base version of Minecraft, rocks and ores are generated through procedural generation algorithms that create a three-dimensional environment composed of voxels (blocks). Each world is generated randomly, with primary block types including stone, deepslate, and various rocks such as granite, diorite, and andesite. These rock types are distributed along a vertical profile, with a significant breakthrough in geological accuracy introduced in version 1.18. This update implemented a new ore distribution algorithm based on uneven probability patterns and the principle of Reduced Air Exposure (RAE), which limits the visibility of certain ores in air-exposed environments.

In the Minecraft coordinate system, the height or depth of a block in the game world is represented by the Y coordinate, with Y = 0 corresponding approximately to sea level. Positive values represent features with positive elevation above sea level, while negative values indicate elevations below sea level. The game's vertical space in current versions (1.18 and above) extends from Y = −64 (the deepest possible layer) to Y = +320 (the maximum build height). There is no direct conversion between in-game Y-levels and real-world depths; however, it is commonly assumed that 1 Y unit roughly corresponds to 1 m. This represents a simplified analogy to real geological depth.

The distribution of individual ores in the game (e.g., coal, iron, copper, gold, diamonds, redstone, emeralds) is defined in newer versions through depth-dependent distribution curves that form so-called triangular occurrence patterns. For example, iron ore is most commonly found in subsurface layers between levels Y = −24 and Y = 54, peaking around Y = 16. Diamonds occur below Y = −16, with their frequency increasing with depth down to bedrock. This pattern reflects a fundamental principle of vertical stratification of ore deposits and introduces a degree of analogy to real geological processes such as magmatic differentiation, hydrothermal mineralization, or metamorphic zoning. The vertical distribution of ores in Minecraft (version 1.18+) by Y-level is illustrated in Figure 2, showing a clear correlation with specific rock environment horizons and a recognizable analogy to the vertical stratification seen in the distribution of mineral resources, depending on the formation mechanisms of their deposits.

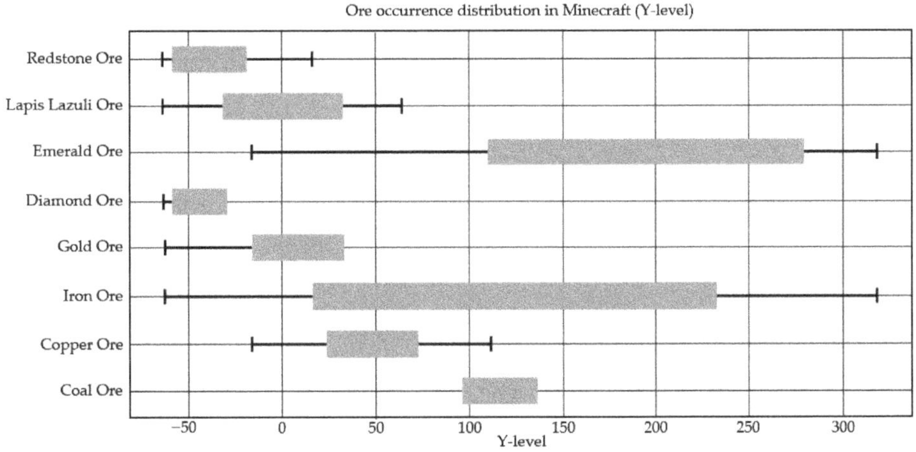

Figure 2. Vertical distribution of ores in Minecraft (Version 1.18+) by Y-level occurrence.

Despite this analogy, the in-game distribution does not account for broader geotectonic relationships, typological classification of ore deposits (e.g., magmatic, sedimentary, or metamorphic), or their association with structures such as fault zones, intrusions, or geochemical anomalies. All ores are dispersed as relatively homogeneous clusters (so-called ore veins), which are generated independently of the genetic type or lithological characteristics of the surrounding rock. In reality, ore deposits exhibit complex geometries, structures, and connections to specific rock complexes and tectonic settings.

From a didactic perspective, however, this represents a significant advancement over earlier versions of the game. The updated algorithm allows players to recognize the basic principle of stratigraphic variability and depth-dependent resource occurrence. Moreover, the Minecraft environment enables straightforward visualization of the vertical profile of a rock mass and allows for the systematic observation of compositional changes—making it particularly suitable for explaining fundamental geological concepts such as layers, horizons, overburden/bedrock, and zonality.

Mods such as BetterGeo (developed by the Geological Survey of Sweden (SGU) with the support of the BetterGeoEdu project funded by the EIT RawMaterials initiative) and Geolosys are significantly more complex in terms of geology and the mineral resource chain. On the other hand, without accompanying explanations or teacher guidance, they are difficult to grasp and complicated for laypersons or the main target group of Minecraft users. In an educational environment (with a teacher or a worksheet), however, they can be very beneficial and understandable. The application of these mods is thus more suited to guided instruction or as part of project-based learning. An important factor in resource acquisition is the abundance of individual minerals that serve as sources of chemical elements or compounds, or more specifically, the relative concentration of these elements within the rock environment. In geochemistry, this aspect is quantified using the Clarke value (or clarke), which represents the average concentration (or relative abundance) of a chemical element in a given geological entity [50].

Since Minecraft does not provide precise values for ore occurrence or elemental content (e.g., in ppm), a simplified approach can be adopted by introducing a quantifier called the Game Occurrence Index (GOI). This index represents the relative abundance of ores as defined by the game's generation algorithms. The GOI can be based on the average number of blocks of a given ore generated in a standard Minecraft world (version 1.18+) per unit of volume, such as a chunk, derived from community-reported world generation data. A higher value indicates greater abundance.

To express relative proportions, both the GOI and real-world Clarke values (in ppm) are normalized to a base value of 1. This allows proportional comparison between the abundance of elements in the Earth's crust, according to Clarke values [51], and their modeled abundance in the game environment. The results are presented in Table 1, taking into account values valid for the main game world, the Overworld, excluding the Badlands biome. Including this biome could distort the results due to the anomalously high occurrence of gold in this specific biome.

Table 1. Quantitative comparison of Clarke values (ppm) for the Earth's crust, with the GOI for selected elements.

Element (Ore)	Real World: Clarke Number [ppm]	Minecraft: Game Occurrence Index [Number of Blocks Per Chunk]	Real World: Relative Abundance (vs. Iron)	Minecraft: Relative Abundance (vs. Iron)	Consistency in Occurrence Rate
Iron (Fe)	56,300	200–300	1	1	Yes (most abundant)
Copper (Cu)	60	150–250	~0.001 (1/938)	~0.7 (1/1.4)	No (significantly higher in Minecraft)
Carbon/Diamond (C)	~200 (total C)	1–5 (max value at Y = −58)	~0.0035 (1/281)	~0.005 (1/200–1/400)	Yes (very low)
Gold (Au)	0.004	5–15 (Overworld, without Badlands)	~0.00000007 (1/14,075,000)	~0.04 (1/10–1/5)	No (significantly higher in Minecraft)

From the perspective of identifying analogies, the values of the GOI indicate that, relatively speaking, iron ranks highest in abundance both in Minecraft and in the real world. Copper is represented by a lower Clarke value in nature (approximately 0.001 relative to iron), yet in Minecraft, its GOI approaches that of iron (0.7), suggesting a prioritization of gameplay mechanics over geochemical realism.

For carbon—represented in the game primarily as diamond—this methodology is more problematic. If we consider total carbon, which corresponds to the Clarke value, a major discrepancy arises due to the relatively low abundance of carbon in nature versus the high abundance of coal in the game. On the other hand, if we consider only diamond, the Clarke value for carbon does not reflect the rarity of this specific allotrope. This example highlights the limitations and nuances involved in mapping natural phenomena and real-world processes into simplified game environments.

Although it is not possible to directly compare the Clarke value of total carbon with the abundance of diamonds in Minecraft, the game successfully simulates the extreme rarity of diamonds, consistent with their exceptional status in nature. The difference in abundance between iron and diamonds in the game (200–300 vs. 1–5 blocks per chunk) serves as a simplified representation of the large disparity between common metals and rare elements or resources in natural settings.

Gold is the rarest of the elements compared in the Earth's crust, with a Clarke value of approximately 0.00000007 relative to iron. While gold is relatively rare in Minecraft, it is still considerably more abundant than in nature (GOI $\approx$ 0.04). This represents the most pronounced deviation from Clarke values in favor of game design balance.

If we express these values in terms of block abundance, we can state the average occurrence of selected ores per chunk. This is visually represented in Figure 3, which also includes error bars illustrating the minimum and maximum number of blocks.

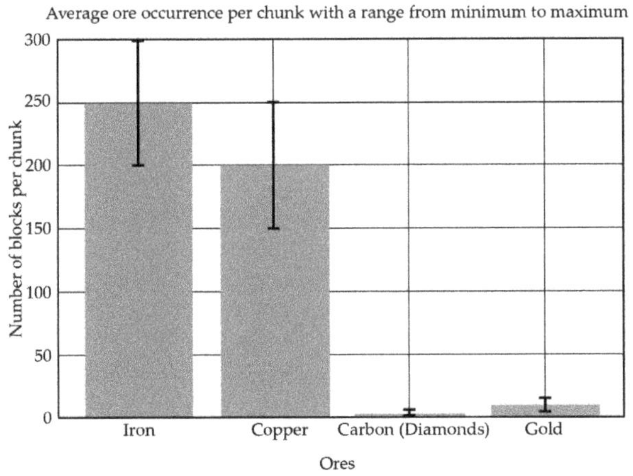

Figure 3. Average occurrence of selected ores/elements per chunk.

Overall, it can be concluded that the in-game representation of geological formation and resource distribution constitutes a simplified yet didactically effective model. It successfully conveys the vertical organization of geological phenomena and the importance of depth for the occurrence of mineral resources. From the perspective of process fidelity, it is a partially accurate model that, however, omits most genetic, structural, and geochemical aspects of real-world geology. The relative abundance of iron and the extreme rarity of diamonds in Minecraft reflect their actual distribution in nature. Nevertheless, the game significantly compresses the abundance range of other ores, particularly copper and gold. This "compression" serves gameplay purposes, as the extreme differences present in real-world geochemistry would otherwise make it excessively difficult for players to access various materials within a reasonable timeframe. Both approaches—realistic and in-game—

demonstrate that Minecraft incorporates principles of mineral and elemental distribution from the real world, albeit with highly adjusted values to support playability.

While the base version of Minecraft offers only highly simplified models of the occurrence and distribution of mineral resources, specialized modifications such as BetterGeo, GeoImmersion, or TerraFirmaCraft already introduce more realistic geological layers, genetic deposit types, and their relationships to specific rock complexes or tectonic environments into the game world. In some cases, they even simulate processes such as metamorphism, hydrothermal mineralization, or sedimentary deposition. These approaches approximate scientific geological simulations and can be used for teaching economic geology and ore deposit models. Our study, however, focused primarily on mainstream game mechanics and easily accessible modifications suitable for developing geoscience awareness among the general population and younger players. Detailed integration of complex geological models into Minecraft would require dedicated research and targeted methodological development. Nevertheless, given increasingly complex technologies that rely on a wide spectrum of materials and resources, even non-specialized mods would benefit from gradually incorporating critical elements such as molybdenum, nickel, lithium, lead and zinc, silver, and some others. The importance of this is also highlighted by the findings published in sources [52–59]. A simplified representation of tectonic features could be integrated into the base version of Minecraft to provide players with a basic understanding of the relationship between geological processes and the distribution of mineral resources. Potential implementations include the following:

Tectonic Map Layers: A simple color-coded overlay indicating different tectonic settings (convergent, divergent, and transform boundaries), which would influence the types and abundances of ores generated.

Fault Lines: Distinctive rock bands visually representing fault zones, signaling potential concentrations of certain minerals or the presence of hydrothermal veins.

Volcanic Zones: Areas with increased volcanic activity (e.g., lava outflows, already partially represented in the game), suggesting the occurrence of ores such as copper, gold, or rare minerals.

Sedimentary Basins: Lowland areas with layered rock formations where deposits of coal and similar resources could be concentrated.

These features would not require complex graphics or advanced simulations, yet they would provide intuitive cues for players, linking ore locations and geological contexts with tectonic processes.

From the perspective of incorporating the perception of geological time, it could be proposed that the distribution of geological formations over time be reflected in the game through phased world generation and the use of special maps simulating ancient geological periods, thereby providing the player with information on the geological evolution of the entire represented environment.

3.2. Geological Exploration Processes

In the real world, geological exploration represents a key phase in the mineral resource acquisition chain, during which geological conditions are systematically mapped, deposits identified, and their economic potential evaluated. This process includes a range of methods, from field observation and geophysical and geochemical analyses to exploratory drilling and deposit modeling. In the base version of Minecraft, however, exploration is not conceived as a distinct phase, but rather as an intuitive activity tied to world discovery—that is, as an integral part of gameplay itself. The player gains knowledge of the environment primarily through visual inspection of the surface and underground, or indirectly through trial-and-error detection of ores while mining. This approach more closely resembles a pre-industrial, "manual" data-gathering technique, lacking systematic tools.

Some player strategies, such as branch mining or strip mining, do represent optimized exploration techniques based on statistical coverage of rock volume and maximizing the chance of ore discovery. However, these are not formal processes, but rather emergent gameplay behaviors shaped by the mechanics of the game. The base game lacks tools for remote ore detection, geophysical measurement, or geological unit mapping. Consequently, it does not provide a realistic representation of geological exploration as a technical-scientific process. This absence limits the understanding of the importance of pre-mining preparation and makes it difficult to communicate essential aspects such as economic feasibility, geological uncertainty, or resource and reserve classification.

A significant advancement is observed with the use of thematic modifications (mods), which explicitly introduce mechanics corresponding to geological exploration. A prime example is the Immersive Engineering mod, which includes a Core Sample Drill that allows players to determine the type of resources located beneath the surface in a given region. This simulates real-world core drilling and deposit modeling and introduces elements of planning, decision-making, and dealing with uncertainty. Similarly, the BetterGeo mod expands the game with basic geological mapping and rock classification, enabling players to predict typical ore associations based on the geological environment.

Other mods such as GeoImmersion or TerraFirmaCraft enrich the game with geochemical parameters of the environment, allowing players to locate deposits based on rock type, depth, or geological domain. These elements significantly enhance the processual realism of the game and allow players to experience geological exploration as a structured phase of the mining cycle. Furthermore, by enabling players to make decisions based on exploration data, the didactic potential of the model increases, creating opportunities to explain principles such as geological prediction, economic efficiency, and mine planning.

From a process-oriented perspective, Minecraft in its base version represents exploration only in a very simplified manner—as random discovery. It is only through specific modifications that this phase may be transformed into a meaningful, structured process analogous to real-world geoscientific and industrial activities. These modifications therefore serve as important tools for education in geology, mineral prospecting, and mining planning.

These statements align well with the findings presented in [52], which indicate that in the field of geology, most teachers are still aware of the potential of games to motivate, enhance, and reinforce the learning of geological content, with digital games being the preferred option. They emphasize the importance of teacher training in this area and the inclusion of game-based applications in school curricula to address various geology-related topics. The presented results suggest a certain lack of consistency in teachers' views on the impact of games on student learning, which creates room for improvement. In the study presented in [60], the authors implemented a virtual course based on Minecraft that achieved both curricular and extracurricular objectives, with its development being relatively simple. They recommend using Minecraft in the future specifically for virtual courses focused on aspects of geological exploration.

To strengthen the geological exploration aspect within the base version of Minecraft, several concepts are proposed:

Observation of Surface Indicators: Implementation of simple "indicator blocks" or rock discolorations on the surface (e.g., small random mineral grains embedded in stone) to suggest the possible presence of ore deposits underground.

Educational effect: Introduces the concept of geological indicators and their role in identifying target exploration sites.

Basic Geological Map: Development of a map displaying rock types in the surrounding area down to a specific depth.

Educational effect: Explains that different minerals are associated with specific geological environments.

Test Pits: Inclusion of small, rapidly deployable exploratory pits revealing surrounding rock layers, with an optional extension to exploratory drilling combined with schematic core analysis.

Educational effect: Demonstrates that exploration activities can include small-scale sampling and subsurface investigation techniques.

Basic Rock Classification: Provision of simplified information about a rock block after mining or examination, linked to the occurrence of specific valuable minerals.

Educational effect: Connects rock types with the typical occurrence of certain ores.

Simple Deposit Potential Assessment: After the discovery of a defined quantity of ore, display an estimate of the deposit's potential (e.g., "rich deposit" vs. "small occurrence").

Educational effect: Introduces the concept of economic viability in mineral resource extraction.

The integration of these elements would enable players to intuitively explore geological concepts such as indicator recognition, geological mapping, sampling, identification of lithological environments, and preliminary economic assessment, while preserving both the entertainment value and the educational potential of the game.

3.3. Mining Techniques and Resource Extraction

The extraction of mineral resources in the real world represents a complex system of technical operations aimed at efficiently, safely, and economically retrieving valuable minerals from the geological environment. These processes involve selecting appropriate mining methods, employing specialized equipment, excavating mine workings, ventilation, dewatering, implementing safety measures, and conducting environmental monitoring. In Minecraft, the mining process is significantly simplified but still exhibits several elements that can be interpreted as abstract models of real-world mining techniques.

In the base version of the game, players extract mineral resources manually using handheld tools—pickaxes—whose efficiency, speed, and durability depend on the material from which they are crafted (wood, stone, iron, diamond, netherite). Netherite is a fictional material that, while occupying the position of the strongest material in the game, has no analog in the real world and is therefore not the subject of our study; it is mentioned here for completeness only. The hierarchy of tools clearly reflects an analogy with the progressive enhancement of material quality and tool design in real mining practices. However, the use of certain materials (e.g., gold) should be interpreted more of a symbolic representation of value rather than a reflection of actual physical and technological properties.

The varying breakability (hardness) of blocks, which determines how long it takes to break them in Minecraft, is represented by the Mining Speed/Time parameter. Each block has an internal hardness value assigned (stone: 1.5, iron ore: 3, diamond ore: 3, obsidian: 50), and each pickaxe type has a base speed modifier (e.g., wooden: 2×, stone: 4×, iron: 6×, diamond: 8×, netherite: 9×). Minecraft also implements the concept of enchantments, which enhance tool performance. The Efficiency enchantment significantly reduces mining time, with each level increasing the mining speed by 25% compared to the tool's base speed (Efficiency I: +25%; Efficiency V: +125%). The Fortune enchantment increases the average quantity of ore dropped as follows: Fortune I: 1–2× (average 1.5×); Fortune II: 1–3× (average 2×); Fortune III: 1–4× (average 2.5×). The Silk Touch enchantment allows for the retrieval of the block itself rather than its fragmented form. This can be understood as a mining process without separation or transformation, representing an analogous equivalent to the extraction of raw material without beneficiation.

The enchantment system can be viewed as analogous to real-world factors that influence the efficiency and yield of mining operations. It reflects an optimization process aimed at maximizing resource gain (material per unit time) relative to investment in tools and upgrades. Figure 4 illustrates the mining time curves for selected blocks (stone, obsidian, diamond ore) depending on tool type and Efficiency enchantment level. It is evident that for harder blocks such as obsidian, the differences in required mining time are more pronounced across tool types, whereas for softer blocks, the differences are less significant. This serves as a meaningful analogy to the impact of tool type relative to mineral hardness or ease of fragmentation in real-world extraction processes.

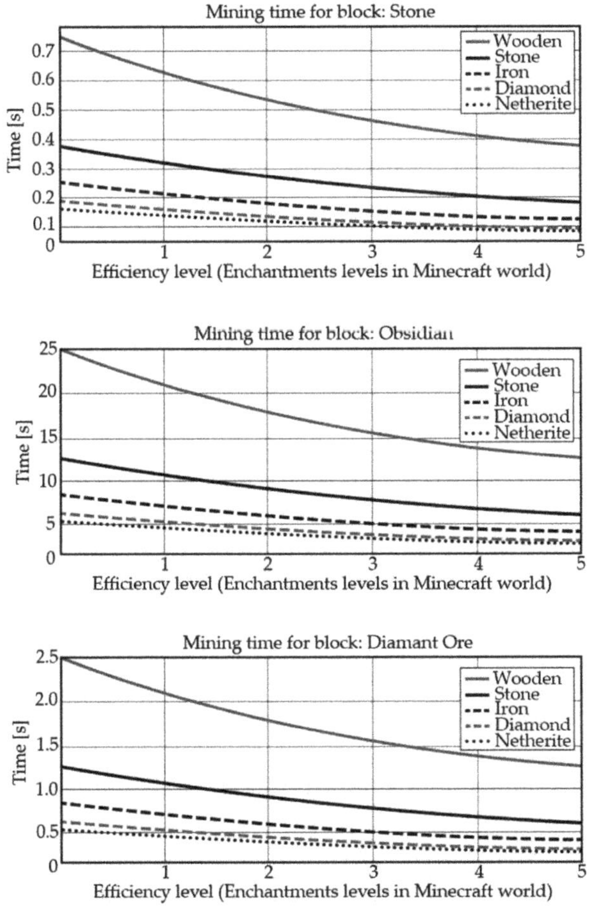

Figure 4. Block mining time (stone, obsidian, diamond ore) depending on tool type and enchantment level.

This system indirectly represents the principle of tool suitability for the hardness of the rock, although it lacks the quantitative accuracy of the Mohs scale or other mechanical parameters of rocks. The player does not have access to mechanized mining or advanced mining technologies (such as continuous miners, drilling rigs, or blasting techniques), and the rock fragmentation process is simplified to repeated clicks simulating the time needed to break a block.

In practice, however, players adopt optimized mining schemes that have emerged as a result of community-driven analysis—most notably branch mining, strip mining, quarry mining, and cave mining. These strategies aim to maximize the volume of explored rock while minimizing tool wear and time consumption. Their existence demonstrates how the game community intuitively replicates the principles of efficient planning of mining panels, space optimization, and the use of natural geological structures.

An important aspect, particularly in underground mining but also in surface excavation and waste dumps, is the stability of mining structures [34,42]. In the base version of Minecraft, apart from specific blocks such as sand, gravel, or red sand—which are affected by gravity and fall if unsupported—most blocks (such as stone, ores, deepslate, etc.) are unaffected by gravity. As a result, when a supporting block is removed beneath them, the blocks above remain suspended regardless of their height or number. Consequently, Minecraft does not simulate natural cave-ins or roof collapses due to structural instability or lack of support.

However, the game does include user-triggered collapses. These occur when a player removes a gravity-affected block (e.g., sand or gravel), which then falls. This mechanic can even be used as a mining technique for such loose materials, presenting an analogy to real-world gravity mining methods. Since blocks (except for sand and gravel) do not obey gravity, Minecraft lacks game mechanics requiring structural reinforcement to prevent collapses. While players may construct walls, ceilings, pillars, or use wooden beams (e.g., spruce/oak logs) as esthetic "supports" in their mines, this is implemented for visual or logistical reasons (e.g., railway support) rather than for functional structural stability.

More advanced tools and techniques are introduced into the game through modifications. For example, the BuildCraft mod adds mining robots (quarry machines) that autonomously excavate large volumes of rock and deposit the extracted materials into containers. This type of automated mining can be considered an analogy to large-scale surface mining operations using conveyor systems and high-capacity excavation machinery. The Immersive Engineering mod, in turn, introduces drills and elevators that improve the efficiency of deep mining and add new technical elements such as power limits, energy requirements, and logistics of extracted resources.

A major limitation of the base version of Minecraft, however, remains the absence of many key aspects of real-world mining processes: there is no ventilation of underground spaces, no monitoring of roof and wall stability, and the player is not exposed to the risks of gas accumulation, collapse, or spontaneous combustion. Basic engineering and environmental measures—such as structural supports, groundwater regulation, or reclamation planning—are also missing. These deficiencies limit the game's capacity to realistically simulate complex mining engineering planning, although it does provide a sufficiently illustrative model of basic excavation and ore sorting concepts.

From an educational perspective, Minecraft can be considered an effective tool for conveying the basic logic of resource extraction—such as the hierarchy of tools, the importance of planning mining routes, the ratio of waste to ore, and the significance of repetitive operations and logistical supply. However, a deeper understanding of these processes requires expansion through specialized modifications capable of integrating advanced technological and safety features into the game. In summary, the game offers a simplified yet comprehensible model of mining, which—especially when enhanced with mods—can be extended into a relatively complex simulation of industrial extraction processes.

3.4. Ore Processing and Refinement

Ore processing and refinement represent the intermediate stage in the process chain between extraction and final utilization of raw materials. Under real-world conditions, this phase comprises a set of technological operations aimed at increasing the concentration of the target element, removing undesirable impurities, and preparing the raw material for further processing—whether metallurgical, chemical, or direct technological applications [36]. This includes processes such as sorting, crushing, grinding, flotation, gravity separation, magnetic separation, smelting, and refining [41]. In Minecraft, these processes are represented only in a very limited and highly simplified form.

In the base version of the game, the entire ore refinement process is reduced to a single step—smelting. After mining ore (e.g., iron ore), the player places the raw block into a furnace or a more advanced device (blast furnace) along with a fuel source to obtain a metal ingot (e.g., iron ingot). This linear and instantaneous conversion of "ore → metal" does not reflect the actual mineralogical or technological reality, where iron ore requires preprocessing (e.g., removal of gangue, drying, calcination), the reduction of iron oxide, and precise control over furnace temperatures and chemical reactions. Furthermore, in the game, each ore block is considered homogeneous and 100% composed of the target element, which is inconsistent with real-world yields, which typically range from several tens to just a few percent.

This shortcoming is partially addressed by thematic modifications that introduce more complex process chains into the game. For example, in the Immersive Engineering mod, players are required to first crush the ore (crusher), then sieve it (screening), and finally separate the gangue from the concentrate (e.g., via magnetic separation), before obtaining a pure material suitable for smelting. This process thus mirrors the basic scheme of physical-mechanical ore processing and reflects real technological steps applied in mineral processing plants. The Tech Reborn mod goes even further, introducing processes such as chemical leaching, electrolytic purification, and fractionation, thereby establishing a more robust and realistic model of industrial mineral processing.

An interesting addition is the inclusion of by-products and waste materials. While the base game uses a binary output system (either a metal is produced or nothing at all), the expansions can generate secondary fractions such as sand, dust, sludge, or secondary metals. These outputs create a more complex balance between input and output streams and offer opportunities to explain the environmental aspects of processing—such as waste production, recycling, or energy consumption.

Process flows in mods also introduce time delays and resource consumption (energy, water, chemicals), thereby enhancing the understanding of the energy demands and complexity of industrial processing. Players are required to plan sequential steps, build interconnected production lines, and optimize material flows, which facilitates an appreciation of the systemic nature of real-world processing industries.

Nevertheless, even the most complex mods rarely cover the entire technological chain, especially regarding the specificity of different metallurgical techniques (e.g., different reducing agents, furnace atmospheres). Mineralogical variability is also not considered—all ores of a given type (e.g., iron) are processed identically, regardless of their origin, structure, or associated minerals. Additionally, processes like quality control, certification, product purity monitoring, or economic feasibility assessments are missing.

From an educational perspective, however, even these simplified processes provide an excellent foundation for explaining linear and parallel processing, yield concepts, the significance of preprocessing, and material flow in technical environments. When supplemented with real-world examples and quantitative data in pedagogical settings, they can greatly enhance the understanding of industrial mineral processing principles. Ultimately, the game's representation of processing can be viewed as a flexible platform that offers a schematic introduction to resource flows and, when combined with mods, serves as a functional simulation tool for teaching technological processes.

3.5. Waste-to-Ore Ratio

One of the key quantitative indicators in the process of mining and mineral processing is the waste-to-ore ratio (W:O). This ratio expresses the amount of accompanying, technologically unutilizable rock (waste) that must be mined, moved, or processed together with the target mineral in order to obtain an economically usable quantity of raw material. In real-world mining, this is a crucial factor affecting costs, efficiency, logistics, and environmental impacts of extraction. Therefore, it is appropriate to analyze how and to what extent this ratio is implicitly or explicitly represented in the environment of Minecraft.

In the base version of the game, each ore block is embedded in a matrix composed predominantly of stone, deepslate, or other rocks, and ore occurrence is statistically rare. Based on the analysis of world generation algorithms in version 1.18 and above, it can be estimated that the occurrence of ores such as iron, copper, or coal ranges approximately from 0.5% to 1.5% of the total rock volume in certain elevation bands. This means that a player must mine approximately 66 to 200 blocks of waste for each block of ore—which aligns with the real-world range of waste-to-ore ratios in underground iron ore mining, where this indicator typically ranges between 10:1 and 100:1 depending on deposit conditions and the mining method.

More precise ratios vary depending on the ore type and the depth at which it occurs. Diamond ore is extremely rare, with an estimated occurrence of approximately 0.084% in regions below $Y = -64$, corresponding to a W:O ratio greater than 1100:1. In contrast, coal can occur in larger clusters and in some areas (e.g., mountain biomes) may comprise up to 2–3% of the rock volume. From a process-analytical perspective, the game environment is thus capable of reflecting the fundamental principle of low ore concentration and the necessity to handle a large volume of waste.

Figure 5 illustrates a comparison of waste-to-ore ratios (W:O) in Minecraft and in real-world mining, based on [44–48]. For comparison with the real world, the analogy was based on determining the waste-to-ore ratio (W:O) expressed as the ratio of waste blocks to useful mineral blocks. Based on data from actual world generation observations derived from [32,49] an effective minable chunk volume was set at approximately 65,000 blocks. Publicly available gameplay and community sources, such as player analyses and empirical tests reported in [32], yielded the following average ore block counts per chunk: diamonds: 1–5 blocks; gold: 5–15 blocks; copper: 150–250 blocks; iron: ~250 blocks; coal: ~500 blocks. These values apply to an unmodified game world generated in version 1.18+ under standard biome conditions. The presented values are given as the mean ± standard deviation, with the 95% confidence interval (CI) shown in parentheses. For example, the average occurrence frequency of iron ore was 247 ± 15 blocks/chunk (95% CI: 240–254). Diamond ore occurred at an average of 3.2 ± 1.1 blocks/chunk (95% CI: 2.9–3.5).

On the other hand, the in-game model does not account for the mass, volume, or logistical costs associated with handling waste material. All blocks are treated equally from the player's perspective—their extraction, transport, and storage are identical regardless of their actual density or volume. There is no incentive to separate waste, dispose of it, or manage it environmentally, which significantly simplifies the gameplay reality compared to the challenges faced in real-world mining operations. In reality, waste rock often represents a major burden, requiring the construction of tailings ponds, spoil heaps, dust control measures, and engineering structures to stabilize terrain.

Some modifications extend this issue. For example, in the Immersive Engineering mod, waste material appears as a byproduct of ore sorting, creating pressure to process or store it. In certain player scenarios, waste even becomes a limiting factor for construction, as it takes up space and complicates logistics. These mechanics enhance process fidelity and open up opportunities for discussing the environmental implications of mining.

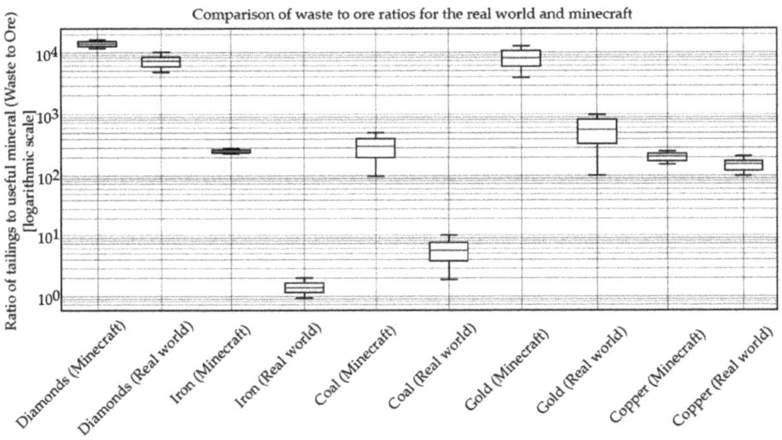

Figure 5. Waste-to-ore ratios (W:O) in Minecraft and in real-world mining.

From an educational standpoint, a quantitative analysis of the waste-to-ore ratio in Minecraft can serve as an excellent tool to introduce concepts such as recovery rate, material sorting, and mining efficiency. It offers a clear illustration that finding an ore is not the end of the task—rather, separating it from a low-concentration system is a technologically and energetically demanding process. The educational value of this analysis is further enhanced when combined with discussions on the environmental and economic impacts of waste disposal in the real world.

3.6. Fluid-Phase Resources

The extraction and processing of liquid resources such as crude oil, natural gas, and various types of liquid hydrocarbons represent a distinct industrial sector within the broader raw material chain in the real world. Unlike solid minerals, which are typically mined through mechanical means, liquid resources are primarily obtained via drilling, vacuum pumping, and subsequent processing in refineries. These operations are technologically complex, conducted under specific pressure and temperature conditions, and involve numerous chemical and separation processes. In terms of process representation, such resources and operations are practically absent in the core version of Minecraft.

The original game includes no form of crude oil or natural gas, and players cannot interact with hydrocarbon-based fluid energy sources. The only liquids present are water and lava, which, while enabling basic simulation of flow and physical properties (e.g., gravity, viscosity), serve no resource or technological function. This absence represents a significant limitation in any attempt to model comprehensive resource cycles—particularly when the aim is to support education in energy systems, refining processes, or petrochemical technologies.

However, this limitation is substantially mitigated through the use of thematic modifications. One of the most significant is the Immersive Petroleum mod, designed as an extension to Immersive Engineering and BuildCraft. This mod introduces a realistic simulation of oil extraction and processing into the game. Oil deposits are procedurally generated beneath the surface, and the player must locate them using a Core Sample Drill, which simulates geological exploration. Once a deposit is identified, the player can install a Pumpjack—a pumping apparatus that continuously extracts crude oil to the surface.

Crude oil in the game has no immediate use on its own. It must first be processed in a structure called the Distillation Tower, which realistically represents a fractional distillation column. Within this unit, crude oil is separated into various fractions—such as diesel, gasoline, heavy oil, and asphalt—each with specific applications in other parts of the game. These fractions serve as fuels, lubricants, or inputs for further chemical processing. This represents an advanced model that captures not only the flow of processes but also the technological interdependence of production stages, the necessity of product storage, and the regulation of throughput.

Additional mods such as BuildCraft, Thermal Expansion, and PneumaticCraft introduce expanded capabilities for fluid handling, including pressure systems, pressurized tanks, pumps, and transport piping. These elements enable players to understand fundamental principles of fluid transport, storage, temperature control, and subsequent utilization. While these are not the exact replicas of real-world industrial technologies, they provide a sufficiently robust framework to explain the logic of fluid management, energy flows, and technological conversion.

From a process analysis perspective, these modifications represent a significant advancement toward a more realistic representation of liquid resources. They allow the player to progress through all key stages—from the discovery and evaluation of deposits, to extraction and transport, to separation and use of products. Furthermore, when incorporating energy requirements, fuel consumption, and output performance, players can create basic material and energy balances and assess operational efficiency—thereby significantly enhancing the game's potential for teaching process engineering.

The use of Minecraft with mods focused on fluid resources thus open up a new dimension in the education of raw material processing. In this case, it offers a more accessible and interactive introduction to the world of petroleum, refining, and liquid energy. These topics are often abstract and difficult to visualize in traditional education, and Minecraft provides a unique didactic bridge between gameplay, simulation, and real-world technological systems.

3.7. Environmental and Systemic Processes

Each real-world mining or processing operation is accompanied by environmental consequences and systemic effects that directly or indirectly impact the surrounding environment, biodiversity, water resources, and air quality. The responsible management of mineral resources therefore requires not only technological efficiency and economic viability but also the minimization of negative externalities and integration of processes into broader environmental and societal frameworks. In the case of Minecraft, environmental aspects are addressed only marginally, and in most gameplay scenarios, there is no ecological feedback or systemic evaluation of the consequences of mining and processing activities.

In the base version of the game, player activities such as mining, smelting, or burning fuels have no environmental repercussions. There is no modeled pollution of air, water, or soil; emissions, mining waste, or noise are not simulated. Players can create large waste piles, destroy vegetation, or drain water sources without any penalty or reaction from the environment. This represents a significant simplification of reality, which prevents players from perceiving resource-related processes as part of a larger ecological system. The game world lacks feedback mechanisms to warn players about overexploitation, the non-renewability of deposits, or landscape degradation.

Despite this absence, several community-driven modifications attempt to bridge this gap. For instance, the Pollution of the Realms mod introduces systems for air, water, and soil pollution. Activities such as metal smelting, coal combustion, or explosions lead to pollution that affects player health, crop productivity, and the appearance of the environment. Such systemic implementations provide a foundation for understanding the ecological consequences of human activity and can simulate the impacts of overmining and unsustainable energy use. Similarly, mods like Better With Mods introduce mechanics for terrain degradation, water contamination, and the need for land reclamation after mining.

From a process-oriented perspective, these additions are significant because they allow the incorporation of concepts such as environmental balance, external costs, and the necessity of environmental management as an integral part of the production cycle. Players must consider not only yield and technical efficiency but also long-term consequences of their actions. This system-based approach reflects the reality of modern resource management, where environmental standards, emission limits, permits, and regulatory frameworks are essential to planning and operation.

Another critical aspect is recycling. While recycling is practically nonexistent in the base game (e.g., tools or weapons cannot be broken down into raw materials), some modifications—such as Tech Reborn or Industrial Foregoing—introduce systems for reprocessing used items, smelting metal tools, and recovering secondary materials. In this way, the game integrates the principle of a closed-loop material cycle, allowing players to optimize resource flows over the long term. Recycling mechanics also emphasize the importance of efficient material use and highlight the environmental advantages of recycling over primary extraction.

From an educational standpoint, the integration of environmental and systemic processes into Minecraft is highly beneficial. It allows learners to understand that technological processes are not isolated but are part of a broader system that impacts nature and society. By linking technical operations to their consequences, the game creates opportunities to discuss sustainability, mining ethics, environmental justice, and the rights of future generations to resources. While these elements are absent from the base game, the possibility of adding them through modifications makes Minecraft a flexible simulation space for exploring not only technical but also social dimensions of resource-related processes.

In Minecraft, it is possible to identify certain processes that reflect characteristics such as material degradation or corrosion. However, these are not implemented universally for all materials and are incorporated in a simplified and schematic form. The phenomenon of corrosion is specifically integrated into the behavior of copper blocks. Copper blocks and their variants (cut copper blocks, copper stairs, and copper slabs) gradually undergo a patination process when exposed to air. They pass through four stages of oxidation: pristine copper → lightly weathered → exposed → fully weathered (green patina). This progression serves as a direct analogy to copper corrosion in the real world.

Beyond copper oxidation, other forms of degradation or material transformation can also be interpreted as geological or environmental processes. Most notably, erosive processes caused by flowing water can transport and destroy certain blocks such as sand, gravel, farmland, or snow. This behavior presents a clear analogy to erosion and environmental degradation induced by natural forces.

The majority of blocks (e.g., stone, dirt, wood, iron, and gold blocks) in Minecraft do not change over time, nor do they disintegrate or corrode. The fact that gold is immune to corrosion in the game accurately reflects the real-world properties of this metal. However, the absence of iron corrosion—unlike copper—in both blocks and items—represents a significant discrepancy between material behavior in Minecraft and in reality. Iron and steel corrosion is a process with major negative implications for the global economy. According to data published in sources such as [52–64], direct global costs of corrosion are estimated to be 3–4% of the global GDP, while total costs, including indirect impacts, may reach 6–8% of GDP.

3.8. Impact of Modification Complexity on Accessibility and Educational Potential

The introduction of advanced modifications (mods) that expand the original game with new technological and geological features has a significant impact on the accessibility and playability of Minecraft, particularly in the context of its use in education. While these mods can substantially increase process accuracy, technological sophistication, and educational content, they can also create barriers for less experienced users or younger students.

From the player experience perspective, complex modifications may reduce the simplicity and intuitiveness that characterize the base version of Minecraft. Installing mods often requires knowledge of file management, the use of launchers (e.g., CurseForge), dependency configuration (e.g., Forge, Fabric), and navigation in an English-language interface. For primary school students or teachers without technical backgrounds, these requirements can be a considerable obstacle. Furthermore, some mods alter the game's economy, resource balance, and progression pace, which can diminish the sense of reward and flow.

An analysis of feedback from user forums (e.g., Reddit: r/feedthebeast, CurseForge discussion threads, Minecraft Forum) suggests that players appreciate advanced mods such as BetterGeo, Immersive Engineering, or GeoImmersion mainly for their realistic content and didactic potential. However, many point out that the complexity of these expansions requires accompanying guides, explanatory materials, or trained instructors. Some comments stress that without guidance, beginners can become "overwhelmed" with information and abandon the game before grasping its educational aim.

From a pedagogical standpoint, it is therefore important to balance the degree of complexity. Ideally, one should begin with simpler mods or specially prepared maps (so-called modpacks) with predefined objectives, simplified interfaces, and limited scope. In later stages, the difficulty can be gradually increased, new elements introduced, and thematic connections deepened. The target age group is also a critical factor—while advanced modifications can be beneficial for high school and university students, they may be counterproductive for younger learners.

Based on these findings, the following is recommended:

- Use thematic mods with regard to age, technical skills, and educational objectives;
- Provide support in the form of explanatory materials or an instructor;
- Combine modified and unmodified environments according to complexity levels;
- Ensure that even advanced game content remains fun, visually engaging, and motivating.

Such an approach will ensure that Minecraft remains not only accurate but also accessible and engaging as a tool for teaching mineral resource processes and environmental education.

Table 2 presents a diagram mapping game mechanics and recommended modifications to specific stages of the mineral resource value chain, taking into account specialized mods aimed at enhancing the educational enrichment potential of each process stage.

Table 2. Comparison of real and simulated processes by category.

Process Phase	Real-World Process	Minecraft Simulation	Level of Correspondence	Educational Potential
1. Geological Formation and Distribution of Resources	Geotectonic processes, deposit typology, association with rock complexes, geochemical anomalies.	Procedural world generation, simple vertical ore stratification, RAE algorithm (version 1.18+), no geotectonic associations.	Medium	Visualizes stratification and depth-dependent resource occurrence.

Table 2. *Cont.*

Process Phase	Real-World Process	Minecraft Simulation	Level of Correspondence	Educational Potential
2. Geological Exploration	Mapping, geophysics, geochemistry, drilling, deposit modeling.	Base game—random discovery; mods—Core Sample Drill (Immersive Engineering), geological mapping (BetterGeo).	Low to medium	Mods allow the simulation of exploration phases and planning.
3. Mining Techniques and Extraction	Surface/underground mining, machinery, ventilation, stability, occupational safety.	Manual mining with tools; hardness and speed parameters; mods—quarry machines, drilling rigs.	Medium	Teaches the principle of tool suitability, route optimization, and basic extraction models.
4. Mineral Processing and Treatment	Sorting, crushing, flotation, separation, smelting, refining.	Base game—direct smelting in a furnace; mods—crushing, magnetic separation, chemical processes (Tech Reborn).	Low to medium	Introduces the concept of yield and material separation.
5. Waste-to-Ore Ratio	Waste rock as accompanying material, handling, stockpiling, environmental measures.	Implicitly high W:O in game; no weight or logistics; mods—byproducts, storage requirements.	Medium	Introduces the concept of yield and material separation.
6. Fluid-State Resources	Oil, natural gas—exploration, pumping, refining, fraction processing.	Base game—absent; mods—Immersive Petroleum (pumping units, fractional distillation).	Medium to high (in mods)	Simulates fluid flow, energy, and technological linkages.
7. Environmental and System Processes	Monitoring, emission control, reclamation, recycling.	Base game—no environmental feedback; mods—Pollution of the Realms, recycling systems.	Low	Enables the explanation of broader mining impacts and sustainability concepts.

4. Conclusions

The analysis of individual processes has shown that Minecraft, while operating with a highly simplified model of mineral resource extraction and processing, is still capable of conveying the basic principles of various stages—from geological exploration, through mining and beneficiation, to technological processing and systemic impacts. Such a game environment offers a useful tool for modeling and understanding the fundamental processes of the mineral resource cycle—from geological exploration, through mining and beneficiation, to technological processing and environmental aspects. In the base game, intuitive, player-optimized approaches dominate without an explicit process structure; however, the inclusion of thematic modifications opens the possibility of simulating more realistic technological and environmental operations. Even with its simplifications, the base game provides a fundamental framework for understanding the logic of mineral resource acquisition and utilization. The included game mechanics support intuitive decision-making, workflow organization, mining planning, and the assessment of inputs and outputs, which are key elements of real-world technological chains. A major contribution to process analysis comes from thematic modifications that enrich the game world with new technological nodes, more realistic mineral processing, more detailed geological exploration, and elements of environmental impact. These expansions enable a more precise tracking of the sequence of operations, illustrate multi-step mineral processing, and introduce feedback between production, consumption, and environmental consequences. This also reinforces the ability to consider sustainability, efficiency, and recycling within the entire resource cycle. Despite the absence of detailed parameters (e.g., yield, contamination, energy balances), the game—especially in targeted education—can convey an understanding of the basic interconnections between inputs, operations, and outputs in the process of resource acquisition. A particular value lies in its ability to link technical operations with their broader impacts, such as waste generation, environmental burden, or recycling. In combination with mods, Minecraft thus provides an effective model space for studying, visualizing, and explaining the individual components of a complex resource system from a process perspective.

From a historical perspective, the time frame reflected in Minecraft's representations of mining and resource needs most closely resembles an era of intensive yet predominantly manual mining—dependent on simple tools and basic processing techniques. The gradual progression of tool quality and the emergence of primitive automation (e.g., Redstone mechanisms) shifts this timeline towards the dawn of the industrial age. The game thus combines elements of mining practice from several centuries, while offering considerable potential for the future integration of contemporary, complex procedures, strategies, and technologies.

From an educational standpoint, the game environment fosters active learning through interaction, experimentation, and visual feedback. Working with game-based models can complement traditional teaching of geology, mining engineering, and environmental science by introducing an experiential dimension that facilitates the understanding of complex phenomena through hands-on simulation. The game also provides a platform for creative thinking, technological problem-solving, and discussions about the interrelations between technical, ecological, and societal aspects of mineral resource extraction.

In conclusion, while not a scientific simulation tool, the virtual environment offers a process-based framework that can be employed not only in science communication but also in formal education about the operations behind mining and processing raw materials. This flexibility and openness make Minecraft an innovative didactic platform with strong potential for further development and research in the domain of digital process modeling.

It is important to distinguish between approaches focused on educational modeling and those aimed at strict scientific simulation. Some of the Minecraft modifications mentioned come close to scientific simulation tools in terms of accuracy, complexity, and emphasis on realistic parameters, and could even be applied in research scenarios. However, our analysis primarily focused on so-called mainstream solutions, which are more accessible, technically less demanding to implement, and better suited for education or raising awareness of raw material processes among the widest possible group of users. This approach prioritizes interactivity, clarity, and player engagement over complete technical precision, thereby aligning with the goals of popularization and educational modeling.

Overall, Minecraft may be regarded as a flexible, modular, and scalable digital model that, with appropriate combinations of modifications, achieves the following:

(1) Covers the majority of key operations in the resource chain with sufficient detail for introductory university instruction;
(2) Provides quantifiable variables (e.g., mining volume, energy consumption, waste-to-ore ratio) suitable for coursework and laboratory exercises;
(3) Enables the integration of environmental and social impacts into technical decision-making.

We therefore recommend using Minecraft as a didactic extension of traditional lectures and lab work. On average, it requires 6–8 teaching hours to build and operate a simplified mining and processing system. Future research should focus on expanding mineralogical variability, improving energy balance accuracy, and connecting the Minecraft platform to real-world deposit databases to enhance scientific validity and support decision-making in the context of sustainable resource management.

The future integration of scientific data sets and educational curricula could significantly enhance the usability and accuracy of models created in the Minecraft environment. Relevant data sources could include geological deposit databases (e.g., USGS Mineral Resources Data System or GeoSciML), environmental data from monitoring networks (e.g., air quality, hydrological parameters), or the results of 3D geological mapping. These data could be directly linked to world generators, allowing players to work within environments that reflect real geological conditions. Equally important is aligning in-game scenarios with official educational curricula, such as Science, Technology, Engineering, and Mathematics (STEM), to create comprehensive educational modules containing tasks, assessments, and feedback. Such integration would provide teachers and instructors with a ready-made framework, enabling them to effectively incorporate game-based simulation into both formal and informal education.

Finally, the authors believe that a thoughtful integration of certain identified concepts and a careful adaptation of game mechanics to real-world processes could deliver significant benefits that go beyond strict science or formal education. This approach has the potential to promote natural public outreach and contribute to a more objective perception of reality through a widely adopted game, thereby helping to shape rational and balanced public attitudes toward key issues concerning the present and future development of society.

Author Contributions: Conceptualization, R.R. and M.B.; methodology, R.R.; formal analysis, M.B.; investigation, M.B.; data curation, M.B.; writing—original draft preparation, R.R.; writing—review and editing, R.R. and M.B.; visualization, M.B.; supervision, L.B.; project administration, R.R.; funding acquisition, L.B. All authors have read and agreed to the published version of the manuscript.

Funding: This research was funded by KEGA: 056TUKE-4/2024, a platform for the effective creation, evaluation, and transfer of innovations and the efficient management of university research outputs with a practical orientation.

Data Availability Statement: The original contributions presented in this study are included in the article. Further inquiries can be directed to the corresponding author.

Conflicts of Interest: The authors declare no conflicts of interest.

References

1. Câmara, R.; Câmara, G.; Andrade Neto, M.; Milosevic, T.; Andrade, J.C.; Rocha, P. Reserve Assessment and Certification of Brazil's Oil and Mineral Sectors and Energy Storage: New Trends, Analysis, and Regulatory Proposals. *Resour. Policy* **2019**, *62*, 347–356. [CrossRef]
2. Trus, F.G.; Braga do Espírito Santo, A.V.; Taniguti de Oliveira, E.Z.; Rimi, L.T.; Dias Parra, M.T. Geoimmersion: Expansion in Geology for Minecraft. *J. Eng. Res.* **2023**, *3*, 2–17. [CrossRef]
3. Rybár, R.; Gabániová, Ľ.; Rybárová, J.; Beer, M.; Bednárová, L. Designing a Tool for an Innovative, Interdisciplinary Learning Process Based on a Comprehensive Understanding of Sourcing: A Case Study. *Processes* **2022**, *10*, 9. [CrossRef]
4. Davies, M. Blockbuster—The Making of Minecraft. Future plc, 10 November 2012. Available online: https://www.pcgamer.com/the-making-of-minecraft/ (accessed on 20 December 2012).
5. Chambers, B. Minecraft Gives Students a Blocky Sandbox to Build Engineering Skills. *Virginia Tech News*. 22 July 2021. Available online: https://news.vt.edu/articles/2021/07/minecraft-ben-chambers-enge-mining-cetl.html (accessed on 29 May 2025).
6. Stuart, K.; MacDonald, K. The 50 Best Video Games of the 21st Century. *The Guardian*. 19 September 2019. Available online: https://www.theguardian.com/games/2019/sep/19/50-best-video-games-of-the-21st-century (accessed on 20 September 2019).

7. Hall, C. Why Minecraft Is the Most Important Game of the Decade. Polygon, 7 November 2019. Available online: https://www.polygon.com/2019/11/7/20952214/minecraft-most-important-game-of-the-decade-2010/ (accessed on 9 November 2019).
8. Spillman, R. Inside the Geeky, Revolutionary World of "Minecraft". *Salon Media Group*. 6 December 2011. Available online: https://www.salon.com/2011/12/06/inside_the_geeky_revolutionary_world_of_minecraft/ (accessed on 23 November 2013).
9. Cellan-Jones, R. The TV Tycoons You May Have Missed. *BBC News*. 10 July 2012. Available online: https://www.bbc.com/news/technology-18709232 (accessed on 10 July 2012).
10. Deterding, S.; Dixon, D.; Khaled, R.; Nacke, L. From Game Design Elements to Gamefulness: Defining "Gamification". In Proceedings of the 15th International Academic MindTrek Conference: Envisioning Future Media Environments, Tampere, Finland, 28–30 September 2011; ACM: New York, NY, USA, 2011; pp. 9–15. [CrossRef]
11. Surendeleg, G.; Murwa, V.; Yun, H.-K.; Kim, Y.S. The Role of Gamification in Education—A Literature Review. *Contemp. Eng. Sci.* **2014**, *7*, 1609–1616. [CrossRef]
12. Squire, K.; Jan, M. Mad City Mystery: Developing Scientific Argumentation Skills with a Place-Based Augmented Reality Game on Handheld Computers. *J. Sci. Educ. Technol.* **2007**, *16*, 5–29. [CrossRef]
13. Baek, Y.; Min, E.; Yun, S. Mining Educational Implications of Minecraft. *Comput. Sch.* **2020**, *37*, 1–16. [CrossRef]
14. Chang, Y.-S.; Wang, Y.-Y.; Tsai, H.-J. The Metaverse in Green Building Concept Learning, Creative Design Performance, and Learning Engagement. *Sustainability* **2024**, *16*, 6264. [CrossRef]
15. Minecraft Education. Sustainability City. Available online: https://education.minecraft.net/worlds/sustainability-city (accessed on 29 May 2025).
16. McNally, B.; de Andrade, B. Altered spaces: New ways of seeing and envisioning nature with Minecraft. *Vis. Stud.* **2022**, *37*, 175–182. [CrossRef]
17. Minecraft Wiki. Ore—Distribution by Version. Available online: https://minecraft.fandom.com/wiki/Ore#Generation (accessed on 29 May 2025).
18. Krappala, K.; Kemppinen, L.; Kemppinen, E. Achievers, Explorers, Wanderers, and Intellectuals: Educational interaction in a Minecraft open-world action-adventure game. *Comput. Educ. Open* **2024**, *6*, 100172. [CrossRef]
19. Furukado, R.; Zhu, Y.; Hagiwara, G. Integrating Minecraft Education in Curriculum Design to Enhance Chemistry Learning. *Procedia Comput. Sci.* **2024**, *246*, 4375–4384. [CrossRef]
20. de Sena, Í.S.; Niroshan, L.; Rosecký, J.; Brůža, V.; Butler, M.; Cocco, C. A Playful Participatory Planning System (P-PPS): A Framework for Collecting and Analyzing Player-Generated Spatial Data from Minecraft Worlds. *ISPRS Int. J. Geo-Inf.* **2025**, *14*, 210. [CrossRef]
21. de Andrade, B.; Poplin, A.; Sousa de Sena, Í. Minecraft as a Tool for Engaging Children in Urban Planning: A Case Study in Tirol Town, Brazil. *ISPRS Int. J. Geo-Inf.* **2020**, *9*, 170. [CrossRef]
22. Nebel, S.; Schneider, S.; Rey, G.D. Mining Learning and Crafting Scientific Experiments: A Literature Review on the Use of Minecraft in Education and Research. *J. Educ. Technol. Soc.* **2016**, *19*, 355–366.
23. EIT RawMaterials. BetterGeoEdu: Teaching Geology with Minecraft. Available online: https://www.eit.europa.eu/news-events/success-stories/bettergeo (accessed on 29 May 2025).
24. Godfirnon, J.; Pirard, E.; Evrard, M.; Barnabé, P.; Shikika, A. BetterGeoEdu: Geology for All through Gaming. SGU Report Series. 2020. Available online: https://www.sgu.se/en/bettergeo (accessed on 29 May 2025).
25. CurseForge. Immersive Geology Mod. Available online: https://www.curseforge.com/minecraft/mc-mods/immersive-geology (accessed on 29 May 2025).
26. Beheshti, M.; Shah, S.A.; Zhang, H.; Barnett, M.; Hira, A. Affordances of Technology for Sustainability-Oriented K–12 Informal Engineering Education. *Sustainability* **2024**, *16*, 6719. [CrossRef]
27. Hobbs, L.; Stevens, C.; Hartley, J. Digging Deep into Geosciences with Minecraft. *Eos (AGU)*. 2018. Available online: https://eos.org/science-updates/digging-deep-into-geosciences-with-minecraft (accessed on 1 July 2025).
28. Minecraft Wiki. Overworld. Available online: https://minecraft.wiki/w/Overworld (accessed on 29 May 2025).
29. CEO Today. Who Created Minecraft? The Game's Origins & Evolution. *CEO Today Magazine*. April 2025. Available online: https://www.ceotodaymagazine.com/2025/04/block-by-block-the-story-of-minecrafts-creator-evolution/ (accessed on 9 June 2025).
30. IGN. Blocks—Minecraft Guide. Available online: https://www.ign.com/wikis/minecraft/Blocks (accessed on 9 June 2025).
31. Minecraft Wiki. Blok—Minecraft Fandom. Minecraft Fandom Wiki (cs). Available online: https://minecraft.fandom.com/cs/wiki/Blok (accessed on 9 June 2025).
32. Minecraft Wiki. Ore—Minecraft Wiki. Available online: https://minecraft.wiki/w/Ore (accessed on 24 June 2025).
33. Cehlár, M.; Rybár, R.; Soušek, R.; Szabo, S. *Surface Mining Technology and Economy*; Institute of Jan Perner: Pardubice, Czech Republic, 2011; p. 331. ISBN 978-80-86530-76-5.

34. Cehlár, M.; Engel, J.; Mihok, J.; Rybár, R. *Surface Mining*, 1st ed.; FBERG, Technical University of Košice: Košice, Slovakia, 2006; p. 347. ISBN 80-8073-608-1.
35. Pfleider, E.P. *Surface Mining*; A Volume in the Seeley W. Mudd Series; American Institute of Mining, Metallurgical, and Petroleum Engineers: New York, NY, USA, 1968.
36. Pheasant Energy. What is Mineral Extraction: Definition, Importance, Methods & Impact. Available online: https://www.pheasantenergy.com/mineral-extraction/ (accessed on 29 May 2025).
37. Bakhtavar, E.; Shahriar, K.; Oraee, K. Mining method selection and optimization of transition from open pit to underground in combined mining. *Arch. Min. Sci.* **2009**, *54*, 481–493.
38. Bakhtavar, E.; Shahriar, K.; Oraee, K. Transition from open-pit to underground as a new optimization challenge in mining engineering. *J. Min. Sci.* **2009**, *45*, 485–494. [CrossRef]
39. Ben-Awuah, E.; Otto, R.; Tarrant, E.; Yashar, P. Strategic mining options optimization: Open pit mining, underground mining or both. *Int. J. Min. Sci. Technol.* **2016**, *26*, 1065–1071. [CrossRef]
40. Leung, R.; Hill, A.J.; Melkumyan, A. Automation and AI Technology in Surface Mining with a Brief Introduction to Open-Pit Operations in the Pilbara. *arXiv* **2023**, arXiv:2301.09771. [CrossRef]
41. Rangefront Mining Services. 8 Steps of Mineral Exploration. Available online: https://rangefront.com/blog/8-steps-mineral-exploration/ (accessed on 29 May 2025).
42. Skrzypkowski, K.; Gómez, R.; Zagórski, K.; Zagórska, A.; Gómez-Espina, R. Review of Underground Mining Methods in World-Class Base Metal Deposits: Experiences from Poland and Chile. *Energies* **2023**, *16*, 148. [CrossRef]
43. Afum, B.O.; Ben-Awuah, E. A Review of Models and Algorithms for Surface-Underground Mining Options and Transitions Optimization: Some Lessons Learnt and the Way Forward. *Mining* **2021**, *1*, 112–134. [CrossRef]
44. Bridge, G. Contested terrain: Mining and the environment. *Annu. Rev. Environ. Resour.* **2004**, *29*, 205–259. [CrossRef]
45. Mancini, L.; Eslava, N.; Traverso, M.; Mathieux, F. Environmental impact of different methods of extraction and processing of raw materials in mining. *Resour. Policy* **2019**, *62*, 347–356. [CrossRef]
46. Prior, T.; Giurco, D.; Mudd, G.; Mason, L.; Behrisch, J. Resource depletion, peak minerals and the implications for sustainable resource management. *Glob. Environ. Change* **2012**, *22*, 577–587. [CrossRef]
47. Werner, T.T.; Mudd, G.M.; Schipper, A.M.; Huijbregts, M.A.J.; Tibbett, M. Environmental pressures of production of platinum group metals: Insights from life cycle assessment. *J. Clean. Prod.* **2017**, *165*, 1167–1176. [CrossRef]
48. Mudd, G.M. Global trends in gold mining: Towards quantifying environmental and resource sustainability? *Resour. Policy* **2007**, *32*, 42–56. [CrossRef]
49. Engineering Research—eduCAPES. Geoimmersion Expansion in Geology for Minecraft. Available online: https://educapes.capes.gov.br/bitstream/capes/720538/1/Geoimmersion%20expansion%20in%20geology%20for%20minecraft.pdf (accessed on 24 June 2025).
50. Égýankosh. Basics of Ore Geology—Definition of Clarke of Concentration as Ratio of Ore Grade to Crustal Abundance. eGyanKosh. Available online: https://www.egyankosh.ac.in/bitstream/123456789/78233/1/Unit-1.pdf (accessed on 1 July 2025).
51. Wikipedia Contributors. Clarke number—the Relative Abundance of Chemical Elements in Earth's Crust; Clarke of Concentration as Index of Elemental Enrichment. Wikipedia, The Free Encyclopedia. Available online: https://en.wikipedia.org/wiki/Clarke_number (accessed on 1 July 2025).
52. John, D.A.; Vikre, P.G.; du Bray, E.A.; Blakely, R.J.; Fey, D.L.; Rockwell, B.W.; Mauk, J.L.; Anderson, E.D.; Graybeal, F.T. *Descriptive Models for Epithermal Gold-Silver Deposits*; U.S. Geological Survey Scientific Investigations Report 2010–5070–Q; U.S. Geological Survey: Menlo Park, CA, USA, 2018; 247p. [CrossRef]
53. Dilles, J.H.; John, D.A. Porphyry and Epithermal Mineral Deposits. In *Encyclopedia of Geology*, 2nd ed.; Alderton, D., Elias, S.A., Eds.; Elsevier: Amsterdam, The Netherlands, 2021; pp. 847–866. [CrossRef]
54. John, D.A.; Ayuso, R.A.; Barton, M.D.; Blakely, R.J.; Bodnar, R.J.; Dilles, J.H.; Gray, F.; Graybeal, F.T.; Mars, J.C.; McPhee, D.K.; et al. *Porphyry Copper Deposit Model; Chapter B of Mineral Deposit Models for Resource Assessment*; U.S. Geological Survey Scientific Investigations Report 2010–5070–B; U.S. Geological Survey: Menlo Park, CA, USA, 2010.
55. Li, Y.; Ke, C.; She, H.; Wang, D.; Xu, C.; Wang, A.; Li, R.; Peng, Z.; Zhu, Z.; Yang, K.; et al. Geology and Mineralization of the Bayan Obo Supergiant Carbonatite-Type REE-Nb-Fe Deposit in Inner Mongolia, China: A Review. *China Geol.* **2023**, *6*, 716–750. Available online: https://doaj.org/article/9e5cc75d3b7c443e9c10f020e0b28339 (accessed on 22 June 2025).
56. Hitzman, M.W.; Selley, D.; Bull, S. Formation of Sedimentary Rock-Hosted Stratiform Copper Deposits through Earth History. *Econ. Geol.* **2010**, *105*, 627–639. [CrossRef]
57. Groves, D.I.; Bierlein, F.P.; Meinert, L.D.; Hitzman, M.W. Iron Oxide Copper-Gold (IOCG) Deposits through Earth History: Implications for Origin, Lithospheric Setting, and Distinction from Other Epigenetic Iron Oxide Deposits. *Econ. Geol.* **2010**, *105*, 641–654. [CrossRef]

58. Bowell, R.J.; Lagos, L.; de los Hoyos, C.R.; Declercq, J. Classification and Characteristics of Natural Lithium Resources. *Elements* **2020**, *16*, 259–264. [CrossRef]
59. Tkachev, A.V.; Vishnevskaya, N.A.; Chesalova, E.I. Lithium Deposits from the Mesoarchean to Present: Their Types, Distribution in Geological Time, Explored Resource Base. *Geol. Ore Depos.* **2024**, *66*, 617–647. [CrossRef]
60. Rader, E.; Love, R.; Reano, D.; Dousay, T.A.; Wingerter, N. Pandemic Minecrafting: An analysis of the perceptions of and lessons learned from a gamified virtual geology field camp. *Geosci. Commun.* **2021**, *4*, 475–492. [CrossRef]
61. Teixeira, I.; Vasconcelos, C. The Use of Educational Games to Promote Learning in Geology: Conceptions of Middle and Secondary School Teachers. *Geosciences* **2024**, *14*, 16. [CrossRef]
62. Kumar, R.; Chaudhary, V.K.; Pramanik, N.; Ghosh, S. Economic and Environmental Impact of Corrosion. In *Sustainability, Safety, and Applications of Nanomaterials-Based Corrosion Inhibitors*; IGI Global: Hershey, PA, USA, 2024; pp. 277–296, ISBN-13: 979-8369376409. Available online: https://www.igi-global.com/gateway/chapter/350982 (accessed on 22 June 2025). [CrossRef]
63. Koch, G.H. Cost of Corrosion. In *Trends in Oil and Gas Corrosion Research and Technologies*; El-Sherik, A.M., Ed.; Elsevier Ltd.: Dublin, Ireland, 2017; pp. 3–30. Available online: https://www.sciencedirect.com/science/article/pii/B9780081011058000012 (accessed on 22 June 2025).
64. NACE International. *International Measures of Prevention, Application, and Economics of Corrosion Technologies Study (IMPACT)*; NACE International: Houston, TX, USA, 2016; Available online: http://impact.nace.org/documents/Nace-International-Report.pdf (accessed on 22 June 2025).

processes

MDPI

Article

Fire Characteristics and Water Mist Cooling Measures in the Coal Transportation Process of a Heavy-Haul Railway Tunnel in Shanxi Province

Wenjin He [1], Maohai Fu [2,*], Lv Xiong [1] and Shiqi Zheng [3]

1 Institute of Rail Transit, Guangdong Communication Polytechnic, Guangzhou 510650, China; hewenjean@gdcp.edu.cn (W.H.); gdcprailh@163.com (L.X.)
2 School of Mechanical Engineering, Southwest Jiaotong University, Chengdu 610095, China
3 Guangzhou Municipal Engineering Design & Research Institute Co., Ltd., Guangzhou 510000, China; zhengsq@gzmedri.com
* Correspondence: gd3987@outlook.com

Abstract: This study investigates the spread patterns of tunnel fires and examines issues related to emergency response. It focuses on the temperature characteristics, spread patterns, conditions leading to multi-source fires, and the efficacy of water mist suppression methods in heavy-haul railway tunnel fires. The research employs theoretical derivations and numerical simulations to achieve its objectives. It was discovered that, during a fire in a heavy-haul railway tunnel, the temperature inside the tunnel can exceed 500 °C. Furthermore, depending on the nature of the goods transported by the train and under specific wind speed conditions, the fire source has the potential to spread to other carriages, resulting in a multi-source fire. Using the numerical simulation software Pyrosim 2022, various wind speed conditions were simulated. The results revealed that at lower wind speeds, the smoke demonstrates a reverse flow phenomenon. Concurrently, when the adjacent carriage on the leeward side of the fire is ignited, the high-temperature reverse flow smoke, along with the thermal radiation from the flames, ignites combustible materials in the adjacent carriage on the windward side of the burning carriage. Through theoretical derivation and numerical simulation, the critical wind speed for the working conditions was determined to be 2.14 m/s. It was found that while a higher wind speed can lead to a decrease in temperature, it also increases the flame deflection angle. When the wind speed exceeds 2.4 m/s, although the temperature significantly drops in a short period, the proximity of combustible materials on the leeward side of the carriage becomes a concern. At this wind speed, the flame deflection angle causes heat radiation on the leeward side, specifically between 0.5 m and 3 m, to ignite the combustible materials on the carriage surface, resulting in fire spread and multiple fire incidents. The relationship between wind speed and the angle of deflection from the fire source was determined using relevant physics principles. Additionally, the relationship between wind speed and the trajectory of water mist spraying was established. It was proposed to optimize the position of the water mist based on its deviation, and the results indicated that under critical wind speed conditions, when the water mist spraying is offset approximately 5 m towards the upwind side of the fire source, it can act more directly on the surface of the fire source. Numerical simulation results show a significant reduction in the maximum temperature and effective control of fire spread. Under critical wind speed conditions, the localized average temperature of the fire decreased by approximately 140 °C when spraying was applied, compared to the conditions without spraying, and the peak temperature decreased by about 190 °C. This modification scheme can effectively suppress the threat of fire to personnel evacuation under simulated working conditions, reflecting effective control over fires. Additionally, it provides theoretical support for the study of fire patterns in tunnels and emergency response measures.

Keywords: tunnel fires; numerical simulation; multiple ignition sources; flame deflection angle; water mist optimization

1. Introductory

Heavy-duty railroads have garnered significant attention in the realm of freight transportation due to their characteristics of large capacity, high efficiency, and low cost. The advantages of heavy-duty railroad transportation are more pronounced compared to other modes of transportation, leading to rapid development in resource-rich countries such as the United States, China, Canada, and Australia. Heavy-duty railroad transportation is particularly common for goods such as coal and ore, which present risks of spontaneous combustion and flammability. As heavy-duty railroads expand their networks extensively, the number of tunnel sections has also increased. Due to the long transportation distances associated with heavy railroads, the conditions within tunnels differ significantly from those of other passenger tunnels [1]. The relatively enclosed environment of tunnels, which contains a substantial amount of electrical equipment, poses unique risks. When heavy trains carrying flammable products traverse these tunnels, they can distribute a considerable quantity of combustible and flammable materials throughout the space. This specific environment increases the likelihood of fire incidents, with the potential hazards being significantly greater than in other heavily loaded sections. Additionally, the number of fire-causing factors within the tunnel is notably elevated, including incidents such as train collisions, electrical fires, and the friction of trains igniting combustible materials [2]. In 2001, a fire was triggered in the St. Gotthard tunnel in Switzerland due to the impact of goods falling off the walls. Additionally, the properties of the cargo can increase the incidence of fire; for instance, coal, which is a primary transport cargo for heavy trains, may accelerate the oxidation process under certain environmental conditions. When heavy trains pass through rail tunnels, the heat generated can accumulate, leading to the spontaneous combustion of coal piles [3–6]. This phenomenon has been observed in the Daqin Railway in China, where a fire was ignited by the spontaneous combustion of coal during transport. When a fire occurs in a tunnel, the unique characteristics of its internal structure can cause temperatures to exceed 1000 °C. The smoke generated by such fires is difficult to eliminate, leading not only to potential casualties but also to significant damage to the tunnel structure due to the extreme heat. Therefore, studying the characteristics of fires in heavy railroad tunnels and implementing relevant measures to control both the fire and the internal temperature can effectively mitigate the impacts of such incidents on both individuals and the environment [7]. This research is crucial for enhancing the overall safety of heavy railroad tunnels.

Scholars from around the globe have extensively studied tunnel fires, primarily through fire experiments, numerical simulations, and other methodologies to enhance the theoretical understanding of tunnel fires. Jae Seong Roh, Seung Shin Yang, et al. [8] utilized scaled-down model fire experiments to examine the heat release rate from fires in various ventilation environments, establishing a positive correlation between wind speed and heat release rate under specific conditions. E. Blanchard, P. Boulet, et al. [9] conducted 4 MW oil pan fire experiments using a tunnel model to investigate heat loss during each phase of the fire, confirming that wind speed is positively correlated with the heat release rate. Furthermore, they analyzed the heat loss associated with each stage of their fire experiments and validated their findings through numerical simulations. McGrattani et al. [10] employed the numerical simulation software FDS to replicate a tanker truck fire in the Howard Street Tunnel. Their findings indicated that the temperature on the surface of the fire source exceeded 1000 °C, while temperatures in the vicinity of the fire source reached over 500 °C. Similarly, Ciro Caliendo, Paolo Ciambelli [11], and colleagues utilized numerical simulation software to examine the temperature distribution within a bi-directional highway tunnel during a vehicle fire. Their results revealed that an increase in the number of fire sources correlates with a heightened risk within the tunnel [12–14]. In heavy railroad tunnel fires, the presence of a significant amount of flammable materials carried by trains makes the study of the temperature field under various boundary conditions particularly crucial. Tunnels tend to accumulate heat and are susceptible to conditions such as ventilation, which can lead to the ignition of unignited combustible materials present in the train, resulting in multiple fire sources. Among the factors influencing temperature distribution, wind speed in the tunnel is one of the most critical. Delichatsios [15] investigated the temperature distribution characteristics of promenade fires in naturally ventilated environments and described the temperature decay equation. Thomas [16,17] analyzed the smoke backward recession phenomenon caused by wind speed in tunnels, examining inertial and buoyancy forces to derive a dimensionless empirical equation for the backward recession length of high-temperature smoke. Hu longhua [18,19] conducted full-scale fire experiments to investigate smoke propagation from the tunnel roof and developed a fire exponential decay model by integrating momentum equations. Additionally, Wu and Bakar [20] performed fire experiments in tunnels with various cross-section shapes, identifying the critical wind speed for their fires and establishing a dimensionless critical wind speed model.

In heavy railroad tunnel fires, the unique characteristics of the goods transported by trains necessitate careful consideration of wind speed. Insufficient wind speed can lead to smoke backflow and may cause flames to be drawn towards the upwind side, exacerbating the fire's spread. Conversely, excessively high wind speeds can reduce the peak temperature within the tunnel [21]. However, they can also alter the flame's trajectory, causing it to deflect excessively downwind. This deflection can result in heat radiation igniting flammable materials in the downwind carriages, further contributing to fire proliferation. Therefore, studying fires in heavy railroad tunnels is of paramount importance. Utilizing numerical simulation software to analyze the impact of wind speed on flame behavior and temperature distribution is essential for enhancing fire prevention and emergency response capabilities in these environments.

2. Numerical Simulation Modeling

2.1. Introduction to Pyrosim Numerical Simulation Software

Pyrosim numerical simulation software, developed by Thunderhead Engineering in the United States, employs computational fluid dynamics methods to simulate flame combustion, heat transfer, and smoke movement in designed scenarios during fire events. It primarily focuses on low-Mach fluid flow and utilizes the mixed fraction method for combustion modeling. The software predominantly implements large eddy simulation and direct numerical simulation techniques for turbulence simulation.

Pyrosim is solved using the Fire Dynamics Simulator (FDS) solver, which is governed by several key equations: the mass conservation equation, the momentum conservation equation, the energy conservation equation, the component transport equation, and the equation of state [22,23]. The mass conservation equation is articulated in Equation (1).

$$\frac{\partial \rho}{\partial t} + \nabla \cdot (\rho u) = 0 \tag{1}$$

where ρ represents the gas density, kg/m^3; t denotes time, s; u is the velocity vector, m/s; and ∇ is the dispersion operator, 1/m. The momentum conservation equation is presented in Equation (2).

$$\frac{\partial u}{\partial t} = \vec{F} - \frac{1}{\rho}\nabla p + v\nabla^2 u - (u \cdot \nabla)u \quad (2)$$

where $\vec{F}$ is the vector of external force acting on the fluid, N/m^3; v is the coefficient of kinematic viscosity. The energy conservation equation is shown in Equation (3).

$$\frac{\partial(\rho h)}{\partial t} + \nabla \cdot (\rho h u) = \frac{Dp}{Dt} + \nabla \cdot (k\nabla T) + q''' - q''_{rad} + q_{chem} \quad (3)$$

where h is the specific enthalpy, J/kg; T is the temperature, K; k is the thermal conductivity, W/(m·k); q''' is the volumetric heat source, W/m^3; q'' is the radiative heat loss, W/m^3; q_{chem} is the chemical reaction exothermic rate, W/m^3. The component transportation equation is Equation (4).

$$\frac{\partial(\rho Z)}{\partial t} + \nabla \cdot (\rho Z u) = \nabla \cdot (\rho D \nabla Z) \quad (4)$$

where Z is the mixing fraction; D is the diffusion coefficient, m^2/s.

At the same time, Pyrosim provides the Smokeview function, which is able to visualize the changes in smoke flow in a fire scenario.

2.2. Fire Full-Scale Modeling

To align the research conditions more closely with the actual project, a heavy-duty railroad transportation tunnel in Shanxi was selected for research visits and investigations. This study is conducted in accordance with the International Union of Railways standard UIC564-2 (International Union of Railways, 2019) [24]. A 500 m long tunnel with a rectangular cross-section measuring 4.8 m in width and 5.5 m in height was established. The wall thickness is 0.2 m, taking into account the wall material used in the actual project. This setup aims to facilitate a more thorough investigation of temperature field variations.

Considering the wall material used in the actual project and to facilitate a more effective study of the temperature field changes, the wall material is selected to be adiabatic. Within the tunnel, a transport carriage is established; this carriage is an open car designed for coal transport. The dimensions of the carriage are as follows: length of 13.5 m, width of 3 m, and height of 2 m. Additionally, several carriages are established with an interval of 0.5 m between them. The coal within the carriage is leveled with the upper end of the compartments. The primary parameters are summarized in Table 1.

Table 1. Geometric parameters and boundary conditions of carriages.

Setting Elements	Parameters
Transport	Coal
Thickness of sidewalls of carriages	3 mm
Thickness of the underside of the carriages	5 mm
Carriage end wall thickness	4 mm
Diameter of carriage wheels	920 mm
Carriage material	Steel

In the fire source, the spontaneous combustion of the coal pile is simulated, with the starting source chosen as the t^2 fire model. This model primarily simulates the combustion process of coal's spontaneous combustion, with the initial combustion point located at the top of the coal in the wagon. The fire model can be expressed as $Q = at^2$, where Q represents the heat release rate measured in kilowatts, kW; while a is the fire growth coefficient in kW/s^2; and t is the time required for the fire to achieve its maximum heat release rate, s.

Based on the combustion characteristics of coal, the fire power in heavy railroad tunnels typically ranges from 5 to 30 MW. Additionally, spontaneous combustion fires in coal piles usually exhibit fire growth coefficients between 0.011 and 0.044. For the purposes of this analysis, Q is established at 10,000 kW, t is set at 300 s, and the fire growth coefficient a is defined as 0.011 kW/s^2. Simultaneously, the coal in the compartment undergoes a pyrolysis reaction. The ignition point of coal typically exceeds 300 °C; therefore, once the temperature of the coal reaches 300 °C, it can ignite. The heat release rate during coal combustion is set at 1000 kW/m^2, achieving its maximum heat release rate after 300 s of combustion. The geometric model diagram of the fire scene is illustrated in Figure 1.

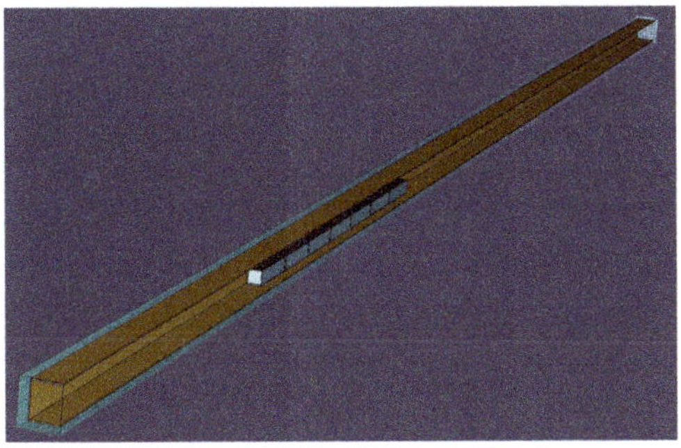

Figure 1. Model diagram of tunnel fire.

Figure 1 illustrates that the central train carriages are designated as the fire carriages. To study the temporal changes in temperature, CO concentration, and other fire characteristics, the center of the fire source in the tunnel serves as a reference point. Thermocouples, smoke concentration sensors, CO concentration sensors, and pressure sensors are installed at vertical heights of 5.5 m, positioned at intervals of 5 m on both the anterior and posterior sides of the fire source. In order to study the fire dynamics at the top of the coal pile within the compartments under varying wind conditions, we examined the temperature field changes at a vertical height of 3.92 m. Utilizing a consistent sensor layout and type, we positioned thermocouples and CO concentration sensors at a height of 1.6 m, which corresponds to the average height of a human eye. These sensors were strategically placed 0.3 m from the walls on both the left and right sides of the carriage. Additionally, we established the fire's center point and deployed monitoring points every 5 m both in front of and behind the carriage. The primary objective of this study is to analyze the escape environment during a fire and to improve rescue operations. The arrangement of the monitoring points is illustrated in Figures 2 and 3.

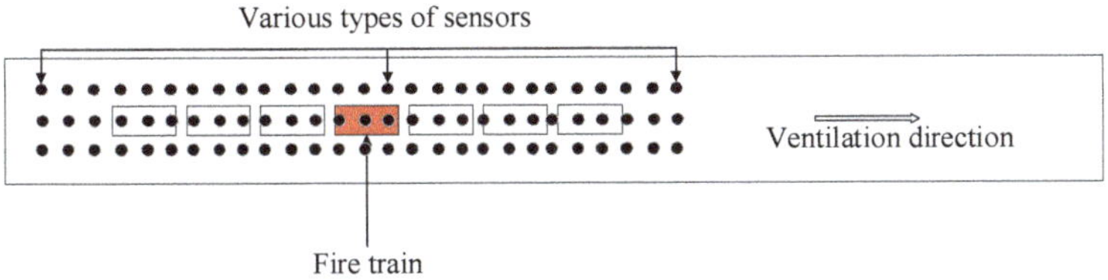

Figure 2. Horizontal arrangement of various types of sensors in the fire tunnel.

In the figure, the black filled circles represent various types of sensors, while the red rectangle indicates the carriage where the initial fire originated. The direction of the arrow illustrates the wind direction, which was analyzed in the simulation by varying wind speeds to observe changes in the temperature field.

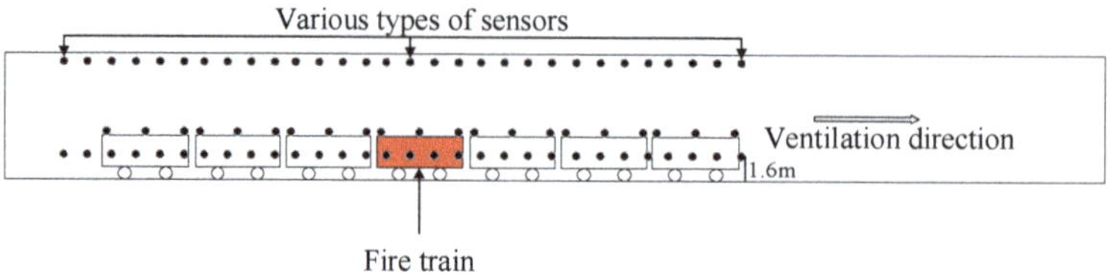

Figure 3. Vertical arrangement of various types of sensors in the fire tunnel.

2.3. Mesh Partitioning and Mesh Irrelevance Test

In Pyrosim, the grid division of the fire model significantly affects the accuracy of fire simulations. Generally, smaller grid sizes lead to more accurate computational results; however, they also increase the computation time required by the computer. Research presented in paper [25] compares temperature trends under various grid size conditions. This analysis is further contextualized by the grid size division standards provided in the FDS user manual, which can be expressed by Equation (5).

$$D^* = \left(\frac{Q}{\rho_0 c_P T_0 \sqrt{g}} \right)^{\frac{2}{5}} \quad (5)$$

where D^* is the characteristic diameter of the fire source, m; ρ_0 is the density of ambient air, kg/m^3; c_p is the specific heat capacity of air, which takes the value of kJ/(kg·K); T_0 is the temperature, K; and g is the acceleration of gravity, m/s^2. The characteristic diameter of the fire source can be calculated to be about 2.34 m, and the results of the calculations are relatively accurate when the characteristic diameter of the fire source is 1/4 to 1/16 times the size of the mesh. Therefore, in order to ensure the accuracy of the fire simulation results, the grid independence test is carried out for the fire simulation.

The temperature trend directly above the fire source for mesh sizes ranging from 0.15 m to 0.5 m is illustrated in Figure 4. It can be observed that the temperature trends across the four grid sizes are consistent, with values remaining largely similar. The maximum error observed does not exceed 15 °C. Therefore, for mesh sizes of 0.15 m, 0.3 m, 0.4 m, and 0.5 m, the fire simulation results are deemed accurate. However, it is important to note that the mesh size significantly affects the simulation time on the computer. To ensure relative accuracy, a mesh size of 0.3 m was found to yield results most similar to those obtained with a mesh size of 0.15 m while also providing relatively high computational efficiency, utilizing a total of 1,066,250 meshs.

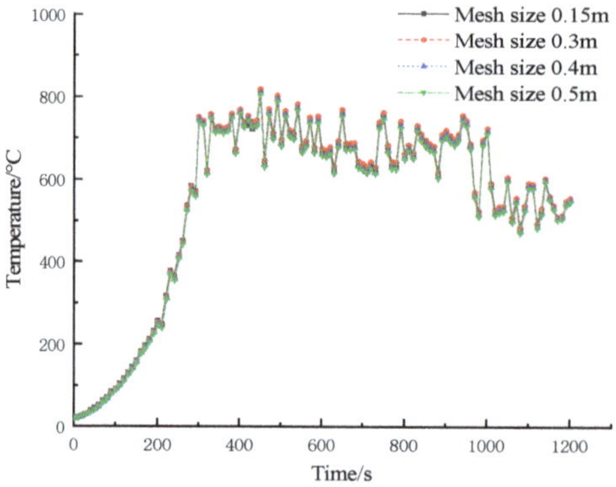

Figure 4. Mesh size 0.15 m, 0.3 m, 0.4 m, 0.5 m temperature comparison chart of the same measurement point.

3. Study on the Fire Characteristics of Heavy Railway Tunnels

3.1. Study on the Fire Spreading Characteristics of Heavy Railroad Trains

In the simulation of coal pile fire within a heavy railroad tunnel train under ventilation conditions of 0 m/s and 1 m/s, it was observed that the combustion behavior and temperature distribution of the fire conformed to general principles in the absence of wind. When a fire occurs, high-temperature smoke impacts the roof, spreading uniformly along the roof walls and forming a roof jet. Given that coal is the primary cargo transported by the train, a low roof height hampers the efficient expulsion of high-temperature smoke, leading to the accumulation of local hotspots. Consequently, the temperature above the train remains elevated, potentially igniting other transport vehicles that are not on fire, thus facilitating a broader spread of the fire. Since the train primarily transports coal, inadequate roof height further delays the timely expulsion of smoke, exacerbating local heat buildup. This causes the temperature above the train to rise significantly, which may lead to the ignition of other coal in the transportation vehicles without an open flame, thereby facilitating the spread of fire and the occurrence of multiple flames. Figure 5 illustrates the fire occurring at different moments, represented by a temperature cloud diagram.

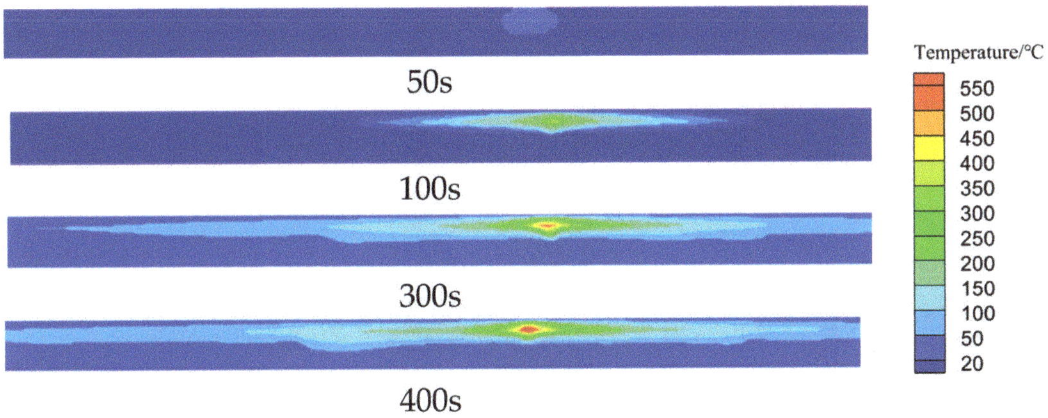

Figure 5. Temperature cloud of horizontal cross-section in the tunnel at different moments.

As illustrated in Figure 5, under windless conditions, when the heat release rate of a single fire source reaches its peak, the temperature of the ceiling is notably elevated, reaching 550 °C. The vertical height is 3.9 m, at which the exposed surface of the coal transported in the carriage has reached a temperature of 230 °C. Given that the general ignition point of coal is approximately 300 °C, the temperature has not yet reached this critical threshold. Consequently, the risk of fire spreading to adjacent carriages remains relatively low. The temperature did not reach the ignition point of the coal, resulting in a relatively low risk of fire spreading to other train carriages. However, in the actual environment of heavy railroad tunnels, there is typically a certain wind speed. After simulating the fire under a wind speed of 1 m/s, the surface temperature of the coal layer in the train carriages is illustrated in Figure 6.

As shown in Figure 6, the temperature in the compartment where the fire occurred is significantly higher and exhibits a fluctuating pattern. This is primarily due to the cooling effect of wind speed on the high-temperature smoke generated by the fire, while the local temperature on the surface of the transported coal exchanges heat with the high-temperature smoke produced by the fire, causing a temporary increase in local temperature. It can be clearly observed that at approximately 1067 s into the fire, the surface temperatures of the adjacent compartments on both the upper and lower wind sides of the burning compartment are significantly above 300 °C, which has reached the basic combustion temperature of coal. Through numerical simulation software Pyrosim, it is evident that the fire spreads to the adjacent compartments, causing additional fires, as illustrated in the flame schematic in Figure 7.

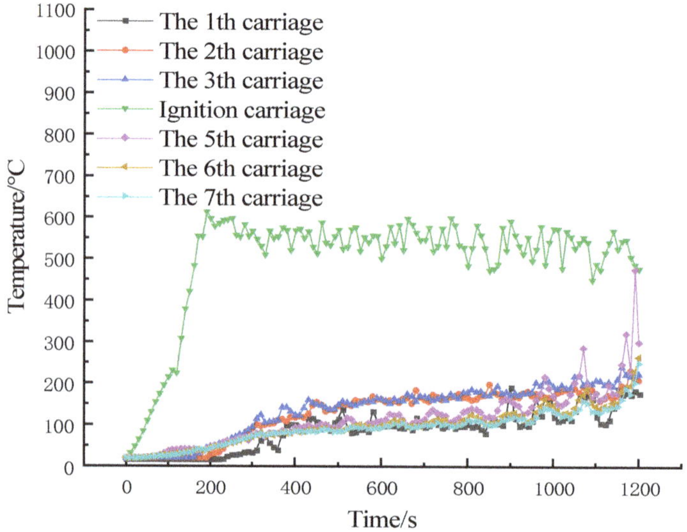

Figure 6. Temperature change diagram of the exposed surface of the coal seam in each carriage with a vertical height of 3.9 m.

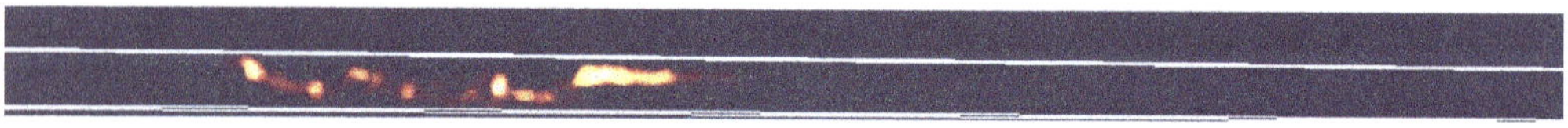

Figure 7. Heavy railroad tunnel fire spread.

Figure 7 clearly indicates that the fire spreads not only to the downwind side of the ignition source but also to the upwind side. As the airflow moves towards the ignition source, it exerts a flexible damping effect on the high-temperature smoke being drawn in. This is primarily due to the increase in temperature near the ignition source, which causes a reduction in local air density on the upwind side of the ignition source. The hot smoke is buoyed upward, while the fresh airflow moving within the tunnel, consisting of air at normal temperature, creates a stratified flow with the high-temperature smoke near the ceiling, resulting in a reverse pressure gradient that suppresses normal air movement. Concurrently, this phenomenon allows the oxygen in the fresh airflow to be utilized more directly by the ignition source, thereby promoting the combustion process and significantly enhancing the heating effect, which may further strengthen the damping effect. As the high-temperature smoke moves towards the upwind side of the ignition source, it is also influenced by the fresh airflow, slowing down its movement and potentially settling locally near the ignition source for a certain period, thus facilitating heat accumulation. When coal is exposed to a high-temperature environment under these conditions for a certain duration, it will ignite. Therefore, this can lead to multi-source fires. The temperature cloud at the height of the compartment at 1067 s is illustrated in Figure 8.

Figure 8 clearly shows that the temperature of the carriage section is higher than that under windless conditions. Additionally, the temperature near the fire source in the carriage is elevated, exceeding 300 °C, which is the ignition point of coal. This increase in temperature contributed to the spread of the fire.

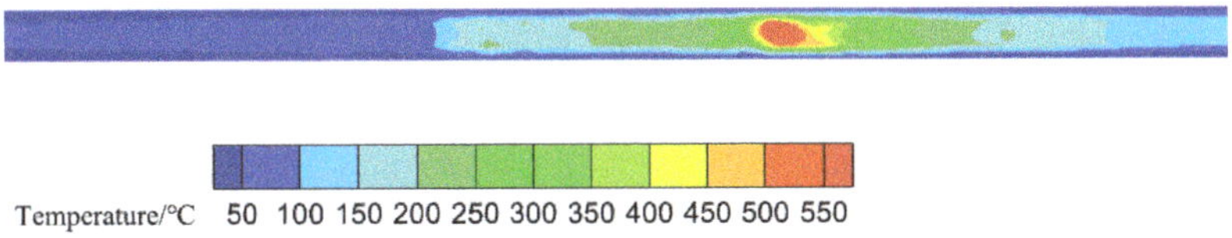

Figure 8. Vertical compartment height temperature cloud map.

Under specific ventilation conditions, although it is possible to control the local temperature decrease in a short period, the oxygen present in the ventilated environment will facilitate the combustion of the fire source, accelerating the rate of heat release from the fire. This not only leads to a rapid increase in fire temperature after a certain period but also causes the fire to spread to other compartments. Meanwhile, the high-temperature smoke that is drawn in will cause the average temperature of the entire tunnel to rise rapidly. This situation severely impacts personnel safety during evacuation and complicates rescue efforts. Additionally, excessive temperatures can inflict damage to the tunnel structure. Therefore, it is essential to determine the critical wind speed for fires in heavy railroad tunnels. At the same time, utilizing wind speed and other measures to prevent the high temperatures from the fire from igniting other carriages is crucial in suppressing the occurrence of multiple fire sources.

3.2. Determination of Critical Wind Speed for Fire in Heavy Railroad Tunnel

The critical wind speed refers to the wind speed condition under which the high-temperature smoke generated by a tunnel fire does not reverse its flow. In this scenario, the critical wind speed is not only crucial for controlling the movement trajectory of the smoke flow and preventing the fire from spreading to the upwind side of the compartment, but it is also an important factor in determining the wind speed conditions.

In the study of critical wind speed, it can be calculated and predicted using mathematical empirical formulas, as demonstrated in Equation (6).

$$v_{cr} = k\left(\frac{gQH}{c_p \rho_0 AT}\right)^{1/3} \tag{6}$$

where v_{cr} is the critical wind speed, m/s; k is a variable constant, which is mainly affected by the location and size of the fire source, and can take the value of 1 in the numerical simulation; H is the height of the heavy railroad tunnel, m; A is the cross-section area, m^2; T is the temperature of the smoke stream, K. Under the condition of no wind, the temperature is about 550 °C. The specific heat capacity of air is approximately 1.3818 kJ/(kg·K), and the critical wind speed is calculated to be around 2.14 m/s. To verify this calculation, a numerical simulation study of fire conditions was conducted at wind speeds of 2.1 m/s, 2.14 m/s, 2.2 m/s, and 2.3 m/s. The study focused on the maximum heat release rate during the fire, as well as the corresponding temperature profiles, which are illustrated in Figure 9.

The temperature cloud in Figure 9 clearly shows that when the heat release rate of the fire source reaches its peak, the fire temperature field stabilizes. Figure 9a illustrates the temperature cloud at a wind speed of 2.1 m/s, where it is evident that the high-temperature smoke generated by the fire is still partially receding backward. However, this backward receding phenomenon completely disappears at a wind speed of 2.14 m/s. This value aligns with the calculated results, allowing us to determine that the critical wind speed for this condition is 2.14 m/s. Furthermore, an observation and analysis of the temperature distribution indicates that wind speed influences both the temperature peak and the average temperature within the fire area of the railroad tunnel, revealing a negative correlation between wind speed and temperature peak.

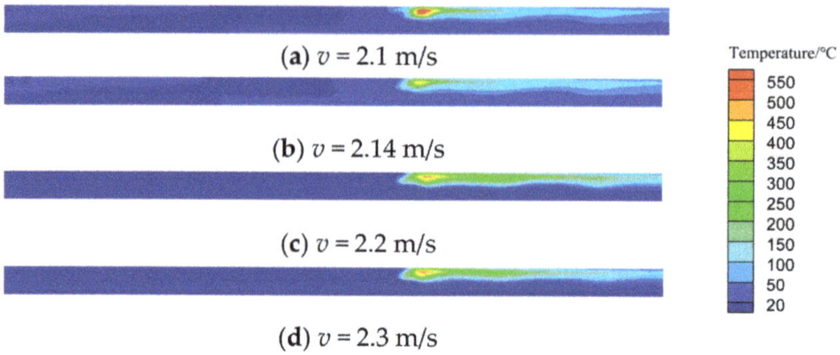

Figure 9. Temperature cloud for different wind speed conditions at 300 s.

3.3. Research on Flame Spread and Wind Speed Control of Fire in Heavy Railroad Tunnel

Different ventilation environments not only affect the temperature of tunnel fires but also lead to certain changes in flame shape. The flame shape is influenced by varying wind speeds, which can cause a certain deflection angle. Due to the special nature of the cargo transported by heavily loaded trains, excessively high wind speeds can increase the deflection angle of the flame. The thermal radiation from the flame can raise the surface temperature of coal on other unburned trains, potentially leading to fire outbreaks and creating multiple fire sources. Therefore, while ensuring that the wind speed in the tunnel is at the critical wind speed, it is also necessary to control the wind speed to prevent the spread of fire. The relationship between the flame deflection angle and wind speed can be calculated using Equation (7).

$$\theta = \arctan(\frac{v}{u_p}) \tag{7}$$

where v is the tunnel ambient wind speed, m/s; u_p is the flame vertical rise velocity, m/s. The flame vertical rise velocity can be described as Equation (8).

$$u_p = C(\frac{Q}{\rho_0 c_p T_0 D^2})^{1/3} \tag{8}$$

where C is an empirical constant, taken as 1; D is the diameter of the fire source, m. After determining its deflection angle, its flame horizontal deflection distance x can be determined by Equation (9).

$$x = H \cdot \tan(\arctan(\frac{v}{u_p})) \tag{9}$$

where H is the height of the flame, m. The distance between the front of the flame and the carriage transporting coal on the downwind side of the fire source can be calculated using Equation (10), which is shown below.

$$d = \sqrt{(L+x)^2 + H^2} \tag{10}$$

where d is the distance from the front of the flame to the coal compartment on the downwind side of the fire source, m. The heat radiation flux to which the coal on its downwind side is exposed can be calculated using Equation (11).

$$q'' = \varepsilon \sigma T_f^4 \cdot \frac{1}{4\pi d^2} \tag{11}$$

where, q'' is the thermal radiation flux kW/m^2; ε is the emissivity, take 0.9; σ is the Stefan–Boltzmann constant; T_f is the flame temperature, K.

Through calculations, it is determined that when the wind speed exceeds 2.3 m/s, the compartments on the downwind side of the fire source, located between 0.5 m and 3 m, have an increased likelihood of ignition. Numerical simulation software was employed to model the tunnel fire at a wind speed of 2.4 m/s. The results indicate that adjacent compartments on the downwind side of the fire begin to ignite at 719 s. Furthermore, by 825 s, the coal within these compartments is ignited, leading to the spread of the fire. The temperature distribution of the fire is illustrated in Figure 10.

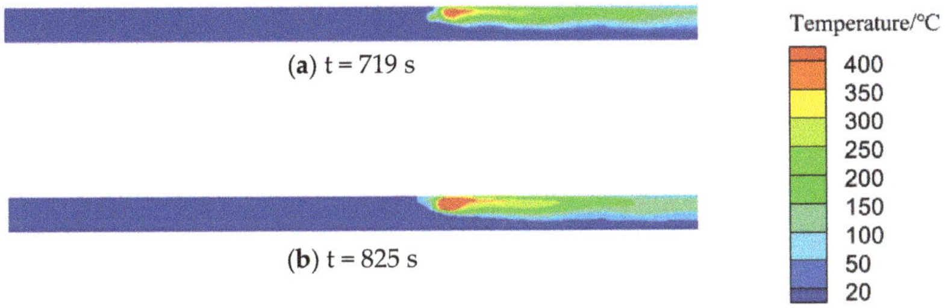

Figure 10. Fire temperature cloud at different moments when v = 2.4 m/s.

In Figure 10, panel (a) illustrates the downwind side of the carriage coal fire at the onset of the fire, where the temperature is relatively low. At this stage, the critical wind speed for the fire remains effective. However, as the heat release rate of the carriage coal gradually increases, the overall heat release rate of the fire rises, leading to the invalidation of the critical wind speed. This phenomenon is further evidenced in Figure 10b, where the high-temperature smoke exhibits the occurrence of inverse recession. This effect is primarily attributed to the increase in fire duration, which correlates positively with the heat release rate. It can be concluded that at 825 s into the fire, the combustion of coal produces an open flame. As the duration of the fire increases, the potential for its spread also rises. The heightened heat release rate exacerbates the smoke countercurrent phenomenon, which may facilitate the fire's spread to the upwind side. Concurrently, factors such as wind speed and thermal resistance may contribute to the fire spreading downwind. Therefore, it is essential to control wind speed and implement additional emergency measures. Currently, the primary fire prevention and suppression strategy in tunnels involves the use of water mist sprinklers. Water mist spraying effectively reduces the high temperatures generated by the fire and helps prevent its spread.

4. Study on the Effect of Water Mist Spraying on Temperature Distribution and Fire Spreading

4.1. Study of the Effect of Water Mist on the Temperature Field at Different Locations

The water mist spraying system not only reduces the temperature generated by a fire but also provides localized cooling to prevent the spread of flames. However, during actual fire scenarios, the thermal buoyancy produced by the fire can significantly affect the trajectory of the spray, preventing it from directly impacting the fire's surface [26–29]. Additionally, the high temperature of the fire plume can cause water mist with lower density to evaporate in such an environment, thereby diminishing its cooling effectiveness. Due to the high temperature of the fire plume, a low density of water mist can lead to evaporation in such an environment, thereby diminishing its cooling effect. Furthermore, wind speed significantly influences the trajectory of the water mist.

This study focuses on working conditions, specifically examining the critical wind speed conditions at the water spray location and the cooling effect of the impact. Additionally, it investigates the fire spread conditions associated with a wind speed of 2.4 m/s, particularly regarding combustible materials, to assess the effectiveness of water spray in preventing fire spread.

In analyzing the motion of water mist particles, their trajectory can be described using mechanistic formulas. Equation (12) represents the horizontal direction, while Equation (13) describes the trajectory in the vertical direction.

$$m_p \frac{du_{px}}{dt} = -\frac{1}{2} C_d \rho_0 A_p |v - u_{px}| (v - u_{px}) \tag{12}$$

$$m_p \frac{du_{py}}{dt} = -\frac{1}{2} C_d \rho_0 A_p |u_{py}| u_{py} - m_p g \tag{13}$$

where m_p is the mass of water mist particles, kg; u_{px} is the horizontal movement speed, m/s; u_{py} is the vertical movement speed, m/s; C_d is the coefficient of resistance, take 1; A_p is the projected area of the particles, m^2. Other water mist spray settings and conditions are shown in Table 2 [30].

Table 2. Water mist spray settings.

Water Mist Spray Related Conditions	Set of Parameters
Water discharge speed	14 m/s
Diameter of water mist particles	100×10^{-6} m
Water mist particle density	1000 kg/m^3
Initial horizontal speed	9.9 m/s
Initial vertical velocity	9.9 m/s
Spraying density	8 L/min·m^2

The path of water mist movement is influenced by wind speed. Under critical wind speed conditions, the mist tends to drift downwind from the fire source. Therefore, Equation (14) can be utilized to calculate its offset distance, as shown below.

$$L = vt \tag{14}$$

where L is the offset distance, m; t is the time for the water mist to fall to the upper surface of the carriage, s.

The calculations indicate that at a velocity of 2.14 m/s, the offset distance of the water mist is 4.1 m, while at a velocity of 2.4 m/s, the offset distance is approximately 4.8 m. Based on these calculations, it is essential to position the water mist sprinkler system on the upwind side of the fire, adjusting for varying wind speeds. This analysis compares the temperature cloud diagrams of a water mist sprinkler system located directly above the fire and one offset by 4.8 m on the upwind side, in contrast to a system without water mist. Figure 11 illustrates the temperature field comparisons before and after the fire occurred at 825 s, highlighting the effects of the water mist spraying system.

According to the research comparison presented in Figure 11, the cooling effect of water mist sprinklers set directly above the fire source is relatively significant, with the lowest temperature peak observed at a distance of 4.8 m from the fire source. This indicates that not only does the sprinkler system significantly suppress the heat release rate of the fire, but its effectiveness is also related to wind speed and the position of the water mist sprinklers. The off-centered sprinklers can directly impact the surface of the fire, while the water mist directly above the fire, although having some cooling effect, cannot directly act on the fire surface and only serves to lower the temperature. Compared to the absence of water mist, the temperature peak decreases less, with a local average temperature drop of approximately 89 °C near the fire source. Under the conditions of the off-centered sprinkler operation, the suppression of the heat release rate results in a temperature reduction, maintaining temperatures below 300 °C at the height of the compartment. This suggests that the local temperatures in adjacent compartments, which did not experience a fire, did not reach the ignition point of coal, and the thermal radiation flux did not reach the critical thermal flux value necessary for coal combustion, leading to the cessation of fire spread. Additionally, the phenomenon of high-temperature smoke receding also disappears. Under wind speed conditions of 2.4 m/s, setting water mist sprinklers 4.8 m on the upwind side of the fire can effectively reduce the average temperature of the fire tunnel, with a temperature peak decrease of approximately 190 °C and an average temperature drop of about 140 °C, thereby suppressing the spread of the fire. At critical wind speed conditions, the likelihood of the fire spreading to other trains is low; therefore, the study primarily focuses on the impact of the optimized off-centered water mist sprinkler conditions on the escape environment at a height of 1.6 m, particularly under wind speed conditions of 2.4 m/s.

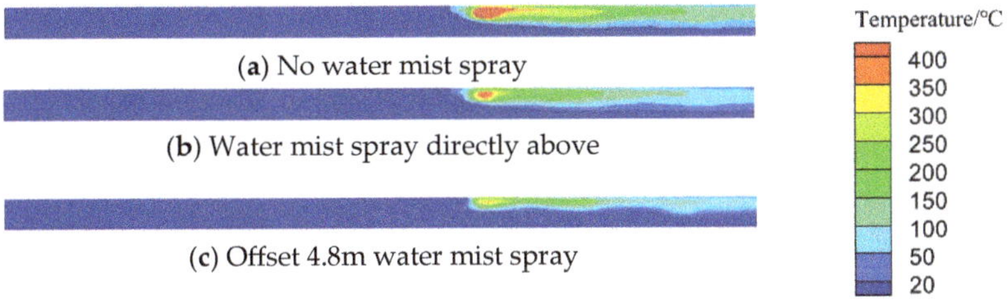

Figure 11. Temperature field comparison of 825 s with and without water mist spray system.

4.2. Study of the Effect of Water Mist Spray Conditions on the Environment for the Escape of People

When the wind speed reaches the critical threshold of 2.14 m/s, the vertical height on the downwind side of the fire corresponds to the characteristic height of the human eye. This parameter is essential for studying the temperature variations across different local areas, as illustrated in the temperature change trend graph shown in Figure 12.

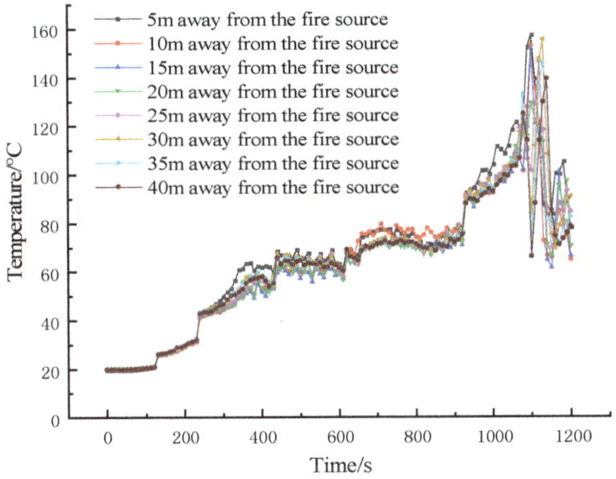

Figure 12. Trend chart of the temperature change of the personnel escape environment at the height of the human eye and the vertical height of 1.6 m.

At the time of the fire, the temperature at a vertical height of 1.6 m on both sides of the compartment was essentially uniform. As illustrated in Figure 12, the temperature change on the left side of the compartment indicates that approximately 430 s after the fire ignited, the local temperature exceeded 60 °C. This poses a significant threat to personnel attempting to escape. Subsequently, the temperature exhibited an upward trend, primarily due to its proximity to the high-temperature smoke layer. The primary reason for the temperature variation with vertical height is its predominant interaction with the high-temperature smoke layer for heat exchange and warming. Given that both ends of the tunnel serve as open exits, the accumulation of the smoke layer remains relatively shallow. Furthermore, the movement of airflow can lead to gradual temperature changes, whether increasing or decreasing, resulting in a slow upward trend in temperature.

The temperatures of each region in the 400 s, both with and without the application of water mist spray, were compared. The resulting comparison graph is presented in Figure 13.

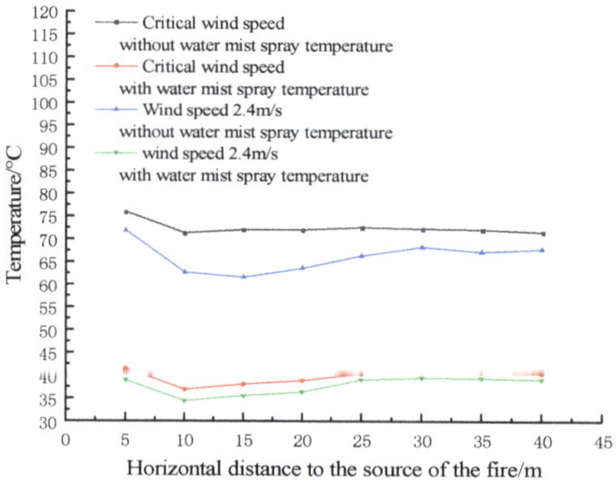

Figure 13. Cooling effect of water mist spraying at vertical height 1.6 m with different wind speeds at 400 s.

Figure 13 illustrates that by optimizing the offset amount of the water spray location, the cooling effect is significantly enhanced. This optimization aligns with personnel escape endurance conditions, thereby providing improved environmental conditions for firefighting and rescue operations.

5. Conclusions

(1) A numerical simulation of a 10 MW heavy railroad tunnel fire, with a peak temperature of 550 °C, was conducted. The results indicated that under ventilation conditions below the critical wind speed, smoke exhibited backward recession. Additionally, the fire spread to both the downwind and upwind sides, igniting adjacent compartments and resulting in multiple fire sources.

(2) This study employs theoretical analysis and numerical simulations to determine a critical wind speed of 2.14 m/s. Various wind speed conditions are simulated to explore the relationship between fire deflection angle and wind speed, as well as the heat radiation flux from the flame front on the downwind side of an adjacent train. Additionally, based on the combustion characteristics of coal, it is deduced that when the wind speed exceeds 2.4 m/s, the heat radiation flux on the surface of the fire source, located 0.5 m to 3 m downwind of the train, is sufficient to ignite the coal. The numerical simulations indicate that a fire event occurs at 719 s, leading to the ignition of coal in the neighboring compartments on the downwind side. This results

in significant burning in these compartments, with high-temperature smoke backing up to the boundary of the upwind side of the fire source by 825 s, ultimately leading to the failure of the critical wind speed.

(3) This study investigates the relationship between the water mist trajectory and wind speed. The trajectory of the water mist was analyzed under both horizontal and vertical initial speeds of 9.9 m/s. It was found that the water mist offset was 4.1 m at the critical wind speed and 4.8 m at a wind speed of 2.4 m/s. Furthermore, the offset was optimized from the initial position of the water mist spray, resulting in a significant cooling effect. Specifically, the peak temperature on the downwind side of the fire source was reduced from 77 °C to 42 °C at a height of 1.6 m, 400 s after the initiation of the spray. At this height, under critical wind speed conditions, the peak temperature on the downwind side of the fire source decreased from 77 °C to 42 °C after 400 s, while at a wind speed of 2.4 m/s, the peak temperature was reduced from 71 °C to 39 °C. The ambient temperature at a height of 1.6 m aligns with the endurance temperature necessary for personnel escaping from the optimized water mist spraying conditions.

(4) The unique characteristics of heavy railroad tunnel fires, along with the specific nature of train cargo transport, facilitate the rapid spread of fire. Low wind speeds can cause high-temperature smoke to reflux, consequently igniting combustion on the upwind side of the carriage. Furthermore, the peak temperature within the tunnel is significantly high, and there exists a negative correlation between wind speed and temperature peaks. Conversely, excessively high wind speeds can lead to an increased angle of flame deflection, which may ignite adjacent combustible materials downwind. The implementation of water spray systems can effectively mitigate fire spread and lower average temperatures, thereby aiding in the evacuation of individuals and enhancing emergency rescue efforts.

6. Discussion

This study investigates the fire characteristics during coal transportation in a heavy-load railway tunnel located in Shanxi Province. In theory, a fire occurring on a train during transport should be quickly detected, and personnel should evacuate promptly. However, the unique properties of coal make it challenging to identify a fire in its early stages without keen inspection. Generally, a freight train with a load of 20,000 to 30,000 tons requires at least 5 min of braking time. During this period, the train's movement continues to supply ample oxygen for coal combustion, potentially accelerating the fire. Therefore, based on the findings of this study, multiple sprinkler systems should be installed in the tunnel in advance, tailored to the properties of different cargo. Additionally, the locations of these sprinkler systems should be marked to indicate optimal stopping points for burning trains and ignition materials, ensuring that the water mist can function effectively. Real-time monitoring of the tunnel's wind speed is also essential; when the wind speed is too low, ventilation equipment should be used to supplement airflow, preventing the backflow of fire smoke within the tunnel and mitigating the risk of secondary disasters.

Author Contributions: W.H. was in charge of the ideas, formulae derivation. He also conducted the first draft of the paper, writing, and theoretical validation; M.F. and L.X. were responsible for the numerical simulation and theoretical analysis; M.F., S.Z. and W.H. were responsible for the numerical simulation and the experimental accuracy validation of the downscaled model; W.H. was responsible for the data visualization component; M.F., L.X. and W.H. were responsible for the collation of articles and correction of errors. All authors have read and agreed to the published version of the manuscript.

Funding: This study was supported by Guangdong Provincial University Innovation Team Project: Technology Innovation Team for Intelligent Application and Maintenance of Urban Rail Transit Equipment (2024KCXTD084); Characteristic Innovation Project of Colleges and Universities in Guangdong Province: Research on Quantitative Evaluation of Urban Rail Transit Safety Risk and Its Maintenance Strategy in the Guangdong-Hong Kong-Macao Greater Bay Area (2024KTSCX382); Guangzhou Philosophy and Social Science Planning Project: Yangcheng Green Beauty Travel Guide: Research on Accessibility Analysis and Path Optimization of "Blue Road Must Arrive" (2024GZGJ120); 2020 Excellent Online Course-Urban Rail Vehicle Structure Overhaul and Maintenance (ZC-AB-05-07-235K).

Data Availability Statement: Data and other information can be obtained by contacting e-mail address gd3987@outlook.com.

Conflicts of Interest: Author Shiqi Zheng was employed by the company Guangzhou Municipal Engineering Design & Research Institute Co., Ltd. The remaining authors declare that the research was conducted in the absence of any commercial or financial relationships that could be construed as a potential conflict of interest.

References

1. Zhao, S.; Liu, F.; Wang, F.; Weng, M. Experimental studies on fire-induced temperature distribution below ceiling in a longitudinal ventilated metro tunnel. *Tunn. Undergr. Space Technol.* **2018**, *72*, 281–293. [CrossRef]
2. Ji, J.; Wan, H.; Gao, Z.; Fu, Y.; Sun, J.; Zhang, Y.; Li, K.; Hostikka, S. Experimental study on flame merging behaviors from two pool fires along the longitudinal centerline of model tunnel with natural ventilation. *Combust. Flame* **2016**, *173*, 307–318. [CrossRef]
3. Liang, K.; Hao, X.F.; An, W.G.; Tang, Y.; Cong, Y. Study on cable fire spread and smoke temperature distribution in T-shaped utility tunnel. *Case Stud. Therm. Eng.* **2019**, *14*, 100433. [CrossRef]
4. Song, Y.J. Research on Key Risk Factors Identification and Control Strategies for Coal Railway Transportation in China. Master's Thesis, North China Electric Power University, Beijing, China, 2024.
5. Zhang, J.R.; Yang, J.F. Environmental impact analysis of coal transportation in Daqin Railway. In Proceedings of the 2020 Annual Conference of the Chinese Society for Environmental Sciences, Yuncheng, China, 18–21 October 2020; Volume 3, pp. 125–129.
6. Chen, C.J. The factors of affecting the effectiveness of coal quality inspection and countermeasures. *Coal Process. Compr. Utilization.* **2025**, *4*, 113–115.
7. Zhong, M.H.; Liu, C.; Yang, Y.X. Study on ventilation mode for train fire in metro tunnel with single-side slope. *J. Saf. Sci. Technol.* **2018**, *14*, 5–13.
8. Jae, S.R.; Seung, S.Y.; Hong, S.R.; Yoon, M.O.; Jeong, Y.T. An experimental study on the effect of ventilation velocity on burning rate in tunnel fire-heptane pool fire case. *Build. Environ.* **2008**, *43*, 1225–1231.
9. Chiara, V.; Bruno, F.; Emilio, P.; Maschio, G. Experimental study on thermal and toxic hazards connected to fire scenarios in road tunnels. *J. Loss Prev. Process Ind.* **2012**, *25*, 718–729.
10. McGrattan, K.; Hamins, A. *Numerical Simulation of the Howard Street Tunnel Fire*; NISTIR6902; National Institute of Standards and Technology: Gaithersburg, MD, USA, 2002.
11. Ciro, C.; Paolo, C. Simulation of fire scenarios due to different vehicle types with and without traffic in a bi-directional road tunnel. *Tunn. Undergr. Space Technol.* **2013**, *37*, 22–36.
12. Jia, Y.; Fan, X.L.; Zhao, X.J.; Deng, Y.L.; Zhu, X.L.; Zhao, W.F. Study on the longitudinal ceiling temperature distribution induced by double pool fires in a tunnel. *Int. J. Therm. Sci.* **2021**, *168*, 107059. [CrossRef]
13. Wang, Q.; Wang, S.; Liu, H.; Shen, J.; Shang, F.; Shi, C.; Tang, F. Characterization of ceiling smoke temperature profile and maximum temperature rise induced by double fires in a natural ventilation tunnel. *Tunn. Undergr. Space Technol. Inc. Trenchless Technol. Res.* **2020**, *96*, 103233. [CrossRef]
14. Li, Y.X. Coal Spontaneous Combustion Characteristics in Low-temperature Oxidation Environment and Anti Spontaneous Combustion Countermeasures for Railway Coal Transportation. *Railw. Logist.* **2024**, *42*, 24–30.
15. Delichatsios, M. Flow of fire gases under a beamed ceiling. *Combust. Flame* **1981**, *43*, 1–10. [CrossRef]
16. Thomas, P.H. The movement of buoyant fluid against a stream and the venting of under ground fires. *Fire Saf. Sci.* **1958**, *351*, 1.
17. Thomas, P.H. *The Movement of Smoke in Horizontal Passages Against Air Flow*; Fire Research Technical Paper; Fire Research Station: Borehamwood, UK, 1968; Volume 7, pp. 1–8.
18. Hu, L.H.; Huo, R.; Chow, W.K. Studies on buoyancy-driven back-layering flow in tunnel fires. *Exp. Therm. Fluid Sci.* **2008**, *32*, 1468–1483. [CrossRef]

19. Yang, D.; Hu, L.H.; Huo, R.; Jiang, Y. Effects of Longitudinal Air Flow on Smoke Stratification in Channel Fire. *J. Combust. Sci. Technol.* **2010**, *16*, 5.
20. Wu, Y.; Bakar, M.Z.A. Control of smoke flow in tunnel fires using longitudinal ventilation systems-a study of the critical velocity. *Fire Saf. J.* **2000**, *35*, 363–390. [CrossRef]
21. Wang, C.; Guo, D.; Xu, B.; Liu, C.; Miao, Z. Experimental and numerical simulation study on fire spreading of urban cable tunnels. *J. Tsinghua Univ. (Sci. Technol.)* **2025**, *65*, 532–546.
22. Zhao, X.; Ni, M.; Wang, W.; Wang, H.; Wang, J. Study on the Fire Characteristics of Dual Fire Sources and the Difference in Power Temperature of Different Fire Sources in Tunnel. *Fire* **2024**, *7*, 273. [CrossRef]
23. Sumic, D.; Males, L.; Rosic, M. An Agent-Based Ship Firefighting Model. *J. Mar. Sci. Eng.* **2021**, *9*, 902. [CrossRef]
24. International Union of Railways. *UIC 564-2: Passenger Rolling Stock—Protection Measures Against Fire and Panic—Part 2: Design Requirements*, 3rd ed.; Union Internationale des Chemins de fer (UIC): Paris, France, 2019.
25. Qin, R.; Li, X.J. Study on the Grid Selection of Block Scale Pyrosim Fire Simulation. *Urban. Archit.* **2021**, *18*, 107–109.
26. Enrico, R. Testing the predictive capabilities of evacuation models for tunnel fire safety analysis. *Saf. Sci.* **2013**, *59*, 141–153.
27. Liu, Y.; Wang, Z.; Zhu, S.; Sun, L.; Wei, Q.; Zhou, X.; Cui, Y.; Wu, D. Research on the Infrared Cooling Characteristics of Composite Nozzle with Water Mist and Water Curtain. *J. Mech. Eng.* **2024**, *60*, 320–328.
28. Kalech, B.; Bouterra, M.; Elcafsi, A. Numerical analysis of smoke flow under the effect of longitudinal airflow in a tunnel fire. *Fire Mater.* **2020**, *44*, 1033–1043. [CrossRef]
29. Rickard, H. Fire behavior of multiple fires in a mine drift with longitudinal ventilation. *Int. J. Min. Sci. Technol.* **2019**, *29*, 245–254.
30. Tian, L. Research on the Evolution Law of Mine Belt Fire and the Prevention and Extinguishing Technology of Fine Water Mist. Master's Thesis, Hunan University of Science and Technology, Xiangtan, China, 2022.

processes

MDPI

Article

Study on Coupled Evolution Mechanisms of Stress–Fracture–Seepage Fields in Overburden Strata During Fully Mechanized Coal Mining

Yan Liu [1], Shangxin Fang [2], Tengfei Hu [1], Cun Zhang [2,3,*], Yuan Guo [4], Fuzhong Li [1] and Jiawei Huang [4]

1 CCTEG Wuhan Engineering Company, Wuhan 430064, China; ly161@126.com (Y.L.); cumthtf@163.com (T.H.); lifz20210503@163.com (F.L.)
2 School of Energy & Mining Engineering, China University of Mining and Technology-Beijing, Beijing 100083, China; fliex0105@163.com
3 School of Mines, China University of Mining and Technology, Xuzhou 221116, China
4 Luotuoshan Coal Mine of Wuhai Energy, Wuhai 016000, China; j589421@163.com (Y.G.); 20082580@ceic.com (J.H.)
* Correspondence: cumt-zc@cumtb.edu.cn

Abstract: Understanding the coupled evolution mechanisms of stress, fracture, and seepage fields in overburden strata is critical for preventing water inrush disasters during fully mechanized mining in deep coal seams, particularly under complex hydrogeological conditions. To address this challenge, this study integrates laboratory experiments with FLAC3D numerical simulations to systematically investigate the multi-field coupling behavior in the Luotuoshan coal mine. Three types of coal rock samples—raw coal/rock (bending subsidence zone), fractured coal/rock (fracture zone), and broken rock (caved zone)—were subjected to triaxial permeability tests under varying stress conditions. The experimental results quantitatively revealed distinct permeability evolution patterns. the fractured samples exhibited a 23–48 × higher initial permeability (28.03 mD for coal, 13.54 mD for rock) than the intact samples (0.50 mD for coal, 0.21 mD for rock), while the broken rock showed exponential permeability decay (120.32 mD to 23.72 mD) under compaction. A dynamic permeability updating algorithm was developed using FISH scripting language, embedding stress-dependent permeability models ($R^2 > 0.99$) into FLAC3D to enable real-time coupling of stress–fracture–seepage fields during face advancement simulations. The key findings demonstrate four distinct evolutionary stages of pore water pressure: (1) static equilibrium (0–100 m advance), (2) fracture expansion (120–200 m, 484% permeability surge), (3) seepage channel formation (200–300 m, 81.67 mD peak permeability), and (4) high-risk water inrush (300–400 m, 23.72 mD stabilized permeability). The simulated fracture zone height reached 55 m, directly connecting with the overlying sandstone aquifer (9 m thick, 1 MPa pressure), validating field-observed water inrush thresholds. This methodology provides a quantitative framework for predicting water-conducting fracture zone development and optimizing real-time water hazard prevention strategies in similar deep mining conditions.

Keywords: overburden rock damage; water hazard control; water-conducting fracture zone; stress disturbance; numerical simulation

1. Introduction

Fossil fuels such as coal and oil and gas remain important energy sources for future human and social development, and their safe and efficient development is of great significance [1]. In the process of coal resource exploitation, the influence of mining response is inevitable [2]. During the mining process of a coal working face, the coal rock,

 https://doi.org/10.3390/10.3390/pr13061753

underground aquifers, and surrounding strata within a certain range undergo deformation, fracturing, and displacement. This process generates extensive fractures and leads to surface subsidence [3,4]. These fractures establish interconnected seepage pathways between the aquifer and the working face, thereby increasing the susceptibility of the working face to water inrush hazards [5–7]. This process involves a coupling change in the stress, fracture, and seepage fields. With the increase in the mining depth, the variation characteristics of the multi-field response of the coal and rock mass are more different than those of shallow mining. The investigation of multi-field coupling evolution mechanisms in mining-induced fractures within overburden strata holds significant importance for both mine water hazard control and water resource protection.

The failure mechanisms of mining-induced overburden strata are significantly influenced by geological conditions, stress states, and engineering contexts [8]. Typically, the overburden strata above the working face are categorized into three distinct zones extending from the coal seam to the ground surface [9–12] (caved zone, fractured zone, and bending subsidence zone). Numerous scholars have validated the above finding through various field detection methods [13] and conducted extensive research in this area. Li et al. [14] analyzed the factors influencing the development height of the water-conducting fracture zone (WCFZ) by using a mathematical model, regression analysis, and field measurement. Zhang et al. [15] predicted the height of the WCFZ based on several factors, including coal seam thickness, proportion of hard rock, floor length, mining depth, and dip angle. Wang et al. [16] investigated the three-dimensional distribution characteristics of porosity in overburden strata during longwall mining, providing critical parameters for permeability distribution and valuable insights for the prevention of coal seam water inrush. Based on the composite structural characteristics of the overburden strata, Jiang et al. [17] characterized the dimensions and distribution patterns of fractures within the strata and determined the height of the water-conducting fracture zone. Wang et al. [18] developed a physical model for a mine and analyzed the dynamic evolution processes and underlying mechanisms of karst roof water inrush under different mining sequences. Through an integrated approach combining physical modeling, theoretical analysis, and field detection, Yang et al. [19] investigated the overburden failure mechanisms in shallow, thick coal seams. Their research demonstrated that fractures within the goaf would eventually close, while those at the goaf boundary serve as the primary pathways for water and sand inrushes into the working face. The numerical simulation method can intuitively observe and study the entire dynamic change process of overburden deformation and failure [20,21], which has significant advantages compared with field measurement. Shi and Zhang [22] investigated the mining fracture morphology in overburden strata above the goaf through an integrated approach combining physical modeling, FLAC3D numerical simulations, and field observations. Based on laboratory tests and PFC numerical simulation, Dong et al. [23] studied the trapezoidal evolution law of overburden fractures in extremely thick coal seams and analyzed the failure forms. Song et al. [24] studied the influence of overburden structural characteristics on WCFZ through FLAC3D numerical simulation, theoretical calculation, and field measurement. Through field measurement and numerical simulation, Chang et al. [25] concluded that the final shape of the WCFZ of the working face was a saddle shape and analyzed the key inhibition layer in the cap layer. Yang et al. [26] obtained the transfer law of mining stress by combining similar simulation with numerical simulation considering fractal geometry theory. He et al. [27] studied the influence of mining parameters on overburden failure and conducted secondary development based on FLAC3D. Cao et al. [28] studied the activation mechanism of water-rich faults and the development characteristics of WCFZ in overlying strata through theoretical analysis, FLAC3D numerical simulation, and field measurement.

Chen et al. [29] comprehensively compared the results of an empirical formula, engineering measurement, a physical simulation model, and UDEC numerical simulation; they studied the height of a water-conducting fracture zone and analyzed the hazard of water inrush and sand flushing on the working face.

However, critical knowledge gaps persist in current methodologies: Existing permeability models predominantly rely on static laboratory measurements, inadequately addressing the dynamic stress–seepage coupling during progressive face advancement, while prevailing numerical frameworks often oversimplify overburden strata as homogeneous media—neglecting the zonal permeability contrasts between caved, fractured, and bending subsidence zones. Such limitations have led to systematic deviations between predicted and observed water-conducting fracture heights, as evidenced by field measurements in deep mining scenarios.

This paper takes the Luotuoshan coal mine fully mechanized mining face as the engineering background. In this paper, different mining influence zones are simulated by coal and rock samples with different damage degrees. The coupling evolution mechanisms of the mining stress–fracture–seepage field in overburden strata were systematically investigated through integrated laboratory experiments and numerical simulations.

2. Experimental Study on Permeability of Coal Rock Sample in Different Overburden Strata Zoning

2.1. Overburden Strata Zoning

During the coal mining process at the working face, the coal seam will be affected by complex mining, which makes the internal fracture structure of the coal seam become complex. In the overburden strata layer, the damage degree of the coal rock body located in different zones varies greatly, which leads to changes in the permeability of gas and water transport. In the process of coal seam mining, the overburden rock structure usually forms a typical "three zone" structure, comprising the caved zone, fractured zone, and bending subsidence zone (Figure 1).

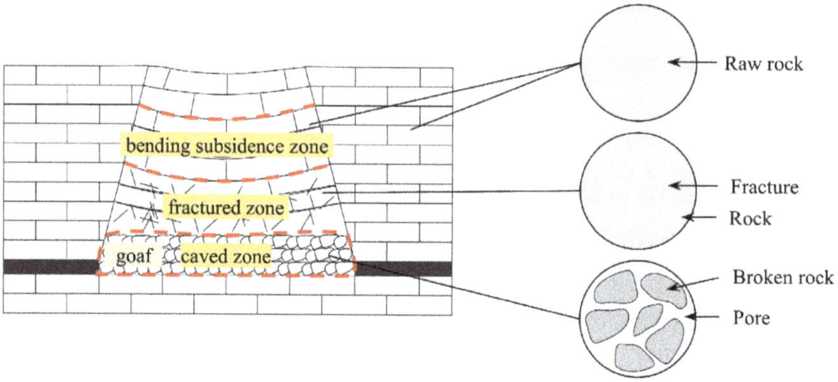

Figure 1. Division of overburden strata areas.

The formation of the "three zones" structure (caved zone, fractured zone, and bending subsidence zone) in overburden strata during coal mining stems from stress redistribution and sequential rock failure mechanisms induced by mining activities. Following coal extraction, the disruption of the original in situ stress equilibrium triggers progressive destabilization of the overlying strata from the bottom upward: the immediate roof, deprived of support, undergoes shear fragmentation and free collapse (caved zone), resulting in disordered accumulation and subsequent compaction of broken rock fragments; the strata above, subjected to gravitational loading and stress concentration, develop tensile–shear composite fractures that evolve into interconnected fracture networks (fractured zone), whose growth is governed by lithological combinations and key strata, while the upper distant strata exhibit continuous elastic bending (bending subsidence zone) due to stress attenuation, coordinating deformation through interlayer friction, and composite beam effects, with only minor bedding separation cracks generated. Coal rock seams located in the bending subsidence zone and above are generally considered to be relatively more complete, with no obvious large fracture, and the coal seams are in the elastic stage, which is defined as a raw coal rock sample in this paper. Located in the overburden fractured zone, the rock seams are generally damaged by shearing or pulling, and the degree of fracture development is large; so, this paper defines this type of coal body as a fracture coal rock sample. The caved zone consists of broken residual coal and rock mass, due to the less residual coal in the mining hollow area of the comprehensive mining face; so, this paper defines the coal rock body in the caved zone as a broken rock sample.

2.2. Permeability Model

Matchstick models are commonly used for standard coal rock seepage characteristics, whereby Seidle et al. [30] proposed an effective stress-based theoretical model of permeability for the particular structure of coal:

$$k_f = k_{f0} e^{-3c_f(\sigma_1 - \sigma_{10})} \tag{1}$$

where k_f is the permeability of cleats and fractures in coal seams, mD; c_f is the fracture compression coefficient, MPa^{-1}; σ_1 is the effective stress, MPa; k_{f0}, σ_{10} are the permeability and effective stress of the initial state, mD, MPa.

According to many scholars, the fracture compression coefficient c_f is not a constant value and varies with the effective stress [31]; it is generally replaced by using the average fracture compression coefficient derived from data fitting:

$$\bar{c}_f = \frac{c_{f0}}{\alpha_f(\sigma_1 - \sigma_{10})}\left(1 - e^{-\alpha_f(\sigma_1 - \sigma_{10})}\right) \tag{2}$$

where c_{f0} is the compression coefficient of the original fracture, and α_f is the change ratio of the fracture compression coefficient with effective stress.

Therefore, combining Equations (1) and (2) can obtain the stress–permeability fitting equation for water, based on which the permeability of the same coal rock sample can be fitted:

$$k_w = k_{w0} e^{-3\frac{c_{w0}}{\alpha_w}(1 - e^{-\alpha_w \sigma_1})} \tag{3}$$

where k_w is the permeability of water in the fractures of the coal rock body, k_{w0} is the permeability of the initial state, c_{w0} is the original fracture compression coefficient, and α_w is the ratio of the change in the fracture compression coefficient with the effective stress when the water is seeping.

2.3. Experimental Program

Figure 2 presents the self-developed small-scale conventional triaxial multi-field coupled experimental system for rock bodies. As shown in Figure 2, the system consists of a cabinet panel and liquid filling pump, a peripheral pressure chamber (including axial pressure and confining pressure loading pumps, with an axial pressure loading range of 0–70 MPa), an autoclave (size φ150 × 150 mm, pressure resistance of 30 MPa), a gas–liquid flowmeter (3 sets of different ranges with automatic switching capability), a hydraulic loading system (peripheral pressure loading range of 0–30 MPa), pressure sensors, and a data acquisition system. This system enables precise control and real-time monitoring of loading, strain, flow, pressure, temperature, and gas components in coal rock bodies under high-pressure and constant-temperature conditions.

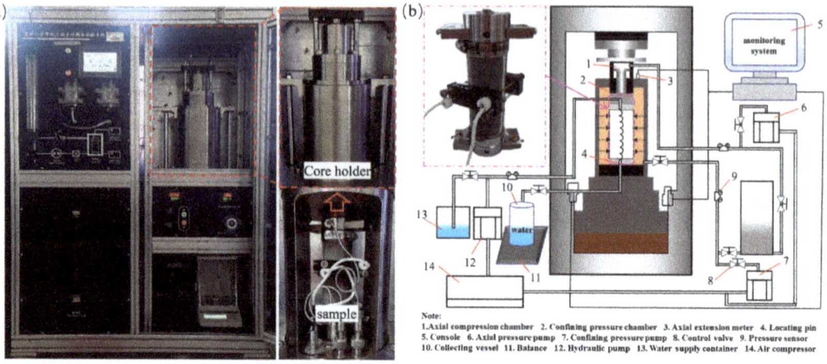

Figure 2. Small-scale conventional triaxial multi-field coupled experimental system for rock bodies. (**a**) Experimental system; (**b**) Schematic diagram.

The experimental samples include raw coal and rock samples, fracture coal and rock samples, and broken rock samples. The specific preparation methods are as follows:

(1) Raw coal rock samples (Figure 3a): The original coal samples and original rock style samples were chosen, and the larger, better integrity, un-weathered coal rock blocks in the coal mining face and overburden strata were directly selected. In the laboratory, according to the 'coal and rock physical and mechanical properties determination method', the provisions were processed into standard raw coal samples, at a size of ϕ50 mm × 100 mm.

(2) Fracture coal rock samples (Figure 3b): The standard coal rock samples were processed into sizes of ϕ50 mm × 100 mm and were subjected to a shear test; after the shear damage occurred, the coal samples in which both halves had more complete shear damage were selected for further processing. After the shear damage, the coal sample was generally divided into two halves, and the integrity of each half of the coal rock sample was better, while the penetrating fractures were perpendicular to the two ends of the sample.

(3) Broken rock samples (Figure 3c): The rock samples were crushed and then sieved using standard sieves. As shown in Figure 4, the experiment required particle size ranges, including <5 mm, 5–10 mm, 10–15 mm, 15–20 mm, 20–25 mm, 25–30 mm, and crushed rock particles > 30 mm.

(**a**) Raw coal rock sample

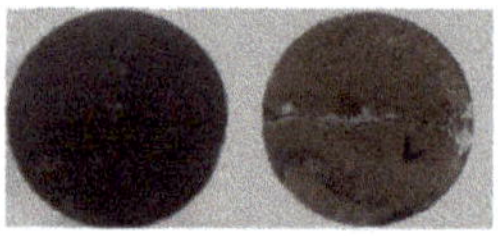

(**b**) Fracture coal rock sample

(**c**) Broken rock sample

Figure 3. Coal rock samples with different levels of damage.

Figure 4. Broken rock samples with different particle size ranges.

Seepage experiments were performed using different selected samples and recorded after the seepage velocity recorded by the instrument was kept constant; in the experiment, the external fractures of the fractured rock samples were sealed with colloid in order to prevent fluid from seeping from the external surfaces of the samples in the subsequent experiments and affecting the experimental results. It should be noted that the experiment was conducted at a room temperature of 22 degrees Celsius. The experimental steps are as follows: (1) Fix the sensor to the processed sample and fix the specimen to the experimental chamber; slowly increase the axial pressure and adjust the axial pressure to 1 MPa after the specimen is compacted and, at the same time, connect the inlet and outlet pipes with the sensing device. (2) Close the experimental chamber and adjust the peripheral pressure to 1 MPa; keep the axial pressure and peripheral pressure the same, load the water pressure step by step, record the seepage data under each water pressure, and unload the water pressure to 0 MPa when the loading of each level of water pressure is completed. (3) Load the axial pressure and peripheral pressure to the next level at the same time; keep the axial pressure and the peripheral pressure the same; repeat the previous step after the pressure is stabilized, until the axial pressure and the peripheral pressure are loaded to the pre-set value and then stop the group of experiments. (4) Unload the axial circumferential pressure in the equipment and replace the pattern; repeat the above process.

The permeability of the different coal rock samples used in the experiment was determined by the following equation:

$$k = \frac{\eta L \Delta Q_i}{A \Delta P \Delta t_i} \tag{4}$$

where k is the average permeability of each measurement point of the rock sample, m^2; η is the fluid viscosity coefficient, taking $\eta = 1 \times 10^{-3}$ Pa·s (water temperature 20 °C); L is the seepage length of the water flow, i.e., the height of the specimen in the test, m; ΔQ_i is the volume of the water flow that seeped through the specimen during the time of Δt_i, m^3; A is the cross-sectional area of the specimen, m^2; ΔP is the difference in the permeation pressure at the upper and lower ends of the specimen, Pa; Δt_i is the recording point interval time, s.

The fitting parameters for the permeability of different coal rock samples are shown in Table 1.

Table 1. Fitting results for different samples.

Type of Sample	Fitting Parameter				Correlation Coefficient R^2
	k_w (mD)	c_{w0} (MPa^{-1})	α_w (MPa^{-1})	c_{w0}/α_w	
Raw coal sample	0.5034	0.3020	0.1105	2.7330	0.9993
Raw rock sample	0.2133	0.3450	0.1247	2.7667	0.9993
Fracture coal sample	28.0314	0.1310	0.1056	1.2405	0.9990
Fracture rock sample	13.5438	0.1104	0.1364	0.8094	0.9980
Broken rock sample	120.3242	0.0303	0.2523	0.1201	0.9930

3. Coupled Simulation of Mining Stress–Fracture–Seepage Fields

3.1. Construction of a Coupled Fluid–Solid Simulation Method

In the above, different areas of overburden strata in the process of working face mining were divided into raw coal rock samples, fracture coal rock samples, and broken rock samples. The evolution laws of permeability under different stress conditions were obtained, respectively, through laboratory seepage experiments. Based on the above foundation, the stress–permeability model of coal rock samples with different damage was established. In FLAC3D numerical simulation, the construction of the coupled simulation method of mining stress–fracture–seepage fields relies on the permeability and stress relationship obtained by fitting based on experimental data.

3.1.1. FLAC3D Fluid Structure Coupling Calculation Method

(1) Seepage calculation method

In FLAC3D, fluid passing through porous media is characterized using the fluid mass balance equation, fluid transport Darcy's law, and constitutive equation for fluid flow. The flow of fluid in FLAC3D is mainly based on Darcy's law:

$$q_{\mathrm{i}} = -k_{\mathrm{il}}\hat{k}(S)\left[p - \rho_{\mathrm{f}}x_{\mathrm{j}}g_{\mathrm{j}}\right]_{,\mathrm{l}} \tag{5}$$

In the formula, q_{i} is the directional fluid velocity; k_{il} is the fluid permeability tensor; $\hat{k}(S)$ is the relative permeability coefficient of the fluid; ρ_{f} is the fluid density; p is the pore pressure; and g_{j} is the component of gravitational acceleration. It is worth noting that k_{il} is the permeability tensor ($\mathrm{m^2/(Pa{\cdot}s)}$) in FLAC3D, which differs in dimension from the permeability k ($\mathrm{m^2}$) obtained in the laboratory. The two can be converted using Equation (6):

$$k_{\mathrm{il}} = \frac{k}{\mu} \tag{6}$$

In the formula, μ is the viscosity coefficient of the fluid (usually taken as 1.01×10^{-3} for water).

When thermal factors are not considered, the constitutive equation for fluid flow is:

$$\frac{1}{M}\frac{\partial p}{\partial t} + \frac{n}{S}\frac{\partial S}{\partial t} = \frac{1}{S}\frac{\partial \varsigma}{\partial t} - \alpha\frac{\partial \varepsilon}{\partial t} \tag{7}$$

The mass equation in the FLAC3D seepage calculation is:

$$q_{\mathrm{V}} - q_{\mathrm{i}} = \frac{\partial \varsigma}{\partial t} \tag{8}$$

In the equation, q_{V} represents the strength of the fluid source. Substituting Equation (8) into Equation (7) yields the expression for the fluid continuity equation:

$$\frac{1}{M}\frac{\partial p}{\partial t} + \frac{n}{S}\frac{\partial S}{\partial t} = \frac{1}{S}(q_{\mathrm{V}} - q_{\mathrm{i}}) - \alpha\frac{\partial \varepsilon}{\partial t} \tag{9}$$

(2) Mechanical calculation methods

In FLAC3D mechanics calculations, when a liquid passes through a unit cell, the forces acting on the unit cell mainly include its weight, buoyancy, and flow resistance. The mechanical equilibrium of a unit cell can be expressed by the relationship between the total stress σ_{ij} and the saturated bulk density ρ_S:

$$\frac{\partial \sigma_{ij}}{\partial x_j} + \rho_S g_i = 0 \tag{10}$$

For a saturated unit cell, its internal materials are solid media and liquid in pores, and its bulk density ρ_S can be expressed by the following equation:

$$\rho_S = \rho_d + nS\rho_f \tag{11}$$

In the formula, ρ_d is the density of the dry solid material.

Based on Terzaghi's theory, the effective stress of saturated porous media in FLAC3D can be expressed by Equation (12):

$$\sigma_{ij} = \sigma'_{ij} - \alpha p \delta_{ij} \tag{12}$$

In the equation, δ_{ij} is the Kronecker tensor, and $\sigma_{ij}{}'$ is the effective stress. It should be noted that the compressive stress acting on the unit cell is negative, while the pore pressure is negative in the tension stress.

By combining Equations (10)–(12), the mechanical equilibrium equation of the saturated unit cell can be obtained:

$$\begin{cases} \frac{\partial \sigma'_{ij}}{\partial x_j} + \rho_d g_i - \alpha(1-n)\frac{\partial p}{\partial x_i} - n\rho_f g \frac{\partial \phi}{\partial x_i} = 0 \\ \phi = \frac{p}{\rho_f g} - \frac{x_k g_k}{g} \end{cases} \tag{13}$$

In the formula, ϕ is the negative pressure head. Among them, $\rho_d g_i$ is the weight of the unit cell, $\alpha(1-n)\frac{\partial p}{\partial x_i}$ is buoyancy, and $n\rho_f g \frac{\partial \phi}{\partial x_i}$ is flow resistance.

3.1.2. Module for Updating the Permeability Dynamics of Coal Rock Bodies

Based on the permeability test results of different coal sample types (raw coal rock samples, fracture coal rock samples, and broken rock samples) under different stress conditions, the relationship between permeability and stress in coal seams was quantitatively described using a fitting formula.

Based on Equation (3) and the experimental results, the fitting equations for different types of coal samples are shown in Table 2.

Table 2. Fitting equations for different styles.

Type of Coal Sample	Fitting Formula	Formula Number
Raw coal sample	$0.5034e^{-8.199(1-e^{-0.1105\sigma_1})}$	RC
Raw rock sample	$0.2133e^{-8.3001(1-e^{-0.1247\sigma_1})}$	RR
Fracture coal sample	$28.0314e^{-3.7215(1-e^{-0.1056\sigma_1})}$	FC
Fracture rock sample	$13.5438e^{-2.4282(1-e^{-0.1364\sigma_1})}$	FR
Broken rock sample	$120.3242e^{-0.3603(1-e^{-0.2523\sigma_1})}$	BR

From the fitting equations, it can be seen that the permeability of the raw coal rock samples decreases exponentially with the increase in effective stress, whereas the permeability of the fracture coal rock samples is significantly higher, showing different permeability behaviors. These permeability formulas were embedded into the FLAC3D numerical model by means of FISH language scripts. These formulas will dynamically update the permeability of the coal rock body according to the different stress conditions of the unit, which becomes a module of the dynamic update of the permeability of the coal rock body, and realizes the coupling effect of stress and seepage in the model. The mechanics and seepage module of FLAC3D will combine the changes in the stress field and fracture field to simulate the expansion and deformation of the coal seam fractures during the mining process. Thus, the seepage characteristics of the aquifer during mining can be accurately reflected.

3.1.3. A Coupled Simulation Method of Mining Stress–Fracture–Seepage Fields

The coupled simulation method of mining stress–fracture–seepage fields is based on the evolutionary characteristics of the overburden strata of the working face and the damage permeability behavior of the coal rock body, and the dynamic coupling of the three fields is realized through experimental–numerical joint modeling. The specific methods are constructed as follows:

(1) Coal rock zoning and permeability modeling

Based on the "three zones" structure of mining overburden (caved zone, fractured zone, and bending subsidence zone), the coal rock body is divided into three categories: raw coal samples (uninjured area, permeability attenuation with the stress index); fracture coal samples (damage zone, permeability jumped due to plastic damage); broken coal rock body (caved zone, permeability attenuation with the compaction process). The stress–permeability modeling of the three types of coal rocks used laboratory seepage experiments.

(2) Permeability dynamic updating mechanism

Three types of coal permeability models are embedded in FLAC3D: Raw coal samples are dynamically corrected by the stress–permeability fitting formula; fracture coal samples trigger a rapid increase in permeability based on a yield criterion; broken coal rock bodies combine the caved zone height equation with compaction theory to simulate the nonlinear evolution of permeability with compaction strain. In the process of calculation, the FISH script is continuously called by FISH callback or the whilestepping command, so as to realize the real-time dynamic update of permeability.

(3) Multi-field coupled numerical simulation flow (Figure 5)

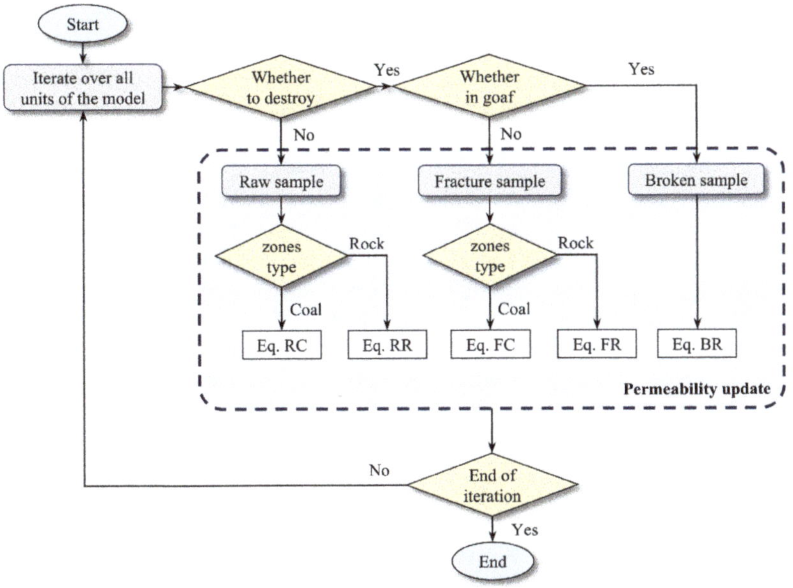

Figure 5. Flow–solid coupling calculation process.

3.2. Engineering Background

In this paper, we chose to select the 110,901 working face (Figure 6) of Luotuoshan coal mine for the study. According to the report of the deep area fine investigation of Luotuoshan well field and the report of the hydrogeological supplementary investigation of the Austrian grey, the water-bearing layer group in the well field is divided into three categories: the Quaternary loose rock pore phreatic water-bearing rock group, clastic rock pore–fracture confined water-bearing rock group, and Ordovician limestone water-bearing rock group.

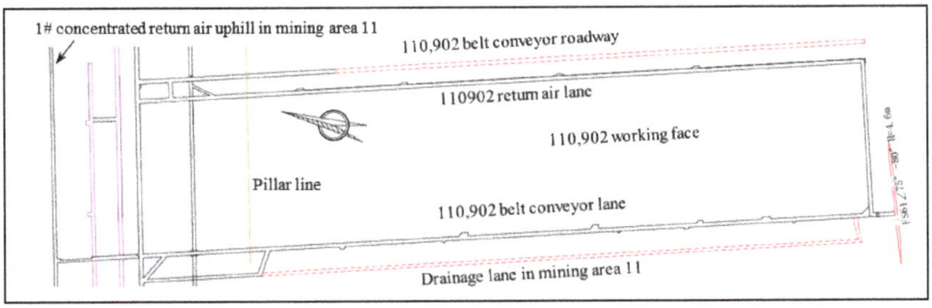

Figure 6. Diagram of the location of the 110,901 working face in Luotuoshan coal mine.

Analyzing the working face mining stress environment, mining layout, surrounding rock aquifer and mechanical properties of coal rock can help to explore the mutual triggering mechanism of a top and bottom water breakout and roof disaster, the occurrence mechanism, and its main controlling factors. Based on this, this paper investigates the distribution and evolution of mining stress–fracture–seepage fields under the condition of a surrounding rock aquifer and reveals the mutual triggering mechanism of roof-bottom water breakout and roof disaster and its main controlling factors in the working face of 110,901 in Luotuoshan coal mine and similar conditions.

3.3. Numerical Model and Parameters

3.3.1. Model Construction

According to the geological exploration data, the FLAC3D numerical model of Luotuoshan was constructed. The calculation range of the model was taken as strike × dip × height = 500 m × 300 m × 170 m. The average burial depth of the coal seam was taken as 308.10 m. The pressure measurement coefficient was taken as 1. The strike × dip of the working face was taken as 400 m × 200 m, and the boundary protection coal pillar was set up on both sides of the coal seam for 50 m. Considering that No.9-2 coal is one of the main mineable coal seams in the well field, the strike of the coal seam is mainly at NW15~20°, and the dip is a monoclinic stratum tilted to the west, with the dip angle ranging from 5~15°, and the average dip angle is 9°; so, the numerical model was set to have a dip angle of 9°. Accordingly, the geological model was constructed by FLAC3D version 6.0 numerical simulation software (Figure 7). Fixed constraints were imposed on the front, back, left, right, and bottom boundaries of the model; the topsoil stratum above the bedrock was converted into an equivalent load, which was uniformly applied to the upper surface of the model.

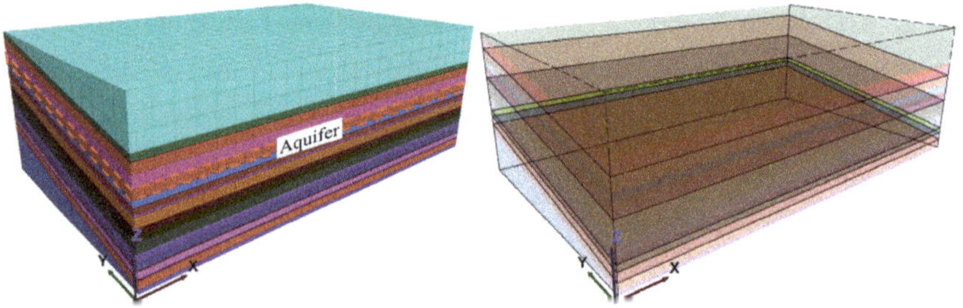

Figure 7. Numerical model of Luotuoshan coal mine.

3.3.2. Parameter Selection

Based on the geological information of the mine's comprehensive borehole column map, strata with similar geological characteristics are classified and simplified to ensure that the model can accurately reflect the actual geological conditions of the mine. In addition, response surface methodology (RSM) is often used to optimize parameters and has been widely applied in many fields [32]. Based on RSM, experimental data, and field measurements, all kinds of rock strata in the model are assigned with corresponding physical and mechanical parameters, and their specific values and layering scheme are shown in Table 3, which is used as the basis for the subsequent simulation and analysis of the evolution law of the mining stress–fracture–seepage fields of the overburden strata in the working face.

Table 3. Physical and mechanical parameters of the reservoir.

Numbering	Lithology	Thickness/m	Density/kg·m^{-3}	Bulk Modulus /GPa	Shear Modulus /GPa	Cohesion /MPa	Angle of Internal Friction/°	Tensile Strength/MPa
1	Fine sandstone	24	2500	1.5	1.4	0.8	22	0.9
2	Mudstone	7	2000	1.0	1.0	0.5	20	0.5
3	Fine sandstone	10	2500	1.5	1.4	0.8	22	0.9
4	Mudstone	12	2000	1.0	1.0	0.5	20	0.5
5	Fine sandstone–aquifer	9	1400	2.5	2.3	2.5	16	1.5
6	Fine sandstone	7	2500	1.5	1.4	0.8	22	0.9
7	Mudstone	6	2000	1.0	1.0	0.5	20	0.5
8	Siltstone	2	2500	1.5	1.4	0.8	22	0.9
9	Mudstone	10	2000	1.0	1.0	0.5	20	0.5
10	Mudstone	6	2000	1.0	1.0	0.5	20	0.5
11	Fine sandstone	7	1800	1.5	1.5	1.2	18	1.2
12	Mudstone	10	2000	1.0	1.0	0.5	20	0.5
13	Siltstone	2	2500	1.5	1.4	0.8	22	0.9
14	Mudstone	5	2000	1.0	1.0	0.5	20	0.5
15	Coal 9#	7	1400	2.5	2.3	2.5	16	1.5
16	Mudstone	1	2000	1.0	1.0	0.5	20	0.5
17	Coal 10#	1	1400	2.5	2.3	2.5	16	1.5
18	Mudstone	6	2000	1.0	1.0	0.5	20	0.5
19	Siltstone	2	2500	1.5	1.4	0.8	22	0.9
20	Mudstone	8	2000	1.0	1.0	0.5	20	0.5
21	Fine sandstone	4	2500	1.5	1.4	0.8	22	0.9
22	Mudstone	24	2000	1.0	1.0	0.5	20	0.5

During the advancement process of the 110,901 working face in Luotuoshan coal mine, the actual measured height of the water-conducting fracture zone based on the monitoring and inversion of the leakage of drilling flushing fluid on site was 58 m. Therefore, in order to ensure the reliability of the numerical simulation results in this paper, the height of the water-conducting fracture zone development during the advancement process of the 110,901 working face was simulated, and the results are shown in Figure 8. According to Figure 8, the development height of the water-conducting fracture zone in the numerical simulation is 55 m, which is close to the development height measured on site. Therefore, it is considered that the numerical simulation results in this paper are reliable.

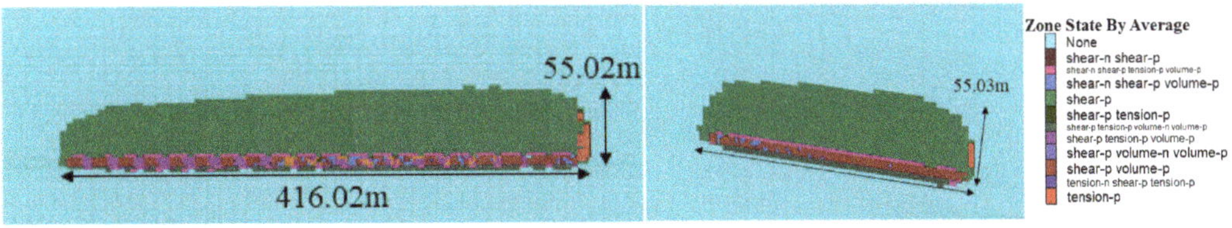

Figure 8. Numerical simulation results of the development height of water-conducting fracture zones.

3.3.3. Measuring Point Arrangement

The stress redistribution of overburden strata and the dynamic expansion of fracture during the mining process are the key factors inducing disasters such as gas transport and water breakout in coal seams. The measurement points are arranged to monitor the mechanical response and seepage parameters of different layers and zones of the coal rock body in real time. They can quantitatively reveal the dynamic evolution characteristics of the "three zones" structure (caved zone, fractured zone, and bending subsidence zone) and provide data support for the coupling analysis of the seepage field and stress field. Monitoring data can effectively reflect the dynamic updating process of coal rock damage permeability modeling, verify the applicability of the graded permeability formula of 'raw coal rock samples—fracture coal rock samples—broken rock samples' in the numerical simulation, and, at the same time, provide a scientific basis for the distribution of over-supporting pressure, the expansion law of the plastic zone, and the determination of the threshold of the sudden change in permeability, which will ultimately serve for the optimization of the mine disaster prevention and control and the mining plan.

The simulated working face is arranged along the dip, with a length of 200 m, and advances along the strike, with an advance length of 400 m. In order to eliminate the boundary effect, a 50 m boundary-protective coal pillar is reserved around the working face; so, it advances along the x-axis from 50 m to 450 m, and five vertical monitoring lines (x = 50 m, 150 m, 250 m, 350 m, 450 m) are arranged along the strike; each line sets multi-level measurement points (Figure 9) on the coal seam roof (z = 35.5 m) and the overlying rocks above it (z = 50 m, 65 m, 80 m, 95 m). The coal seam roof (z = 35.5 m) and the overlying rocks above it (z = 50 m, 65 m, 80 m, 95 m) are set up with multi-layer measuring points (Figure 9), covering the dip range of the working face from y = 50 to 250 m. The main points are as follows: (1) Measuring points of coal seam roof (z = 35.5 m): they are set up along the dip at 10 m intervals, focusing on capturing the migration patterns of vertical stress concentration zones (the peak of the pressure of the over-abutting support). (2) Overlying fracture zone measurement points (z = 50~95 m): the stress and permeability of the overburden strata are monitored and distinguished from the permeability decay in the raw phase and the jumping behavior after plastic damage. In order to effectively capture the relevant data and evolution pattern of each measurement point, the monitoring interval of the model measurement points is set to 100 steps.

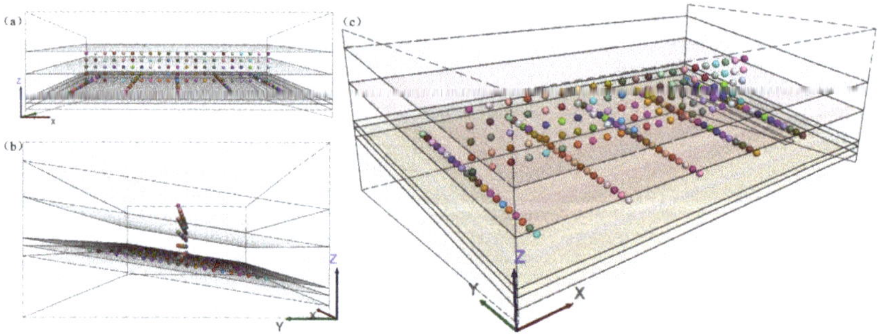

Figure 9. Measuring point arrangement. (**a**) Front; (**b**) Side; (**c**) Three-dimensional.

In this paper, the concept of damage degree is additionally introduced to monitor in real time the evolution of the damage degree of the overburden strata above the coal seam and the roof plate below the coal seam in the process of working face mining; the seepage mechanics data are integrated in all the measuring points to form a three-dimensional monitoring network of "vertical stratification comparison, stepwise tracking along the strike and dip coverage", which provides comprehensive parameters for the study of the dynamic evolution of "mining stress field, fracture field and seepage field".

4. Seepage Evolution Mechanisms of the Mining-Disturbed Overburden Aquifer

4.1. Characteristics of Overburden Permeability Mining Evolution

In order to deeply study the evolution law of the overburden seepage field during the advancement of the coal mine working face, to improve the ability of mine water damage prevention, this paper further analyzes the change characteristics of overburden permeability under the mining stress–fracture evolution. The redistribution of overburden strata stress after mining leads to the development and expansion of fractures, which in turn changes the permeability of the coal and rock body, resulting in significant changes in groundwater transport characteristics. Numerical simulation can be used to construct a dynamic evolution model of the seepage field, combine the stress–fracture–seepage coupling mechanism, and quantify the changes in coal rock permeability during the mining process.

Figure 10 shows the evolution pattern of the seepage field at X = 50 m, 150 m, and 250 m slice locations of the modeled strike during advancement of the working face along the strike.

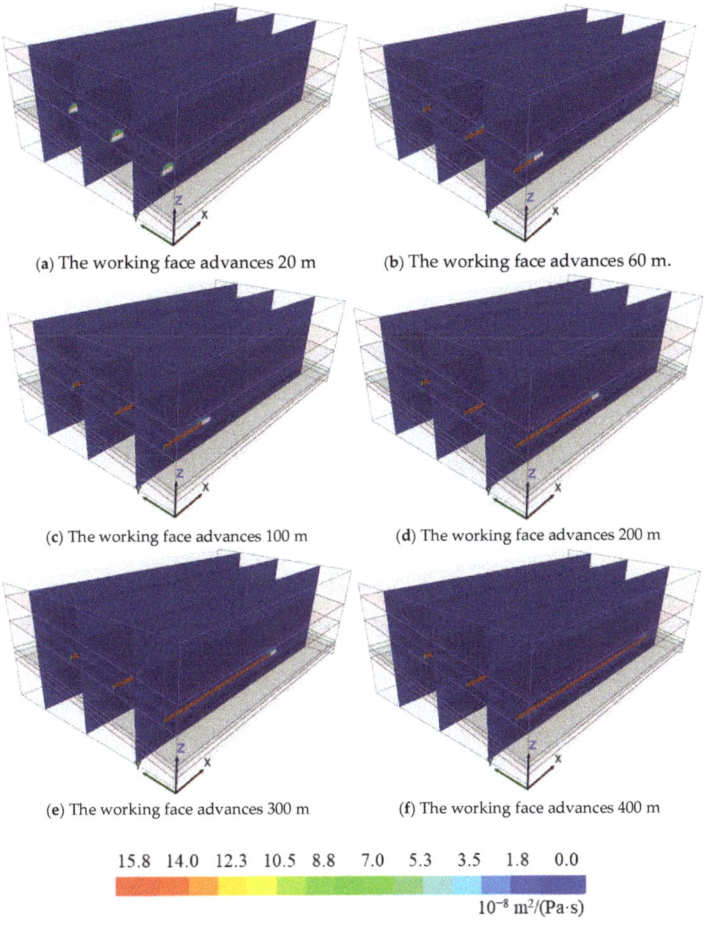

Figure 10. Distribution characteristics of seepage field along the strike of the working face.

Figure 11 shows the evolution of the seepage field at the position of the X = 150 m slice of the model dip during the advancement of the working face along the strike. When the working face advances to 20 m (Figure 11a), the seepage field is mainly concentrated near the coal seam roof, and the permeability change area is relatively small, indicating that the initial mining influence is limited, and only a local permeability enhancement phenomenon is generated in the front of the working face and the boundary of the mining area. After advancing to 60 m (Figure 11b), the high permeability area expands upward, and the permeability of the fracture zone increases, indicating that the stress adjustment of the overburden strata of the coal seam has caused the initial formation of seepage channels. When advancing to 100 m (Figure 11c), the influence range of the seepage field is obviously enlarged, the rupture degree of the roof is intensified, the fracture network is gradually penetrated, and the permeability is increased as a whole. When advancing to 200 m (Figure 11d), the high permeability zone further develops upwards, the scale of the seepage channel expands, the connectivity of the fractures increases, and the transport capacity of the water and gas improves, indicating that the development of mining fractures has entered a mature stage. With advancement to 300 m (Figure 11e) and 380 m (Figure 11f), the distribution of the seepage field tends to stabilize, and the areas of high values of permeability remain at the boundaries of the fracture zones and goaf, which further verifies the controlling effect of overburden fractures on the permeability performance. Overall, the evolution law of the seepage field indicates that with the advancement of the working face, the formation and expansion of the fractures directly lead to the increase in permeability, and the permeability channel between the goaf and the fracture zone is gradually established and eventually tends to be stable. This study provides a scientific basis for coal mine water inrush risk assessment and seepage disaster prevention and control.

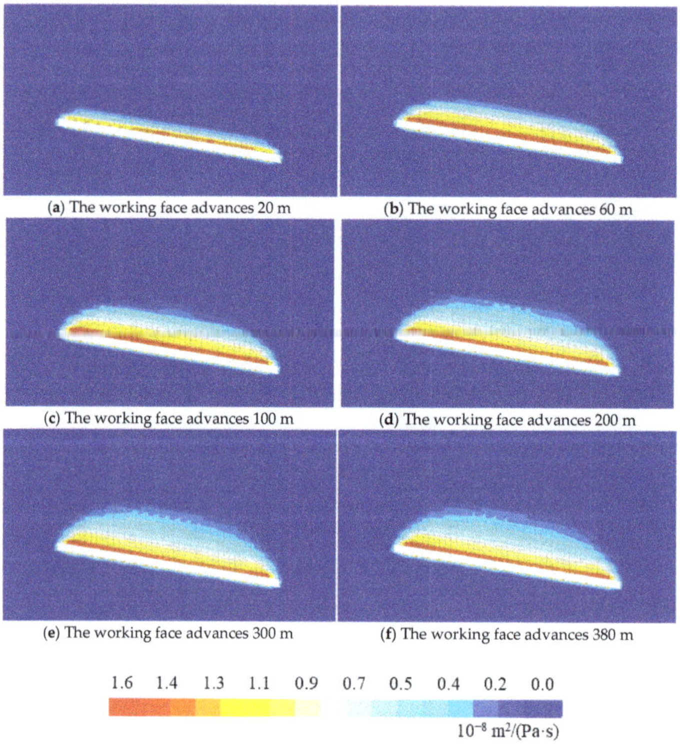

Figure 11. Distribution characteristics of seepage field along the dip of the working face.

According to permeability monitoring data at the y = 150 m strike line (z = 35.5 m) during face advancement (Figure 12a), the spatial–temporal evolution of mining-induced permeability exhibits significant differentiation, reflecting the multi-physical coupling mechanisms of stress–fracture–seepage fields. When the working face advanced to 50 m, permeability ahead of the measuring line (x = 50 m) surged from 4.62 mD to 26.99 mD (484% increase), indicating plastic damage initiation in the roof strata under abutment pressure. At 200 m advancement, permeability at x = 250 m reached 81.67 mD—an order of magnitude higher than adjacent undisturbed areas—marking fracture network penetration. As advancement progressed to 400 m, permeability at x = 150 m peaked at 15.14 mD (228% increase from the initial point), demonstrating stress concentration migration with face progression. Post-advancement, permeability in the goaf rear (x ≥ 350 m) stabilized at 23.72 mD (4.1 × initial value), confirming persistent seepage channels through plastic zone fractures.

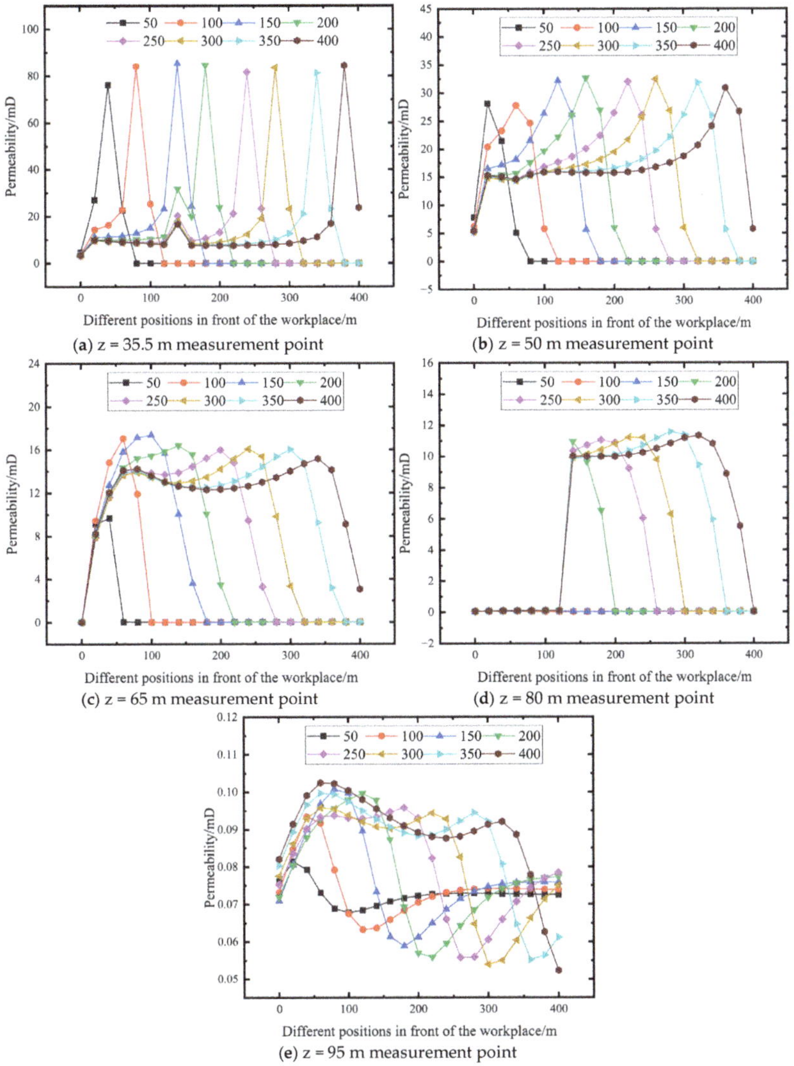

(**a**) z = 35.5 m measurement point (**b**) z = 50 m measurement point

(**c**) z = 65 m measurement point (**d**) z = 80 m measurement point

(**e**) z = 95 m measurement point

Figure 12. Evolution of overburden permeability of different working face advance distances.

The permeability has the following evolutionary pattern: (1) Within the range of overshooting influence (50–80 m in front of the working face), the permeability shows an exponential growth, synchronized with the migration of the peak support pressure, which conforms to the three-stage evolution model of "stress compression-fracture emergence-permeability leap"; (2) the lateral coal pillar area (x = 50 m) tends to stabilize after a sharp increase in permeability at the beginning of the advancement (0.0356 mD), reflecting the formation of a low permeability barrier after the protective coal pillar undergoes elastic–plastic deformation; (3) the middle part of the goaf (x = 250 m) shows a sudden change in permeability in the middle and late stages of advancement, which corresponds to the formation of the dominant water-conducting zone of the off-stratum fracture triggered by the breakage of the key layer. This law provides data support for revealing the differences in permeability characteristics among the "three vertical zones" of the overburden strata and indicates that the evolution of the seepage field is regulated by both the opening degree of the mining fracture and the effect of stress recovery.

4.2. Seepage Migration Characteristics of Overburden Aquifer

The hydrogeological conditions of the Luotuoshan coal mine are complex, and the distribution of overburden aquifers has an important impact on mine mining. At a distance of approximately 55 m above the working face (Figure 7), there exists a layer of fine-grained sandstone aquifer, which has a water pressure of about 1 MPa and thus belongs to the medium level of water pressure; the thickness of the aquifer is 9 m, which has strong water storage capacity and permeability, and it is one of the main sources of water influx in the mine. The existence of fine-grained sandstone aquifer makes it necessary to pay special attention to waterproofing measures in the process of working face mining in order to prevent the occurrence of sudden water accidents. Therefore, Luozhoushan coal mine needs to strengthen the monitoring and management of this aquifer in the process of mining to ensure the safe production of the mine.

Figure 13 shows the saturation evolution characteristics at the modeled strike Y = 150 m slice location during the advancement of the working face along the strike.

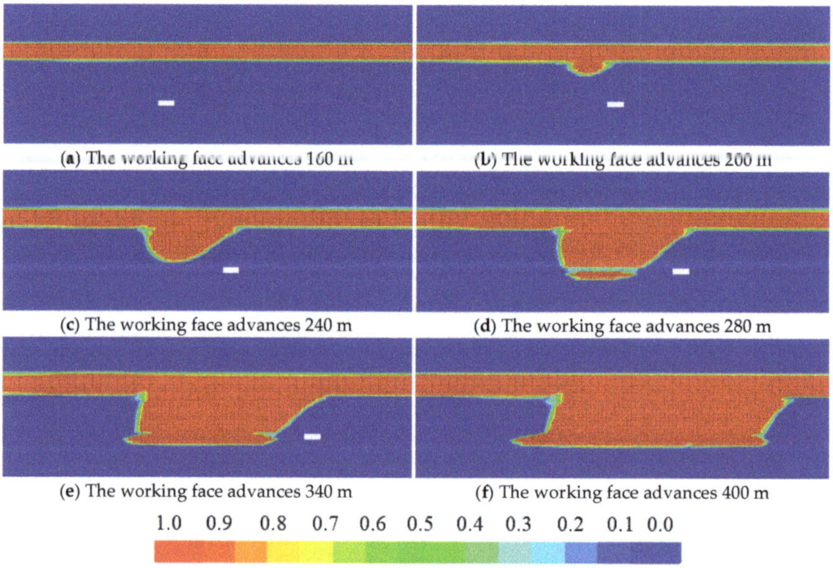

Figure 13. Distribution characteristics of saturation along the strike of the working face.

Figure 14 shows the evolution characteristics of the pore water pressure field at the location of the Y = 150 m slice of the modeled strike during the advancement of the workface along the strike. The change in pore water pressure shows a clear distribution trend as the working face advances. In the early stages of advancement (before 160 m), the pore water pressure was low and no significant permeability occurred. When advancing to 200 m, the pore water pressure at the measuring point began to rise significantly, especially at the measuring point close to the mining area; the pressure gradually increased, indicating that the water in the aquifer began to infiltrate into the mining area. To more than 250 m, drastic changes appeared in the pore water pressure at a number of measurement points; the pressure peak was obviously related to the concentration of the goaf, indicating that the seepage formed a stable channel and that the working face was affected by water intensification. As it advances to 400 m, the seepage path expands further, and the risk of water damage rises significantly. Therefore, in the actual mining process, we should focus on the evolution of pore water pressure after advancing to 200 m and strengthen the over-drainage and reinforcement measures in order to reduce the risk of water breakout and guarantee the safety of the working face.

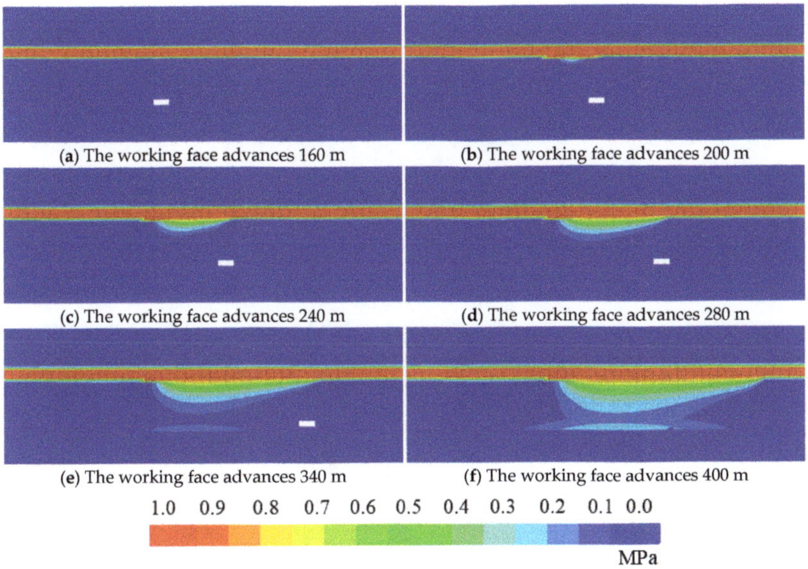

Figure 14. Distribution characteristics of pore water pressure field along the strike of the working face.

Figure 15 shows the saturation evolution characteristics at the location of the model dip X = 250 m slice during the advancement of the working face along the strike. From the simulation results, the plastic zone of the overburden strata gradually expands upwards as the working face advances and eventually connects with the upper aquifer, leading to water seepage. When advancing to 160 m, the plastic zone develops into the aquifer, meaning that the water-conducting fracture zone starts to affect the stability of the aquifer. At this time, the water flow seeps along the fracture to the goaf, forming a local seepage channel and posing a potential threat to the safety of the working face. When advancing to 250 m, due to further development of the fracture, water in the aquifer seeps into the goaf on a large scale, indicating that the pressurized water in the aquifer may have broken through the overburden barrier and entered the working face area. This process shows that the stability of the goaf is strongly influenced by aquifer seepage, which may lead to increased risk of a water breakout disaster.

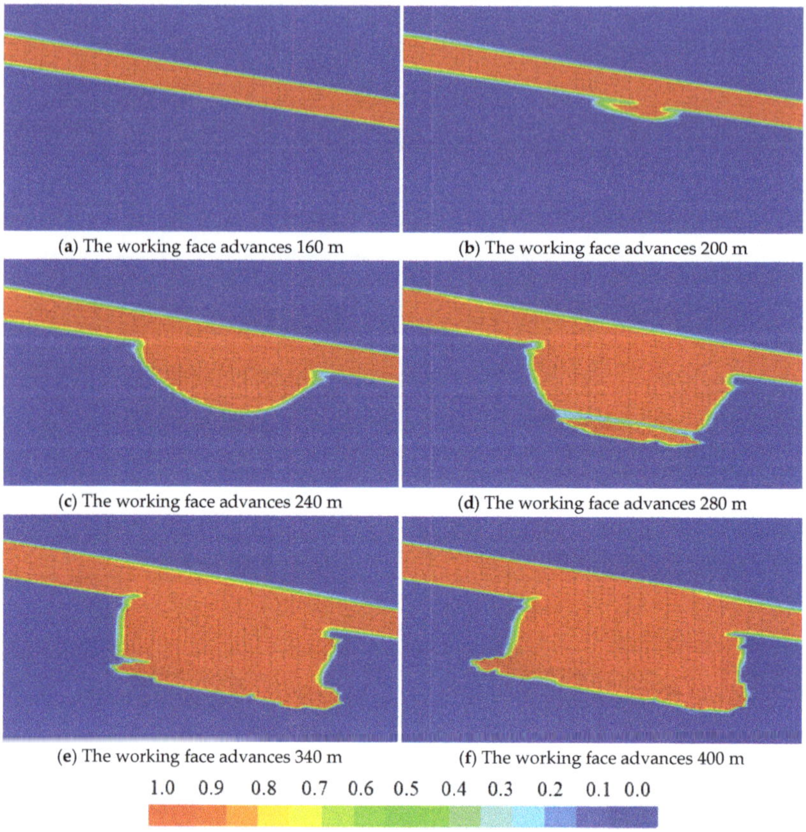

(a) The working face advances 160 m
(b) The working face advances 200 m
(c) The working face advances 240 m
(d) The working face advances 280 m
(e) The working face advances 340 m
(f) The working face advances 400 m

Figure 15. Distribution characteristics of saturation along the dip of the working face.

Figure 16 shows the evolution characteristics of the pore water pressure field at the location of the model dip X = 250 m slice during the advancement of the working face along the strike. From the evolutionary characteristics of pore water pressure, with the increase in advance distance the pore water pressure decreases significantly at the junction of the aquifer and fracture zone, which indicates that the water flow moves along the fracture to the low-pressure zone and forms a seepage field. This seepage may not only accelerate the collapse of the overburden strata but may also lead to a significant decline in the volume of water in the aquifer, with implications for the surrounding ecosystem. Therefore, during the advancement of the working face, The monitoring of the water-conducting fracture zone must be strengthened, and measures such as grouting reinforcement and pre-pumping of hydrophobic water must be taken to reduce the risk of water disaster and ensure the safe production of the mine.

The pore water pressure at different heights of the overburden strata shows an obvious trend at different advancing distances (Figure 17). In the early stage of advancement (0–100 m), the pore water pressure at all measurement points is close to zero, indicating that the overburden strata has not been disturbed by mining at this time, and the pore water pressure in the aquifer is kept in static equilibrium. However, as it advances to 120 m and above, especially after 140 m, the pore water pressure at some measurement points starts to increase, indicating that the WCFZ gradually expands upwards and starts to affect the aquifer area. At this stage, the pressure change of the measurement point at a low height (35.5 m) is not obvious, while the pressure of the measurement point at a higher height (50 m, 65 m) gradually increases, which reflects that the initial development of the fracture mainly affects the permeability of the higher area and that the water pressure of the aquifer has not been fully released. Entering the 260–320 m stage, there is an obvious surge in pore water pressure, especially in the low-level measurement points near the coal seam roof, indicating that the redistribution of stress in the goaf has drastically altered the structure of the overburden strata, allowing the water from the aquifer to penetrate into the overburden strata. At this stage, the pressure of the aquifer at 80 m decreased, reflecting the gradual formation of the trend of downward infiltration of water, and the further development of the fracture also increased the potential risk of mine water breakout. When advancing to 260 m and above, water damage monitoring must be strengthened, focusing on the fracture connectivity, and taking over-drainage measures to reduce the threat of water breakout. When the working face advances to 340 m and above, the pore water pressure shows drastic changes at each measuring point, indicating that the fracture expansion has basically connected the aquifer and the goaf, forming a more stable seepage channel.

The pore water pressure at different heights of the overburden strata experienced four stages: static equilibrium, fracture expansion, seepage formation, and high risk of water breakout. The evolution process of pressure directly reflects the development state of fracture. To ensure safe mining, it is suggested that over-drilling be carried out when advancing beyond 200 m to assess fracture expansion. At the same time, pre-pumping hydrophobic, grouting reinforcement and other measures should be combined to reduce the aquifer pressure in advance to reduce the risk of water breakout, and to strengthen the drainage management when advancing to more than 300 m to cope with possible water breakout problems and ensure the safe production of the mine.

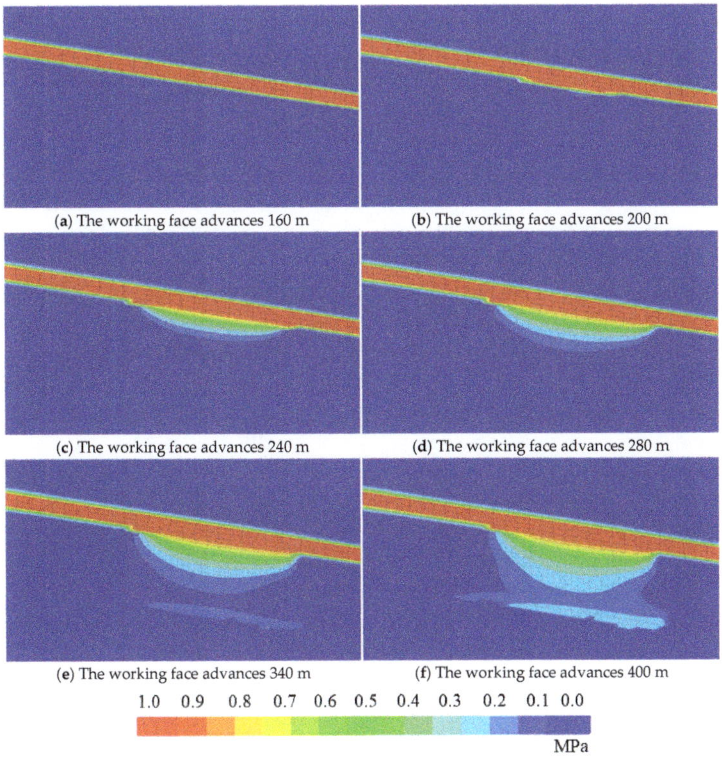

Figure 16. Distribution characteristics of pore water pressure field along the dip of the working face.

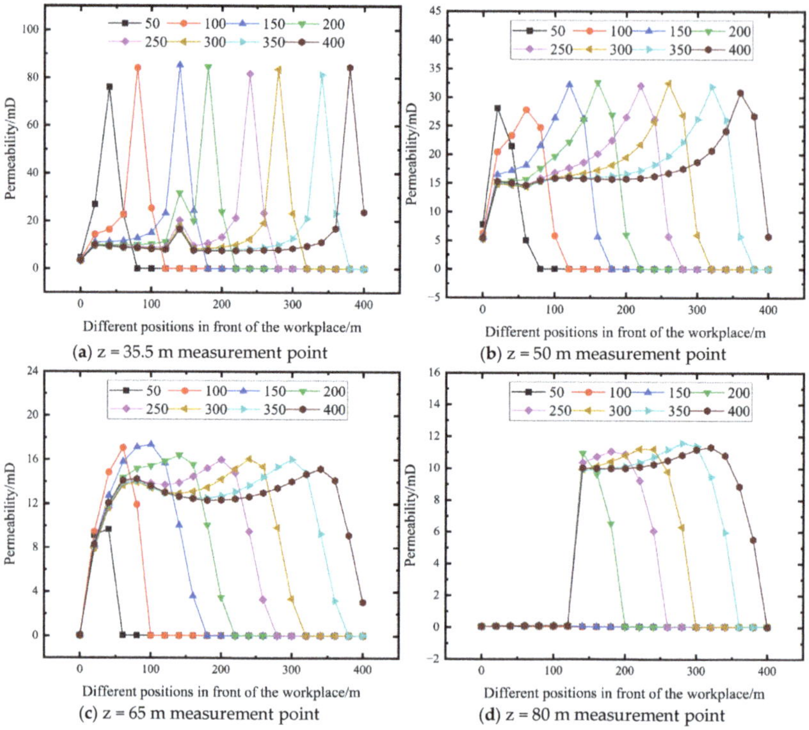

Figure 17. Evolution of pore water pressure in the overburden strata under different advancement distances of the working face.

5. Conclusions

(1) This study systematically investigated the coupled evolution mechanisms of stress–fracture–seepage fields in overburden strata through integrated experimental and numerical approaches. Laboratory triaxial permeability tests on three types of damaged coal rock samples (raw, fractured, and broken) revealed distinct seepage behaviors: fractured samples exhibited 23–48 times higher initial permeability (28.03 mD for coal, 13.54 mD for rock) compared to intact samples, while broken rock permeability decayed exponentially from 120.32 mD to 23.72 mD under compaction. These findings established quantitative permeability models ($R^2 > 0.99$) that effectively characterize zonal differences in mining-induced damage, providing a critical experimental basis for multi-field coupling simulations.

(2) A dynamic coupling methodology was developed by integrating FISH scripting with FLAC3D, enabling real-time permeability updates based on stress states and failure zone identification. The algorithm embedded three mechanisms: stress-dependent exponential decay for intact zones, yield-triggered permeability jumps for fractured regions, and compaction-driven nonlinear evolution for caved zones. This approach bridged the gap between static laboratory models and dynamic mining conditions, achieving automatic multi-field coupling during face advancement simulations.

(3) Numerical simulations of the Luotuoshan 110,901 working face identified critical seepage thresholds and spatial patterns. High-permeability zones (5–15 × background values) persistently localized at goaf boundaries and fracture zones, with peak permeability (81.67 mD) occurring at 200 m advancement. The simulated fracture zone height reached 55 m, directly connecting with the overlying aquifer (9 m thick, 1 MPa

pressure), consistent with field-measured water inrush events. These results demonstrated the method's capability to predict water-conducting fracture development and optimize drainage strategies.

(4) The stress–fracture–seepage coupling process evolved through four quantifiable stages: static equilibrium (0–100 m, <5 mD), fracture expansion (120–200 m, +484% permeability), seepage channel formation (200–300 m, aquifer connectivity), and high-risk inrush (300–400 m, 23.72 mD stabilized flow). These findings provide actionable thresholds for staged prevention measures—grouting before 200 m, targeted dewatering at 260–320 m, and enhanced drainage beyond 340 m. The methodology offers a predictive framework for deep mining water hazards, with future research recommended for multi-aquifer systems and real-time monitoring integration.

Author Contributions: Conceptualization, Y.L. and C.Z.; methodology, C.Z.; software, S.F.; validation, T.H., Y.G. and F.L.; formal analysis, S.F.; investigation, J.H.; resources, Y.L.; writing—original draft preparation, S.F.; writing—review and editing, C.Z.; visualization, S.F.; supervision, Y.L.; project administration, Y.G. All authors have read and agreed to the published version of the manuscript.

Funding: This research was funded by the 2022 First Batch of Science and Technology Projects of China Energy Group (GJNY-22-68), the Innovation Project of CCTEG Wuhan Research Institute (ZZYF202318), and the the National Key R&D Program of China (2023YFC3012101).

Data Availability Statement: Some or all data, models, or codes that support the findings of this study are available from the corresponding author upon reasonable request.

Conflicts of Interest: Authors Yan Liu, Tengfei Hu, and Fuzhong Li were employed by CCTEG Wuhan Engineering Company. Authors Yuan Guo and Jiawei Huang were employed by Luotuoshan Coal Mine of Wuhai Energy. The remaining authors declare that the research was conducted in the absence of any commercial or financial relationships that could be construed as a potential conflict of interest.

References

1. Li, Q.; Li, Q.; Wu, J.; Li, X.; Li, H.; Cheng, Y. Wellhead stability during development process of hydrate reservoir in the Northern South China Sea: Evolution and mechanism. *Processes* **2024**, *13*, 40. [CrossRef]
2. Han, P.; Wang, K.; Pang, J.; Ji, X.; Zhang, C. Response properties of geometries of coal penetrating fracture on seepage behavior. *Int. J. Min. Sci. Technol.* **2025**, *35*, 191–211. [CrossRef]
3. Bai, E.; Guo, W.; Tan, Y.; Guo, M.; Wen, P.; Liu, Z.; Ma, Z.; Yang, W. Regional division and its criteria of mining fractures based on overburden critical failure. *Sustainability* **2022**, *14*, 5161. [CrossRef]
4. Feng, D.; Hou, E.; Wang, S.; Xie, X.; Yuan, F.; Guo, L.; Wang, G.; Xie, Y.; Chen, Z. Research on 3D development characteristics of water-conducting fractured zone based on field measurement. *Front. Earth Sci.* **2022**, *10*, 808931. [CrossRef]
5. Wang, L.; Zhang, W.; Cao, Z.; Xue, Y.; Liu, J.; Zhou, Y.; Duan, C.; Chen, T. Effect of weakening characteristics of mechanical properties of granite under the action of liquid nitrogen. *Front. Ecol. Evol.* **2023**, *11*, 1249617. [CrossRef]
6. Xu, J.; Yang, D.; Zhang, Z.; Sun, Y.; Zhao, L. Study on fracture evolution and water-conducting fracture zone height beneath the sandstone fissure confined Aquifer. *Sustainability* **2024**, *16*, 6006. [CrossRef]
7. Zhang, C.; Wang, X.; Han, P.; Zhang, T.; Zhang, L.; Wang, F. Acoustic emission and splitting surface roughness of sandstone in a Brazilian splitting test under the influence of water saturation. *Eng. Geol.* **2024**, *329*, 107369. [CrossRef]
8. Wang, G.; Wu, M.; Wang, R.; Xu, H.; Song, X. Height of the mining-induced fractured zone above a coal face. *Eng. Geol.* **2017**, *216*, 140–152. [CrossRef]
9. Liu, Q.; Huang, J.; Guo, Y.; Zhong, T. Research on the temporal-spatial evolution of ground pressure at short-distance coal seams. *Arab. J. Geosci.* **2020**, *13*, 1014. [CrossRef]
10. Lu, C.; Xu, J.; Li, Q.; Zhao, H.; He, Y. Research on the development law of water-conducting fracture zone in the combined mining of jurassic and carboniferous coal seams. *Appl. Sci.* **2022**, *12*, 11178. [CrossRef]
11. Su, B.Y.; Yue, J.H. Research of the electrical anisotropic characteristics of water-conducting fractured zones in coal seams. *Appl. Geophys.* **2017**, *14*, 216–224. [CrossRef]
12. Wang, S.; Li, X.; Wang, D. Mining-induced void distribution and application in the hydro-thermal investigation and control of an underground coal fire: A case study. *Process Saf. Environ. Prot.* **2016**, *102*, 734–756. [CrossRef]

13. Wang, H.; Zhang, D.; Wang, X.; Zhang, W. Visual exploration of the spatiotemporal evolution law of overburden failure and mining-induced fractures: A case study of the Wangjialing coal mine in China. *Minerals* **2017**, *7*, 35. [CrossRef]
14. Li, X.; Li, Q.; Xu, X.; Zhao, Y.; Li, P. Multiple Influence Factor Sensitivity Analysis and Height Prediction of Water-Conducting Fracture Zone. *Geofluids* **2021**, *2021*, 8825906. [CrossRef]
15. Zhang, D.; Gao, Y.; Zhang, G.; Tang, Z.; Ding, F.; Gao, M. Failure characteristics of overlying strata and mechanism of strong ground pressure during the large-scale and continuous mining of deep multi working faces. *Energy Sci. Eng.* **2024**, *12*, 4950–4964. [CrossRef]
16. Wang, S.; Liu, K.; Wang, S.; Liang, Y.; Tian, F. Three-dimensional stochastic distribution characteristics of void fraction in longwall mining-disturbed overburden. *Bull. Eng. Geol. Environ.* **2022**, *81*, 414. [CrossRef]
17. Jiang, D.; Tang, Y.; Huang, W.; Hou, K.; Luo, Y.; Liu, J. Research on the height of the water-conducting fracture zone in fully mechanized top coal caving face under combined-strata structure. *Sustainability* **2022**, *14*, 13781. [CrossRef]
18. Wang, W.; Li, Z.; Du, F.; Cao, Z.; Li, G. Study of roof water inrush control technology and water resources utilization during coal mining in a Karst area. *Mine Water Environ.* **2023**, *42*, 546–559. [CrossRef]
19. Yang, D.; Guo, W.; Tan, Y. Study on the Evolution Characteristics of Two-Zone Failure Mode of the Overburden Strata under Shallow Buried Thick Seam Mining. *Adv. Civ. Eng.* **2019**, *2019*, 9874769. [CrossRef]
20. Ghabraie, B.; Ren, G.; Zhang, X.; Smith, J. Physical modelling of subsidence from sequential extraction of partially overlapping longwall panels and study of substrata movement characteristics. *Int. J. Coal Geol.* **2015**, *140*, 71–83. [CrossRef]
21. Wang, C.; Zhang, C.; Li, Z.; Zhou, J. Using discrete element numerical simulation to determine effect of persistent fracture morphology on permeability stress sensitivity. *Int. J. Numer. Anal. Methods Geomech.* **2023**, *47*, 570–584. [CrossRef]
22. Shi, X.; Zhang, J. Characteristics of overburden failure and fracture evolution in shallow buried working face with large mining height. *Sustainability* **2021**, *13*, 13775. [CrossRef]
23. Dong, F.; Yin, H.; Ren, H.; Cheng, W.; Tai, S.; Miao, T.; Zhang, Y. Trapezoidal Failure Behavior and Fracture Evolution Mechanism of Overburden in Extra-thick Coal Mining in Weakly Cemented Strata. *Rock Mech. Rock Eng.* **2024**, *57*, 11239–11260. [CrossRef]
24. Song, S.; Ruan, H.; Wei, J.; Niu, R.; Cheng, X.; Chen, B. The Influence of the Key Characteristics of Overburden Rock Structure on the Development Height of Water-Conducting Fracture in Yushenfu Coal Mine Area, China. *Appl. Sci.* **2024**, *14*, 10537. [CrossRef]
25. Chang, X.; Wang, M.; Zhu, W.; Fan, J.; Liu, M. Study on height development characteristics of water-conducting fracture zone in fully mechanized mining of shallow thick coal seam under water. *Sustainability* **2023**, *15*, 7370. [CrossRef]
26. Yang, B.; Yuan, S.; Liang, Y.; Liu, J. Investigation of overburden failure characteristics due to combined mining: Case study, Henan Province, China. *Environ. Earth Sci.* **2021**, *80*, 143. [CrossRef]
27. He, X.; Zhang, C.; Han, P. Overburden Damage Degree-Based Optimization of High-Intensity Mining Parameters and Engineering Practices in China's Western Mining Area. *Geofluids* **2020**, *2020*, 8889663. [CrossRef]
28. Cao, Z.; Yang, X.; Li, Z.; Du, F. Evolution mechanism of water-conducting fractures in overburden under the influence of water-rich fault in underground coal mining. *Sci. Rep.* **2024**, *14*, 5081.
29. Chen, K.; Ge, Y.; Liu, Z.; Chen, L.; Zhang, Q. Overburden failure associated with slicing mining in a super thick coal seam under special weak aquifers. *Water* **2022**, *14*, 3882. [CrossRef]
30. Seidle, J.P.; Jeansonne, M.W.; Erickson, D.J. Application of matchstick geometry to stress dependent permeability in coals. In Proceedings of the SPE Rocky Mountain Regional Meeting, Casper, WY, USA, 18–21 May 1992; SPE: Richardson, TX, USA, 1992; p. 24361.
31. Li, S.; Tang, D.; Pan, Z.; Xu, H.; Huang, W. Characterization of the stress sensitivity of pores for different rank coals by nuclear magnetic resonance. *Fuel* **2013**, *111*, 746–754. [CrossRef]
32. Ahmadi, S.; Khormali, A.; Razmjooie, A. Experimental investigation on separation of water in crude oil emulsions using an oil-soluble demulsifier. *Iran. J. Chem. Chem. Eng.* **2023**, *42*, 2332–2343.

processes

MDPI

Article

Study on the Mechanisms of Floor Heave in Roadways and Control Strategies via Grooving and Pressure Relief

Tongxi Li *, Bingyuan Hao and Chaoyao Shi

College of Mining Engineering, Taiyuan University of Technology, Taiyuan 030024, China
* Correspondence: 17200620521@163.com or lidongxi1221@link.tyut.edu.cn

Abstract: With the increasing depth of mining operations, roadways experience higher ground stress and pronounced deformation. Elevated in situ stress leads to a continuous rise in vertical stress on surrounding rock. The excavation of roadways and working faces further redistributes the stress field, resulting in more frequent and severe floor heave. To address this issue in the 1232 transportation roadway of Shuguang Coal Mine, a numerical model was developed using the discrete element method (PFC2D) to systematically analyze the impacts of floor stiffness, strength, and joint distribution on heave mechanisms. A mechanical device for underground slotting operations was designed, and the optimal slotting depth was determined through simulation. The results indicate that floor heave stems from progressive failure of composite strata, governed by stiffness, strength, and moment of inertia (influenced by strata thickness and joint development). Effective suppression requires slotting depths to penetrate the shallow fractured zone and reach the load-bearing structure. Stress arching effects significantly mitigate deformation at 2.5 m depth, providing a theoretical basis for optimal slotting design.

Keywords: floor heave mechanism; discrete element modeling; slotting control; composite strata; pressure relief by slotting

1. Introduction

Statistics show that floor deformation control accounts for 65–70% of roadway surrounding rock stability management. Floor heave is one of the most common manifestations of floor failure. It not only narrows the roadway cross-section—hindering pedestrian access, transportation, and ventilation—but also compromises the stability and integrity of support systems, severely impacting safe mine operations [1]. Therefore, in depth investigation of the mechanisms of floor heave and its control methods holds significant theoretical and practical value for improving roadway floor stability. Numerous scholars, both domestic and international, have conducted extensive research on this issue. Kang et al. [2,3], He et al. [4–6], and others concluded that roadway floor heave is primarily caused by various mechanisms including elasto-plastic deformation, dilation, water-induced swelling, rheological behavior, and flexural fracturing of the floor strata. Li et al. [7] studied the stress distribution patterns in the surrounding rock of retreat roadways under mining influence using theoretical analysis and FLAC3D numerical simulation, focusing on the effect of different pillar widths and spacing. Indraratna [8] from Australia employed simulation software to analyze fracture development in the floor, providing valuable insights into the mechanisms and predictive modeling of crack propagation. Xu et al. [9], in response to floor heave deformation in Caojiatan Mine, developed a series of core technologies for roadway protection through roof cutting and pressure relief, based on detailed studies of overburden movement and coal seam variation. Song et al. [10] proposed mitigation measures for both reinforced and pressure-relief-type floor heave based on extensive field observations, theoretical analysis, and numerical simulations. Wang et al. [11] integrated practical engineering scenarios with floor heave mechanisms and proposed a terminal anchor cable bundle reinforcement method for floor stabilization using theoretical and numerical approaches. Zhu et al. [12], using the disturbed intake roadway in the west wing of Yuwu Coal Mine as the research context, proposed an optimized combined support system of "inverted U-shaped steel arches + grouted anchor bolts", supported by theory and simulation. The scheme's effectiveness was validated through industrial-scale trials. Li et al. [13] systematically examined the mechanisms of floor heave under mining disturbances and proposed a method combining floor slotting for pressure relief with additional bolt support at the base of both sidewalls. Guo et al. [14] demonstrated the effectiveness of slotting and pressure relief in controlling roadway floor heave. Their in depth analysis revealed that when the slotting depth exceeds 0.5 times the roadway width, the mitigation effect is significantly enhanced. Moreover, changes in slot angle alter the stress distribution in the floor, offering further control of floor deformation. Hou et al. [15] identified floor heave and creep as the two major challenges in enhancing roadway stability in underground coal mines, and proposed a technology to address both issues based on theoretical analysis. S. Mo [16] showed that floor bolts can effectively reduce surrounding rock deformation and thereby mitigate floor heave. Ren et al. [17] developed a novel block particle discrete element simulation method that matches the dual media of overlying rocks (OLR) and loose layers (LSL). Han et al. [18] studied the permeability characteristics of five sets of penetrating fractures (CPF) with different joint roughness coefficients (JRC). Based on 3D shape scanner testing and hydraulic coupling testing, the method of evaluating effective geometric parameters on the surface of the crack was improved under various constraint pressures, and the relationships between effective geometric parameters and constraint pressures were established. Zhang et al. [19] concluded that discrete element method (DEM) simulation is crucial for bridging macro and micro studies, summarized the research status, and discussed related issues. Han et al. [20] applied slotting pressure-relief techniques to control floor heave in deep, high-stress roadways. Their study revealed that a slotting depth of 2 m results in the most uniform heave distribution. Wei et al. [21] and others used FLAC3D simulation to investigate the deformation characteristics and energy transfer of the floor and surrounding rock after slotting in deep-buried roadways. Li et al. [22] addressed severe floor heave caused by the stress concentration and weak floor lithology using slotting pressure relief techniques. Zhou et al. [23] studied variations in stress distribution and the displacement of surrounding rock before and after slotting pressure relief through numerical simulation. Hao et al. [24] and colleagues validated the predictive reliability of their proposed model by comparing it with numerical simulation results. Many researchers have shifted from single-aspect approaches to comprehensive control strategies for managing large deformations in the rock surrounding roadways, and have successfully applied these solutions in real mining environments.

2. Project Overview

The fully mechanized mining face #1232 is located in the Loess Plateau region, characterized by low hills and beam-shaped loess tablelands. The surface elevation ranges from +951 m to +1142 m, resulting in a relative elevation difference of 91 m. The western boundary of the working face coincides with the edge of the mining district. To the east of the #1232 face lies the central track roadway; its western side borders the mine boundary. The southern side is adjacent to the active #1230 mining face, while the northern boundary connects to the yet-to-be-mined #1234 face. The mining layout of the working face is shown in Figure 1.

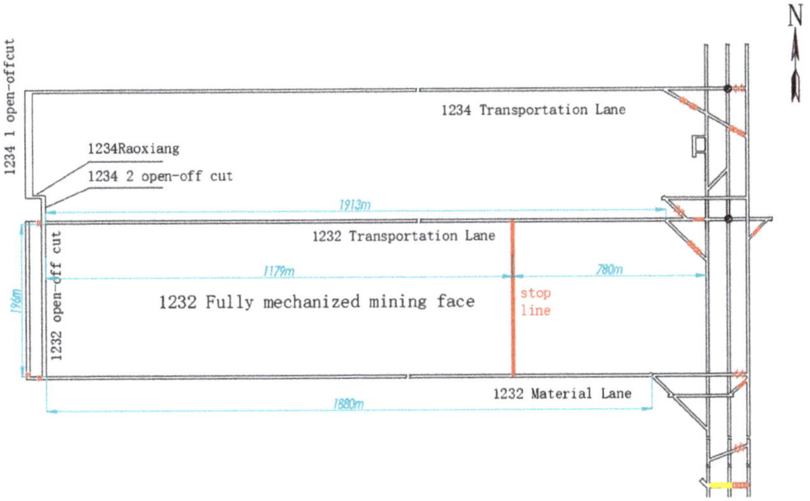

Figure 1. Comprehensive mining face mining layout plan for #1232.

Shuguang Coal Mine currently extracts the No. 2 coal seam. The buried depth of coal seam No. 2 in 1232 Transportation Lane is greater than 600 m, which is considered a deep buried coal seam, where deformation of the roadway's surrounding rock is predominantly characterized by floor heave. This issue has become more pronounced since the adoption of gob-side entry retaining without coal pillars, which has led to intensified deformation in the retained roadways. As a result, roadway reconstruction costs—such as bottom re-excavation and reinforcement support—have significantly increased. Additionally, the complexity of ensuring roadway stability has greatly increased, posing serious challenges to safe and efficient mine production. As shown in Figure 2, floor heave in the 1232 transportation roadway is evident, with a heave magnitude of no less than 30 cm.

Figure 2. On site photo of the bottom drum in the roadway of Shuguang Coal Mine.

Mechanical Parameter Testing of Rock Surrounding the Roadway

The determination of the physical and mechanical properties of rock forms the foundation of rock mass engineering analysis in mining. These test data serve not only as key indicators for rock mass quality classification systems but also as critical input parameters for constructing multi-scale models of rock mass stability. The experimental results indicate that parameters such as uniaxial compressive strength and Poisson's ratio have a decisive influence on the design and control of rock surrounding a roadway. In view of the floor heave observed in the 1232 transportation roadway of the No. 2 coal seam at Shuguang Coal Mine, combined with the comprehensive bar chart of #1232 as shown in Figure 3, a study on control strategies was conducted. In agreement with the mine operator, core sampling for rock mechanical testing was carried out approximately 200 m from the roadway entrance. The physical and mechanical parameters of each rock stratum are listed in Table 1.

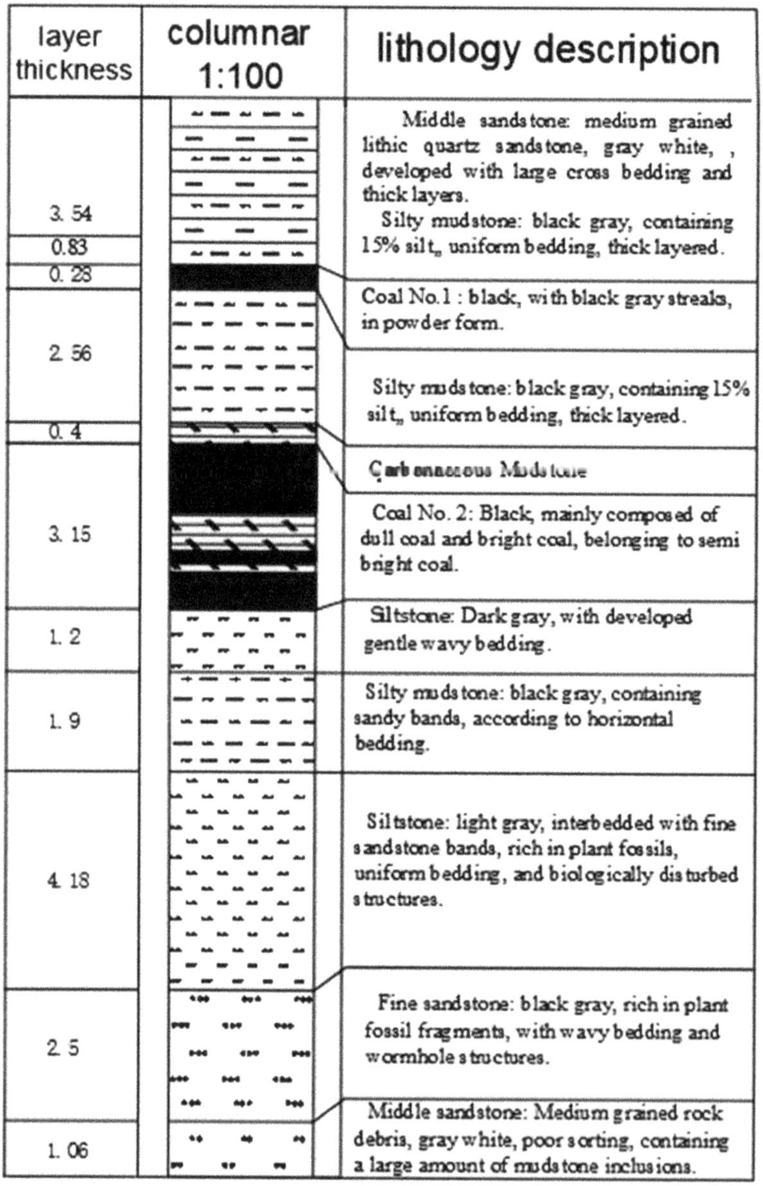

Figure 3. Coal seam top and bottom slate layer column chart of 1232 Transportation Lane.

Table 1. Physical and mechanical parameters of rock formations.

Lithology	Thickness/m	Shear Modulus /GPa	Bulk Modulus /GPa	Cohesion /MPa	Tensile Strength /MPa	Internal Friction Angle/(°)	Density/ (kg m^{-3})
Middle sandstone	3.54	3.35	5.08	4.88	3.1	31.3	2500
Silty mudstone	0.83	1.92	3.58	3.02	2.13	27.3	2400
Coal No. 1	0.28	0.62	1.35	1.03	0.81	23.8	1400
Silty mudstone	2.56	1.92	3.58	3.02	2.13	27.3	2400
Carbonaceous mudstone	0.4	1.53	2.88	3.57	1.46	25.6	2400
Coal No. 2	3.15	0.62	1.35	1.03	0.81	23.8	1400
Siltstone	1.2	4.63	6.21	5.83	4.2	33.4	2600
Silty mudstone	1.9	1.92	3.58	3.02	2.13	27.3	2400
Siltstone	4.18	4.63	6.21	5.83	4.2	33.4	2600
Fine sandstone	2.5	4.63	6.21	5.83	4.2	33.4	2600
Middle sandstone	1.06	3.35	5.08	4.88	3.1	31.3	2500

3. Discrete Element Analysis of Floor Heave Mechanism

3.1. Model Construction

Due to its depth (≈600 m) and structural background, vertical stress dominates in Shuguang Coal Mine. On site monitoring shows that vertical stress accounts for about 85% of the total stress concentration in the bottom strata. Therefore, this study considered vertical stress and applied horizontal displacement constraints to establish a model.

Based on the comprehensive stratigraphic column and the physical–mechanical parameters of the rock layers at Shuguang Coal Mine's #1232 working face, along with the mining and excavation layout, a two-dimensional model (24 m long × 21.6 m high) was established to study the floor heave mechanism of the 1232 transportation roadway, as shown in Figure 4. The 1232 transportation roadway is centrally positioned within the model, with both sides located 9.75 m from the model boundaries. The model constrains horizontal displacement at the left and right boundaries, applies vertical displacement constraints at the base, and imposes a uniformly distributed load of 15 MPa on the top surface. The model is constructed using periodic block elements, with initial in situ stresses applied. The load-bearing layer is represented by converting the applied stress into force loads directly on the top particles of the model. Once the model reaches equilibrium under the applied in situ stresses, excavation of the 1232 transportation roadway is simulated.

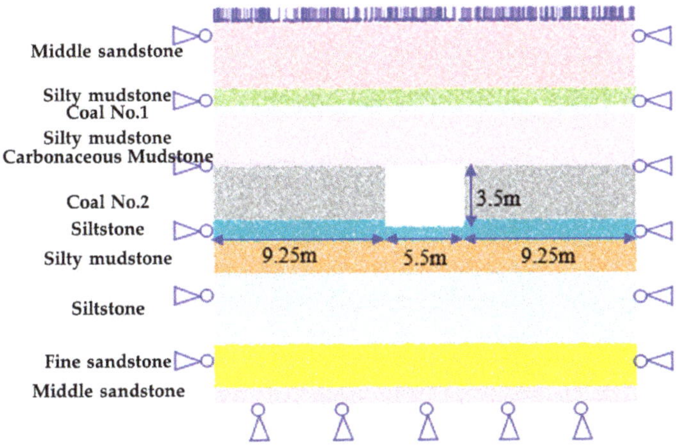

Figure 4. Discrete element model of bottom drum in 1232 Transportation Lane of Shuguang Coal Mine.

The numerical model employs particles with a uniform radius distribution across all simulated strata, comprising a total of 53,418 particles to represent 11 distinct rock layers. To address the computational inefficiency associated with excessive particle generation using the compaction–rejection method, the model was constructed using periodic brick elements. Each brick element consists of a granular assembly within a periodic spatial domain, pre-equilibrated to ensure mechanical stability. This approach eliminates the need for additional computational time to achieve equilibrium during the model initialization phase, thereby streamlining the model establishment process. Upon assembling the periodic brick elements, initial in situ stresses were applied to the system. Specifically, the overburden load was converted into equivalent force vectors and directly applied to the top layer of particles. After achieving equilibrium under the initial in situ stresses, the excavation of the 1232 transportation roadway was simulated to investigate the dynamic response of the surrounding rock mass.

The Particle Flow Code (PFC) method, rooted in discontinuous medium mechanics theory, employs assemblies of discrete rigid particles to simulate the macroscopic mechanical behavior of materials. In this framework, particle overlaps are permitted to characterize contact forces governed by Newtonian dynamics, while the material's macro-scale mechanical properties dynamically emerge from evolving interparticle contact states. The transition from pre-peak linear behavior to post-peak nonlinear characteristics is triggered by contact failures (e.g., shear or tensile failure modes), eliminating the need for predefined macroscopic constitutive relationships. Instead, micro-mechanical parameters—including particle gradation, stiffness coefficients, frictional properties, and bond strengths—indirectly govern the material's macroscopic response. The term "particles" in PFC serves as an abstract representation of material behavior rather than literal granular constituents, enabling applications ranging from crystalline rock mechanics to amorphous material analysis. The collective mechanical complexity of particle assemblies (e.g., nonlinear deformation, failure patterns) manifests through contact mechanics: contact strength parameters dictate fundamental stress-bearing capacities (tension, compression, shear), while contact failure initiates internal crack propagation. Crucially, the stress-strain relationships under complex loading conditions are automatically derived through computational algorithms tracking microstructural evolution. This bottom-up modeling paradigm allows macroscopic material properties to naturally emerge from the progressive reconfiguration of microscopic contact networks, bypassing traditional phenomenological assumptions inherent in continuum-based approaches.

3.2. *Analysis of Simulation Results*

Analysis of Actual Floor Heave Deformation in the 1232 Transportation Roadway

The characteristics of surrounding rock stress manifestation after excavation of the 1232 transportation roadway are shown in Figure 5. Following excavation, stress is released into the roadway, causing fractures to develop in the sidewalls, floor, and shoulder areas of the roof. These fractures exhibit a butterfly-shaped distribution. The force chain distribution indicates the formation of an elliptical stress-relief zone (loosening zone) in the surrounding rock, with an average extent of approximately 1.5 m in both sidewalls. The thick, silty mudstone in the floor is almost entirely fractured, with a failure depth reaching up to 2.5 m. Under the lateral load transfer from the sidewalls, the broken mudstone is compacted, forming a load-bearing structure. The above analysis shows that the extent of the loosening zone in the surrounding rock plays a controlling role in floor heave deformation. Its distribution in the floor determines both the termination position and the depth of the heave. As this study focuses on the influence of floor stratigraphy on floor heave, the effect of roadway support parameters was not considered in the model. If support parameters are taken into account, the extent of the loosening zone in the surrounding rock will be reduced, increasing the load transferred from the sidewalls to the floor, and consequently intensifying the floor heave deformation.

Crack distribution	force chain distribution	bottom bulge deformation

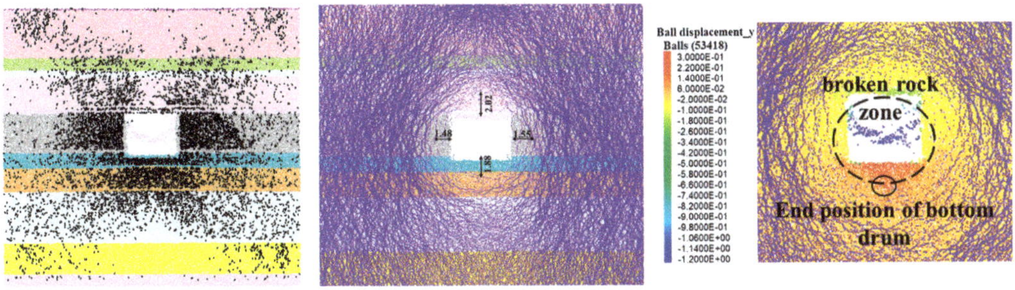

Figure 5. Rock pressure characteristics of rock surrounding 1232 Yunxiang of Shuguang Coal Mine.

3.3. Analysis of Influencing Factors on Floor Heave

Roadway excavation and retreat mining operations lead to the redistribution of stress within the surrounding rock. When the floor loses displacement constraints, the combined effects of horizontal and vertical stresses cause internal failure in the floor strata, resulting in upward deformation and the formation of floor heave.

The butterfly-shaped fracture distribution arises from stress redistribution following roadway excavation. Vertical stress is released toward the free surface of the roadway, while horizontal stress is transferred laterally into the surrounding rock mass, resulting in compressive stress concentrations at the roof–floor corners and tensile stress zones along the sidewalls and floor. High deviatoric stress regions trigger diagonal shear fractures, whereas low-stress zones generate vertical tensile cracks. The interplay of these failure mechanisms forms the characteristic butterfly-shaped fracture pattern. Geometric symmetry of the roadway and lithological homogeneity promote stress redistribution through a self-stabilizing "stress arch", redirecting stresses deeper into the strata while creating a symmetric loosening zone near the excavation. Progressive failure in composite strata further reinforces this pattern: weak layers (e.g., mudstone) experience rapid tensile fracturing due to low tensile strength, while stiff layers (e.g., sandstone) exhibit delayed shear failure. The spatiotemporal heterogeneity of failure across strata ultimately defines the distinct butterfly-shaped fracture geometry. This mechanism aligns with field observations in deep coal mines and classical theories of excavation-induced damage (e.g., Kang et al., 2015 [3]; He et al., 2009 [4]), validating its universality in geomechanical engineering.

During this process, the vertical stress acting on the floor surface significantly decreases, while the horizontal stress transferred from both sides gradually increases, shifting from compressive to tensile stress. As the stress redistributes, it transfers toward the goaf boundary, leading to instability and failure in the floor of the retreat roadway due to the increasing horizontal stresses transmitted from the sidewalls. Under high tensile stress, once the stress exceeds the rock's ultimate tensile strength, progressive failure occurs, becoming a primary factor in the failure of floor strata in retreat roadways.

After roadway excavation during retreat mining, the loss of vertical constraint in the floor allows stress to be released toward the roadway, manifested as floor heave. From a mechanical perspective, floor heave is a progressive failure process that develops layer by layer from the shallow floor to deeper strata, until it reaches a competent bearing layer.

This stable layer restricts further development of the heave. Therefore, floor heave in retreat roadways is governed by the layered failure characteristics of composite floor structures. According to composite plate theory, the flexural deformation of each floor rock layer is primarily influenced by three factors: stiffness, strength, and moment of inertia. The moment of inertia is largely affected by the thickness of the rock layers and the degree of joint development. By varying the stiffness, strength, and moment of inertia of the rock layers, the influence of each factor on floor heave can be systematically studied.

(1) Influence of Floor Stiffness and Strength on Floor Heave Deformation

To investigate the influence of floor strength on floor heave deformation, the stiffness and strength of the first-layer siltstone, second-layer silty mudstone, and third-layer siltstone in the floor were increased by a factor of 2.5, while the parameters of the other strata in the model remained unchanged. The stress distribution characteristics after excavation with the strengthened floor are shown in Figure 6.

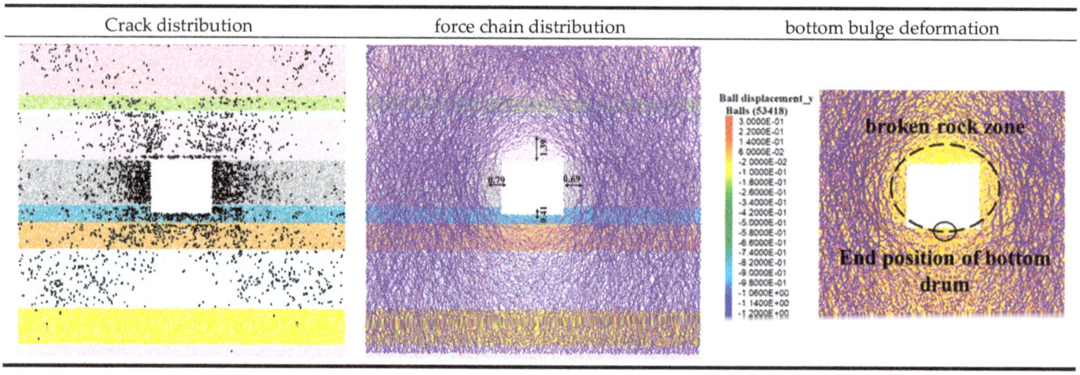

Figure 6. Characteristics of rock pressure in the surrounding rock after strengthening the bottom plate of Shuguang Coal Mine.

After excavation with the strengthened floor, fewer fractures developed, primarily concentrated in the sidewalls. Fracture development in both the floor and roof significantly decreased, with most of the floor cracks confined to shallow depths. The force chain distribution indicates that the stress-bearing arch in the surrounding rock shifted closer to the roadway, and the extent of the loosening zone was significantly reduced. The loosening zone in the sidewalls decreased from 1.5 m to 0.7 m, the stress-reduced zone in the roof narrowed from 2.02 m to 1.39 m, and the failure depth in the floor was reduced from 2.5 m to 0.41 m. Ultimately, the floor heave was limited to no more than 4 cm.

Increasing the stiffness and strength of the floor generally enhances its stability. A stable floor can withstand greater stress and more effectively distribute loads, thereby reducing flexural deformation and minimizing floor heave. After floor reinforcement in the 1232 roadway, deformation and failure of the roof and sidewalls were restrained, the load-bearing capacity of the sidewalls was improved, the loosening zone was reduced, and floor heave deformation was effectively mitigated.

(2) Influence of Floor Layer Thickness and Joint Development on Floor Heave Deformation

The floor strata can be regarded as a composite beam structure. Under vertical stress disturbances from the roof and coal pillar, the beam ends are subjected to loads exceeding their bearing capacity, while simultaneously experiencing horizontal compressive forces due to stratum movement. This results in flexural deformation and upward arching of the beam, eventually leading to failure through a layer-by-layer failure process, as illustrated in Figure 7. The location and extent of joint development within the floor strata affect the composite beam's resistance to deformation. Consequently, the degree of jointing directly influences the extent of floor heave.

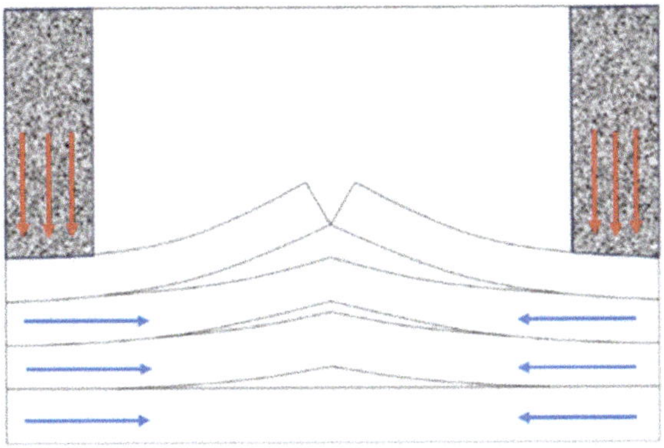

Figure 7. The composite beam of the bottom plate gradually bends and bulges, layer by layer.

1) Single Joint Located in the Middle of the Second Floor Stratum

When a single joint is located in the middle of the second floor stratum, the stress characteristics of the surrounding rock after excavation of the 1232 transportation roadway are shown in Figure 8. Following excavation, stress is released into the roadway, leading to fracture development in the sidewalls, floor, and shoulder areas of the roof, exhibiting a butterfly-shaped distribution. According to the force chain distribution, the loosening zone in the surrounding rock of the roadway is elliptical in shape, with an average extent of approximately 1.5 m on both sides. The thick, silty mudstone in the floor is almost completely fractured, with a failure depth of up to 2.5 m. The fractured mudstone is compacted under lateral load transfer from the sidewalls, forming a load-bearing structure. The floor heave exceeds 30 cm and terminates at the upper boundary of the load-bearing structure formed by the broken silty mudstone. This boundary also corresponds to the edge of the loosening zone in the 1232 roadway's surrounding rock.

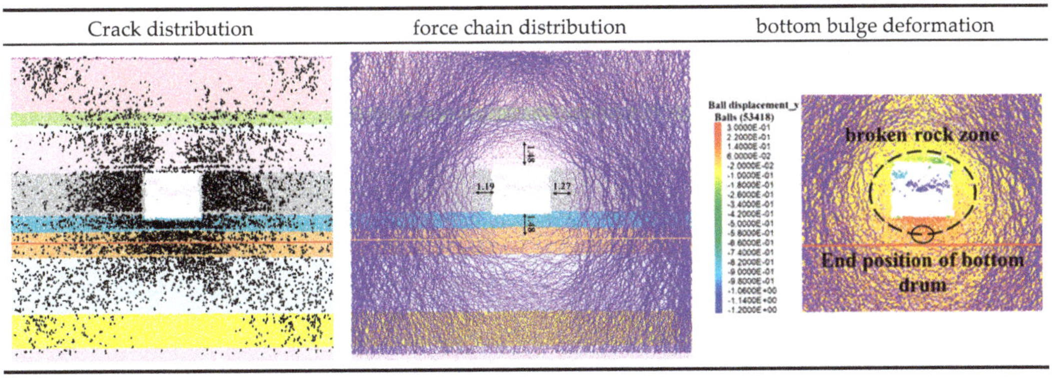

Figure 8. A single joint is located in the middle of the second layer of rock strata on the floor, and the rock pressure characteristics of the rock surrounding the 1232 transportation roadway.

The second layer of the floor stratum is completely damaged under the combined effects of stress release due to excavation and lateral load transfer from the sidewalls. The presence of a single joint in the middle of this layer does not significantly alter the composite beam's resistance to floor heave deformation. Therefore, when a single joint is located in the middle of the second floor layer, the stress characteristics of the surrounding rock in the 1232 roadway do not exhibit significant changes.

2) Single Joint Located 1 m Below the Top of the Third Floor Stratum

When a single joint is located 1 m below the top of the third floor stratum, the stress characteristics of the surrounding rock after excavation of the 1232 transportation roadway are shown in Figure 9. the original 4.18-m-thick siltstone layer is divided into two layers with thicknesses of 1 m and 3.18 m. This segmentation reduces the overall resistance of the composite floor beam to floor heave deformation, leading to increased fracture development, expansion of the loosening zone, and intensified floor heave. According to the simulation results, when the joint is located 1 m below the top of the third floor stratum, the surrounding rock of the 1232 roadway exhibits the following stress characteristics: fractures develop in the sidewalls, floor, and roof shoulders, forming a butterfly-shaped pattern; the number of fractures increases compared to previous cases. The loosening zone in the surrounding rock remains elliptical, with an average width of 1.6 m on both sides. Floor failure extends to the stratification layer at a depth of 3.5 m.

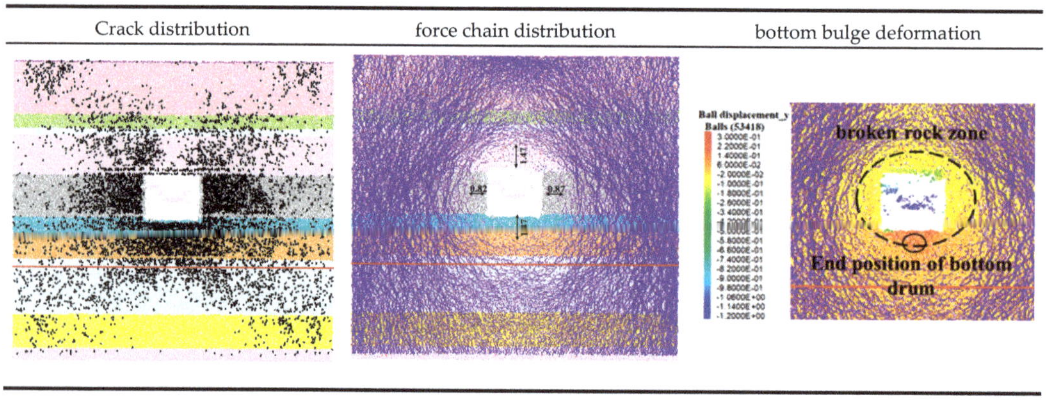

Figure 9. The rock pressure characteristics of the rock surrounding the 1232 transportation roadway, where a single joint is located 1 m below the top of the third rock layer on the bottom plate.

(3) Two Joints Located 1 m and 2 m Below the Top of the Third Floor Stratum

When two joints are located 1 m and 2 m below the top of the third floor stratum, the stress characteristics of the surrounding rock after excavation of the 1232 transportation roadway are shown in Figure 10. the original 4.18-m-thick siltstone is divided into three layers of 1 m, 1 m, and 2.18 m in thickness. This further reduces the composite floor beam's resistance to floor heave deformation, and theoretically leads to more extensive fracturing, larger loosening zones, and increased floor heave. However, simulation results show that the stress characteristics of the 1232 roadway's surrounding rock under the condition of two joints at 1 m and 2 m depth are consistent with those under a single joint at 1 m depth, showing no significant changes. The force chain distribution reveals that the horizontal compression between the two joints located at 1 m and 2 m depth forms a hinged structure. This increases the floor's bearing stress and effectively limits the downward propagation of floor heave, resulting in a maximum heave depth.

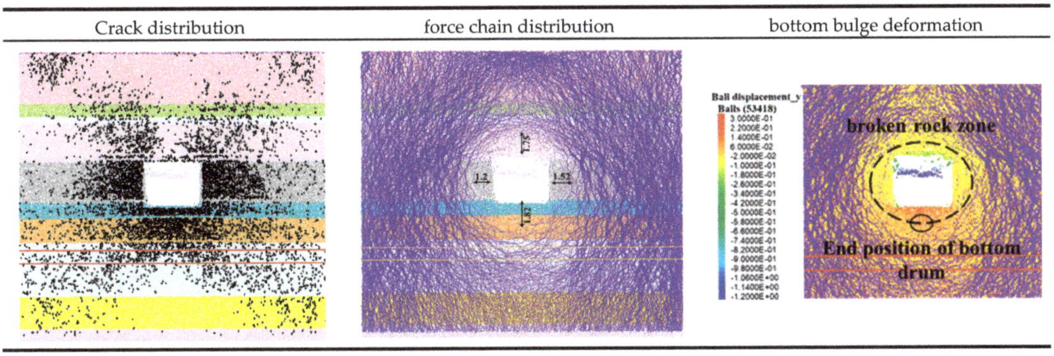

Figure 10. Double joints are located at 1 m and 2 m below the top of the third rock layer on the bottom plate. Characteristics of rock pressure in the rock surrounding the 1232 transportation roadway.

As a result, the extent of floor heave shows little difference compared to the case with a single joint at 1 m depth. The fractured rock mass between the two joints forms a compacted structure that constrains further heave, defining the maximum depth range of floor heave, as illustrated in Figure 11.

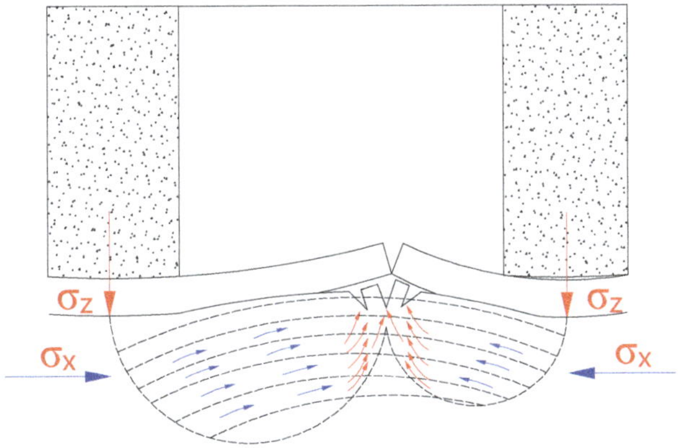

Figure 11. The compression structure of the fractured rock layer on the bottom plate restricts the deformation of the floor bulge.

As the joints extend deeper into the floor strata, when a joint is located in a layer close to the immediate floor, its influence on floor heave is minimal, since the first floor layer is already completely damaged. However, when the joint is located in deeper strata farther from the immediate floor, and the fractured rock fails to form a compacted, stable load-bearing structure, the extent of floor heave increases. If the fractured rock mass forms a load-bearing structure, it can reduce floor bending and deformation, effectively preventing further downward propagation of the heave. As a result, the maximum depth of floor heave is limited to the upper boundary of the bearing structure formed after rock failure.

4. Research on Slotting and Pressure Relief Control Technology

4.1. Floor Heave Control Scheme

Slotting and pressure relief in the floor can effectively alleviate stress concentration, improve the mechanical state of the surrounding rock, reduce support costs and maintenance frequency, and adapt flexibly to complex geological conditions. In the 1232 retreat roadway of Shuguang Coal Mine, the floor is primarily composed of mudstone, which has relatively low strength. Therefore, it is proposed that slotting and pressure relief be implemented under the original support conditions in the 1232 transportation roadway of Shuguang Coal Mine. To this end, a technologically advanced, safe, and efficient floor-slotting machine was designed and developed for field industrial trials, aiming to effectively mitigate floor heave and achieve engineering control objectives in the roadway.

4.2. Key Parameters for Slotting Control of Floor Heave in the 1232 Transportation Roadway of Shuguang Mine

To understand the control mechanism of slotting depth on floor heave deformation, three numerical models were developed: one where the slot penetrates the first floor layer, one that penetrates the second, and one that reaches the third. In all models, the slot width was set to 20 cm, as shown in Figure 12. The slotting width should be determined based on roadway production and technical constraints, and generally should not exceed 30 cm.

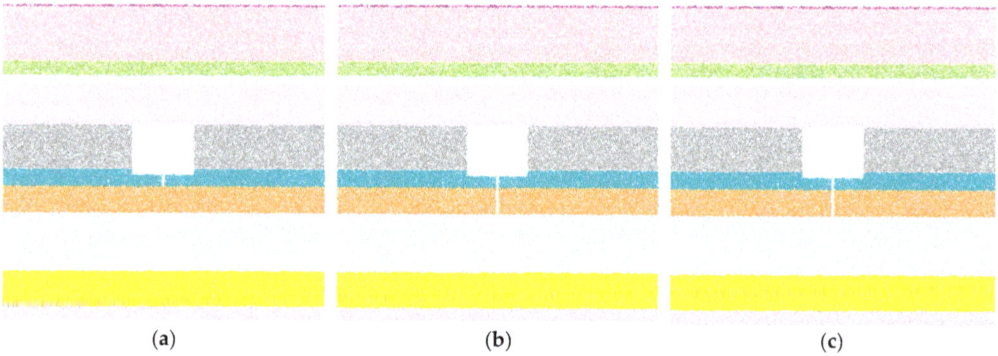

Figure 12. Numerical model for determining the depth of groove cutting in the bottom drum controlled by groove cutting control. (**a**) Slotting through the first floor stratum; (**b**) Slotting through the second floor stratum; (**c**) Slotting to 2 m below the top of the third floor stratum.

(1) Shallow Slotting (Penetrating the First Floor Layer)

When the slot penetrates the first floor layer, the stress characteristics of the surrounding rock after excavation of the 1232 transportation roadway are shown in Figure 13. Compared to the condition before slotting, stress release causes a similar number and pattern of fractures, with butterfly-shaped distributions primarily concentrated in the sidewalls and floor. The force chain distribution indicates that both cases share similar loosening zone characteristics: the sidewall loosening zone extends about 1.5 m, and the floor loosening zone reaches a depth of 2.5 m. Slotting does not alter the depth or degree of floor failure; the second layer of thick silty mudstone in the floor is completely damaged both before and after slotting. Prior to slotting, the maximum floor heave deformation exceeds 30 cm. After slotting, the created slotting space allows for the movement of fractured floor material, thereby reducing the maximum floor heave deformation to approximately 10 cm.

When the slot penetrates the first floor stratum, it lies entirely within the shallow fractured zone. While it contains the floor heave deformation of the shallow layers within the slotting space, it does not alter the movement characteristics of deeper fractured strata. Therefore, its overall effect on controlling floor heave deformation is limited.

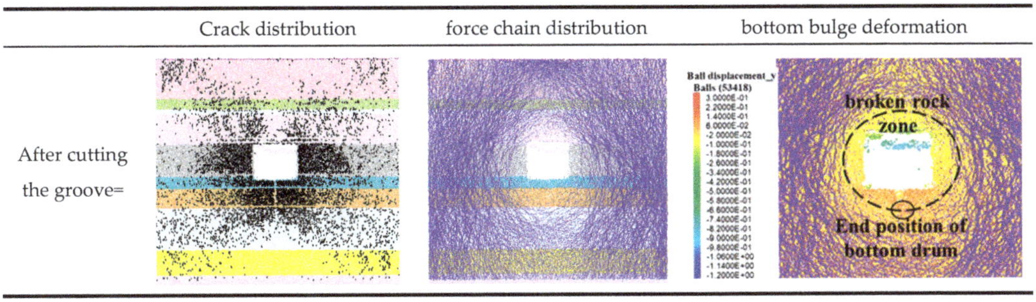

Figure 13. Cut through the first layer of rock layer 1232 of the bottom plate by slotting and the characteristics of rock pressure in the rock surrounding the transportation roadway.

(2) Intermediate Slotting (Penetrating the Second Floor Stratum)

When the slot penetrates the second floor stratum, the stress characteristics of the surrounding rock after excavation of the 1232 transportation roadway are shown in Figure 14. Compared to the pre-slotting condition, the fracture distribution still exhibits a butterfly-shaped pattern, mainly concentrated in the sidewalls and floor, with an increased number of fractures in the floor. The loosening zones in the sidewalls and roof of the 1232 roadway remain largely unchanged, while the loosened zone in the floor expands. As the slot cuts through the second floor stratum, this layer bears little to no load. The unloading range of the floor increases, and the primary load-bearing structure shifts from the second to the third floor stratum. The maximum floor heave deformation is reduced to approximately 10 cm.

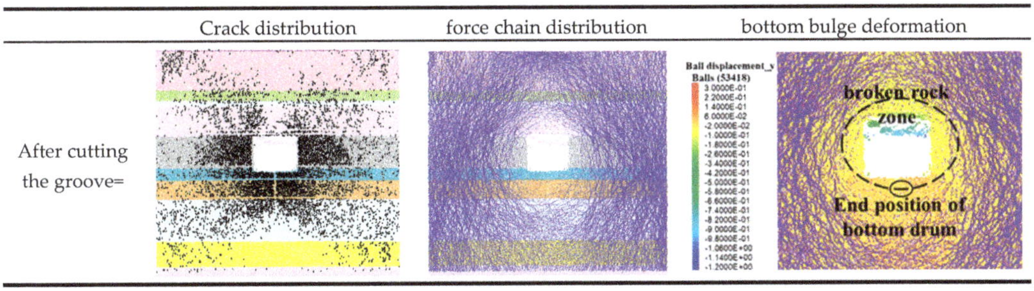

Figure 14. Cut through the bottom plate of the second layer of rock layer 1232 and the characteristics of rock pressure in the rock surrounding the transportation roadway.

The identical floor heave reduction observed under both shallow and intermediate slotting conditions arises because these methods partially relieve stresses and modify the load-bearing structure. However, due to inherent limitations in stratum strength and load-bearing structure distribution, they fail to fully eliminate deformation. In contrast, deep slotting achieves superior control by inducing more comprehensive stress redistribution and a downward shift of the load-bearing structure. This deeper intervention transfers loads to stable, less disturbed strata, thereby minimizing residual deformation through enhanced mechanical stability.

After the third floor stratum in the 1232 transportation roadway fails, a compacted load-bearing structure is formed, which effectively limits the further downward development of floor heave deformation. When the slot penetrates the second floor stratum, it provides sufficient space for floor deformation, thereby effectively controlling floor heave.

(3) Deep Slotting (2 m Below the Top of the Third Floor Stratum)

When the slot penetrates to 2 m below the top of the third floor stratum, the stress characteristics of the surrounding rock after excavation of the 1232 roadway are shown in Figure 15. Compared to the pre-slotting condition, the fractures still exhibit a butterfly-shaped distribution, primarily concentrated in the sidewalls and floor, with a noticeable increase in the number of floor fractures. The loosening zones in both sidewalls and the floor are significantly enlarged. Due to the deeper slotting, the floor unloading zone expands, enlarging the stress-reduction area in the sidewalls and shifting the primary floor bearing structure further downward. As a result, the maximum floor heave deformation is reduced to approximately 6 cm.

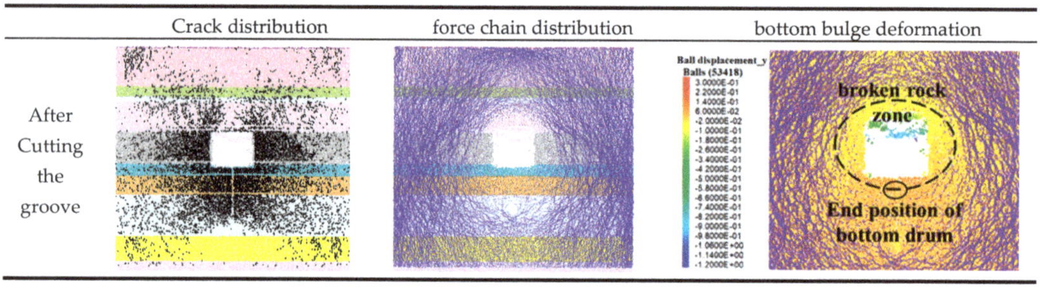

Figure 15. Cut through the top 2 m of the third layer of rock strata on the bottom plate, and analysis of the rock pressure characteristics of the surrounding rock in the 1232 roadway.

Greater slotting depths can alter the stress arch distribution in the sidewalls and floor, shifting the arches into deeper surrounding rock, which is beneficial for controlling floor heave. However, considering mining technical conditions, slotting efficiency, and equipment capabilities, excessively deep slots are not advisable. A reasonable depth should reach just above the main load-bearing structure in the floor to balance effectiveness and feasibility.

5. Summary and Conclusions

(1) The floor strata can be conceptualized as a composite beam structure. When subjected to vertical stress disturbances from the roof and coal pillars, the beam ends experience loads exceeding their bearing capacity, coupled with horizontal compressive forces induced by stratum movement. This results in flexural deformation and upward arching of the beam until failure, manifesting as layer-by-layer progressive failure. Floor heave in retreat roadways is governed by the stratified failure characteristics of composite floor strata. According to composite plate theory, the flexural deformation of individual floor layers depends primarily on three factors: stiffness, strength, and moment of inertia. The moment of inertia is predominantly influenced by stratum thickness and joint development.

(2) Enhanced strength and stiffness of floor strata constrain the failure and displacement of the sidewalls and roof, thereby improving the load-bearing capacity of the sidewalls, reducing the extent of the surrounding rock loosening zone, and mitigating floor heave deformation. The joint distribution within the floor strata critically affects the composite rock mass's resistance to heave. Shallow joints exhibit negligible impact due to the complete failure of shallow strata under stress redistribution. However, as joints extend into deeper strata, the composite system's resistance to heave diminishes, leading to amplified deformation. When joints penetrate beyond a critical depth, fractured strata form a compacted load-bearing structure that confines heave propagation, with maximum heave depth localized above this stabilizing structure.

(3) Shallow slotting within the fractured zone confines heave deformation to the slotting space but fails to alter the migration characteristics of deeper fractured strata, resulting in limited control efficacy. In contrast, deeper slotting modifies the stress arch distribution in the sidewalls and floor, redirecting stresses into deeper rock masses and significantly improving heave suppression. Considering practical constraints—such as mining technical conditions, slotting efficiency, and equipment capabilities—excessively deep slotting is not advisable. The optimal slotting depth should terminate immediately above the primary load-bearing structure, balancing control effectiveness and engineering feasibility.

(4) The PFC model with smooth joints successfully reproduced the full floor heave process—"deformation–separation–fracture–compaction"—overcoming the limitations of traditional continuum methods in simulating discontinuous failure. Comparative analyses of crack propagation and force chain evolution under varying joint distributions and slotting parameters confirm the advantages of discrete element methods in revealing rock mass micro-failure mechanisms, providing methodological insights for similar engineering studies.

Author Contributions: Conceptualization, T.L.; Formal analysis, T.L and B.H.; Software, T.L. and C.S. All authors have read and agreed to the published version of the manuscript.

Funding: This research received no external funding.

Data Availability Statement: The data presented in this study are available on request from the corresponding author.

Conflicts of Interest: The authors declare no conflicts of interest.

References

1. Xiong, Y.; Kong, D.Z.; Yang, S.L.; Wu, G.; Zuo, Y.; Cheng, Z. Comprehensive identification of coal wall stability in steeply inclined working faces using cloud model. *China Saf. Sci. J.* **2022**, *32*, 144–151.
2. Kang, H.P. *Mechanism and Prevention of Floor Heave in Soft Rock Roadways*; China Coal Industry Publishing House: Beijing, China, 1993.
3. Kang, H.P.; Fan, M.J.; Gao, F.Q.; Zhang, H. Deformation characteristics and support technology of roadways in ultra-deep coal mines over 1000 meters. *Chin. J. Rock Mech. Eng.* **2015**, *34*, 2227–2241.
4. He, M.C.; Zhang, G.F.; Wang, G.L.; Xu, Y.L.; Wu, C.Z.; Tang, Q.D. Control mechanism and application of floor heave in deep coal roadways. *Chin. J. Rock Mech. Eng.* **2009**, *28*, 2593–2598.
5. He, M.C.; Zou, Z.S.; Zou, Y.F. *Introduction to Soft Rock Roadway Engineering*; China University of Mining and Technology Press: Xuzhou, China, 1993.
6. He, M.C. *Theory and Practice of Soft Rock Roadway Support in Chinese Coal Mines*; China University of Mining and Technology Press: Xuzhou, China, 1996.
7. Li, S.Y. *Study on the Mechanism and Control of Floor Heave in Soft Rock Withdrawal Roadway*; Taiyuan University of Technology: Taiyuan, China, 2020.
8. Indraratna, B.; Nemcik, J.A.; Gale, W.J. Review and interpretation of primary floor failure mechanism at a longwall coal mining face based on numerical analysis. *Geotechnique* **2000**, *50*, 547–557. [CrossRef]
9. Xu, X.D. *Mechanism of Strong Ground Pressure Manifestation and Pressure Relief Control in Gob-Side Roadways of Extra-Thick Coal Seams in Caojiatan Mine*; Jilin University: Changchun, China, 2022.
10. Song, H.H.; Zhao, Y.X.; Jiang, Y.D.; Zhang, X.Z. Influence of coal heterogeneity on failure characteristics under uniaxial compression. *J. China Coal Soc.* **2017**, *42*, 3125–3132.
11. Wang, X.Q.; Kan, J.G.; Jiao, J.K. Mechanism and control practice of floor heave in high-stress soft rock roadways. *J. Min. Saf. Eng.* **2017**, *34*, 214–220+227.
12. Zhu, X.X. *Study on Control Mechanism of Floor Heave in Dynamic Pressure Roadways of Yuwu Coal Mine*; China University of Mining and Technology: Xuzhou, China, 2019.
13. Li, Y.J. Analysis of floor heave mechanism and control technology in mining-affected roadways. *Coal Chem. Ind.* **2017**, *40*, 31–33.
14. Guo, B.H.; Lu, T.K. Mechanism of floor heave in deep roadways and slotting control technology. *J. Min. Saf. Eng.* **2008**, *1*, 91–94.
15. Hou, C.J. Key technologies for controlling surrounding rock in deep roadways. *J. China Univ. Min. Technol.* **2017**, *46*, 970–978.

16. Mo, S.; Tutuk, K.; Saydam, S. Management of floor heave at Bulga Underground Ope rations—A case study. *Int. J. Min. Sci. Technol.* **2019**, *29*, 73–78. [CrossRef]
17. Ren, Z.; Zhang, C.; Wang, Y.; Yang, S.; Li, Q. Collaborative movement characteristics of overlying rock and loose layer based on block–particle discrete-element simulation method. *Int. J. Coal Sci. Technol.* **2025**, *12*, 17. [CrossRef]
18. Han, P.; Wang, K.; Pang, J.; Ji, X.; Zhang, C. Response properties of geometries of coal penetrating fracture on seepage behavior. *Int. J. Min. Sci. Technol.* **2025**, *35*, 191–211. [CrossRef]
19. Zhang, C.; Chen, Y.; Wang, Y.; Bai, Q. Discrete element method simulation of granular materials considering particle breakage in geotechnical and mining engineering: A short review. *Green Smart Min. Eng.* **2024**, *1*, 190–207. [CrossRef]
20. Han, L.; Hou, S.Y.; Zhang, X.K. Mechanism and control technology of floor heave in steeply inclined coal seam roadways. *Saf. Coal Mines* **2018**, *49*, 209–212+217.
21. Wei, J.Q.; Xu, Y.; Xu, Z.Q.; Li, D.X.; Cui, J.Q.; Wang, X.L. Energy transfer law and control during floor heave in roadways. *Met. Mine* **2021**, *2*, 29–35.
22. Li, L.; Duan, X.B. Study on slotting control technology for floor heave in Yitang coal mine roadways. *Shanxi Coal* **2020**, *40*, 1–4.
23. Zhou, R. Pressure relief support technology for return air roadways in deep mine with soft surrounding rock. *Shandong Coal Sci. Technol.* **2020**, *4*, 79–82.
24. Yang, R.; Hao, B.Y. *Pressure Relief Control Technology for Floor Heave in Mining Roadways*; Taiyuan University of Technology: Taiyuan, China, 2022.

processes MDPI

Article

Study on Rational Roadway Layout and Air Leakage Prevention in Shallow Close-Distance Coal Seam Mining

Ying Liu

School of Management Science and Engineering, Beijing Information Science and Technology University, Beijing 100083, China; liuyingquality@163.com

Abstract: To address the issues of roadway instability and severe air leakage in goaf areas during overlapping coal pillar mining in shallow multi-seam coalfields, this study takes the 22,209 working face of Huojitu Shaft in the Shendong Daliuta Mine as the research object. Using the discrete element method (DEM), the optimal layout of roadways in the lower coal seam and the corresponding evolution of overburden fractures were simulated. In addition, the effectiveness of goaf backfilling in controlling overburden air leakage channels was analyzed and verified. The results indicate that the width of coal pillars in the upper seam should be greater than approximately 23 m to ensure that roadways remain in a stress-stable zone. Roadways in the lower seam should be horizontally arranged within a range of 35–55 m from the center of the overlying coal pillar. This layout effectively avoids placing the roadway beneath the high-stress concentration zone or the pressure-relief area of the goaf. After mining the upper coal seam, the overburden collapse zone takes on a "trapezoidal" shape, and mining-induced fractures develop upward to the surface, forming vertical and inclined fracture channels that penetrate to the surface, resulting in severe air leakage in the goaf. Following the mining of the lower seam, the interlayer strata are completely fractured, leading to secondary development of fractures in the overlying old goaf. This results in the formation of a connected fracture network spanning from the surface through the seam goaf linkage. Implementing goaf backfilling measures significantly reduces the vertical settlement of the overburden, prevents the formation of through-layer air leakage channels, and effectively mitigates interlayer air leakage problems during lower-seam mining.

Keywords: shallow multi-seam; roadway layout; overburden fracture; air leakage; goaf backfilling; roadway instability; DEM

1. Introduction

Shallow coal resources are widely distributed in northern China. Their extraction frequently triggers large-scale surface subsidence, leading to ecological degradation, frequent geological hazards, and environmental deterioration [1–3]. Multi-seam mining operations in shallow regions are particularly challenging due to narrow inter-seam spacing and secondary mining disturbances, which complicate the stress distribution of overlying strata and exacerbate air leakage dynamics. Under these conditions, lower-seam mining faces are prone to hazards such as roadway instability and collapse, frequent hypoxia in working faces, and spontaneous combustion in goaf areas [4–6]. Consequently, investigating the optimal positioning of mining roadways and the fracture development characteristics in overlying strata during shallow multi-seam extraction is critical for ensuring roadway stability and preventing gas-related disasters.

Extensive research has been conducted on the rational layout of close-distance coal seam roadways from mechanical perspectives, including overburden stress fields, roadway deformation, and spatial arrangement [7,8]. Wang et al. [9] introduced the Dempster–Shafer evidence theory to develop a system optimization model for coal mine roadway layout, establishing a comprehensive factor framework for spatial arrangement. Zhu et al. [10] analyzed roadway deformation patterns at the 2112 working face in Liangjia Coal Mine and proposed a BBSCS method to reinforce roadway roofs and coal seams. Wei et al. [11] optimized the rational layout of lower-seam roadways by comparing stress distribution characteristics in roadways at varying positions. Zhang et al. [12] elucidated the pressure-relief mechanism of roadways beneath goaf areas and investigated stress compositions in the roof, floor, and surrounding strata. Wei et al. [13] analyzed stress variations in lower seams influenced by residual coal pillars and explored optimal roadway layouts for lower-seam mining. Simultaneously, understanding the initiation, distribution, and evolution of overburden fractures during mining is essential for characterizing air leakage pathways [14,15]. Researchers have extensively studied fracture development patterns in shallow multi-seam mining overburden from diverse perspectives, yielding valuable insights [16,17]. Pan et al. [18] investigated the vertical progression of overburden fractures in shallow coal seams and numerically simulated their spatial evolution during mining. Wang et al. [19] spatially analyzed fracture propagation, surface subsidence, and strata collapse, clarifying the transmission mechanisms of overburden fractures. Xu et al. [20] used 3DEC software to simulate the spatiotemporal evolution of overlying strata fractures during coal mining, studied the fractal characteristics of the overlying strata caused by mining, and revealed different time series of fractal properties and spatial fracture patterns of the overlying rock mass. Wang et al. [21] simulated continuous multi-seam mining processes, revealing that the dynamic interaction between collapsed zones and surrounding rock dominates the evolution of overburden load-bearing structures. Zhang et al. [22] proposed a Space Sky Surface (3S) integrated system that effectively monitors the movement of the cover layer in high-intensity mining areas in real time. However, comprehensive research on the reasonable layout of shallow multi seam mining tunnels and the corresponding development of air leakage cracks in overlying rock mining is not yet complete [23,24].

Therefore, this study establishes a DEM model for shallow multi-seam mining based on the 22,209 working face in Huotuji Mine. First, post-mining stress distributions in the upper seam were analyzed. The stress characteristics of lower-seam roadways at varying horizontal positions relative to overlying residual coal pillars were investigated, enabling the proposal of optimal roadway layouts under these conditions. Subsequently, the overlying strata collapse and fracture development patterns during contiguous multi-seam extraction were simulated, identifying the evolutionary mechanisms and spatial distributions of primary air leakage channels. Finally, the efficacy of goaf backfilling in mitigating strata subsidence and suppressing fracture development was comparatively evaluated.

2. Review

The Huojitu mine, part of the Shendong-Daliuta coal mining area, is situated in Shenmu City, Shaanxi Province. The mine contains two coal seams: #12 and No. 22, with the latter being the current primary mining target. The #22 coal seam has an average burial depth of 120.28 m and a thickness of 4–5 m. Overlying this, the #12 coal seam exhibits an average burial depth of 99.73 m and a thickness of approximately 6 m. The inter-seam spacing between the #22 and #12 coal seams averages 30.64 m, and they can thus be classified as a close-distance coal seam group. As the #12 seam has been largely mined out, abundant residual coal pillars remain in the goaf areas. This configuration poses two critical engineering challenges: (1) optimizing roadway layout in the #22 seam and (2) preventing surface air leakage. This study focuses on the #22 coal seam's 22,209 fully mechanized mining face. The burial depth of this working face ranges from 67.2 to 116.9 m (average: 70.43 m), with a coal seam thickness of 4.2 m and dip angles of 1–3°. The immediate roof and floor of the working face consist of siltstone. Above the face lie the mined-out areas of the #12 coal seam's 12,201 and 12,203 working faces, with an inter-seam spacing of approximately 24.5 m. The stratigraphic column of the 22,209 working face is illustrated in Figure 1.

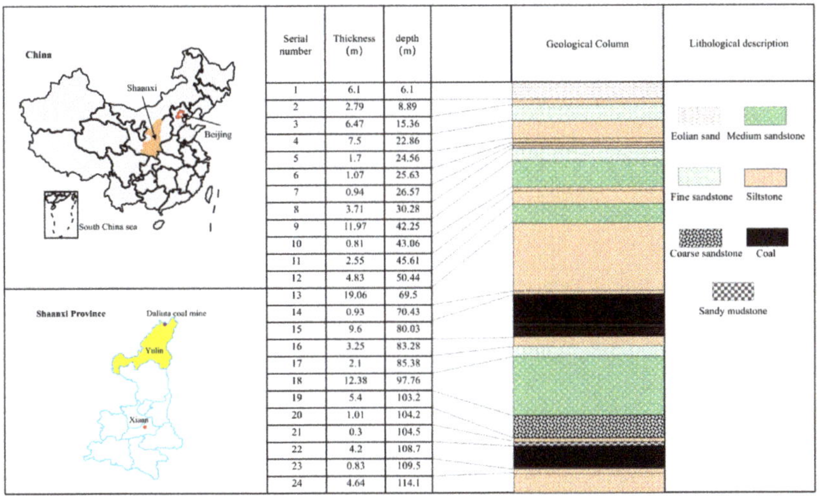

Figure 1. Column chart of coal-bearing strata in the 22,209 working face of Huojitu Well.

3. Numerical Modeling

A two-dimensional discrete element method (DEM) numerical model for shallow multi-seam mining was developed using PFC^{2D} (5.0) software, based on stratigraphic data from the 22,209 working face in Huojitu Mine, Shendong mining area (Figure 2). The model spans 350 m in length and 100 m in height, comprising 22 strata layers. Particles were generated with a minimum diameter of 0.35 m and a maximum-to-minimum size ratio of 1.66, totaling 43,585 discrete elements. Particle movement was confined within the cuboid boundaries of the model. A full-seam-height mining method was implemented, sequentially extracting the upper and lower coal seams from left to right. The model was solved at 1 m mining increments.

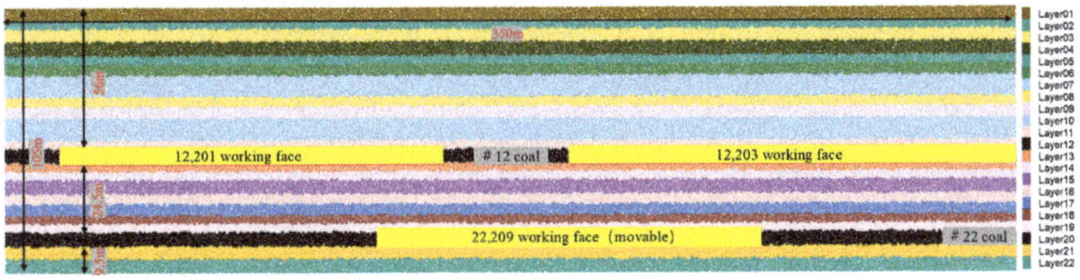

Figure 2. DEM model of the coal measure strata for the 22,209 working face.

To accurately capture material mechanical behavior, mesoscopic parameters in the particle flow model were calibrated to match macroscopic rock properties. The calibration process followed Wang et al. [25], focusing on two parameter categories: (1) particle properties: density, friction coefficient, Young's modulus, and normal-to-shear stiffness ratio; (2) parallel bonds: Young's modulus, normal-to-shear stiffness ratio, tensile strength, shear strength, and friction angle. Calibration was validated through agreement between simulated and experimental stress–strain curves and failure modes, ensuring DEM parameters effectively represented macroscopic coal-rock behavior. The calibrated mesoscopic parameters for each stratum are summarized in Table 1, Figure 3 shows the calibration results of coal and fine-grained sandstone.

Table 1. Microparameters of calibrated DEM model strata.

Property	Parameter	Mudstone	Sandy Mudstone	Fine-Grained Sandstone	Medium-Grained Sandstone	Coarse-Grained Sandstone	Siltstone	Coal
particle	Density (Kg/m^3)	2300	2400	2600	2500	2400	2400	1650
	R_{min}				1.2			
	R_{max}/R_{min}				1.3			
	E_c (GPa)	7.23	11.25	15.87	13.42	9.75	18.88	3.34
	μ	0.5	0.5	0.6	0.5	0.5	0.55	0.45
	k_n/k_s	2.0	2.0	1.8	1.8	2.0	1.7	2.15
bond	$\overline{E_c}$ (GPa)	7.23	11.25	15.87	13.42	9.75	18.88	3.34
	$\overline{\sigma_c}$ (MPa)	11.05	14.25	24.63	16.62	7.51	18.82	4.21
	$\overline{\tau_c}$ (MPa)	11.15	14.25	24.53	16.83	7.51	18.92	4.31
	$\overline{k_n/k_s}$	2.0	2.0	1.8	1.8	2.0	1.7	2.15
	φ	35	35	40	35	35	40	42

Note: D is the density; E_c is the Young's modulus of particle contact; $\overline{E_c}$ is the parallel bonding Young's modulus; μ is the coefficient of friction of particles; k_n/k_s is the particle stiffness ratio; $\overline{\sigma_c}$ is the parallel bonding tensile strength; $\overline{\tau_c}$ stands for Parallel Bond Shear Strength; φ is the parallel bonding friction angle.

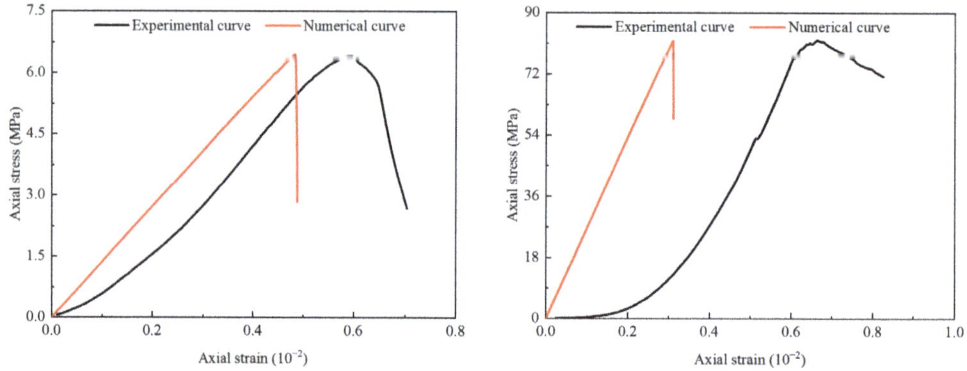

(**a**) Coal. (**b**) Fine-grain sandstone.

Figure 3. The contrast of stress–strain curves between numerical simulations and the laboratory experiment.

4. Results

4.1. Optimal Roadway Layout

4.1.1. Rational Width of Section Coal Pillars in Upper-Seam Mining

Figure 4 illustrates the vertical stress distribution in the stope following extraction of the #12 coal seam's 201 working face. Since our research focuses on the stress distribution during the mining process of the 201 working face under shallow buried multi-coal-seam geological conditions, without involving specific external factors, we assume that the rock is isotropic. During mining of the 201 working face, the original stable stress state was disrupted, leading to stress redistribution and localized concentration around the excavated area. Post mining of the upper seam, stress release occurred in the floor strata beneath the goaf, forming a low-stress zone. Conversely, stress concentration developed near the lateral coal pillars of the lower close-distance seam due to overlying goaf pressure, propagating downward along the seam floor. Simultaneously, distinct stress concentration zones emerged on both sides of the 201 goaf, with vertical stress gradually recovering to in situ levels as distance from the goaf increased. As shown in Figure 4, vertical stress in the outer coal seam decreased progressively with distance from the goaf boundary, stabilizing at in situ stress levels approximately 23 m from the goaf edge. Thus, sufficient pillar width must be reserved between roadways in the same seam, and new roadways should avoid stress concentration zones induced by adjacent mining activities. For the Huojitu Mine, section coal pillars should exceed 23 m in width. Current design for Huojitu Mine sites adopts 30 m-wide pillars, which satisfies this criterion.

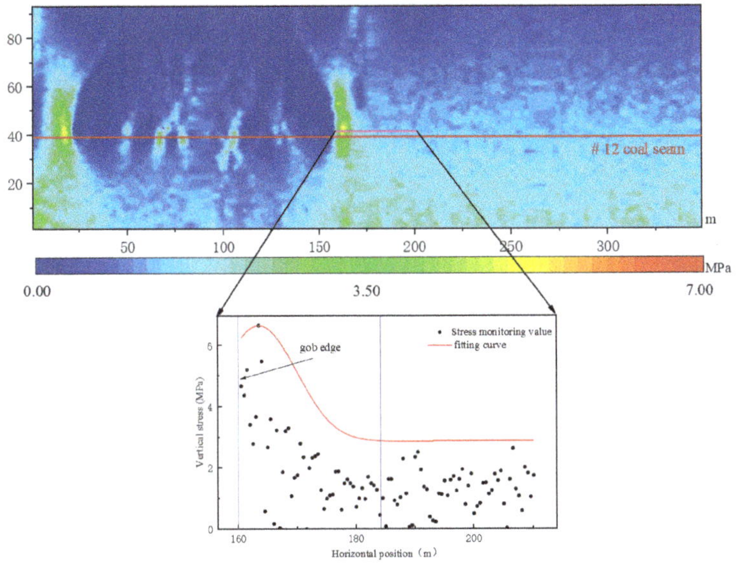

Figure 4. Stress distribution in mining area after 201 working face mining.

4.1.2. Optimal Positioning of Lower-Seam Roadways

As shown in Figure 5, mining of the overlying #12 coal seam induced significant stress concentration in section coal pillars. This resulted in a trapezoidal stress distribution pattern in the underlying strata, where the spatial extent of stress concentration expanded with depth while its magnitude diminished. Stress distribution in the #22 coal seam beneath the #12 seam was monitored (Figure 6). The stress magnitude peaked directly below the pillar center, exhibiting a symmetrical decline toward both sides. At 35 m from the pillar center, stress levels fell below in situ values due to pressure relief from the overlying goaf.

Figure 7: Vertical stress variations in the roadway surrounding rock at different distances from the #12 coal pillar center. At 5 m from the pillar center (directly beneath the section coal pillar), both left- and right-side surrounding rock exhibited elevated vertical stresses. At 15 m (adjacent to the pillar's right edge), the left-side strata experienced marginally increased vertical stress, while the right-side strata, located within the pressure-relief zone of the 203 goaf, showed a significant stress reduction. At 35 m (the critical boundary of the #12 pillar's influence zone), roadway surrounding rock stress decreased substantially. Beyond this distance, stress continued to decline slightly but at a diminishing rate. Notably, excessive proximity to the upper goaf interior risks overburden stress accumulation, significantly increasing roadway loads. Therefore, roadways should be positioned horizontally within 35–55 m of the residual upper pillar center to balance stress mitigation and operational safety.

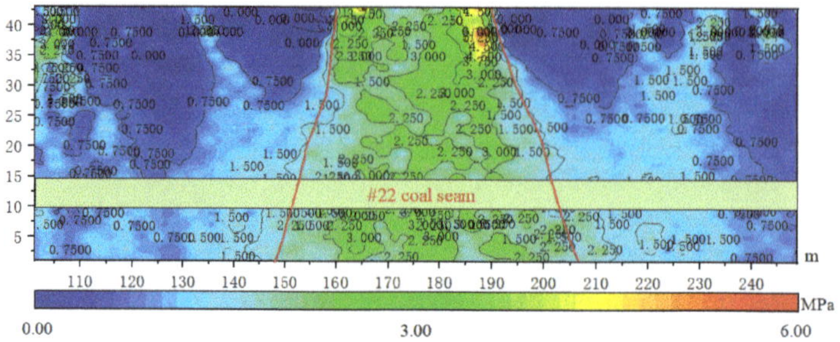

Figure 5. Stress distribution in the stope after mining of 201 and 203 working faces.

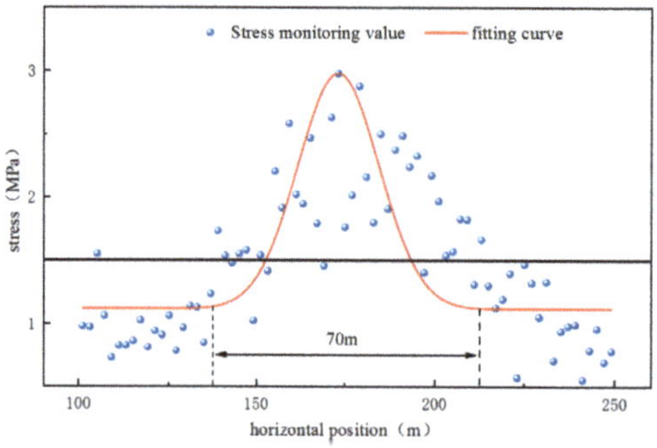

Figure 6. Stress distribution in the #22 coal seam after mining of the 201 and 203 working faces.

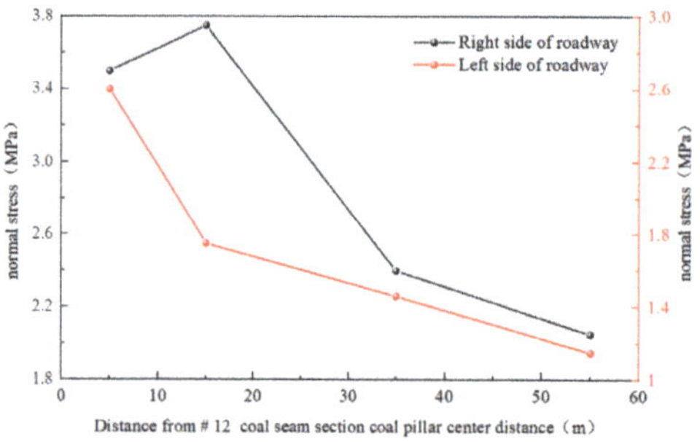

Figure 7. Vertical stress variations in roadway surrounding rock at distances from the #12 pillar center.

4.2. Fracture Development Patterns

4.2.1. Upper-Seam Mining

Figure 8: Simulated overburden collapse during mining of the #12 coal seam's 201 working face. As shown in Figure 8a, when the 201 working face advanced to 56 m, the immediate roof (Layer 11) experienced initial collapse, triggering the first weighting event. Mining-induced fractures propagated upward, with collapsed zones exhibiting higher porosity. The main roof (Layer 10) remained largely unaffected, retaining lower porosity. At 74 m advancement, failure of the key stratum (Layer 10) induced the first periodic weighting of the main roof, with a weighting interval of 18 m. Collapsed zones adopted a trapezoidal geometry, accompanied by pronounced lateral fractures. Vertical fractures emerged in central goaf areas, extending upward to Layer 07, while high-porosity zones and bed separation cracks dominated goaf margins. During the third periodic weighting at 105 m, the key stratum (Layer 07) prevented fracture propagation to overlying strata (Layers 07–05). Cyclic collapse occurred between Layers 11 and 08, expanding horizontally rather than vertically. The fourth periodic collapse at 121 m caused vertical expansion of collapsed zones through Layers 07–05. Microscopic fractures proliferated in Layers 04–02, with porosity maps revealing interconnected vertical fractures along goaf margins, forming inclined fracture channels extending to the bed separation zone between Layers 05 and 04. At 138 m advancement, pronounced bending subsidence occurred in upper Layer 05. Minor bed separation between Layer 02 and surface Layer 01 connected preexisting inclined fracture channels, enabling fracture propagation to the surface. This suggests potential air leakage pathways from the #12 goaf to the surface, allowing atmospheric air ingress. Throughout the mining process, periodic weighting events recurred at intervals of 13–20 m.

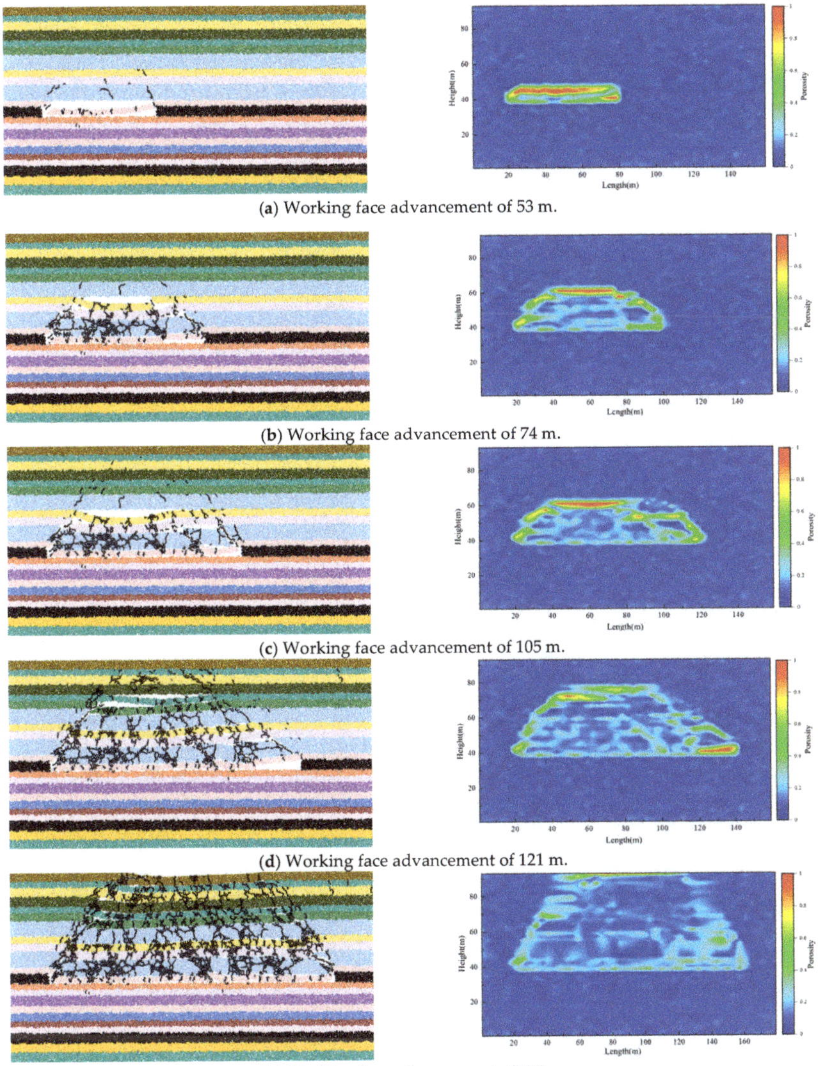

(a) Working face advancement of 53 m.

(b) Working face advancement of 74 m.

(c) Working face advancement of 105 m.

(d) Working face advancement of 121 m.

(e) Working face advancement of 138 m.

Figure 8. Overburden collapse characteristics during excavation of Panel 201 in #12 coal seam.

Figure 9 illustrates the comprehensive overburden collapse pattern, fracture development, and air leakage pathways following mining operations at Panels 201 and 203 in #12 coal seam. Based on previous analysis of rational roadway layout, a 30 m inter-panel coal pillar was established between Panels 201 and 203. The overburden collapse mechanism at Panel 203 followed patterns similar to those observed at Panel 201 (as detailed in prior excavation analysis) and will not be reiterated here. As shown in Figure 9, mining-induced fractures propagated to the ground surface following completion of extraction operations in #12 coal seam.

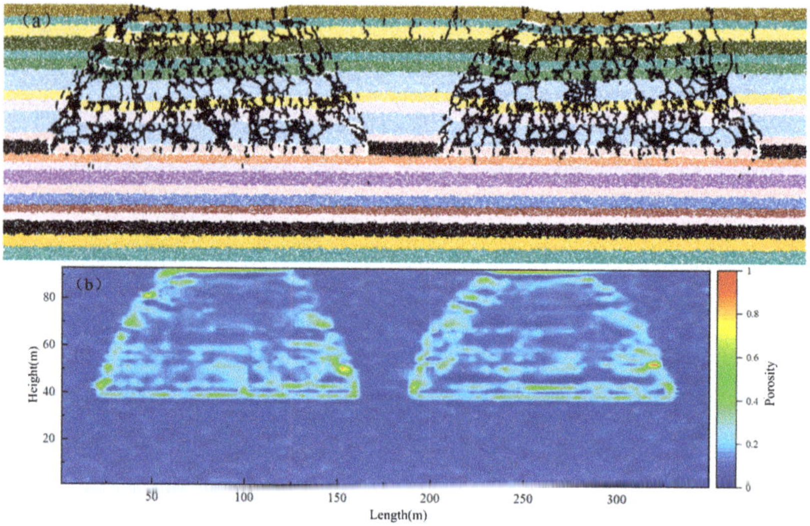

Figure 9. Mining-induced fractures and air leakage pathways in overburden after extraction of #12 coal seam. (**a**) Overburden collapse after the mining of the 201 and 203 faces; (**b**) The development of air leakage channels after the mining of the 201 and 203 faces.

4.2.2. Lower-Seam Mining

Mining of the #22 coal seam's 209 working face commenced after completion of the overlying #12 seam. Based on optimal roadway layout analysis, the 209 working face roadway was positioned 45 m horizontally from the center of the upper section coal pillar. Figure 10 depicts the overburden collapse during mining of the 209 working face. At 52 m advancement, Layers 19–17 fractured and collapsed, while mining-induced cracks developed in Layers 16–13. High-porosity "oblique zones" formed along both sides of the goaf, yet porosity did not extend to Layer 16 or the interburden strata, indicating no interconnected air leakage pathways between seams at this stage. At 70 m advancement, the face traversed the overlying section coal pillar. Excessive interburden stress caused complete fracture of the interlayer strata. Vertical cracks propagated through the central and peripheral zones of the goaf, connecting to the abandoned #12 goaf. This established cross-seam air leakage pathways, allowing oxygen-depleted gas from the #12 goaf to flow downward into the 209 working face goaf. Porosity distribution maps confirmed that high-porosity zones extended to the #12 goaf. Notably, the high-porosity oblique zones on both sides of the section pillar became interconnected, enabling surface air leakage to penetrate both upper and lower goaf areas through this channel. Cumulative mining thickness increased overburden subsidence, inducing a "reactivation effect" on the abandoned upper goaf. Fractured strata in the upper goaf underwent secondary crack development, with microscopic fractures proliferating and overall porosity rising significantly. At 87 m and 110 m advancements, the interburden experienced its 2nd and 4th periodic fractures, respectively. The influence zone of lower-seam mining expanded progressively, with porosity increases spreading across the abandoned upper goaf. Simultaneously, strata above the section coal pillar fractured extensively, accompanied by marked porosity elevation.

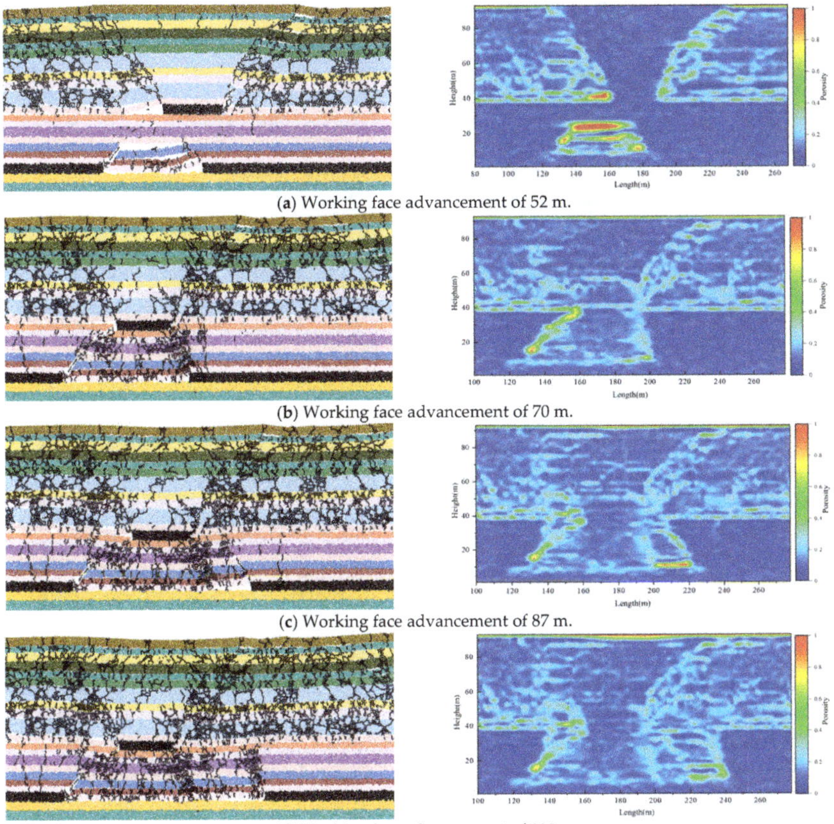

(a) Working face advancement of 52 m.

(b) Working face advancement of 70 m.

(c) Working face advancement of 87 m.

(d) Working face advancement of 110 m.

Figure 10. Overburden collapse characteristics during excavation of Panel 209 in #22 coal seam.

Figure 11 illustrates the comprehensive fracture development and air leakage pathway formation in stabilized overburden following extraction at Panel 209 of #22 coal seam. Severe overburden damage occurred after mining the lower #22 coal seam. Comparative analysis with Figure 9 reveals a significant increase in microfracture density and overall porosity within the overburden. Numerous, intricate air leakage pathways developed between the ground surface and the goaf area of #22 coal seam. The primary leakage channels were identified as oblique fractures along the goaf margins and zones between periodic overburden breakages.

Following the completion of lower-coal-seam extraction, interconnected vertical fracture zones were formed in the interlayer region, exhibiting an average width of 35 cm and a maximum vertical development height of 82 m. These fracture zones penetrate into the overlying goaf area. Symmetrically distributed oblique fractures were observed along the goaf margins, with a width of 25–30 cm on each side, extending over 80 m along the coal seam strike.

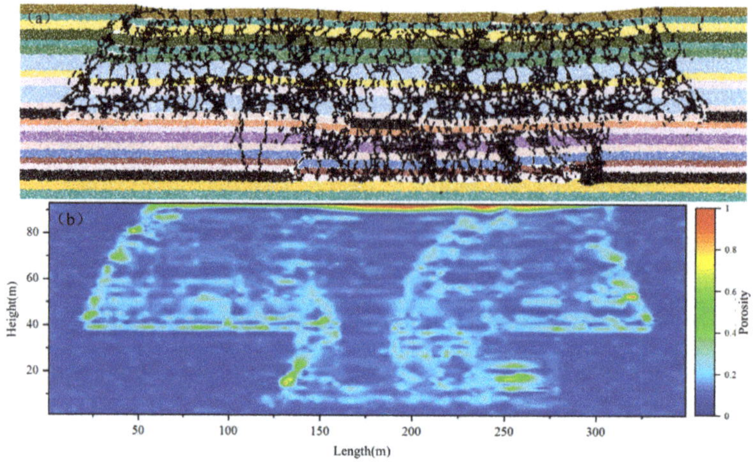

Figure 11. Distribution of overburden fractures and air leakage pathways after completion of Panel 209 excavation in #22 coal seam. (**a**) Conditions of overburden collapse after the mining of the lower coal seam is completed; (**b**) The development of the air leakage channel after the mining of the lower coal seam is completed.

5. Discussion

5.1. Fracture Development Control Measures

As established, thin interburden layers and substantial cumulative mining thickness in shallow, closely spaced coal seams induce complete interlayer failure and secondary activation of overlying abandoned goafs. This process intensifies overburden fragmentation, accelerates fracture/leakage pathway development, and amplifies surface-to-goaf air leakage. The utilization of solid waste materials for underground backfilling has emerged as a pivotal strategy in sustainable coal mining practices. Unfilled goafs created by coal extraction generate substantial voids, triggering sequential subsidence, fracturing, and collapse of overlying strata from lower to upper levels, accompanied by drastic stress redistribution and displacement. Fractured strata reorganize during collapse, creating extensive bed separations, voids, and fractures. These unstable structures pose critical safety risks while severely damaging overlying aquifers and surface ecosystems. Solid waste backfilling significantly modifies overburden deformation patterns and structural responses, effectively reducing strata damage and fundamentally altering the spatiotemporal evolution of fractures. A numerical model simulating fracture development in overlapping shallow coal seams was implemented, incorporating backfill blocks (5 m intervals, 3 m height, 60% backfill ratio) during lower-seam panel advancement. This simulation investigated overburden failure mechanisms, fracture propagation, and leakage pathway formation under backfilled conditions.

Figure 12 presents simulated overburden collapse patterns and fracture generation processes during backfill mining at Panel 22,209 of Huojitu Mine.

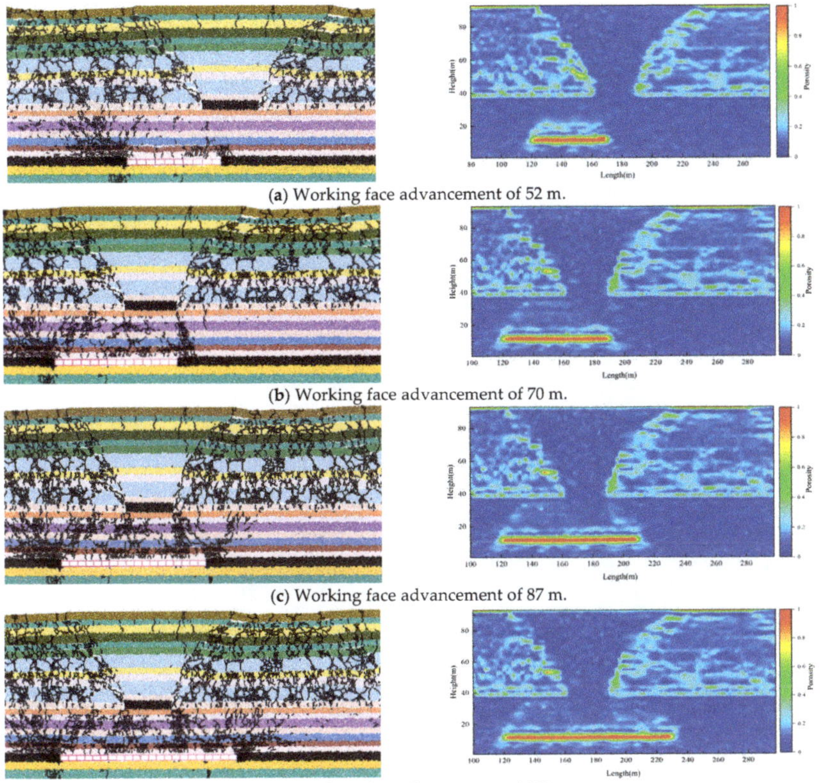

(a) Working face advancement of 52 m.

(b) Working face advancement of 70 m.

(c) Working face advancement of 87 m.

(d) Working face advancement of 110 m.

Figure 12. Overburden collapse characteristics during excavation of Panel 209 in #22 coal seam.

Figure 12a reveals that initial overburden fracture and collapse occurred at a panel advance of 52 m, with pronounced subsidence observed in Layers 19 to 17. The presence of backfill blocks substantially mitigated deformation severity. Comparative analysis with Figure 10a demonstrates significant reduction in interlayer bed separations and oblique fractures along goaf margins under backfilled conditions. Microscopic fractures propagated upward to Layers 16–14. At 70 m advancement, microfractures extended to the floor of Coal Seam #12, resulting in complete interlayer failure. Backfill blocks effectively restricted settlement height of fractured overburden. Vertical high-porosity pathways were constrained, with only multiple transverse interlayer channels developing (porosity ~0.2). Comparison of Figures 10b and 12b indicates markedly reduced porosity in fractured overburden and non-interconnected leakage pathways between upper and lower goafs under backfilling. Porosity remained stable at ~0.2 during subsequent advances (87 m and 110 m), though spatial propagation occurred in both vertical and lateral directions. Oblique leakage pathways along abandoned upper goafs progressively interfaced with lower channels. Nevertheless, compared to unfilled conditions, both pathway development intensity and leakage magnitude were significantly reduced.

Figure 13 illustrates the spatial distribution of overburden fractures and air leakage pathways following mining completion at Panel 209 in coal seam #22 under backfilled goaf conditions. Figure 13a reveals sustained high fragmentation intensity in the backfilled goaf overburden, particularly within periodic weighting zones where microfracture clustering is pronounced. However, backfill blocks significantly reduced overburden settlement height, enabling recompaction of fractured strata. Consequently, overall porosity remained stable, with limited connectivity to overlying goafs through low-porosity leakage pathways (high airflow resistance). Comparative analysis of Figures 8, 10 and 12 demonstrates that lower-seam backfilling effectively mitigates reactivation effects on overlying abandoned goafs during mining. Fracture density and porosity distribution in overlying legacy goafs showed negligible variation under lower-seam backfilled mining conditions. The specific changes in porosity are shown in Table 2.

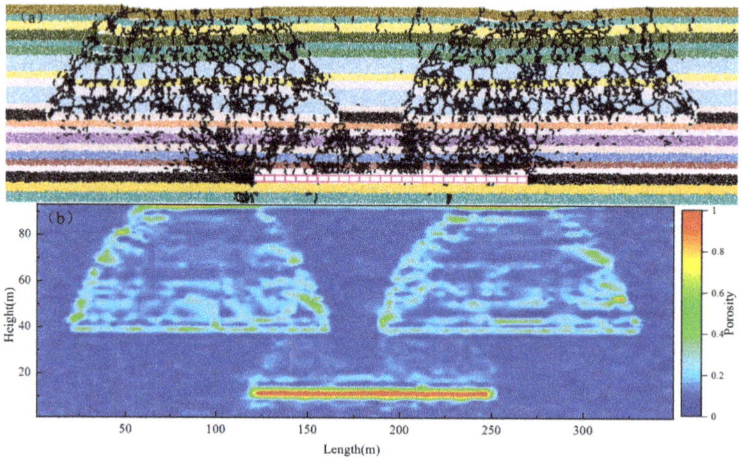

Figure 13. Distribution of overburden fractures and air leakage pathways after completion of mining under backfilled goaf conditions at Panel 22,209, Huojitu Mine. (**a**) Conditions of roof collapse after backfilling of the goaf; (**b**) Development of air leakage channel after backfilling of the goaf.

Table 2. Table of changes in porosity before and after filling.

Advancing Distance	Observation Indicators	Before Filling	After Filling	Drop
52 m	Interlayer fissure	0.25–0.30	0.15–0.18	40–42%
0 m	Vertical through passage	0.35–0.40	≤0.20	43–50%
	Horizontal interlayer channel	0.28–0.32	0.18–0.22	33–36%
87 m	Periodic break zone	0.38–0.45	0.20–0.25	43–47%
110 m	Activation area of old goaf	0.40–0.50	0.18–0.22	55–60%

5.2. Settlement Characteristics

Vertical displacement of marker particles at varying elevations above upper- and lower-seam roofs was monitored pre- and post-mining. Figure 14 presents settlement curves for distinct overburden strata following lower-seam extraction. As shown in Figure 14a, post-mining settlement curves exhibit a concave profile with steep flanks. Proximal strata experienced deeper subsidence and more severe collapse, attenuating with increasing distance from the seam. Distant strata displayed reduced subsidence magnitude and spatial extent due to load-bearing effects from collapsed lower layers. Collapse completeness correlated inversely with vertical distance from the mined seam.

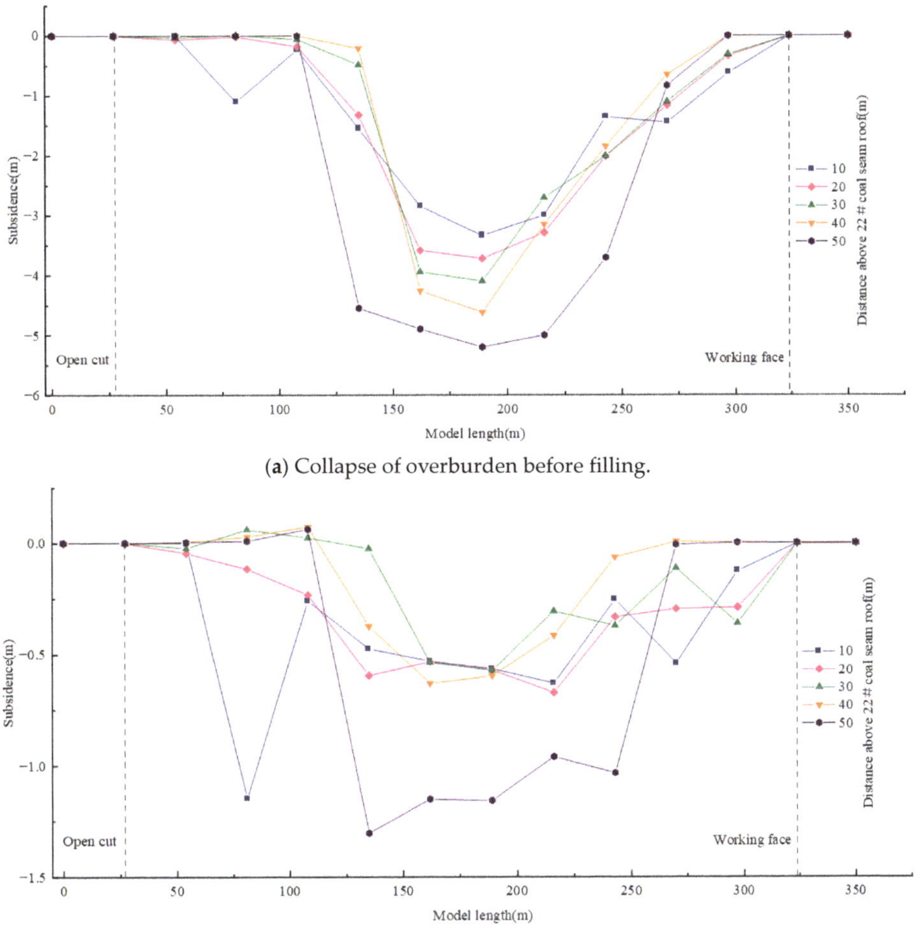

(a) Collapse of overburden before filling.

(b) Collapse of overburden after filling.

Figure 14. Settlement curve of coal seam mining strata before and after filling.

Figure 14b depicts settlement patterns under backfilled conditions, revealing substantial modifications compared to unfilled scenarios in shallow multi-seam mining. First, backfilling reduced maximum settlement magnitudes across all strata through enhanced recompaction, flattening the concave profile. Second, settlement curves exhibited minor fluctuations rather than abrupt discontinuities at profile edges. Notably, near-surface strata maintained minimal deformation, demonstrating backfilling's efficacy in controlling surface subsidence and leakage pathway development.

5.3. Field Application

Based on research findings, the following technical guidelines are proposed for adjusting mining plans: (1) Pillar width optimization: strictly maintain the residual pillar width in the upper seam ≥23 m to ensure lower-seam roadways remain within the stress-stable zone outside the pillar projection. (2) Roadway layout selection: lower-seam roadways should be horizontally arranged within 35–55 m on both sides of the overlying pillar centerline. Real-time monitoring using borehole stress gauges during construction is required to avoid entering high-stress concentration zones or goaf pressure-relief areas. (3) Backfill reinforcement: prioritize paste backfill technology for upper-goaf treatment, requiring a backfill rate exceeding 85% and backfill strength $\geq$5 MPa to suppress mining-induced fracture development. This study provides methodologies for addressing roadway layout in shallow coal seams and overburden fracture development. However, due to variations in the properties of different coal seams, field applications require case-specific analysis. The proposed solutions should be adapted to formulate appropriate prevention measures based on actual geological conditions.

6. Conclusions

(1) Analysis of vertical stress distribution outside the upper-seam goaf revealed a gradient attenuation pattern with increasing distance from the goaf boundary. Vertical stress recovered to the in situ stress threshold at 23 m, necessitating inter-panel coal pillars exceeding 23 m to ensure roadway placement within stable in situ stress zones. (2) When the lower-seam roadway was 5 m from the overlying panel pillar center, both sides exhibited high vertical stresses due to direct superposition. At 15 m spacing (adjacent to the pillar's right edge), left-side stresses increased marginally under concentrated loading, while right-side stresses dropped sharply within the goaf pressure relief zone. Stress decay became critical at 35 m (stress influence boundary), with further distance (>35 m) yielding gradual stress reduction. Optimal roadway positioning is recommended within 35–55 m of legacy pillar centers. (3) Post mining of upper seams, mining-induced fractures propagated to the surface, causing severe air leakage under atmospheric pressure differentials. Lower-seam extraction intensified overburden damage, with microfracture density and overall porosity increasing markedly. Complex leakage networks formed between surface and goaf, dominated by oblique fractures along goaf margins and zones of periodic overburden failure. Thin interburden and high cumulative extraction thickness in shallow multi-seam mining triggered complete interlayer failure and secondary activation of abandoned goafs, exacerbating overburden fragmentation and leakage pathway development. Backfilling in lower-seam goafs significantly reduced reactivation effects on overlying abandoned goafs. Fracture density and porosity distribution in legacy goafs remained largely unchanged under backfilled mining. Settlement curves demonstrated pronounced modifications, including reduced maximum subsidence magnitudes and collapse intensity.

Funding: This work was supported by the Humanities and Social Science Project of the Chinese Ministry of Education (23YJC630229).

Data Availability Statement: The original contributions presented in this study are included in the article. Further inquiries can be directed to the corresponding author.

Conflicts of Interest: The author declares no conflicts of interest.

References

1. He, Y.; Zhang, J.; Yang, T.; Wu, J.; Gao, S.; Sun, J. Study on the Bearing Structure of Key Strata and the Linkage Evolution Mechanism of Surface Subsidence in Shallow Coal Seam Mining. *Appl. Sci.* **2024**, *14*, 9608. [CrossRef]
2. Ding, L.; Liu, Y. Three-dimensional physical simulation on overlying strata's motion rule in shallow seam. *Fresenius Environ. Bull.* **2017**, *26*, 5314–5322.
3. Bo, H.; Guo, G.; Li, H.; Wang, Y.; Jiang, Q.; Hu, S.; Zhang, F. Study on surface subsidence prediction method of shallow coal seam backfill-strip mining under the hard roof. *Bull. Eng. Geol. Environ.* **2023**, *82*, 281. [CrossRef]
4. Jilo, N.Z.; Assefa, S.M.; Assefa, E. Numerical analysis of underground tunnel deformation: A case study of Midroc Lega-Dembi gold mine. *Sci. Rep.* **2024**, *14*, 7964. [CrossRef]
5. Li, M.; Ye, L.; Feng, J.; Fang, Y.; Wen, H.; Wu, X. Failure mechanism and control technology of soft-rock roadways subjected to high structural stress. *Front. Earth Sci.* **2024**, *12*, 1473108. [CrossRef]
6. Sun, Y.; Li, G.; Zhang, J.; Xu, J. Failure mechanisms of rheological coal roadway. *Sustainability* **2020**, *12*, 2885. [CrossRef]

7. Wu, G.; Chen, W.; Jia, S.; Tan, X.; Zheng, P.; Tian, H.; Rong, C. Deformation characteristics of a roadway in steeply inclined formations and its improved support. *Int. J. Rock Mech. Min. Sci.* **2020**, *130*, 104324. [CrossRef]
8. Xu, X.; He, F.; Li, X.; He, W. Research on mechanism and control of asymmetric deformation of gob side coal roadway with fully mechanized caving mining. *Eng. Fail. Anal.* **2020**, *120*, 105097. [CrossRef]
9. Wang, J.; Zhao, M.; Feng, L.; Chen, X. Optimization model of coal mine roadway layout system based on evidence theory and its application. In Proceedings of the 16th International Conference on Industrial Engineering and Engineering Management, Beijing, China, 21–23 October 2009; pp. 185–189. [CrossRef]
10. Zhu, W.; Wang, F.; Chen, S.; Yin, D.; Zhou, J.; Jie, Z.; Zou, Y. Asymmetric Deformation Mechanism of Roadway with Continuous Mining and Continuous Backfilling. *Rock Mech. Rock Eng.* **2023**, *57*, 2125–2142. [CrossRef]
11. Wei, M.; Zhu, Y.; Wang, P.; Luo, Y.; Ren, H.; Li, P. Failure characteristics and stress field distribution law of roadway in downward mining of deep close distance coal seam. *Sci. Rep.* **2024**, *14*, 28735. [CrossRef]
12. Zhang, J.; Zhang, Y.; Han, Y.; Fan, W.; Song, Z.; Yu, H.; Zhang, J.; Zhang, H.; Liu, J. Study on stress distribution law of surrounding rock of roadway under the goaf and mechanism of pressure relief and impact reduction. *Eng. Fail. Anal.* **2024**, *160*, 108210. [CrossRef]
13. Wei, X.; Zhao, Z.; Wang, Y.; Wang, X.; Hui, Z. Study on the law of stress distribution in the presence of remaining coal pillar in a close-distance coal seam and the reasonable location of the roadway. *Sci. Rep.* **2025**, *15*, 8737. [CrossRef]
14. Zhang, J.; He, Y.; Yang, T.; Bai, W.; Wu, J.; Zhuo, Q.; Gao, S. Coevolution mechanism and branch of pillar-overburden fissures in shallow coal seam mining. *Energy Sci. Eng.* **2023**, *11*, 1630–1642. [CrossRef]
15. Ling, C.; Liu, B.; Zhang, C.; Teng, T.; Zhang, K.; Sun, B.; Zhou, J. Fractal characteristics of overburden rock fractures and their impact on ground fissures in Longwall coal mining. *Fractal Fract.* **2023**, *7*, 699. [CrossRef]
16. Chen, Z.; Zou, Q.; Gao, S.; Zhan, J.; Chen, C.; Ran, Q.; Sun, X. Long-term evolution of overlying rock fractures in mined-out areas and its effect on gas flow conductivity. *Fuel* **2023**, *353*, 129213. [CrossRef]
17. Zhu, H.; Fang, S.; Huo, Y.; Guo, J.; Wu, Y.; Hu, L. Study of the Dynamic Development Law of Overburden Breakage on Mining Faces. *Sci. Rep.* **2020**, *10*, 6555. [CrossRef]
18. Pan, W.; Jiang, P.; Li, B.; Yang, Y. The Spatial Evolution Law and Water Inrush Mechanism of Mining-Induced Overburden in Shallow and Short Coal Seam Group. *Sustainability* **2022**, *14*, 5320. [CrossRef]
19. Wang, K.; Zhang, X.; Wang, Y.; Li, K.; Sun, J.; Zhou, A.; Du, F.; Guo, Y. Progressive failure analysis and fractures space model construction of overlying strata in shallow multi-seam overlapping mining. *Eng. Fail. Anal.* **2025**, *173*, 109462. [CrossRef]
20. Xu, C.; Wang, K.; Guo, L.; Zhao, C.; Wang, Y.; Yang, Y.; Wu, S. Fractal characteristics of mining-induced overburden fracture development during fully mechanized coal mining. *Environ. Earth Sci.* **2024**, *83*, 113. [CrossRef]
21. Wang, G.; Wu, M.; Wang, R.; Xu, H.; Song, X. Height of the mining-induced fractured zone above a coal face. *Eng. Geol.* **2016**, *216*, 140–152. [CrossRef]
22. Zhang, C.; Zhao, Y.; He, X.; Guo, J.; Yan, Y. Space-sky-surface integrated monitoring system for overburden migration regularity in shallow-buried high-intensity mining. *Bull. Eng. Geol. Environ.* **2020**, *80*, 1403–1417. [CrossRef]
23. Shan, R.; Li, Z.; Wang, C.; Wei, Y.; Tong, X.; Liu, S.; Shan, Z. Study on the distribution characteristics of stress deviator in the surrounding rock when mining closely spaced coal seams. *Environ. Earth Sci.* **2021**, *80*, 602. [CrossRef]
24. Wu, B.; Wang, X.; Bai, J.; Wu, W.; Zhu, X.; Li, G. Study on crack evolution mechanism of roadside backfill body in GOB-Side entry retaining based on UDEC trigon model. *Rock Mech. Rock Eng.* **2019**, *52*, 3385–3399. [CrossRef]
25. Wang, K.; Zhang, X.; Du, F.; Guo, Y.; Li, K.; Deng, Y. Evolution of the Overlying-Strata-Bearing structure and the permeability distribution of the broken interlayer in shallow multiseam mining. *Int. J. Geomech.* **2023**, *23*, 04023111. [CrossRef]

processes

MDPI

Article

Mechanism and Application of Static Stress Intervention for Controlled Directional Roof Caving in Fully Mechanized Mining Faces

Hao Shi [1,*], Bingyuan Hao [1], Xingyun Ren [2] and Ji Zhang [1]

[1] College of Mining Engineering, Taiyuan University of Technology, Taiyuan 030024, China
[2] Department of Resources and Mechanical Engineering, Lyuliang University, Lyuliang 033001, China
* Correspondence: shihao0824@link.tyut.edu.cn

Abstract: To address roof overhang hazards (e.g., rock bursts and gas accumulation) in high-gas coal mines, this study proposes a static stress intervention method for controlled directional roof collapse. Using the 150110 fully mechanized face at Yiyuan Coal Mine as a case study, we investigate the mechanical mechanism of static stress intervention-induced roof collapse through theoretical modeling and FLAC3D simulations in the absence of pre-cracks. The study reveals that advanced boreholes filled with static expansion agents generate stress concentration zones along the drilling array. When superimposed with mining-induced stresses, this configuration induces tensile failure preferentially at borehole locations, thereby achieving controlled directional roof collapse. Theoretical calculations indicate that roof fracturing occurs at predetermined locations when expansion pressure reaches ≥9.11 MPa. FLAC3D simulations analyzed stress redistribution and plastic zone evolution under combined static and mining-induced stresses, demonstrating the method's efficacy in optimizing roadway stability. Field trials implement spaced boreholes (65 mm diameter, 16 m depth, 1 m spacing) with alternating expansion agent charging, achieving a 6 m reduction in roof collapse intervals, effectively mitigating overhang hazards. Results confirm that static stress intervention reshapes the roof stress field, inducing tensile failure along predetermined paths without relying on pre-cracks. The findings provide theoretical and technical insights for roof stability control in high-gas coal mines.

Keywords: static stress intervention; swelling pressure; high-gas mine; roof stability

1. Introduction

In underground coal mines, the timing and spatial distribution of roof collapse in fully mechanized mining faces directly influence the stability of the mining site and overall mining efficiency. However, due to the influence of complex geological conditions, the roof often forms a large-span unsupported roof due to excessive exposure, further triggering disasters such as rock burst and gas accumulation, which severely affect coal mine safety production. To address this challenge, researchers have proposed techniques such as dense drilling roof cutting [1,2], blasting pre-cracking [3–5], hydraulic fracturing [6–8], and mechanical fracturing [9] for roof pressure relief. However, each of these techniques has limitations. Traditional drilling roof cutting requires densely spaced boreholes, which is expensive and leads to uncontrollable disturbance ranges. Explosive blasting, although effective in regulating roof pressure, poses significant safety risks in high-gas mines and is therefore limited in application. Hydraulic fracturing equipment is precise, but crack

 https://doi.org/10.3390/10.3390/pr13051552

propagation is constrained by the continuity of the rock mass. Mechanical fracturing incurs high costs and poses potential air leakage risks [10].

In recent years, static stress intervention technology has gained attention due to its mild reaction characteristics and operational safety. Li et al. [11] determined the optimal composition ratio of the breaking agent through experimental analysis, providing strong support for roof cutting and pressure relief. Huang [12], Huang et al. [13], and others used the expansion pressure of static expansion agents to pre-crack the roof by adjusting the water-to-cement ratio. Researchers such as Hao et al. [14–17] further verified its engineering applicability. Compared to traditional methods, static expansion agents are ordinary building materials that do not require special approval. Additionally, the process is controllable, avoiding the risk of explosive blasting. However, existing research primarily focuses on pre-crack-guided collapse, with little research on the collaborative effect of static and mining-induced stresses in the roof cutting mechanism.

Numerical simulation, as an effective tool for simulating mining processes, has gradually become an effective means of solving coal mining problems. For example, Tien Trung Vu et al. [18] used UDEC software to simulate the mining process of the 11th coal seam in the Nuibao coal mine in Vietnam, analyzed the displacement zone of the surrounding rock, and determined the affected area on the terrain surface; Dmytro Babets et al. [19] used RS3 and RS2 FEM simulation software to establish a three-dimensional model to analyze the stress–strain distribution of surrounding rock and study the deformation of tunnels under different parameter combinations. Li et al. [20] used FLAC3D software to simulate different fracturing schemes for key layers and analyzed the stress distribution pattern. Li et al. [21] demonstrated through comparative analysis of strain and deflection values obtained from load tests on a PCB bridge that finite element analysis can be effectively employed to validate engineering applications.

To address the problem of difficult roof collapse at the end of the fully mechanized mining face, the 150110 working face at Yiyuan Coal Mine is used as the engineering background. We propose a novel method to collaboratively utilize static and mining-induced stresses for roof cutting. Through theoretical modeling, the reshaping effect of static stress intervention on the roof stress field is analyzed. Combined with numerical simulations and field tests, the mechanical mechanism of collaborative stress driving directional roof collapse without pre-cracks is revealed. This study breaks through the traditional reliance on pre-cracks in roof-cutting technology, providing a new theoretical foundation and technical pathway for precise control of the roof in mining areas.

2. Project Overview

Jinneng Holding Group and Shun Yiyuan Coal Industry Co., Ltd. (referred to as "Yiyuan Coal Industry") has a designed production capacity of 1.8 million tons per year. It is a high-gas mine, mining the No. 15 coal seam with an average seam thickness of 5.85 m, an average dip angle of 5°, and using the fully mechanized mining method for roof cutting and pressure relief. Based on the mining pressure data from the adjacent working faces, the initial collapse step distance for the immediate roof is 9.2 m, the first pressure step distance for the old roof is 32.2 m, and the periodic pressure step distance is 15 m. The roof and floor lithology mainly consist of mudstone, sandstone, and limestone. Figure 1 illustrates the stratigraphic column.

According to the mining sequence plan, the 150108 working face is the continuation of the 150110 working face, with a pillar width of 30 m. The relative position relationship is shown in Figure 2. Affected by the mining activities of the 150110 working face, the mining pressure in the transportation roadway of the 150108 working face is significant, and the significant surrounding rock deformation, requiring the expansion and repair of secondary support. At the same time, during the advancement of the 150110 working face, untimely roof collapse caused a large area of roof hanging at the end. To prevent the accumulation of corner gas and air leakage in the mined-out area, frequent construction of gob bag isolation walls at the working face end is necessary. The actual site conditions are shown in Figure 3. The large deformation of the surrounding rock and large-area roof hanging in the roadway pose certain safety risks to production operations, seriously affecting production progress and increasing labor and material input.

Histogram	Layer thickness/m	Lithology
	4.03	sandy mudstone
	5.14	mudstone
	7.61	limestone
	0.35	14# coal
	4.70	Siltstone
	14.12	mudstone
	5.85	15# coal
	6.85	mudstone

Figure 1. Comprehensive stratigraphic histogram.

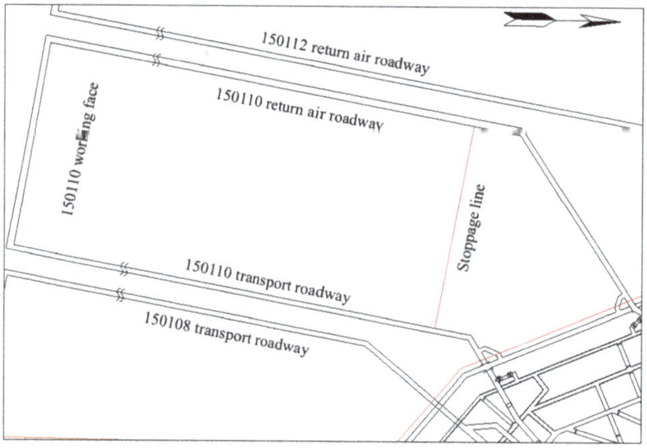

Figure 2. 150110 working face mining layout plan.

Figure 3. Construction of gangue bag separation wall.

3. Mechanism of Static Stress Intervention in Roof Cutting

During the roof cutting and pressure relief process, if the roof rock mass has a free surface and the stress inside the hole reaches or exceeds its tensile strength, cracks will form; If the rock mass has no free surface and is under triaxial compression inside the hole, even if the expansion stress reaches the tensile strength of the rock, it will only change the stress state of the surrounding rock, thus achieving static stress intervention.

The principle of collaborative roof cutting between static stress and mining-induced stress is as follows: In the advanced working face, boreholes are drilled in the roof of the recovery roadway. Static expansion agents are filled into the holes, and the hydration reaction of the expansive agents generates expansion pressure that acts on the hole wall, creating a high-pressure state inside the hole and forming stress concentration zones along the borehole line. When the working face advances to the static stress intervention position, under the combined effect of mining-induced stress, the stress concentration zone is induced to become the location of roof fracture, causing the roof of the mining area to collapse along the borehole line in a timely manner. This achieves the engineering goal of roof cutting and pressure relief while optimizing the stress environment around the roadway. The 3D schematic diagram is shown in Figure 4.

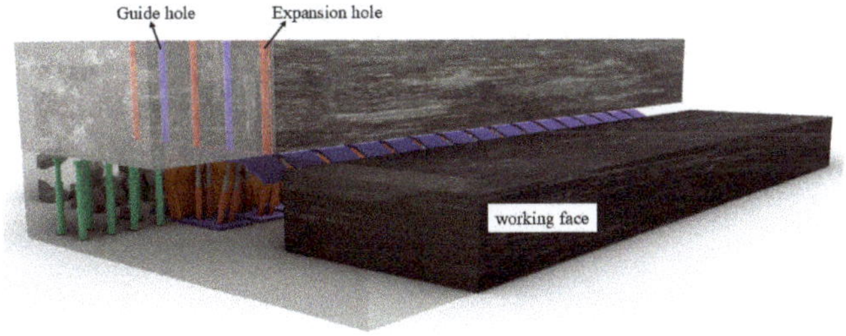

Figure 4. Static stress intervention structure model diagram.

As the working face advances, the roof loses the support of the coal seam, and the collapsed gangue cannot provide timely support for the roof, causing the roof at the end of the working face to form a cantilever beam structure. Without manual control of the roof, the roof fractures and collapses when it reaches the critical bending moment. Due to the overhanging roof structure, the upper section is subjected to tensile stress, while the lower section experiences compressive stress. Since the tensile strength of the rock is much lower than its compressive strength, the failure first occurs on the upper surface of the roof. If pressure is applied through drilling, causing the stress at the borehole location to reach the ultimate tensile strength in advance, it is possible to intervene in the location of the roof collapse.

The constructed stress intervention mechanical model is shown in Figure 5.

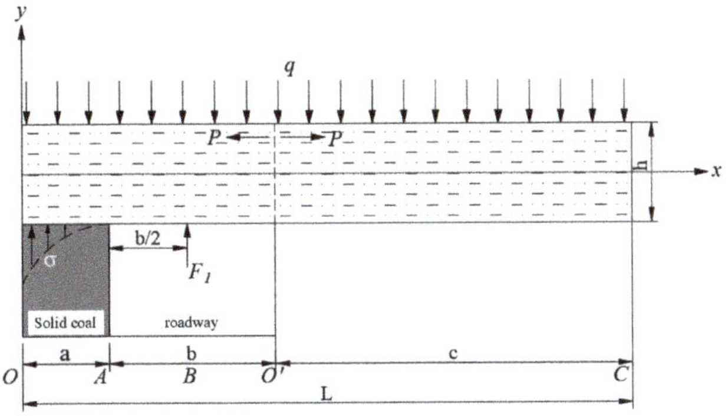

Figure 5. Stress intervention mechanics model.

In the figure, σ represents the support force provided by the coal body in the plastic state of the solid coal pillar to the roof rock layers, in MPa; F_1 represents the support force provided by the roadway support to the roof rock layers, in kN; q represents the average load of the immediate roof and the overlying rock layers, in kN/m; a represents the width at which the solid coal enters the plastic state, in meters; b represents the width of the roadway, in meters; L represents the overhang length of the immediate roof, in meters, i.e., the periodic pressure step distance of the immediate roof; h represents the thickness of the immediate roof, in meters.

When the roof fractures at point O, the width of the ultimate equilibrium zone inside the solid coal can be calculated using the following formula [22]:

$$a = \frac{m}{2\xi \tan \varphi} \ln \frac{K\gamma H + Ccot\varphi}{\xi(P_x + Ccot\varphi)} \tag{1}$$

In the equation, m represents the mining thickness of the coal seam, in meters; ξ is the triaxial stress coefficient, $\xi = (1 + \sin \varphi)/(1 - \sin \varphi)$; φ is the internal friction angle of the coal body, in degrees; c is the cohesion of the coal body, in MPa; K is the stress concentration coefficient; γ is the average unit weight of the overlying rock layers, in kN/m^3; H is the burial depth of the roadway, in meters; P_x is the support resistance of the coal pillar, in MPa.

The support force of the coal body in the plastic state on the roof can be calculated using the following formula [23]:

$$\sigma = \left(\frac{C}{\tan \varphi} + \frac{P_x}{\xi}\right) e^{\frac{2x \tan \varphi}{m\xi}} - \frac{C}{\tan \varphi} \tag{2}$$

According to the principles of material mechanics, without applying stress intervention (p), the cantilever beam will fracture at point O, and the ultimate bending moment at the point of fracture is

$$\begin{aligned} M_{\max} &= M_O = -\tfrac{1}{2}qL^2 + \left(a + \tfrac{b}{2}\right) \times F_1 \\ &+ \textstyle\int_0^a \sigma(a - x)dx \end{aligned} \tag{3}$$

The maximum stress is

$$\sigma_{\max} = \frac{M_{\max} z}{I_z} \tag{4}$$

In the equation: $\sigma_{\max}$ is the maximum stress, in KPa; $M_{\max}$ is the maximum bending moment, in kN·m; I_z is the moment of inertia of the section, in m^4.

If the unit width of the beam is taken, the moment of inertia of the section is

$$I_z = \frac{1}{12} h^3 \tag{5}$$

In the equation: h is the thickness of the immediate roof, in meters.

At the upper surface ($z = h/2$), the ultimate tensile stress at point O is:

$$\sigma_{\max} = \frac{6\left(\int_0^a \sigma(a - x)dx + F_1\left(a + \frac{b}{2}\right) - \frac{1}{2}qL^2\right)}{h^2} \tag{6}$$

The advanced working face drills a hole at position O′ in the roof and fills it with static expansion agents, forming continuous expansion pressure along a specific direction. The expansion force is symmetrically distributed within the hole, and the resulting bending moments cancel each other out. Stress is formed radially within the borehole, and the expansion pressure is assumed to be p.

When no expansion force is applied, the stress at point O′ is

$$\sigma_{O'} = \frac{M_{O'}z}{I_z} = \frac{-3q(L-a-b)^2}{h^2} \tag{7}$$

After applying the expansion force, the total stress at point O′ is

$$\sigma'_{max} = \sigma_{O'} + \sigma_x = p + \frac{-3q(L-a-b)^2}{h^2} \tag{8}$$

When the total stress at point O′ reaches the ultimate tensile strength faster than at the fixed end, it causes the roof to collapse at the borehole location in advance. Therefore, the critical expansion force p must satisfy the following equation, i.e.:

$$p \geq \frac{6\left(\int_0^a \sigma(a-x)dx + F_1\left(a+\frac{b}{2}\right) - \frac{1}{2}qL^2\right)}{h^2} + \frac{3q(L-a-b)^2}{h^2} \tag{9}$$

Based on the actual geological conditions of the 150110 transportation roadway, the required parameters in the formula are assigned as follows: b = 5.5 m, P_x = 0.153 MPa, φ = 30°, c = 1 MPa, m = 2.8 m, F_1 = 200 kN, H = 300 m, K = 3, L = 15 m, γ = 25 kN/m^3, and the basic roof load $q = \gamma H$. Substituting into Equations (1), (2), and (9), the results are a = 6.78 m, σ = 2.77 MPa, and the force required to cause the roof rock layer to collapse at the specified location O′ is: $p \geq 9.11$ MPa. The existing static expansion agent can meet the requirement of applying 9.11 MPa of expansion pressure inside the borehole.

In summary, by applying the expansion force, the roof is induced to reach the tensile strength limit at the borehole location first. After the working face recovers, under the combined effect of stress intervention and mining-induced stress, the roof is driven to initiate tensile failure along the borehole line, achieving the goal of roof cutting. Theoretical calculations show that, for the 150110 transportation roadway, when the expansion pressure exceeds 9.11 MPa, the roof will initiate tensile failure at the borehole location first, intentionally shifting the collapse location.

4. Analysis of Roof–Rock Pressure Under Static Stress Intervention

4.1. Numerical Simulation Model

Based on the actual geological conditions of the 150110 transportation roadway at Yiyuan Coal Industry, a FLAC3D numerical model is established to analyze the influence of static stress intervention on roof–rock pressure of the working face. The dimensions of the numerical model are 115 m × 30 m × 42 m (X × Y × Z), and the excavation size of the roadway is 5 m × 30 m × 3.3 m. The model's perimeter and bottom are displacement-constrained, with fixed boundaries. The 150110 working face is buried at a depth of 300 m, and a vertical stress of 7.5 MPa is applied at the top of the model to simulate the pressure from the overlying rock layers, thirty-meter coal pillars are retained on both sides to mitigate boundary effects using the Mohr–Coulomb constitutive model. The model comprises 1,146,539 grids and 194,948 nodes. A gradient meshing strategy is adopted, where tetrahedral meshes are employed to locally refine critical research areas such as roadways and boreholes, achieving a continuous transition in mesh density from dense to sparse through a progressive gradient approach. The numerical model is shown in Figure 6, and the mechanical parameters of each rock layer are presented in Table 1.

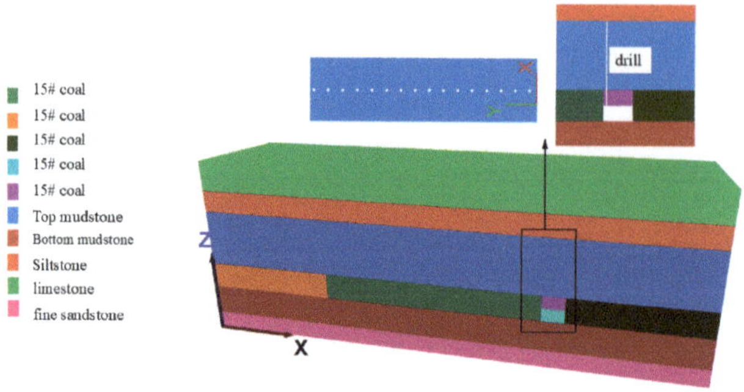

Figure 6. Schematic diagram of the model.

Table 1. Mechanical parameters of rock layer.

Lithologic	Density /(g/cm^{-3})	Bulk Modulus /GPa	Shear Modulus /GPa	Cohesion /MPa	Internal Friction Angle/(°)	Tensile Strength /MPa
Limestone	2.89	4.57	3.53	4.92	36	2.13
Siltstone	2.46	9.1	5.0	2.75	38	1.84
Mudstone	2.61	5.08	2.47	3.7	30	0.44
15# coal	1.4	0.85	0.38	1.0	28	0.5
Mudstone	2.61	5.08	2.47	3.7	30	0.44
Fine sandstone	2.64	4.0	2.7	3.8	35	1.97

The borehole diameter is 65 mm, the height is 16 m, and the spacing is 1 m, arranged vertically to the roof. The radial expansion pressure inside the borehole is set to 30 MPa [12, 13]. The borehole is located 800 mm from the coal pillar and arranged in a series along the tunnel direction, with expansion pressure applied to every other borehole. A stepwise excavation method is adopted, with an excavation step of 5 m. Excavation is carried out sixtimes, with a total excavation length of 30 m along the tunnel direction. The excavation plan is shown in Figure 7.

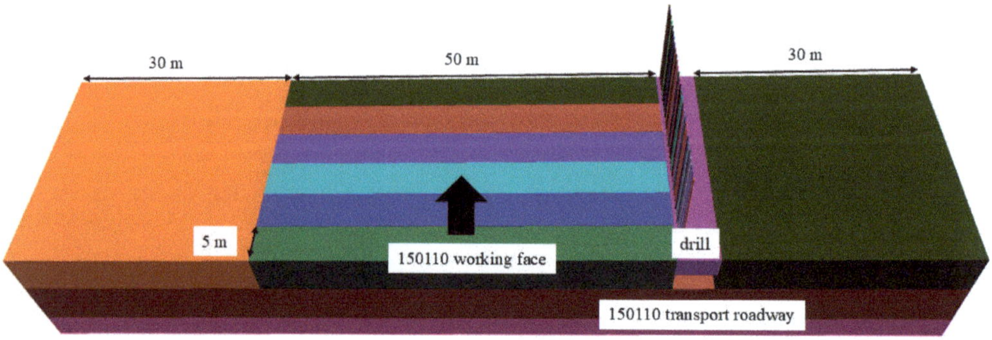

Figure 7. Excavation plan.

4.2. Roof Mining Pressure Distribution Pattern

After the calculations, analyze the failure characteristics of the roadway roof on sections A-A and B-B during excavation advancement to verify the roof cutting effect under the combined effects of static stress and mining-induced dynamic stresses.

Figure 8 shows the plastic zone distribution in the roadway surrounding rock and the A-A and B-B cross-sections after static stress intervention in the advanced working face.

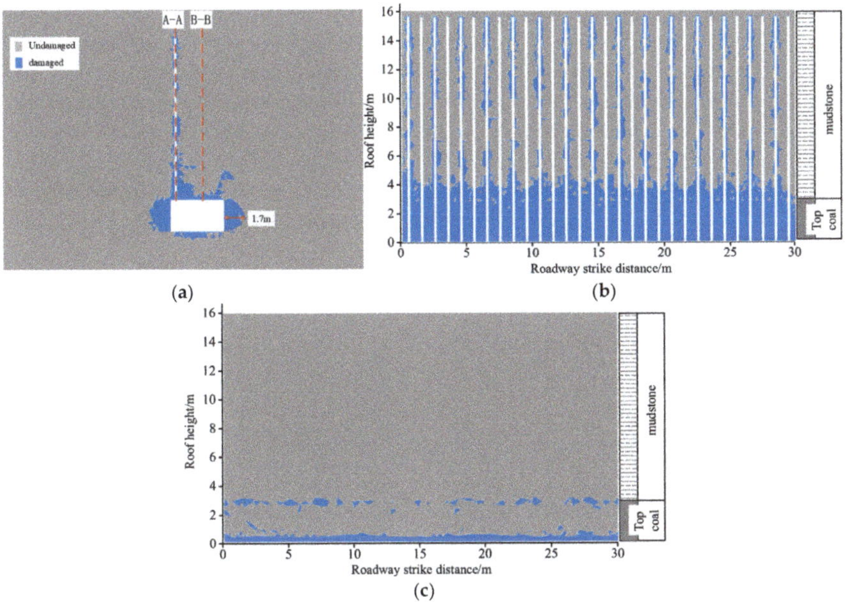

Figure 8. (**a**) Plastic zone distribution of roadway surrounding rock; (**b**) Section A-A; (**c**) Section B-B.

From Figure 8, it can be seen that the coal body on both sides of the roadway has low strength, resulting in small-scale damage, with the maximum damage depth reaching 1.7 m. From the roof borehole profile A-A, it can be seen that, after static stress intervention, failure initially occurred within a 3 m zone of the roof coal with lower strength; the strength of the immediate roof is higher, and no continuous damage occurred, thus maintaining its integrity. From the tunnel roof profile B-B, it can be seen that the tunnel roof has not undergone significant damage.

Figure 9 shows the distribution of stress in the tunnel roof after static stress intervention in the advanced working face.

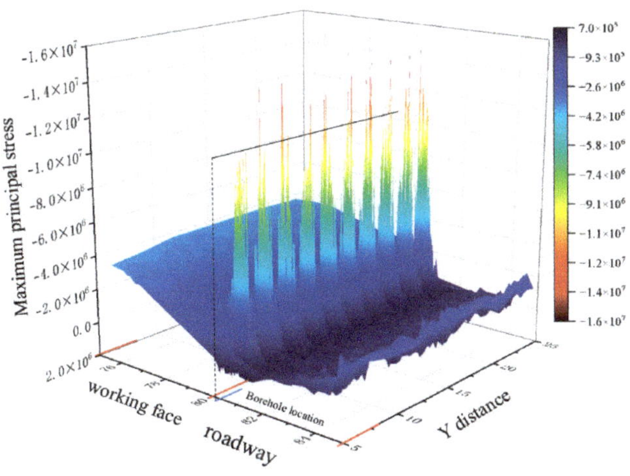

Figure 9. Stress distribution in the roof of the tunnel after drilling expansion.

Figure 9 reveals the three-dimensional stress field characteristics of the tunnel roof under static stress intervention. The tunnel roof on the right side shows a significant stress difference compared to the left side, where the working face has not yet been mined. The overall roof is in a low-stress pressure relief zone. However, a significant stress anomaly occurs at the position where static stress intervention is applied, with a peak stress of 15.5 MPa and a stress concentration coefficient of 2.01. The roof of the working face in the non-cutting region gradually returns to the original rock stress state at greater distances.

Figure 10 shows the distribution of plastic failure in the A-A cross-section of the roadway roof under varying advance distances.

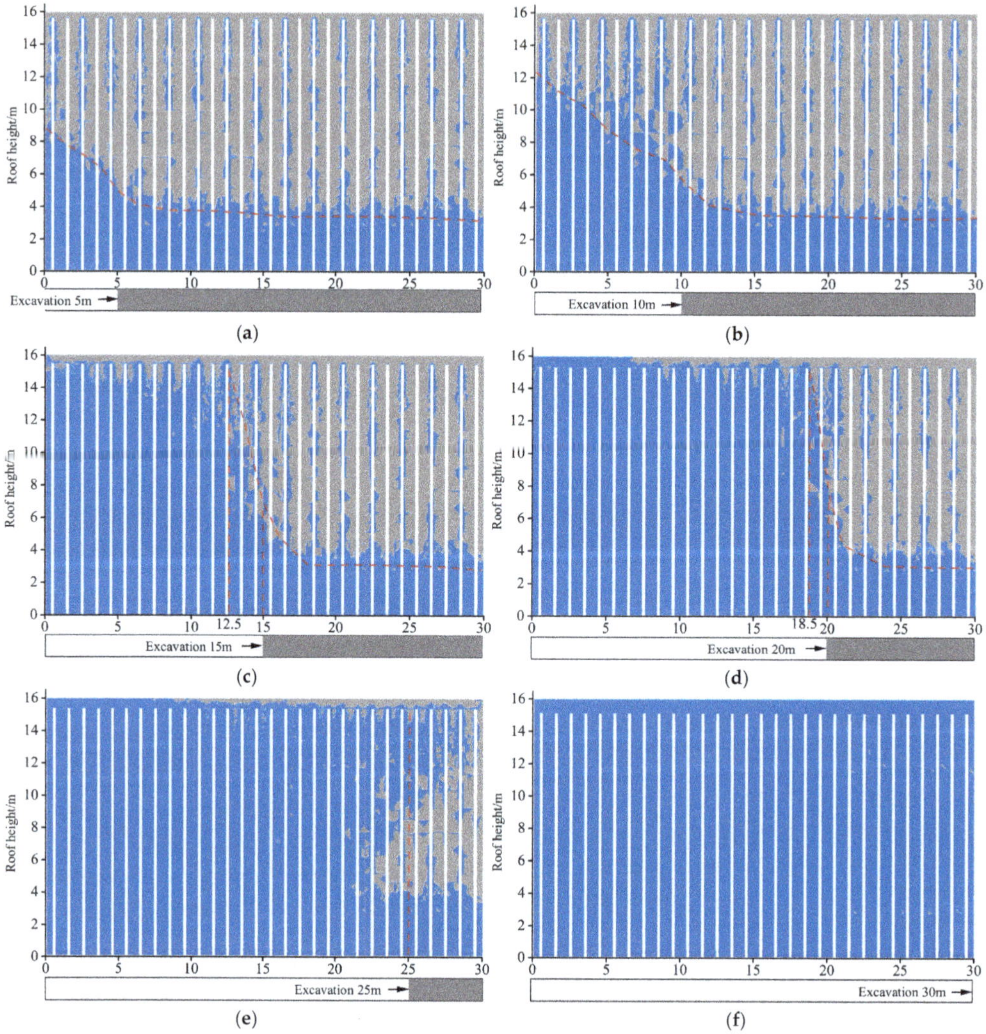

Figure 10. Failure characteristics of section A-A at different excavation lengths: (**a**) excavation 5 m: (**b**) excavation 10 m; (**c**) excavation 15 m; (**d**) excavation 20 m; (**e**) excavation 25 m; (**f**) excavation 30 m.

From Figure 10, it can be seen that, after 5 m of excavation, the roof of the mined-out area begins to undergo plastic damage, with a maximum damage depth of 8.5 m. At this point, the area of the roof in the mined-out area that has undergone plastic damage accounts for 35.25%. After 10 m of excavation, the degree of plastic damage to the roof of the mined-out area increases further, with a maximum damage depth of 12.2 m. The area of the roof in the mined-out area that has undergone plastic damage accounts for 54.75%. After 15 m of excavation, the entire roof in the mined-out area within 12.5 m undergoes plastic damage, the periodic pressure step distance shortens, and the area of the roof in the mined-out area that has undergone plastic damage accounts for 84.2%. After 20 m of excavation, the entire roof in the mined-out area within 18.5 m undergoes plastic damage. After 25 m of excavation, nearly all of the roof in the mined-out area behind the excavation undergoes plastic damage. It can be seen that the roof drilling above the top coal is complete and no conduction crack is formed during the mining process.

Figure 11 shows the distribution of plastic failure in the B-B cross-section of the roadway roof under varying advance distances.

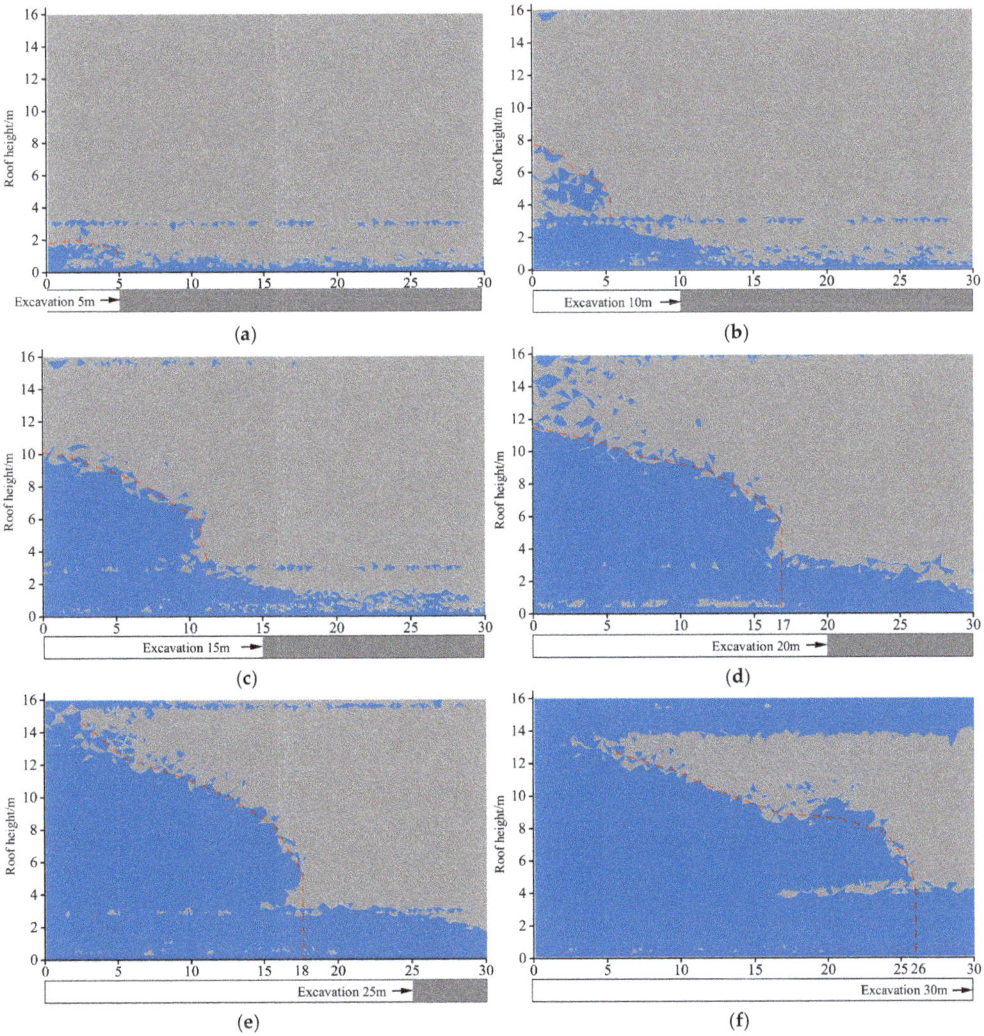

Figure 11. Failure characteristics of section B-B at different excavation lengths: (**a**) excavation 5 m; (**b**) excavation 10 m; (**c**) excavation 15 m; (**d**) excavation 20 m; (**e**) excavation 25 m; (**f**) excavation 30 m.

From Figure 11, it can be seen that, after 5 m of excavation, partial damage occurs in the roof coal, with a maximum damage depth of 2 m. The integrity of the upper roof remains relatively good. After 10 m of excavation, damage starts to occur in the tunnel roof within the 0–5 m range, with a maximum damage depth of 8 m. At this point, the roof damage lags behind the working face by 5 m. After 15 m of excavation, the roof in the 11 m range behind the tunnel experiences damage, with a maximum damage depth of 10.1 m. At this point, the roof damage lags behind the working face by 4 m. After 20 m of excavation, the roof in the 17 m range behind the tunnel experiences damage, with a maximum damage depth of 11.8 m. At this point, the roof damage lags behind the working face by 3 m. After 25 m of excavation, the roof in the 18 m range behind the tunnel experiences damage. At this point, the roof damage lags behind the working face by 7 m. After 30 m of excavation, the roof in the 26 m range behind the tunnel experiences damage. At this point, the roof damage lags behind the working face by 4 m. As the working face advances, the roof coal has relatively low strength and experiences plastic damage first. However, the strength of the immediate roof above is relatively high, causing it to lag behind the collapse of the mined-out area roof, which helps maintain the stability of the tunnel roof before the working face recovery.

The simulation results show that (1) the static stress intervention will form a high stress region near the borehole; (2) when the working face is pushed to the high stress area, the roof of the mined-out area is plastic damaged along the borehole layout face; (3) the roadway roof can remain stable before mining, which verifies the feasibility of static stress intervention in roof cutting.

5. On-Site Application and Effect Analysis

In actual engineering practice, continuous loading may lead to premature roof fractures. To provide a free surface and play a guiding role, a spaced loading method is used. Boreholes filled with static expansion agents are called charging holes, while boreholes without static expansion agents are called guiding holes. The spacing between charging holes and guiding holes is mainly determined by the performance of the static expansion agent and the strength of the roof rock. Based on previous engineering experience, three spacing options are available for the charging and guiding holes: 800 mm, 1000 mm, and 1200 mm. Through preliminary tests, a spacing of 1000 mm can effectively achieve roof cutting while maximizing material savings. The diagram of spaced loading is shown in Figure 12, and the scene of expansion borehole construction is shown in Figure 13.

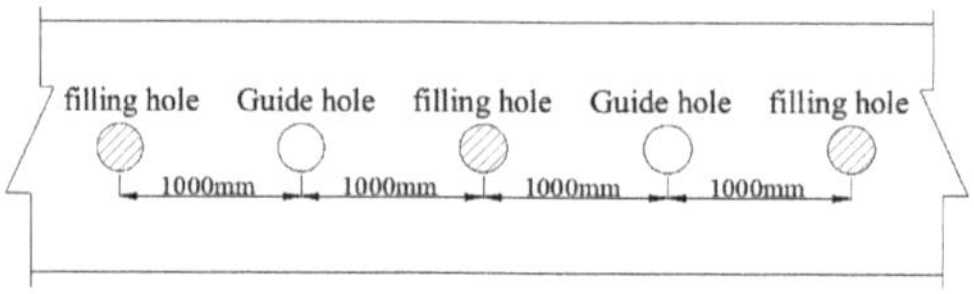

Figure 12. Schematic diagram of spacer charging.

Based on the rock fragmentation theory [24], the borehole depth is determined to be 16 m. In the 150110 working face transportation roadway, a row of boreholes is drilled along the direction of the working face. Based on field tests and construction conditions, the borehole diameter is set to 65 mm, the borehole spacing is 1 m, the depth is 16 m, the distance from the coal pillar is 800 mm, and the boreholes are arranged vertically to the roof with expansion agents loaded into every other hole.

Figure 13. Photographs of expansion holes in the field.

When the working face advances to the range of the roadway intervention, the working face continues to advance about 9 m, and the roof of the mined-out area collapses almost completely, with the collapse step distance reduced by about 6 m. As the temporary support with individual props and roof beams at the retreating end is removed, the roof of the side roadway in the mined-out area collapses in a timely manner. The roof collapse situation is shown in Figure 14.

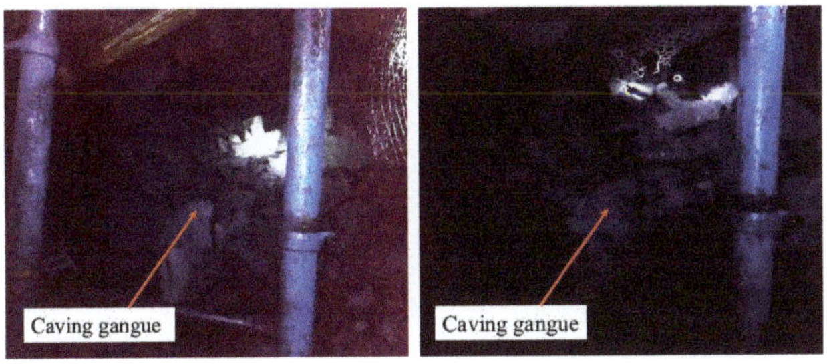

Figure 14. Top slab collapse at the end of the roadway.

6. Conclusions

A method combining static expansion stress with mining-induced stress was proposed to achieve directional roof collapse in high-gas coal mines, addressing roof overhang hazards. Through theoretical modeling, FLAC3D numerical simulations, and field tests, the mechanical mechanism of static-mining stress synergy driving directional roof failure was revealed, and its engineering applicability was validated.

1. Mechanism Analysis: The expansion pressure ($\geq$9.11 MPa) generated by static expansion agents within boreholes induces stress concentration zones at borehole locations. Superimposed with mining-induced stresses, tensile failure preferentially initiates along the borehole trajectory, achieving directional collapse. Theoretical calculations confirm that roof fracturing occurs at predetermined positions when the expansion pressure reaches the critical threshold.

2. Stress Redistribution: Static stress intervention significantly alters roof stress distribution, forming high-stress concentration zones around boreholes (peak stress: 15.5 MPa, stress concentration factor: 2.01) while maintaining borehole integrity. As the working face advances, plastic failure propagates along the borehole trajectory, validating the feasibility of stress synergy-driven collapse.

3. Field Application: Field implementation, employing a spaced charging configuration (1 m spacing, 65 mm borehole diameter, 16 m depth), reduced roof collapse intervals by 6 m, achieving effective mitigation of face-end overhang hazards.

4. Innovation and Significance: This study overcomes the limitations of conventional pre-cracking techniques by actively regulating the roof stress field through static stress intervention, enabling directional collapse without pre-existing fractures. The findings establish a novel theoretical and technical framework for roof stability control in high-gas coal mines.

Author Contributions: Conceptualization, B.H.; methodology, H.S. and X.R.; software, H.S. and J.Z.; validation, H.S., X.R. and J.Z.; writing—original draft preparation, H.S. and X.R.; writing—review and editing, B.H.; funding acquisition, B.H.; All authors have read and agreed to the published version of the manuscript.

Funding: This research received no external funding.

Data Availability Statement: The data presented in this study are available on request from the corresponding author.

Conflicts of Interest: The authors declare no conflicts of interest.

References

1. Li, X.; Liu, S.; Fu, M.; Peng, B.; He, Y. Research and application of influencing factors of key parameters of roof cutting and pressure relief by dense drilling. *Coal Sci. Technol.* **2023**, *51*, 243–253.
2. Ma, J.Q.; Li, X.H.; Yao, Q.L.; Xia, Z.; Xu, Q.; Shan, C.H.; Sidorenko, A.; Aparin, A. Numerical simulation on mechanisms of dense drilling for weakening roofs and its application in roof control. *J. Cent. South Univ.* **2023**, *30*, 1865–1886. [CrossRef]
3. Zhang, B.; Wang, P.; Cui, S.; Fan, M.; Qiu, Y. Mechanism and surrounding rock control of roadway driving along gob in shallow-buried, large mining height and small coal pillars by roof cutting. *J. China Coal Soc.* **2021**, *46*, 2254–2267.
4. Wang, M.; Zheng, H.; Ma, Z.; Mu, H.; Feng, X. Control Technology of Roof-Cutting and Pressure Relief for Roadway Excavation with Strong Mining Small Coal Pillar. *Sustainability* **2023**, *15*, 2046. [CrossRef]
5. Zhang, K.; Liu, C.; Zhang, H.; Yue, X.; Liu, H. Research on Roof Cutting Pressure Relief of the Gob-Side Entry Retaining With Roadside Backfilling. *Front. Earth Sci.* **2022**, *10*, 835497. [CrossRef]
6. Feng, G.; Fan, Y.; Wang, P.; Guo, J.; Gao, R.; Wen, X.; Zhang, P.; Zhu, L.; Qian, R.; Zhang, J. Fracture propagation law of hydraulic fracturing of rock-like materials based on discrete element method. *J. China Coal Soc.* **2024**, *49*, 2231–2246.
7. Zhang, H.; Liu, C.; Chen, Z.; Yu, X.; Zhang, K.; Liu, H. Analysis and Application of Hydraulic Fracturing to Control Hard and Stable Roof in Initial Mining Stage. *Sustainability* **2023**, *15*, 10518. [CrossRef]
8. Cheng, L.; Zhang, Z.; Jiang, P.; Yang, J. Research and application of stress field response mechanism based on roof hydraulic fracturing pressure relief. *J. Min. Saf. Eng.* **2023**, *40*, 722–729.
9. Tai, Y.; Yu, B.; Xia, B.; Li, Z.; Xia, H. Research on stress release for the gob-side roadway using the roof-cutting technology with a chainsaw arm. *R. Soc. Open Sci.* **2020**, *7*, 191663. [CrossRef]
10. Feng, S.; Chen, X.; Wang, L.; Li, L.; Li, X. Air Flow Movement Law in Working Face and Gob with Roof-Cutting and Pressure-Releasing Mining: A Numerical Simulation and Engineering Verification. *ACS Omega* **2023**, *8*, 25960–25971. [CrossRef]
11. Li, S.; Li, Z.J.; Luo, M.K. Experimental Study on Reasonable Composition and Proportion of Coal and Rock Static Cracking Agent. *Blasting* **2018**, *35*, 137–141+153.
12. Huang, H. *Application and Research of Static Broken Used in High-Gas Coal Mine Roadway End Caving Roof Control*; Taiyuan University of Technology: Taiyuan, China, 2014.
13. Huang, X.P. *Research on Optimization of Static Roof Cutting and Stress Releasing Parameters at the Side of Mining Roadway*; Taiyuan University of Technology: Taiyuan, China, 2018.
14. Wang, F.Q.; Li, Z.X.; Wu, J.; Gao, C.; Wang, J.N.; Ren, X.Y.; Ma, X.C.; Zhang, T. Study on Static Expansion and Roof Cutting Technology at End of Large Mining Height Working Face. *Coal Technol.* **2023**, *42*, 95–99.
15. Hao, B.Y.; Li, Z.H.; Huang, H.; Zhao, G.L. Research and Application of Static Crushing Technology Handled Roadway Roof Overhang in High-gas Coal Mine. *Blasting* **2015**, *32*, 109–114.
16. Xiang, H.T.; Yan, G.C.; Ding, H.Q.; Kou, K.B.; Li, G. Research and Application of Static Precrack Technology of Roof Before Initial Mining in High Gas Mine. *Min. Res. Dev.* **2022**, *42*, 14–19.

17. Ding, H.; Yan, G.; Kou, K.; Dong, J.; Xiang, H. Study on the technology of roof cracking caused by static crushing agent inXiaochang Coal Industry. *Min. Saf. Environ. Prot.* **2023**, *50*, 100–104+110.
18. Vu, T.T.; Do, S.A. Determination of the rock mass displacement zone by numerical modeling method when exploiting the longwall at the Nui Beo Coal Mine, Vietnam. *Min. Miner. Depos.* **2023**, *17*, 59–66. [CrossRef]
19. Dmytro, B.; Olena, S.; Hapieiev, S.; Prykhodchenko, V. Multifactorial analysis of a gateroad stability at goaf interface during longwall coal mining—A case study. *Min. Miner. Depos.* **2023**, *17*, 9–19.
20. Yang, X.; Yang, G.; Huang, R.; Wang, Y.; Liu, J.; Zhang, J.; Hou, S. Comprehensive study on surrounding rock failure characteristics of longwall roadway and control techniques. *Appl. Sci.* **2021**, *11*, 9795. [CrossRef]
21. Li, P.; Yang, C.; Fu, X.; Li, J.; Jin, D. Reinforcement of Insufficient Transverse Connectivity in Prestressed Concrete Box Girder Bridges Using Concrete-Filled Steel Tube Trusses and Diaphragms: A Comparative Study. *Buildings* **2024**, *14*, 2466. [CrossRef]
22. Hou, C.J.; Ma, N.J. Stress in in-seam roadway sides and limit equilibrium zone. *J. China Coal Soc.* **1989**, *14*, 21–29.
23. Huang, W.P.; Gao, Y.B.; Wen, Z.J.; Gao, L. Technology of gob-side entry retaining using concrete-filled steel tubular column as roadside supporting. *J. China Univ. Min. Technol.* **2015**, *44*, 604–611.
24. He, M.C.; Gao, Y.B.; Yang, J.; Guo, Z.B.; Wang, E.Y.; Wang, Y.J. The energy-gathered roof cutting technique in non-pillar mining and its impact on stress evolution of surrounding rocks. *Chin. J. Rock Mech. Eng.* **2017**, *36*, 1314–1325.

processes

Article

Distribution Characteristics of Mining-Induced Stress Fields and Surrounding Rock Control Technology in Adjacent Working Faces Within Fold Structure Zones

Jingya Wang [1,2,3,4], Gao Li [5], Wencai Wang [1,2,3,4,*], Hu Liu [6], Rui Wang [1,2,3,4], Hao Zhang [1,2,3,4] and Shengxiao Yuan [1,2,3,4]

1 Institute of Mining and Coal, Inner Mongolia University of Science and Technology, Baotou 014010, China; 2021011025@stu.imust.edu.cn (J.W.)
2 Inner Mongolia Key Laboratory of Mining Engineering, Baotou 014010, China
3 Inner Mongolia Research Center for Coal Safety Mining and Utilization Engineering and Technology, Baotou 014010, China
4 Inner Mongolia Cooperative Innovation Center for Coal Green Mining and Green Utilization, Baotou 014010, China
5 China Energy Trading Group Co., Ltd., Beijing 100089, China
6 Sunsangou Coal Mine of Inner Mongolia Hongyuan Coal Group Co., Ltd., Ordos 010300, China
* Correspondence: wencai@imust.edu.cn

Abstract: Mining operations in fold structure zones are often subject to dynamic disasters due to the influence of tectonic topography. To explore the interaction between the tectonic stress field and the mining-induced stress field throughout the entire mining process of adjacent working faces in fold structure zones, this study adopts a comprehensive research methodology that integrates field investigations, theoretical analysis, numerical simulations, and industrial experiments. The stress distribution characteristics before and after mining in fold structure zones are systematically analyzed to elucidate the evolution laws of stress and displacement in coal seams, reveal the mechanisms of surrounding rock instability, identify high-risk locations for roof collapse, and propose targeted surrounding rock control strategies for practical application. The key findings of this research are as follows: (1) In fold structure zones, the horizontal stress is significantly influenced by tectonic factors, whereas the vertical stress is predominantly affected by mining activities. (2) The evolution of the mining-induced stress field in fold structure zones is jointly governed by the initial tectonic stress and the mining-induced stress. The advancing position of the working face determines the specific locations of stress concentration, while the tectonic stress regulates the intensity of stress concentration across different regions. (3) The mechanism of surrounding rock failure and instability in fold structure zones is irreversible, with the stress field being a superposition of tectonic and mining-induced stresses. The extent of failure depends on the combined stress concentration at specific locations, which is directly correlated with the distribution of the initial tectonic stress field. (4) Based on the failure patterns of surrounding rock in fold structure zones, a coordinated control strategy incorporating supplementary roof support was developed, along with detailed parameter specifications. The practical implementation of this strategy ensured the stability of surrounding rock during mining through fold structure zones, effectively preventing incidents of roof collapse or rib spalling.

Keywords: fold structure zones; working face; stress field; surrounding rock control

1. Introduction

In the current global energy landscape, coal remains a crucial foundational energy resource in ever-increasing demand. However, we must confront the reality that as easily accessible shallow coal deposits gradually deplete, coal mining operations are inevitably extending to deeper strata and geologically complex regions. Fold structures, as common geological phenomena resulting from crustal movements, are widely distributed in coal-bearing strata. These fold structures significantly influence the occurrence state of coal seams, leading to frequent dynamic disaster incidents in such mining areas, including roof collapse, rib spalling, and floor heave [1,2]. The formation process of fold structures involves intense stress interactions, resulting in markedly heterogeneous and complex stress distributions within coal seams and surrounding rock masses. During the extraction of folded coal seams, the distribution characteristics of the stress field directly govern the stability of the working face surrounding rock [3,4]. Therefore, analyzing the impact of underground mining activities on stress environments in fold structure zones and implementing targeted prevention measures against mining-induced dynamic disasters under such geological conditions holds significant theoretical value for guiding safe and efficient coal mining operations in structurally complex terrains.

Scholars both domestically and internationally have conducted extensive research on the stress field distribution characteristics in fold structure zones through theoretical analysis, numerical simulation, and field measurement methods, yielding a series of significant research outcomes. Based on a large number of measured data, He Hu et al. [5] monitored seismic events during coal seam mining in folded structure zones and studied the evolution patterns of these events. They found that intense seismic events are more likely to occur in the syncline axis, with a higher probability of impact hazards compared to other areas. Wang Cunwen et al. [6] used a combination of mining pressure theory and numerical simulation to analyze the mechanism of rockburst caused by folded structures. They also provided case studies. The research concluded that folded structures can be divided into five zones based on stress conditions, with the syncline axis, anticline axis, and fold limbs being more prone to impact disasters. Wang Zhanling et al. [7] investigated the changes in roadway support under the influence of folded structure stress and explored methods to improve roadway support. They found that using high-prestress bolts for roadway support can ensure the integrity of the roof coal and rock and significantly reduce roadway deformation. Jing Guangcheng et al. [8] used theoretical analysis and numerical simulation methods combined with engineering examples to summarize the evolution of mining-induced stress and energy consumption patterns in folded structure working faces. Chen Guoxiang et al. [9] studied the relationship between horizontal stress in folded structure coal seams, stress changes due to mining disturbances, and rockburst. They also discussed the causes of rockburst hazards. Kang Hongpu et al. [2] conducted in situ measurements and numerical simulations to analyze the stress distribution around large-scale syncline structures. They found that syncline structures significantly affect the stress distribution underground, especially in the syncline axis, where horizontal stress increases sharply. In contrast, the syncline limbs have relatively lower horizontal stress, with the increase in horizontal stress being much higher than that of vertical stress. Wang Shengben et al. [10] theoretically analyzed the relationship between geological structure zones and the direction of stress. They used numerical simulation to establish a numerical model of folded structures and studied the characteristics and patterns of overburden movement and stress distribution in folded structure zones. Guo Changbao et al. [11] used ANSYS software for numerical simulation. They designed two excavation schemes with different excavation locations and directions, as well as two tunnel types: horseshoe-shaped and circular arch with straight walls. They studied the stress redistribution characteristics of surrounding rock in tunnels at different angles under the influence of folded structures and the probability of impact hazards occurring in tunnels at different angles. Shi Qiang et al. [12] investigated the stress distribution characteristics of coal seams in complex folded structures. They found that a large amount of energy is accumulated during the formation of folded structures, and this accumulated energy is an important factor leading to instability and failure in working faces. Zhou Ye et al. [13] used FLAC to conduct numerical simulation experiments on two-dimensional single-layer folds, mainly discussing the evolution patterns of maximum principal stress and horizontal strain during the deformation process of single-layer folds. Wang Jinsheng [14] studied the impact of folded structures on coal mine mining. He found that folded structures bring difficulties to roadway layout and roof management and that gas accidents are more likely to occur around the axes of anticlines and synclines. Cao Anye et al. [15] used in situ measurement data and numerical simulation to study the stress evolution patterns of the roof, floor, and coal seam during mining in folded structure zones. They found that the vertical and horizontal stress field characteristics vary in different folded zones, with horizontal stress playing a dominant role. Chen Ming [16] analyzed the causes of folds theoretically and studied their impact on coal seam mining. He found that the internal accumulated stress in folded structures creates numerous fractures, making the coal seam roof unstable. During mining, the release of structural stress increases the probability of roof collapse accidents. Wu Yun et al. [17] used the UDEC numerical software to analyze the destruction and deformation patterns of the overlying strata above the working face and the development characteristics of fractures in the overlying strata during coal mining, with different mining steps and stages. Bu Wankui et al. [18], based on the theory of curved beams and using geological data such as the folded structure of the working face, conducted research on the changes in overburden and stress distribution patterns in folded structures through theoretical analysis and numerical calculations. Liu Botao et al. [19] used numerical simulation to analyze the deformation and destruction characteristics of the overlying strata under different folded structures during mining. The results showed that different folded structures lead to different forms of overburden deformation and destruction. Anticlines inhibit the subsidence of the overlying strata, while synclines have the opposite effect. Tian Chen et al. [20] used three-dimensional similarity simulation to explore the patterns of overburden breakage and mining pressure manifestation during the mining process of large-height working faces in folded structure zones. Tang Long et al. [21] obtained the stress evolution laws and distribution characteristics of the surrounding rock of fold structures through numerical simulation and field measurement methods. Lu Cai et al. [22] studied the stress evolution during coal mining in fold areas and analyzed the stress distribution characteristics of coal and rock in different positions of working faces in fold areas.

While the aforementioned scholars have investigated the stress environment in fold structure zones, research on the interaction between the tectonic stress field and mining-induced stress field throughout the entire mining process of adjacent working faces in such zones remains limited and requires further elaboration. This study, based on the mining conditions of adjacent working faces in the 12# coal seam at the Bulianta Coal Mine in Shendong, addresses critical practical challenges, including intense strata pressure manifestations and frequent roof collapse incidents in fold structure working faces. By adopting an integrated research methodology that combines field investigations, theoretical analysis, numerical simulations, and industrial experiments, this study systematically analyzes the stress distribution characteristics before and after mining in fold structure zones. The objectives are to elucidate the stress and displacement evolution laws of coal seams in these zones, uncover the mechanisms of surrounding rock instability during mining, pinpoint risk-prone locations for roof collapse, and develop and implement targeted surrounding rock control strategies.

2. Case Study Overview

The Bulianta Coal Mine is located in the northeastern part of the Ordos Plateau, a typical loess highland region. The 12515 working face of the Bulianta Coal Mine has a total length of 327.7 m. The fully mechanized mining face is arranged along the strike of the coal seam, with a strike length of 2941.7 m and a burial depth ranging from approximately 270 to 290 m. The mining method employed is the fully mechanized longwall mining with a single-pass full-seam height extraction and the caving method for goaf management. A 30-m-wide protective coal pillar is left between the 12515 working face and the 12514 goaf. Both the tailgate and headgate of the 12515 working face are driven along the floor with a rectangular cross-section design (width × height = 5400 mm × 4300 mm). The fold structure zones are prominently distributed within the mining area of the 12515 working face. The geological overview of the 12515 working face is illustrated in Figure 1.

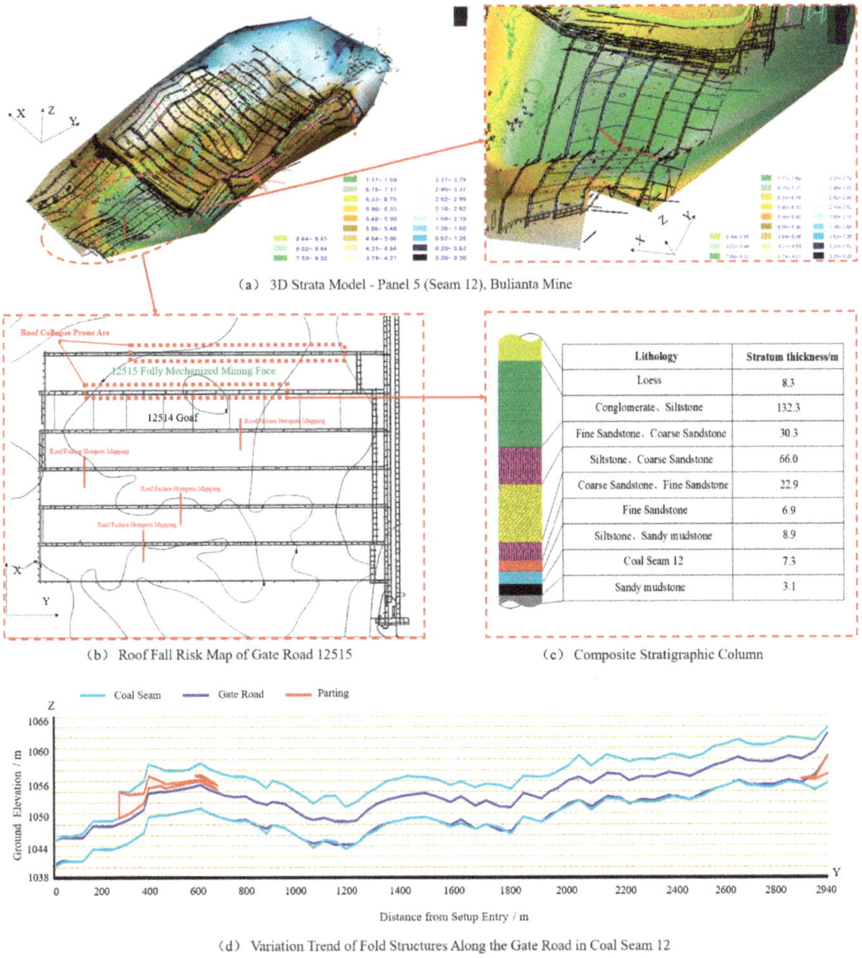

Figure 1. Geological Overview of Face 12515.

As shown in Figure 1a,d, the 12 coal working face is significantly distributed in the folded structure area. Among them, the 12515 working face is an unmined working face. It is affected by the folded structure and the mining of adjacent working faces at the same time, and there is a high risk of roof fall. Combined with the distribution of roof fall positions in other working faces, it can be known that the roof fall risk area is located in the position with large undulations of the folds. Therefore, the risk location is determined as shown in Figure 1b.

It can be known from the engineering data that the roof strata of the 12515 working face are mainly composed of sandstone, sandy mudstone, and siltstone. To investigate the surrounding rock structure in the easily caving zones of the roof at the Bulianta 12515 working face, one borehole each was drilled in three high-risk caving areas along the return airway and the belt roadway. The boreholes had a diameter of 75 mm and a depth of 15 m, with a 60-degree angle to the roof of the working face. The design of the borehole inspection scheme and the layout of the boreholes are illustrated in Figure 2, and the results of the borehole inspection are presented in Figure 3.

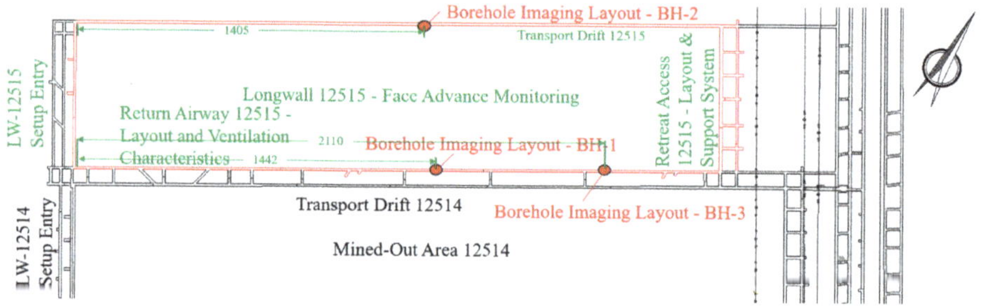

Figure 2. Borehole Camera Observation Points.

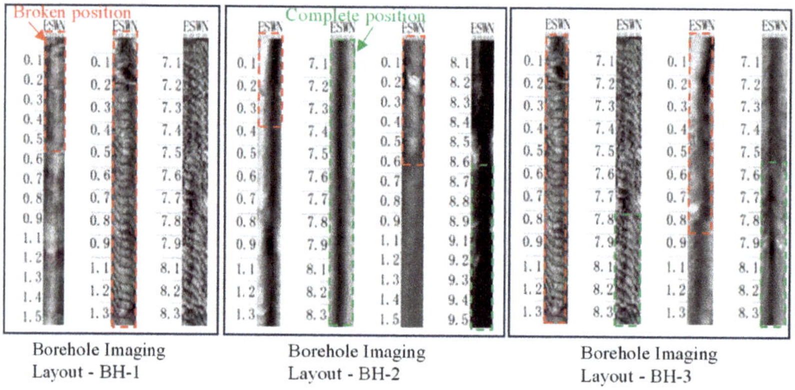

Figure 3. Borehole Digital Imaging.

As shown in the figures, analysis of the borehole inspection video and the borehole unfolding diagram reveals the following observations: Within the range of 0–1.5 m from the roof, the fractures are densely interwoven, and the rock mass is severely fragmented. The coal body, after being cut, is broken into blocks of varying sizes, with a significant accumulation of coal fragments around the borehole wall. Partial collapse of the borehole is observed, showing a tendency to connect with the overlying goaf. In the range of approximately 7–8 m from the roof, the surrounding rock exhibits multiple horizontal, vertical, and inclined fractures. These fractures are large in scale and widely extended, containing a small amount of fine debris. Within the range of about 8–9.5 m from the roof, only a few small-scale fractures are locally present. The borehole wall is relatively smooth, and the surrounding rock maintains good integrity.

A one-month observation of mine pressure was conducted in the area where the 12514 working face passed through the fold structure. The pressure changes at different support positions under various advancement degrees were statistically analyzed. The mine pressure manifestation surface diagram of the 12514 working face passing through the fold area is shown in Figure 4. A total of 24 periodic pressure arrivals were statistically analyzed in the horizontal and fold areas. The monitoring data of the working face head, middle, and tail sections were respectively statistically analyzed. The average values of periodic pressure arrivals, dynamic load coefficients, and pressure duration at the three positions were plotted into curves, as shown in Figure 5.

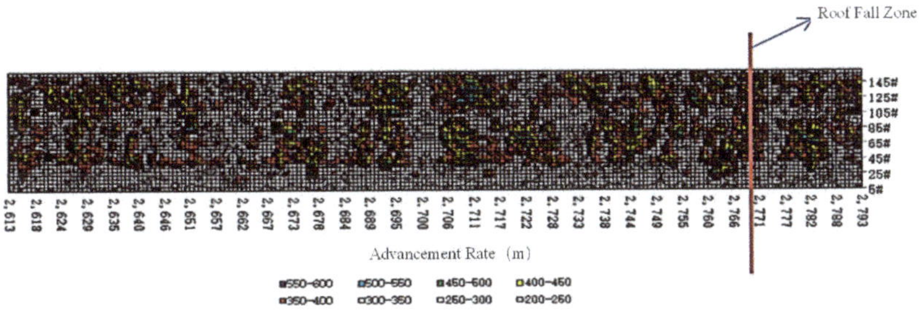

Figure 4. Strata Pressure Distribution Map Across Fold Zone in Face 12514.

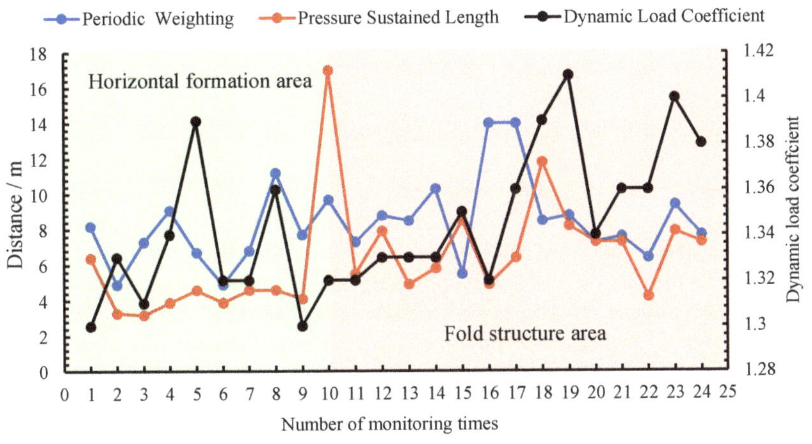

Figure 5. Monitoring results of periodic weighting.

As shown in Figure 4, within a certain range before and after the working face passes through the fold zone, the range of roof pressure increases, the pressure is obviously irregular, the pressure step distance decreases, and there is a significant continuous pressure, with the pressure distance increasing. According to the statistics of periodic pressure in Figure 5, the manifestation of mine pressure can be divided into two stages. The first stage is when the working face is being mined in the horizontal area, within which the maximum periodic pressure step distance is 11.2 m, the minimum is 4.9 m, and the average periodic pressure step distance is 7.4 m; during the pressure period, the maximum average dynamic load coefficient is 1.39, the average is 1.30, and the average pressure distance is 4.3 m. The pressure intensity is moderate, and the overall pressure change is not significant. The second stage is when the working face enters the fold influence area. Within this range, the maximum periodic pressure step distance is 14 m, the minimum is 5.5 m, and the average periodic pressure step distance is 8.9 m. During the pressure period, the maximum average dynamic load coefficient is 1.41, the maximum pressure distance is 17 m, the average pressure distance is 7.7 m, and the average dynamic load coefficient is 1.40. This stage is characterized by strong pressure intensity. Whether from the pressure distance or the pressure intensity, there is an increase to varying degrees after entering the fold influence area. The working face is in a long-term pressure state. Under the circumstances of increased mine pressure intensity and duration, the roof is prone to a roof fall accident.

3. Stress Distribution Characteristics and Dynamic Evolution Laws of Adjacent Working Faces in Fold Structure Zones

Due to the unique coal seam occurrence conditions in fold structures, the stress field distribution within the surrounding rock of the working face exhibits significant variability. Therefore, analyzing the in situ stress characteristics before mining in fold-affected working faces, as well as the dynamic stress evolution and distribution patterns of the surrounding rock during the entire mining process of adjacent working faces, is crucial for ensuring the efficient and safe extraction of fully mechanized working faces.

3.1. Establishment of Numerical Model

Based on the actual layout of adjacent working faces and the real distribution of folds, a FLAC^{3D} numerical model was established using the Mohr-Coulomb criterion. The model dimensions are 700 m × 630 m × 150 m (length × width × height), with fold parameters set to a wavelength of 500 m, an amplitude of 40 m, and a limb angle of 150°. The coal seam thickness is 7.3 m, with a burial depth ranging from 270 m to 290 m, and the dip length of the two working faces is 330 m. There exists a 110-m-thick overlying rock stratum from the ground surface to the top of the model. According to geological data, this stratum has an average volume force of 25 kN/m^3. Therefore, a vertical load of 2.75 MPa is applied to the upper part of the model to simulate the self-weight of the overlying loose rock mass. Fixed constraints are applied to the bottom and surrounding sides of the model. The numerical simulation model is illustrated in Figure 6. Based on the engineering geological data, the maximum horizontal principal stress of the surrounding rock at the working face of the 12514 fold structure was measured to range from 1.61 to 1.93 times the vertical stress. Given the complexity and risk associated with the structural stress environment of the 12515 fold structure, the lateral pressure coefficient was iteratively adjusted to ensure that the stress distribution in the fold model aligns with the actual conditions. Consequently, a maximum lateral pressure coefficient of 2.1 was established for the numerical simulation. The physical and mechanical parameters of the rock and coal are listed in Table 1.

Table 1. Rock mass mechanical properties.

Name of Stratum	The Angle of Internal Friction/(°)	Shear Modulus of Elasticity/GPa	Bulk Modulus /GPa	Cohesion/MPa	Tensile Strength/MPa	Density Kg/m^{-3}
Siltstone	38	8.1	10.8	3.5	3.20	2480
Fine Sandstone	42	13.5	18.0	15.3	3.50	2490
Sandy Mudstone	29	1.7	2.3	2.0	1.30	2350
Gritstone	34	2.9	4.2	5.0	1.90	2560
Coal	26	1.5	1.8	1.3	1.22	1930

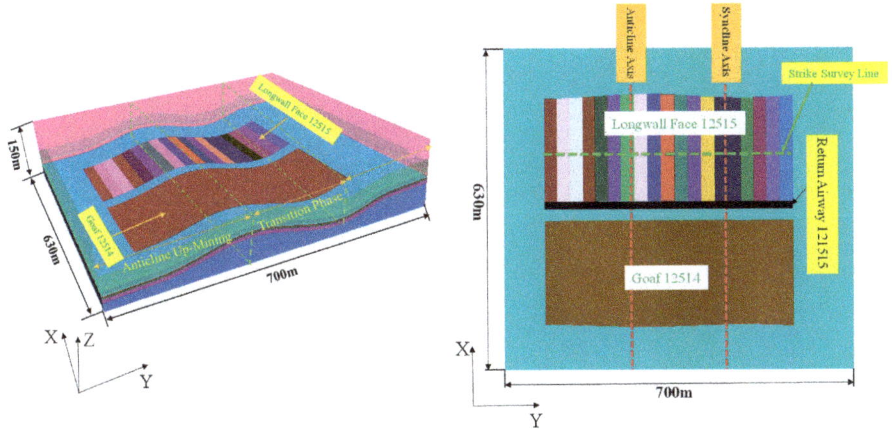

(a) Plan View of the 3D Geological Model (b) 3D Geological Model

Figure 6. Numerical Computation Model.

The horizontal stress field in the fold-affected zone is composed of both the tectonic stress and the self-weight stress components. To establish a reliable numerical model that accurately reflects the initial in situ stress state of the coal and rock mass in the fold-affected zone, the horizontal displacement constraints on the x-direction sides of the model were removed, and horizontal stress boundary conditions were applied. A horizontal tectonic stress of 5.5 MPa was applied at the top of the model, and 15.5 MPa was applied at the bottom, creating a linear gradient characteristic of fold structures. The boundary load application method for the model is illustrated in Figure 7.

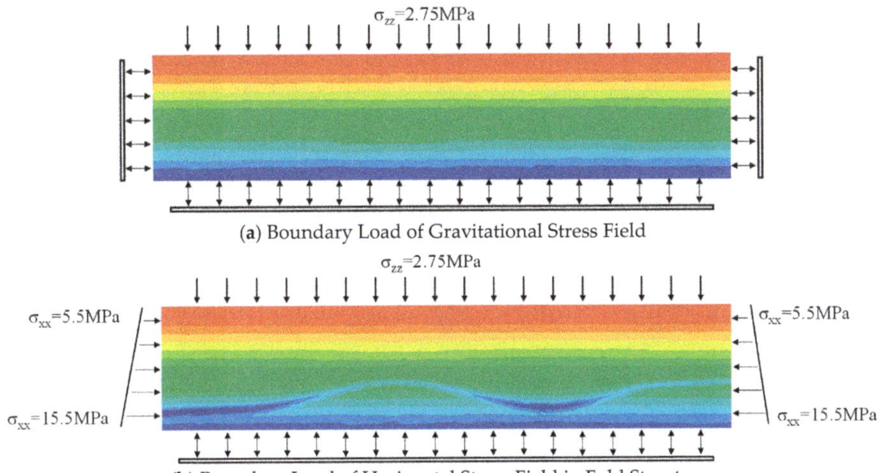

Figure 7. Boundary Load Application Methods in Numerical Models.

3.2. Initial In Situ Stress Distribution Characteristics in Fold Structure Zones

To analyze the stress distribution characteristics of surrounding rock during mining activities in fold structure areas, it is essential to first examine the initial stress distribution within these regions. The stress nephogram of the surrounding rock in the fold structure area, following the excavation of the working face, is depicted in Figure 8.

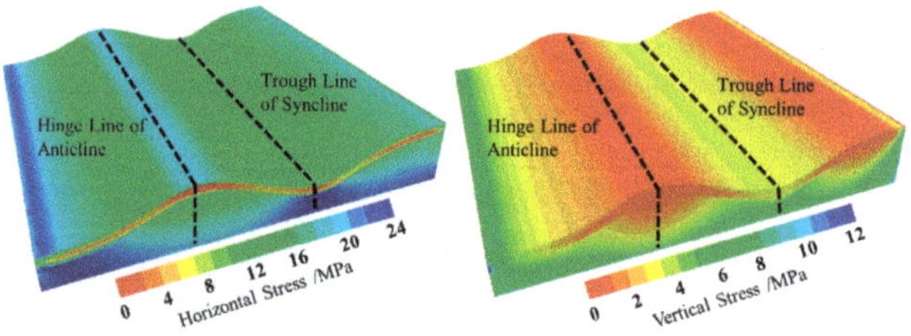

(**a**) In situ Horizontal Stress (σ_{H_0}) (**b**) In situ Vertical Stress (σ_{V_0})

Figure 8. Initial Stress Contour Map in Folded Zone.

The initial horizontal stress in the fold structure area is generally greater than the vertical stress. As shown in Figure 8a, the horizontal stress of the surrounding rock in the fold structure area is concentrated at the top of the anticlinal axis coal seam and the bottom of the synclinal axis coal seam. The horizontal stress in the coal seam is generally relieved, while the horizontal stress in other areas is relatively uniform. As illustrated in Figure 8b, the vertical stress of the surrounding rock in the fold structure area exhibits gradient relief behind the anticlinal axis and beyond the synclinal axis. The stress relief zones on both flanks of the anticlinal axis display a relatively symmetrical arch-shaped distribution. Overall, the vertical stress of the surrounding rock in the uplifted terrain transfers to the downward concave terrain, with significant gradient changes in vertical stress on both flanks of the syncline, indicating poor stability of the coal-rock mass.

3.3. Stress Distribution Characteristics in Adjacent Gob Areas of Fold Structure Zones

Under the layout conditions of adjacent working faces in fold structure zones, the stress redistribution within the surrounding rock after the completion of mining in the upper working face and the formation of the gob area directly influences the stress field distribution in the current working face area. Therefore, based on the initial stress field, the post-mining state of the upper working face was simulated to further investigate the stress field distribution characteristics of the surrounding rock prior to mining in the current working face. The stress nephogram of the surrounding rock in the adjacent gob area of the fold structure zone is shown in Figure 9.

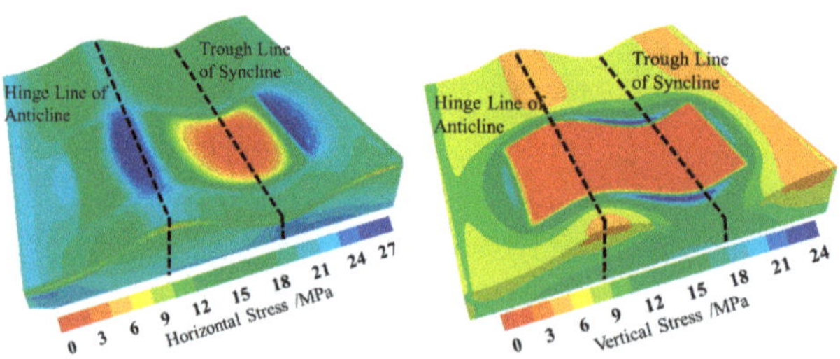

(**a**) Horizontal Stress (**b**) Vertical Stress

Figure 9. Surrounding Rock Stress Contour Map Near Gob Areas in Folded Zones.

As illustrated, after the collapse of the upper working face, the horizontal stress in the fold structure is concentrated in the anticlinal axis region, while the horizontal stress in the synclinal flanks is released. The vertical stress in the fold structure, influenced by the superposition of lateral abutment pressure from the gob area, is concentrated laterally in the synclinal flanks of the gob area. At a lateral distance of 30 m from the gob area, which corresponds to the boundary of the 12515 working face, the horizontal stress distribution in the coal seam roof is not significantly affected by the gob area, whereas the vertical stress is notably influenced by the lateral stress concentration from the gob area.

In summary, the horizontal stress in the fold structure zone is significantly influenced by tectonic factors, with stress concentration locations remaining consistent before and after mining. Post-mining, the horizontal stress in the synclinal axis and its flanks is gradually released. Conversely, the vertical stress in the fold structure zone is more affected by mining activities. Upon completion of the working face, internal stress is released, and stress concentration becomes evident at the synclinal axis and its flanks outside the working face. The anticlinal axis, due to tectonic influences, experiences stress release, resulting in a lower stress concentration compared to the synclinal axis. Therefore, prior to mining in the current working face, the surrounding rock in the synclinal flanks near the coal pillar is subjected to stress concentration due to the combined effects of the fold structure and adjacent gob area, posing a threat to the stability of the surrounding rock. Further investigation into the stress field conditions during the mining of this working face is urgently needed.

3.4. Stress Distribution Characteristics During Mining in the Current Working Face of Fold Structure Zones

The stress field during the mining of the current working face represents a further evolution based on the pre-mining stress field. Therefore, to investigate the dynamic evolution of the surrounding rock stress field during the mining process in the fold structure zone, the mining conditions of the current working face were simulated on the pre-mining model. A stress monitoring line was installed in the immediate roof of the coal seam at a position 20 m from the coal pillar. The stress distribution nephograms throughout the mining process and the dynamic evolution curves of stress at different advancing distances were extracted, as shown in Figure 10.

As illustrated, the mining process is divided into three distinct zones based on the distribution of fold structures and stress variations: the anticline mining zone, the anticline-syncline transition zone, and the syncline mining zone.

As shown in Figure 10a, in the anticline mining area, the horizontal stress ahead of the working face gradually increases as the face advances. Influenced by the superposition of fold structure-induced horizontal stress, the horizontal stress ahead of the working face reaches its peak value of 25.8 MPa when the face advances to the anticlinal axis. As the working face continues to advance and the gob area forms behind it, the vertical stress also peaks ahead of the working face. However, closer to the anticlinal axis, the vertical stress is released due to the fold structure, resulting in a reduced concentration of vertical stress in the roof ahead of the working face. When the face reaches the anticlinal axis, the peak vertical stress in the immediate roof ahead of the working face decreases to 14.3 MPa.

As shown in Figure 10b, in the anticline-syncline transition area, the horizontal stress ahead of the working face increases and then returns to the in situ stress state. Within the range from the anticlinal axis to the synclinal axis, the closer the working face is to the synclinal axis, the more pronounced the horizontal stress relief in the gob area due to the influence of the initial tectonic stress field. When the working face advances to 375 m from the setup entry, located at the synclinal axis, the horizontal stress ahead of the working face reaches its minimum value of 12.1 MPa in this region. As the gob area expands behind the working face, vertical stress concentration occurs ahead of the face, with the concentration increasing closer to the synclinal axis. When the working face advances to 300 m from the setup entry, the vertical stress reaches its peak value of 23.2 MPa.

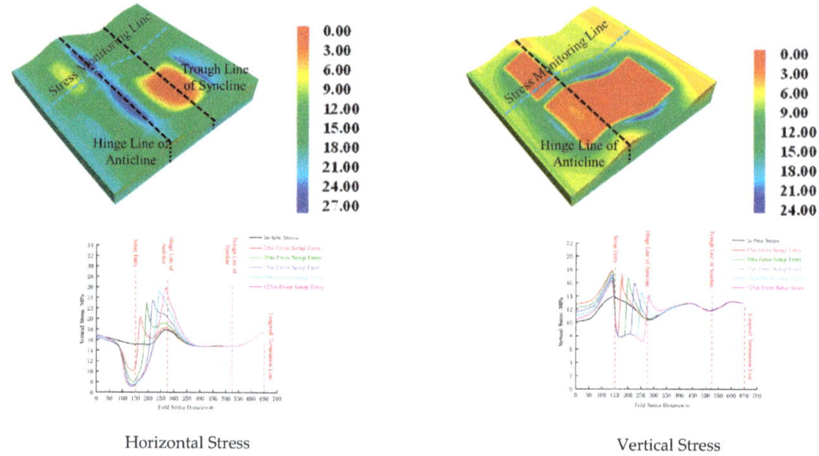

(**a**) Stress Distribution Contour Map in Up-Dip Mining Area of Anticline Structure

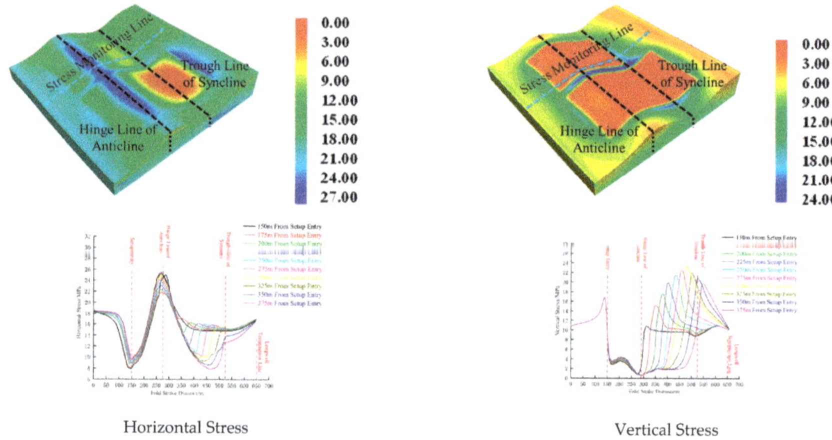

(**b**) Stress Distribution Cloud Diagram in Anticline-Syncline Transition Zone

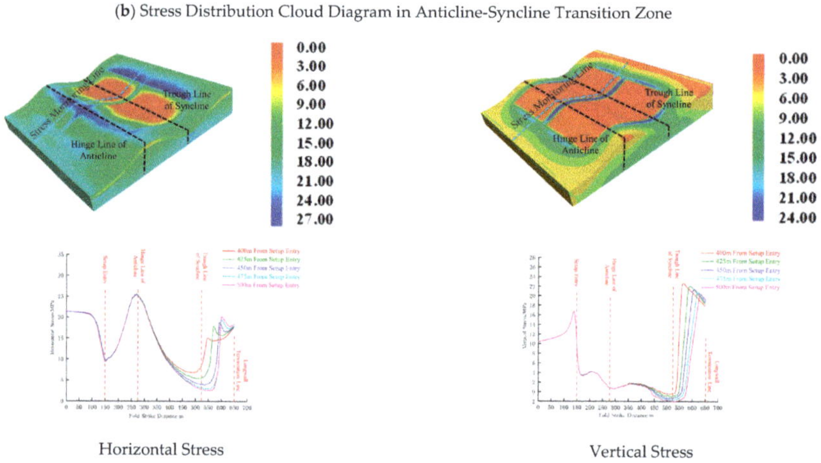

(**c**) Stress Distribution Nephogram of the Upward Mining Panel in a Syncline

Figure 10. Characteristic Stress Distribution Diagrams During Full-Seam Mining in Folded Zones.

As shown in Figure 10c, in the syncline mining area, the horizontal stress distribution trend is similar to that in the anticline mining area. The stress ahead of the working face initially increases and then decreases, with another increase in horizontal stress near the stopping line. The farther the working face is from the synclinal axis, the greater the horizontal stress concentration. When the working face advances to 500 m from the setup entry, the peak horizontal stress is 21.5 MPa. As the working face continues to advance, influenced by topographic conditions, the vertical stress in the surrounding rock gradually enters a relief zone. The farther the working face is from the syncline, the lower the vertical stress concentration ahead of the face. When the working face advances to 400 m from the setup entry, the maximum vertical stress is 22.6 MPa, and when it advances to 500 m, the vertical stress ahead of the face is 20.5 MPa.

In summary, the evolution of the mining-induced stress field in the current working face of the fold structure zone is jointly influenced by the initial fold structure stress and the mining-induced stress. The advancing position of the working face determines the location of stress concentration, while the fold structure stress controls the degree of stress concentration in different regions. Therefore, when the working face advances to the synclinal flanks, the stress in the roof of the gob area behind the face is released, and stress concentration ahead of the face increases significantly, leading to a higher risk of roof fracture and instability.

4. Evolution Patterns and Mechanism Analysis of Surrounding Rock Failure and Instability During Mining in Fold Structure Zones

4.1. Evolution Patterns of Immediate Roof Displacement in Fold Structure Working Faces

Based on the stress distribution characteristics of the surrounding rock during mining in adjacent working faces within the fold structure zone, a displacement monitoring line was installed in the immediate roof of the coal seam at a position 20 m from the coal pillar. This setup was used to study the subsidence and instability process of the immediate roof in the current working face, revealing the evolution patterns of immediate roof displacement in the fold structure zone. The vertical displacement distribution curve is shown in Figure 11.

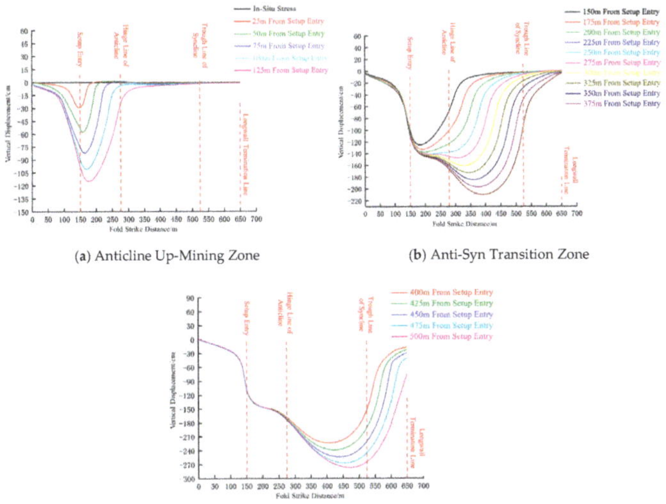

(**a**) Anticline Up-Mining Zone (**b**) Anti-Syn Transition Zone

(**c**) Syncline Highwall Zone

Figure 11. Roof Vertical Displacement in Fold Structure Zone (Face 12515).

As shown in Figure 11a, in the anticline mining zone, the immediate roof of the working face remains stable with no displacement before excavation. As the working face advances and the gob area forms behind it, significant vertical displacement occurs in the immediate roof near the setup entry. The overall displacement curve exhibits an initial increase followed by a decrease between the setup entry and the advancing face, with no displacement occurring at the face itself. The closer the advancing distance is to the anticlinal axis, the more severe the subsidence of the immediate roof. When the working face advances to 125 m from the setup entry, the maximum vertical displacement in the gob area behind the face is 115 cm, while the maximum vertical displacement ahead of the face is 32 cm.

As shown in Figure 11b, when the working face advances into the anticline-syncline transition zone, the gob area behind the face continues to expand, and the immediate roof near the setup entry undergoes further subsidence. The farther the face advances, the greater the subsidence displacement in the gob area. After passing the anticlinal axis, influenced by the fold topography, the closer the face is to the synclinal axis, the greater the vertical stress, leading to further subsidence of the immediate roof in this region. The closer the face is to the synclinal axis, the more the overall displacement curve exhibits a bimodal distribution. When the working face advances to 375 m from the setup entry, the maximum vertical displacement in the roof of the gob area behind the face occurs near the synclinal axis, reaching 211 cm, while the vertical displacement ahead of the face is 118 cm.

As shown in Figure 11c, when the working face advances into the syncline mining zone, the overall displacement curve of the gob area follows a similar trend, exhibiting a bimodal distribution with peaks located ahead of the setup entry and behind the synclinal axis. The peak behind the synclinal axis is significantly larger than that near the setup entry. The farther the face advances, the greater the displacement peaks. When the working face advances to 500 m from the setup entry, the maximum vertical displacement in the gob area behind the face is 282 cm, while the vertical displacement ahead of the face decreases to 73 cm.

In summary, during the mining process of the current working face in the fold structure zone, the subsidence of the immediate roof in the gob area gradually increases as the working face advances. When the face is in the anticline mining zone, the influence of fold topography is minimal, and the maximum displacement of the immediate roof occurs ahead of the setup entry. After the face enters the anticline-syncline transition zone, the influence of fold topography becomes significant. The closer the face is to the synclinal axis, the subsidence of the immediate roof in the gob area exhibits a bimodal distribution, with the maximum subsidence occurring at the synclinal axis. Therefore, based on the evolution patterns of immediate roof displacement in the fold structure working face, the key areas for surrounding rock failure and instability during mining are the synclinal flanks of the fold structure zone.

4.2. Mechanism of Surrounding Rock Failure and Instability During Mining in Fold Structure Working Faces

The failure and instability of the working face during mining in the fold structure zone are directly related to the stress environment distribution characteristics under different mining conditions. Therefore, to further analyze the failure and instability characteristics at each mining stage, it is necessary to conduct an in-depth analysis of the relationship between the tectonic stress field and the mining-induced stress field. Figure 12 provides a schematic diagram illustrating the mechanism of surrounding rock failure and instability during mining in the fold structure working face.

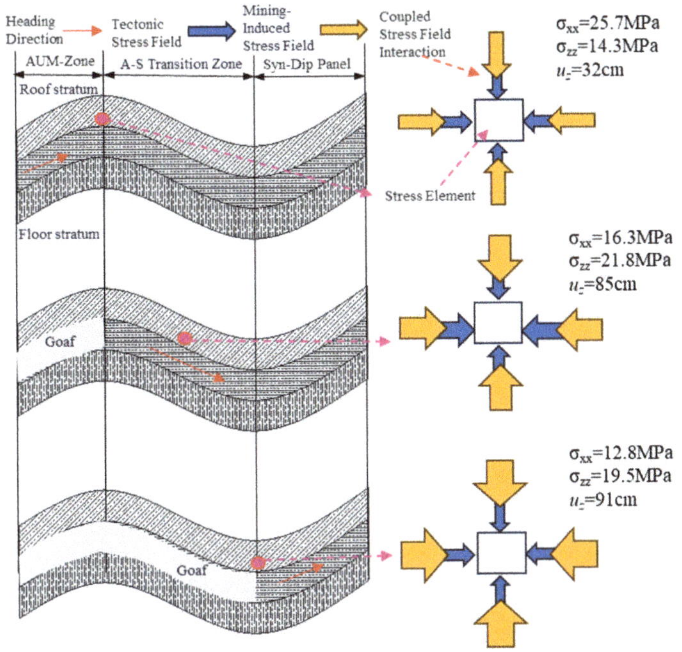

Figure 12. Schematic Diagram of Surrounding Rock Failure Mechanisms in Longwall Mining Through Fold Structures. (AUM = Anticline Up-Mining, A-S = Anticlinal-Synclinal, Syn-Dip = Syncline + Dip).

As shown in Figure 12, the mining process in the fold structure zone exhibits zonal distribution, with the overall stress field being a superposition of the tectonic stress field and the mining-induced stress field. When the working face advances into the anticline mining zone, stress concentration occurs in the roof due to the generation of the advanced mining-induced stress field. However, influenced by the tectonic stress field, horizontal stress concentrates at the anticline, while vertical stress is in a relieved state, resulting in relatively small vertical displacement in this area. In this zone, the influence of the tectonic stress field outweighs that of the mining-induced stress field. When the working face advances into the anticline-syncline transition zone, the mining-induced stress gradually increases as the gob area behind the face expands. In this zone, the concentration of horizontal stress in the tectonic stress field intensifies, and vertical stress also increases with decreasing vertical distance, leading to a significant rise in vertical displacement. As the working face advances into the syncline zone, the gob area behind the face gradually collapses, and under the combined influence of the mining-induced and tectonic stress fields, vertical displacement further increases. Therefore, during the anticline-syncline transition and syncline mining stages, the mining end is significantly affected by the superposition of the tectonic and mining-induced stress fields, resulting in substantial roof displacement and an increased risk of roof collapse.

In summary, the mechanism of surrounding rock failure and instability during mining in the fold structure working face can be summarized as follows: the damage caused by mining is irreversible, and the stress field is influenced by the superposition of the tectonic and mining-induced stress fields. The extent of damage depends on the degree of stress concentration resulting from the superposition of these two fields at a given location. The degree of damage in different structural positions of the working face is directly related to the distribution of the fundamental tectonic stress field.

5. Surrounding Rock Control Technology for Mining in Fold Structure Zones

Based on the analysis of the surrounding rock stress field distribution characteristics during the entire mining process of adjacent working faces in fold structural zones, it is evident that the risk areas for roof fall accidents are primarily concentrated in regions of high fold structural stress, specifically at the transition phase of anticlines to synclines and the mining ends during the syncline uplift phase. Drawing from the engineering practices at the 12515 working face in the 5th panel of the Bulianta Coal Mine, this study proposes principles and strategies for controlling the surrounding rock to ensure the stability of the roof and roadway walls during the mining of fold structural zones, thereby facilitating efficient extraction in fully mechanized mining faces.

5.1. Principles of Surrounding Rock Control in Mining Faces Within Fold Structural Zones

In the high-stress-gradient fold flanks, the return air roadway of the 12515 working face is subjected to the combined effects of tectonic stress from the fold terrain, lateral stress from the goaf of the 12514 working face, and advanced mining-induced stress from the current working face, resulting in severe damage to the surrounding rock. To address this, a reinforcement support measure using anchor cables and steel straps was implemented for the roadway roof. The anchor cables were installed into the stable rock layer above the roof, allowing the cables and the anchored rock mass to jointly bear the load, thereby enhancing the support strength. Coupled with W-shaped steel straps, this approach effectively suppresses roof delamination, ensuring the integrity of the roadway roof, preventing roof collapse, and reducing mesh bagging phenomena.

Additionally, to improve the strength of the coal-rock mass, high-molecular chemical grout was uniformly injected into the weak structures of the coal rock mass, such as pores, fractures, and joints, using hydraulic and pneumatic principles. The grout fills, penetrates, and compacts the coal-rock mass, binding it into a cohesive unit with high strength, excellent anti-seepage properties, and enhanced stability. This improves the physical and mechanical properties of the coal-rock mass, increasing its integrity and strength, thereby mitigating roof leakage, collapse, and coal wall spalling to a certain extent.

During mining through fold flanks, the roof pressure on the working face can be significant, especially when the roof rock beams fracture parallelly, leading to intense pressure that may trigger roof falls or even crush the supports. Hydraulic supports, as the primary equipment ensuring mining space height, divide the stope into the goaf and the controlled roof area. Increasing the working resistance of hydraulic supports can control roof subsidence in high-mining-intensity faces, but blindly increasing support resistance often yields limited results. Therefore, adopting pressure control technologies in both the controlled roof area and the goaf can effectively mitigate the impact of periodic weighting on the supports, ensuring the safety of the stope space and the stability of the surrounding rock structure.

In summary, to address the challenges of roof and surrounding rock control in working faces affected by intense mining in fold structures, a coordinated control scheme is proposed. This includes pressure control technologies in the controlled roof area and goaf, high-molecular chemical grouting reinforcement, and reinforced support for the roadway roof. This approach leverages the synergistic advantages of the "working face-support" and "roadway-support" structures to ensure the stability of the surrounding rock and the safe, efficient production of the working face during mining through fold structures.

5.2. Surrounding Rock Control Strategy for Roadways in Fold Structural Zones

To ensure safe production in the working face, pre-grouting reinforcement should be implemented in the transition zone of the anticline and the flanks of the syncline, as well as within a 20-m range before and after these areas. This measure aims to further control roof falls and rib spalling. Specifically, when the working face advances beyond the anticlinal axis and reaches the flanks of the syncline, grouting reinforcement should be applied to both the roof and the coal wall side.

For areas severely affected by folding, where the roof is fractured and rib spalling is prominent, a combination of roof grouting, fiberglass bolt reinforcement, and coal wall grouting should be adopted. Two rows of grouting holes are arranged on the coal wall of the working face. The first row of grouting holes is positioned 2.0 m below the roof, with holes drilled at an angle of 60–70° upward from the horizontal direction. The depth of these holes should be no less than 10 m, with a spacing of 3 m between holes. After drilling, fiberglass bolts with a diameter of 18 mm and a length of 4 m are inserted into the grouting holes, followed by the injection of high-molecular materials for reinforcement. The second row of grouting holes is positioned 2.0 m above the floor, with holes drilled at a 15° upward angle from the horizontal direction. The depth of these holes should also be no less than 10 m, with a spacing of 3 m. Detailed parameters of the grouting hole arrangement are shown in Figure 13.

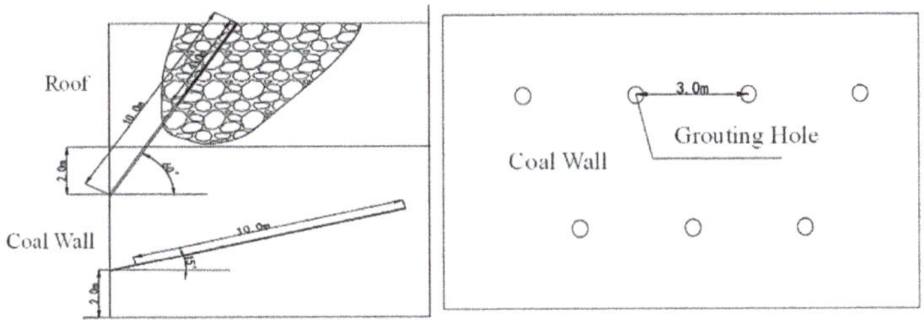

(**a**) Side View (**b**) Plan View

Figure 13. Grouting Borehole Layout Schematic in the Working Face.

When the specific area is located at the headgate or tailgate region, high molecular materials are injected from the solid side of the roadway towards the roof side of the working face within one entry. Boreholes are drilled at the intersection of the roof and the solid side, inclined upwards at a 30° angle to the horizontal, with a depth of no less than 10 m and spaced 3 m apart. The construction plan, including borehole location, depth, and angle, can be adjusted based on field conditions. The layout of the grouting holes is illustrated in Figure 14.

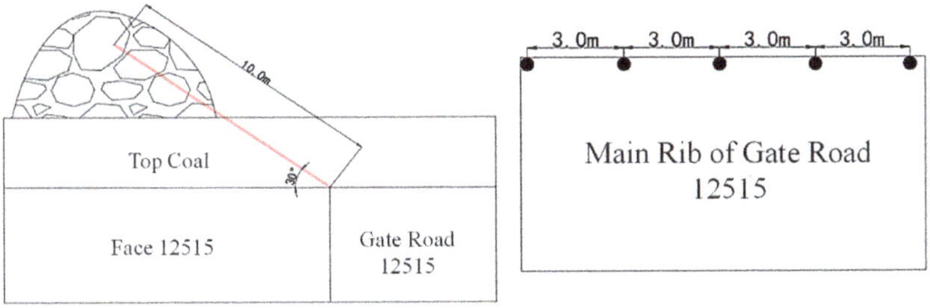

(**a**) Side View (**b**) Front view

Figure 14. Schematic Layout of Grouting Holes in Stope Drift.

Within the range from the open-off cut to the anticlinal axis, the stress gradient exhibits minimal variation, with low stress concentration and vertical displacement. In this region, the original roadway support scheme is maintained. In the transition zone between the anticline and syncline, where the stress gradient is significant, and within the syncline's uplifted fold limb, additional reinforcement measures are implemented for the extraction roadway. Roof support is enhanced using $\Phi 20.8 \times 5150$ mm anchor cables combined with steel straps. Solid wooden cribs are erected in the entry crosscuts to support the roof, and ventilation barriers are installed at the crosscut entrances. Within the fold limb area, one row of two $\Phi 21.6 \times 5150$ mm anchor cables with $1500 \times 300 \times 12$ mm PVC straps is installed at 3.0 m intervals to prevent rib spalling. Based on field conditions, if the roof-to-canopy distance at the working face is large (exceeding 1.0 m after extending the canopy) and the roof remains relatively intact, $\Phi 20.8 \times 5150$ mm anchor cables are used to reinforce the roof in the gaps between supports and the unsupported areas ahead of the canopy. This prevents roof failure and collapse due to inadequate support in areas with large canopy-to-roof distances (the specific locations for anchor cable installation are determined based on actual conditions). The roadway support cross-section is illustrated in Figure 15.

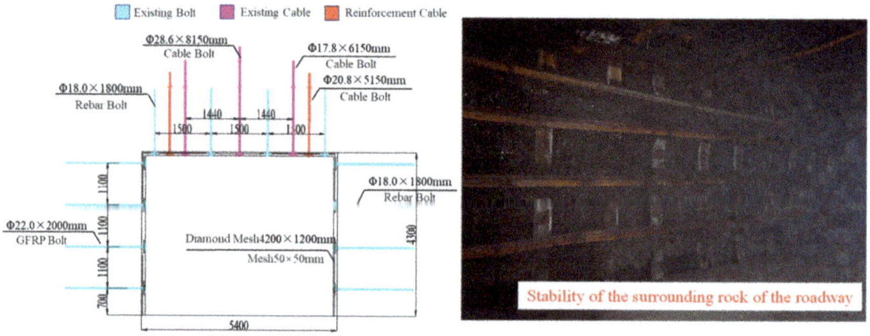

Figure 15. Cross-Sectional View of Roadway Support Structure.

Based on the aforementioned research findings and the practical production conditions of the 12515 working face, the proposed ground control strategy was implemented on-site. During the mining process through the fold structure zone, the surrounding rock of the 12515 working face remained stable and secure, with no incidents of roof collapse or rib spalling occurring.

6. Conclusions

(1) The pre-mining stress field in fold structure zones is characterized by significant horizontal stress influenced by the structure, with stress concentration locations remaining consistent before and after mining. Post mining, the horizontal stress in the syncline axis and its flanks is gradually released. Vertical stress is more affected by mining activities; after mining, internal stress is released, and stress concentration is evident at the syncline axis and its flanks. The stress concentration at the anticline axis is less pronounced due to structural influences.

(2) The evolution of the mining-induced stress field in fold structure zones is jointly influenced by the initial tectonic stress and mining-induced stress. The advancing position of the working face determines the stress concentration locations, while the tectonic stress controls the degree of stress concentration in different regions. As the working face

advances into the syncline flanks, stress is released in the goaf behind the face, and stress concentration increases ahead of the face, significantly raising the risk of roof failure.

(3) The mechanism of surrounding rock failure and instability in fold structure zones is irreversible, with the stress field being a superposition of tectonic and mining-induced stresses. The degree of failure depends on the combined stress concentration at specific locations, directly related to the distribution of the initial tectonic stress field.

(4) For the control of roof and surrounding rock in roadways under strong mining influence, a coordinated control strategy was developed, incorporating roof pressure control, goaf pressure control, polymer chemical grouting reinforcement, and supplementary roof support in roadways. Application of this strategy in the 12515 working face ensured the stability of surrounding rock during mining through fold structure zones, with no incidents of roof collapse or rib spalling.

Author Contributions: Software, R.W. and H.Z.; Investigation, H.L.; Data curation, S.Y.; Writing—original draft, J.W. and G.L.; Writing—review and editing, W.W. All authors have read and agreed to the published version of the manuscript.

Funding: This research was supported by the National Natural Science Foundation of China (Grant No. U24A2085, 52264010), Natural Science Foundation of Inner Mongolia (Grant No. 2024LHMS05017), Inner Mongolia Autonomous Region First-class Discipline Scientific Research Special Program (Grant No. YLXKZX-NKD-036), Development Project of Young Scientific and technological talents in Inner Mongolia Autonomous region (Grant No. NJYT23065) and Inner Mongolia Autonomous Region Directly-Affiliated Universities Basic Scientific Research Operating Funds Program (Grant No. 2023CXPT003, 2023QNJS096).

Data Availability Statement: The original contributions presented in this study are included in the article. Further inquiries can be directed to the corresponding author.

Conflicts of Interest: The author Gao Li was employed by China Energy Trading Group Co., Ltd. The author Hu Liu was employed by Sunsangou Coal Mine of Inner Mongolia Hongyuan Coal Group Co., Ltd. The remaining authors declare that the research was conducted in the absence of any commercial or financial relationships that could be construed as a potential conflict of interest.

References

1. Wang, L.; Cao, A.; Guo, W.; Dou, L.; Wen, Y.; Xue, C.; Hu, Y.; Lv, D. Rock burst mechanism and characteristics of roadway in "fault-fold" area. *J. Min. Saf. Struct. Eng.* **2023**, *40*, 69–81+90.
2. Kang, H.; Wu, Z.; Gao, F.; Ju, W. Effect of geological structures on in-situ stress distribution in underground coal mines. *Chin. J. Rock Mech. Eng.* **2012**, *31* (Suppl. S1), 2674–2680.
3. Wang, H.; Deng, D.; Jang, C.; Shi, R.; Yan, X. Study on evolution characteristics of uneven stress field in mining induced fold tectonic area. *Coal Sci. Technol.* **2020**, *48*, 59–69.
4. Hu, G.-D.; Cui, H.-Q.; Guan, J. Numerical simulation of stress distribution in small foldof coal seam. *J. Saf. Environ.* **2016**, *16*, 54–57.
5. He, H.; Dou, L.; Gong, S.; Zhou, P.; Xue, Z.; Jiang, H. Research on Seismicity in High Tectonic Stress Zones. *J. China Univ. Min. Technol.* **2011**, *40*, 7–13.
6. Wang, C.; Jiang, F.; Liu, J. The Controlling Effect of Structures on Rockburst and Case Analysis. *J. China Coal Soc.* **2012**, *37* (Suppl. S2), 263–268.
7. Wang, Z.; Kang, H.; Lin, J. Research on the Influence of Folded Structure Stress on Roadway Support. *Coal Sci. Technol.* **2011**, *39*, 25–28.
8. Jing, G. Research on Stress Evolution and Rockburst Incubation in Thick Coal Seam Mining in Folded Structure Zones. Master's Thesis, China University of Mining and Technology, Xuzhou, China, 2017.
9. Chen, G. The Mechanism of the Effect of Maximum Horizontal Stress on Rockburst and Its Application. Ph.D. Thesis, China University of Mining and Technology, Xuzhou, China, 2009.
10. Wang, S.; Zhang, X. The Relationship Between Geological Structures and Stress in Coal Mine Roadways. *J. China Coal Soc.* **2008**, *33*, 738–742.

11. Guo, C.; Zhang YJiang, L.; Sun, D.; Xiong, T.; Du, Y.; Zheng, G. Rockburst Mechanism and Suitable Tunnel Section in Deep-Buried Tunnels in Folded Structures. *Chin. J. Rock Mech. Eng.* **2012**, *31* (Suppl. S1), 2758–2766.
12. Shi, Q.; Pan, Y.; Li, Y. Typical Cases of Rockburst in China and Analysis. *Coal Min.* **2005**, *10*, 13–17.
13. Zhou, Y.; Lin, G.; Gong, F.; Liu, S.; Zhang, D. Changes in the Maximum Principal Stress and Horizontal Strain During the Deformation of a Single Fold and Influencing Factors. *Geotecton. Metallog.* **2007**, *31*, 37–43.
14. Wang, J. The Impact of Folds and Flexures on Coal Seam Mining in Yangmei No. 5 Mine. *Coal Min.* **2015**, *20*, 20–22.
15. Cao, A.; Xue, C.; Wu, Y.; Wang, S.; Guo, W. Mechanism and Prevention of Rockburst in Coal Mining in Folded Structure Zones. *Coal Sci. Technol.* **2021**, *49*, 82–87.
16. Chen, M. The Impact of Folded Structures on Coal Seam Mining. *Inn. Mong. Coal Econ.* **2017**, *10*, 5–6.
17. Wu, Y. Research on the Stress Field Distribution and Evolution of Overburden Spatial Structure in Folded Structure Zones. Master's Thesis, China University of Mining and Technology, Xuzhou, China, 2018.
18. Bu, W.; Xu, H.; Zhao, Y. Analysis of Overburden Deformation and Stress in Folded Structures Based on Curved Beam Theory. *J. Min. Saf. Eng.* **2019**, *36*, 827–833.
19. Liu, B.; Li, X.; Xie, C. Numerical Experimental Analysis of Coal Mining Subsidence in Folded Structure Zones. *Jiangxi Coal Sci. Technol.* **2016**, *3*, 12–15.
20. Tian, C.; Yang, Z.; Li, G. Three-Dimensional Similarity Simulation of Mining Pressure in Large-Height Working Faces in Folded Structure Zones. *J. Heilongjiang Univ. Sci. Technol.* **2023**, *33*, 861–868.
21. Tang, L.; Tu, S.; Tu, H.; Zhang, L.; Li, W.; Miao, K.; Zhao, H.; Ma, J. Mining stress evolution law and control technology of surrounding rock in fold structure area. *J. Min. Saf. Eng.* **2024**, 1–11. Available online: http://kns.cnki.net/kcms/detail/32.1760.TD.20240412.1751.002.html (accessed on 5 May 2025).
22. Lu, C.; Zhang, X.; Xiao, Z.; Wang, C.; Wang, B.; Zhou, T.; Li, H.; He, Z. Study on controlling law of fold structure on evolution of mining stress in deep mines. *Coal Sci. Technol.* **2020**, *48*, 44–50.

processes

MDPI

Article

Deformation and Pore Structure Characteristics of Lignite Pyrolysis with Temperature Under Triaxial Stress

Feng Zhang [1], Shiwei Niu [1,2,3,*], Jiawei He [1], Kai Zhang [1] and Zhongcheng Qin [1]

1 College of Energy and Mining Engineering, Shandong University of Science and Technology, Qingdao 266590, China; zhangfeng08838@163.com (F.Z.); a17311253116@163.com (J.H.); 13105118050@163.com (K.Z.); qinzhch@126.com (Z.Q.)
2 National Demonstration Center for Experimental Mining Engineering Education, Shandong University of Science and Technology, Qingdao 266590, China
3 Xinwen Mining Group Co., Ltd., Taian 271200, China
* Correspondence: nshiwei1123@sdust.edu.cn

Abstract: As people pay increasing attention to the clean and efficient mining and utilization of coal resources, efforts to improve the utilization rate of coal, modify coal resources, and carry out coal gasification have become more and more important. The deformation characteristics of lignite, the most appropriate coal type for underground coal gasification, are intricately linked to its mechanical properties, permeability characteristics, and mining efficiency throughout the extraction process. The deformation and pore structure characteristics of lignite from room temperature to 650 °C have been studied through high-temperature triaxial penetration testing systems, NMR, and X-CT. As the temperature increases, the porosity of lignite rises, its mechanical strength decreases, and significant deformation occurs, and high temperatures promote pore development in lignite. The axial deformation of lignite pyrolysis is divided into three stages: the dehydration and degassing at room temperature to ~200 °C, the slow deformation between 200 °C and 300 °C, and the pyrolysis deformation from 300 °C to 650 °C. Significant deformation occurs during both the dehydration degassing and pyrolysis deformation stages. Between 250 °C and 650 °C, a large number of highly interconnected pore networks form. Investigating the deformation and pore structure characteristics of lignite is crucial for elucidating its mechanical and permeability features under varying temperature and pressure conditions.

Keywords: lignite; deformation characteristics; temperature; triaxial pressure; NMR; X-CT

1. Introduction

Lignite, the principal type of low-rank coal, is a significant energy resource. Lignite is challenging to store because of its elevated water and volatile matter content [1]. At present, the majority of lignite is combusted directly, resulting in considerable pollution from the emission of CO and CO_2 into the atmosphere. The advancement of underground coal gasification (UCG) technology presents a significant potential to address the constraints of lignite [2–6].

Lignite experiences significant physical and chemical transformations under the influence of temperature and pressure [7]. The internal structure of lignite can be readily determined by permeability tests. Lignite experiences intricate chemical and physical transformations, including dehydration and pyrolysis, due to temperature and stress. A distinction exists from traditional coal rock deformation [8,9]. The in situ steam injection

https://doi.org/10.3390/10.3390/pr13051444

process of lignite demonstrates that lignite deformation under stress modifies both the direction and magnitude of the stress on the coal seam, as well as alters the material transmission pathway, thereby indicating a correlation between permeability and deformation.

Recent studies have examined the fluctuations of mechanical parameters in both conventional and triaxial circumstances, considering confining pressure and loading path [10–14]. The impact of temperature on the mechanical properties of coal under conditions of temperature and pressure is confined to its effect on coal strength [15]. Research indicates that all metamorphic coal demonstrates a propensity to diminish in strength with rising temperatures, and the pattern varies slightly among coal of different ranks. The rise in temperature augments the flexibility of coal, resulting in alterations to its mechanical behavior and lithology.

The deformation and failure characteristics of coal as a unique organic rock at elevated temperatures significantly differ from those at room temperature [16]. Research conducted by Bustin [17], Jiang Bo [18,19], Liu Junlai [20], and others has demonstrated that elevated temperatures significantly affect the deformation characteristics of both organic (coal) and inorganic rocks. The degree of metamorphism and the presence of volatile gases in coal are the primary factors contributing to its structural deformation. The primary deformation mechanisms of coal are high-rank deformation and low-rank coal toughness. V.L. Shkuratinik integrated uniaxial and triaxial compression tests with acoustic emission to investigate the internal fracture characteristics of coal samples during compression [21–23]. Consequently, they conducted further research on the acoustic emission memory effect of anthracite subjected to cyclic loading [24]. A.S. Voznesenskii, Wang Enyuan, and Nie Baisheng [25–27] investigated acoustic emission events during various stages of coal sample deformation and the frequency spectrum characteristics of coal fracture acoustic emission, discovering that alterations in the frequency spectrum characteristics of coal acoustic emission are intricately linked to the processes of coal deformation and fracture. Zou Yinhui [28] employed damage theory and defect distribution characteristics to formulate the theoretical model of acoustic emission during coal rock failure.

X-ray computed tomography (X-CT) and nuclear magnetic resonance (NMR) technologies have been extensively employed in experimental investigations of the mesoscopic properties of coal and rock in recent years [29–33]. Zhu Qingzhong et al. [34] employed CT technology to investigate the fracture patterns and permeability of high-rank coal matrices at varying degrees of expansion. The decrease in permeability was mostly attributed to the diminishment of pore size and interconnected porosity. Cheng Ming et al. [35] employed nuclear magnetic resonance (NMR) and X-ray computed tomography (X-CT) methods to investigate the compressibility of pores and fractures in low-rank coal, examining its correlation with heterogeneity and the impact of fracture morphology and mineral composition on compressibility. Zhao Yang et al. [36,37] employed NMR to analyze the pore structures of coal samples. They examined the correlations among pore structure, physical attributes, and the fractal dimension derived from NMR T_2 spectra.

Based on the above research, most studies on the pore structure and deformation characteristics of lignite pyrolysis have been conducted under uniaxial stress or low-temperature conditions. There is limited research on triaxial stress or high-temperature triaxial coupling conditions. Therefore, a triaxial stress high-temperature gas heating reaction system was designed and developed. This system allowed us to study the deformation characteristics of lignite pyrolysis under triaxial stress conditions. In addition, we combined NMR technology and X-CT scanning technology to investigate the pore structure characteristics of lignite pyrolysis under triaxial stress conditions. The deformation and pore structure characteristics of lignite pyrolysis with temperature under triaxial stress were obtained, providing a theoretical support for underground coal gasification.

2. Materials and Methods

2.1. Sample Preparation

The lignite specimens were sourced from the No. 5 coal seam at the Yuanbaoshan open-pit coal mine located in Chifeng City, Inner Mongolia Autonomous Region, China. The samples were treated and polished to produce cylinders (Φ7 × 10 mm and Φ50 × 100 mm), with both ends ground flat. The lignite specimens are illustrated in Figure 1.

The lignite is located in the lower cretaceous Fuxin formation of the Mesozoic erathem. Proximate and ultimate analyses of lignite were conducted, and the results are listed in Table 1. It is a low-ash and medium-sulfur moderately clean coal, with poor calorific value, poor thermal stability, ash content of 9.17%, moisture of 18.56%, and volatile matter content of 54.6%. It exhibits good chemical reactivity with CO_2 and a relatively high tar yield.

Figure 1. Lignite samples.

Table 1. Proximate and ultimate analyses of lignite.

Items	Proximate Analysis (wt%)			Ultimate Analysis (in daf. Basis, wt%)				
	M_{ad}	A_d	V_{daf}	C	H	O	N	S
lignite	18.56	9.17	54.6	67.44	4.48	1.04	0.81	26.23

M_{ad}: moisture; A_d: ash; V_{daf}: volatile matter content.

2.2. Experimental Equipment and Methods

2.2.1. High-Temperature Triaxial Experiment

A critical aspect for the successful implementation of underground coal gasification of lignite is a comprehensive investigation of the deformation features and affecting factors of lignite. Current testing apparatus typically operate at below 200 °C [9] and are incapable of systematically assessing the deformation features of lignite during the pyrolysis process at room temperature to 650 °C. The high-temperature triaxial penetration testing system utilized in this experiment, depicted in Figure 2, comprises a power system, a test control system, a temperature measuring and control system, a cooling system, and a permeation chamber. It can heat the sample to 650 °C and also offer insulation and pressurization.

The test bench can assess the deformation of coal and rock within a temperature range of 25 °C to 650 °C. The lateral pressure of the test bench utilizes a solid pressure transfer mechanism, in which the load is conveyed by sodium chloride composed of very small particles. The axial and confining pressure loads are regulated independently during loading. The sample's axial deformation is quantified using a dial indicator, while the heating temperature is monitored with a thermocouple. The testing machine has an axial and confining maximum load of 10,000 kN. The maximum axial and confining pressure of the specimen is 25 MPa, and the maximum thermal stability temperature of the sample is 650 °C.

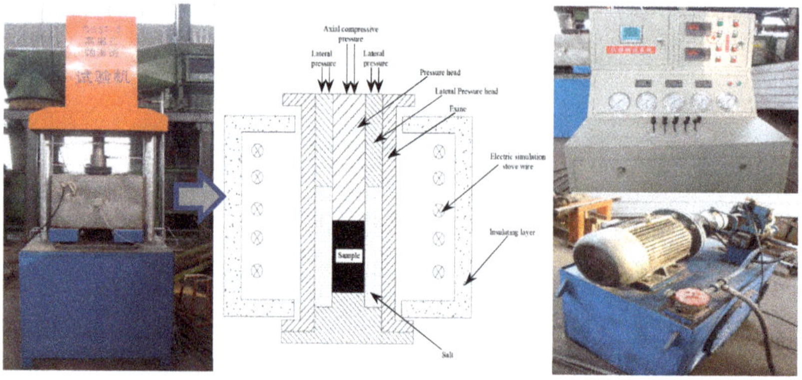

Figure 2. High-temperature triaxial testing system.

This experiment aims to simulate the temperature and deformation of lignite under their original rock stress state. The burial depth of lignite sample is 300 m, the axial pressure is set to 7.5 MPa, and the lateral pressure is 1.2 times the axial pressure, which is 9 MPa. First, weigh and install the specimen, and purge the internal air by introducing nitrogen. Then, set the axial and confining pressure to 7.5 MPa and 9 MPa, raise the temperature to 25 °C, and maintain the pressure for 4 h. Finally, while keeping the set axial and confining pressures constant, heat the specimen. The applied temperatures are 50 °C, 100 °C, 150 °C, 200 °C, 250 °C, 300 °C, 350 °C, 400 °C, 450 °C, 500 °C, 550 °C, 600 °C, and 650 °C at a rate of 10 °C/min, with each temperature stage maintained for 4 h. During the experiment, measure the specimen's deformation every 10 min

2.2.2. Experiment of NMR [32]

NMR is among the most swiftly progressing non-destructive testing methods in recent years. This study employed an NMR analyzer (MicroMR-12-025V, Newman Technology, Suzhou, China) to investigate the pore structure and fluid saturation of coal rock. The device's primary frequency, probe diameter, and minimum echo time were 12 MHz, 25.4 mm, and 60 μs, respectively.

Lignite samples were initially analyzed using NMR at 25 °C, subsequently positioned on the reaction table, and heated to a predetermined temperature (150 °C, 250 °C, 350 °C, or 450 °C) at a rate of 10 °C/min. The temperature was maintained for one hour, and the samples were let to cool naturally under the protection of nitrogen gas. The naturally cooled samples were subsequently extracted and promptly sealed with wax, followed by testing the pertinent parameters using NMR. The specific process is shown in Figure 3.

2.2.3. Experiment of X-CT [30]

This experiment utilized the μCT225kVFCB high-precision X-ray computed tomography testing and analysis equipment was developed by taiyuan university of technology and china academy of engineering physics in taiyuan, china. The test system consists of a micro-focus X-ray apparatus, a digital flat panel detector, a high-precision turntable fixture, a machine base, a horizontal movement mechanism, a data acquisition and processing system, and additional components. This technology is applicable for 3D X-CT scanning of diverse metallic and non-metallic materials. The magnification range spans from 1 to 400, the sample diameter runs from 1 to 50 mm, and the resolution of the scanning unit varies from 0.5 to 194 μm. The apparatus can identify pores measuring 0.5 μm in diameter and fractures with a breadth of 1 μm.

The lignite samples were subjected to heating at 150 °C, 250 °C, 350 °C, and 450 °C utilizing a high-temperature gas heating apparatus, maintaining a temperature deviation within 5 °C. The samples were maintained for one hour and subsequently allowed to cool naturally to below 25 °C. The samples were subsequently positioned on a rotating test bench of the μCT225kVFCB high-precision X-CT testing and analysis equipment for scanning, and the results were reconstructed in three dimensions. The specific process is shown in Figure 3.

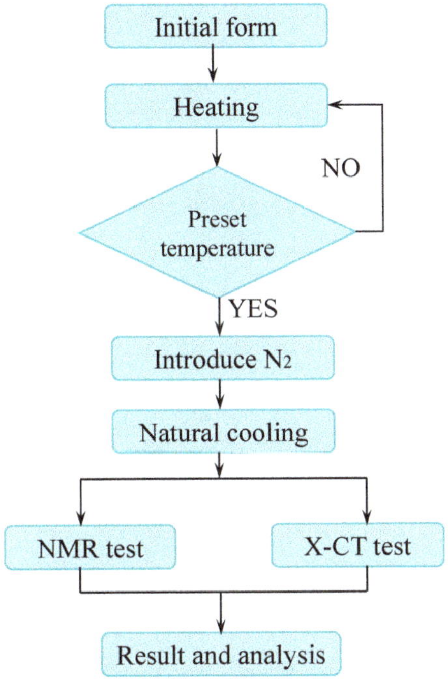

Figure 3. Experimental flowchart.

3. Results

3.1. Deformation Characteristics of Lignite in the High-Temperature Triaxial Experiment

The lignite was heated from room temperature to 650 °C for 4116 min. The relationship curve between the length and strain of the lignite sample and time and temperature during the entire temperature change process was obtained, as shown in Figures 4 and 5. The sample's maximum heating temperature was 650 °C. The pyrolysis process was categorized into three stages based on the interplay of strain, temperature, and time. The process consisted of three stages: the initial stage involved heating from room temperature to 200 °C, the second stage entailed slow deformation from 200 °C to 300 °C, and the last stage was the pyrolysis deformation from 300 °C to 650 °C. The dial indication revealed variations in the sample's axial length. The lignite sample was compressed from an initial length of 9.7896 cm to 7.1971 cm, resulting in an overall deformation of 2.5989 cm. The deformation ratio was 26.55 percent.

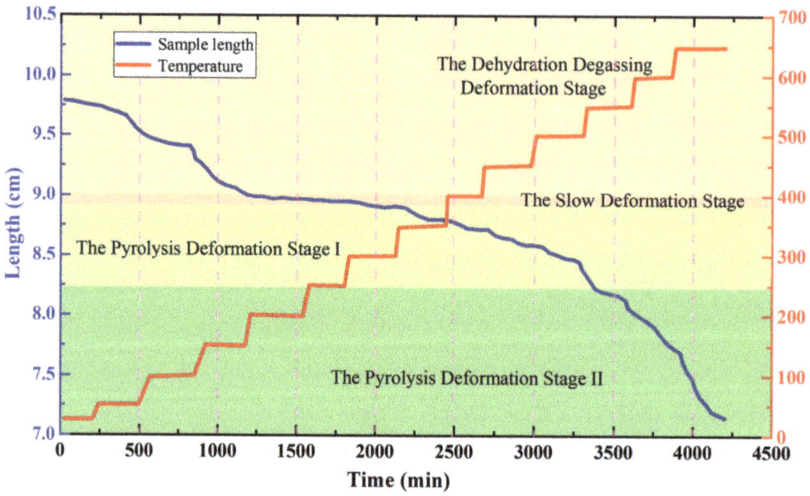

Figure 4. Relationship between sample length, temperature, and time during heating.

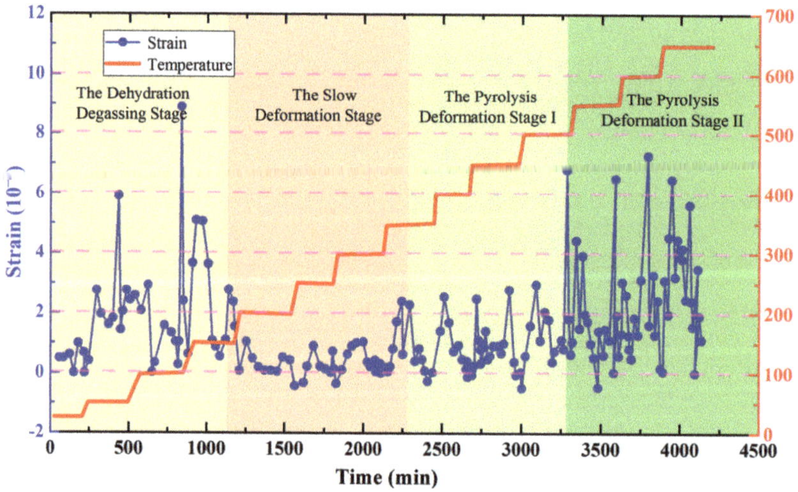

Figure 5. Relationship between sample deformation strain, temperature, and time during heating.

(1) The Dehydration Degassing Deformation Stage

The initial stage involved dehydration, degassing, and deformation from the onset of temperature elevation to 1181 min, namely from room temperature to 200 °C. The sample contracted from 9.7896 cm to 8.9981 cm, resulting in a deformation length of 0.7979 cm and a deformation ratio of 8.15%. At this stage, the free water in the lignite evaporated rapidly at around 100 °C, while the bound water was also destroyed as the temperature increased. The extent of lignite deformation was quite pronounced, especially at approximately 100 °C.

(2) The Slow Deformation Stage

This phase mostly pertained to a temperature range of 200 °C to 300 °C, with a corresponding duration of 1181 min to 2167 min. In comparison to a temperature of 200 °C, the axial length of lignite decreased by 0.096 cm, resulting in a deformation ratio of 0.98%. The strain–time curve indicates that the sample underwent expansion during the initial phase of insulation at 250 °C and 300 °C. At this stage, no chemical reactions occurred in the lignite. Under certain confining pressure conditions, the hard particles within the lignite itself are sufficient to resist the external load, preventing the overall coal matrix skeleton from being damaged. The deformation of lignite occurred gradually, with both the overall deformation and the rate of deformation being minimal.

(3) The Pyrolysis Deformation Stage

Upon reaching a temperature of 300 °C, the sample's deformation transitioned to a new phase, with future deformation mostly influenced by the pyrolysis of the lignite. The deformation process was categorized into two stages: 300 °C to 500 °C and 500 °C to 650 °C. The relevant time intervals were 2367 min to 3298 min and 3298 min to 4116 min, respectively. The ultimate length of the specimen at 650 °C was 7.1971 cm. The total deformation length of the sample was 1.705, and the deformation ratio was 17.42%. The deformation degree of the sample between 300 °C and 500 °C was markedly less than that observed at 500 °C. Between 300 °C and 500 °C, the lignite underwent initial pyrolysis. The carboxyl groups and some oxygen-containing functional groups within the lignite broke, accompanied by the release of various gases. As the temperature rose further to between 500 °C and 650 °C, the lignite underwent further pyrolysis. The macromolecular structure of lignite decomposed, releasing CO_2 and CH_4. During the initial pyrolysis stage, the lignite skeleton retained relatively high strength and a certain load-bearing capacity, with minimal deformation. During secondary pyrolysis, as the lignite underwent reactions such as polycondensation and flow, the lignite skeleton softened, its load-bearing capacity diminished, and deformation became more pronounced.

3.2. Evolution in Porosity Fissure Structure of Lignite by NMR [32]

Figure 6 displays the T_2 spectra of lignite subjected to various pretreatment temperatures, as acquired through NMR analysis. Significant disparities in signal intensity and T_2 time are evident under varying temperature settings. The fundamental idea of NMR T_2 spectral analysis of lignite pore fissures is that the lateral relaxation time of the 1H nuclei in the pore fissure water of lignite correlates with the pore radius. Consequently, an increase in pore size correlates with an extended relaxation time, and conversely.

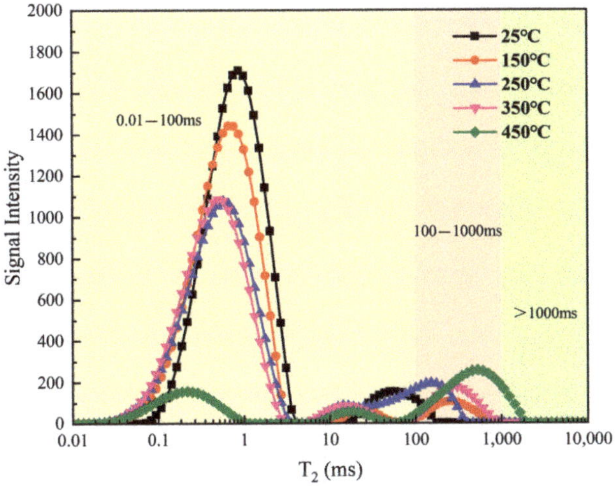

Figure 6. The T2 spectra of lignite.

While the T_2 spectrum can indicate the general architecture of the pores and fractures within lignite, it requires processing to more correctly characterize the distribution of pore sizes. Consequently, by analyzing the data acquired from NMR, the NMR T_2 spectrum is converted into a pore size distribution graph. Figure 7 illustrates the pore size distribution of lignite derived from the lignite T_2 spectrum presented in Figure 6.

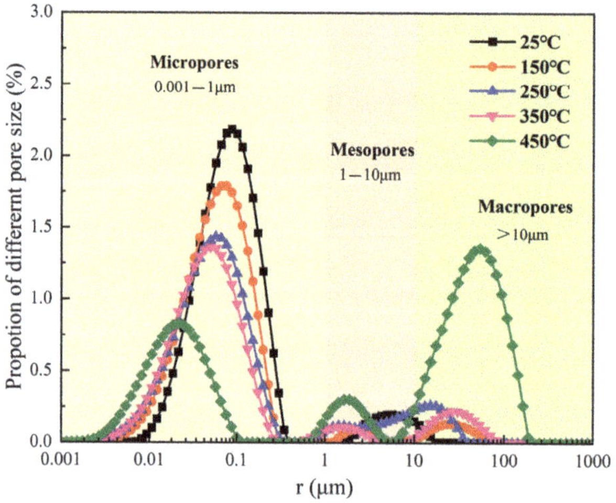

Figure 7. Pore size distribution diagram of lignite.

Based on the pore fracture features of lignite's aperture distribution, the pore cracks in lignite can be classified into three size groups. Pores from 0.001 to 1 μm in diameter are classified as micropores, those measuring 1 to 10 μm are designated as mesopores, and those >10 μm are referred to as macropores or crevices [38]. Figure 8 illustrates a histogram depicting the pore dispersion across varying temperature settings. Figures 7 and 8 illustrate that the lignite was predominantly characterized by small pores at room temperature (25 °C), comprising up to 93.34% of the total pore volume, with a minor presence of mesopores being seen. At a pretreatment temperature of 150 °C, significant cavities emerged in the lignite, although the quantity of mesopores was reduced compared to 25 °C. At pretreatment temperatures of 250 °C and 350 °C, although the pore fissures remain mostly characterized by micropores, the proportion of mesopores to macropores progressively rose. As the temperature increased to 450 °C, macropores supplanted micropores, becoming the predominant feature of the pore fissures in the lignite, constituting over 55%.

The research indicates that the micropores in the lignite predominantly fell within the range of 0.01 to 0.5 μm. As the pretreatment temperature progressively rose, the diameters of the micropores diminished and the proportion of macropores gradually increased. The research indicates that the lignite underwent dehydration and degassing at 150 °C, leading to the formation of many pore fractures. As the pretreatment temperature increased to approximately 250 °C, the volatilization of the lignite generated numerous micropores, while the alteration in mesopores remained minimal. Heat induced intricate physical and chemical interactions inside the lignite, leading to the expansion and amalgamation of micropores, resulting in the formation of new mesopores and macropores, while concurrently reducing the quantity of micropores. The quantity of macropores escalated with warmth, and their openings persisted in enlarging. At around 450 °C, the volatilization of lignite occurred in substantial amounts, leading to the constant increase in pore volumes. At temperatures beyond 300 °C, the lignite became obstructed as it softened and generated tar, leading to a substantial decrease in the overall pore count.

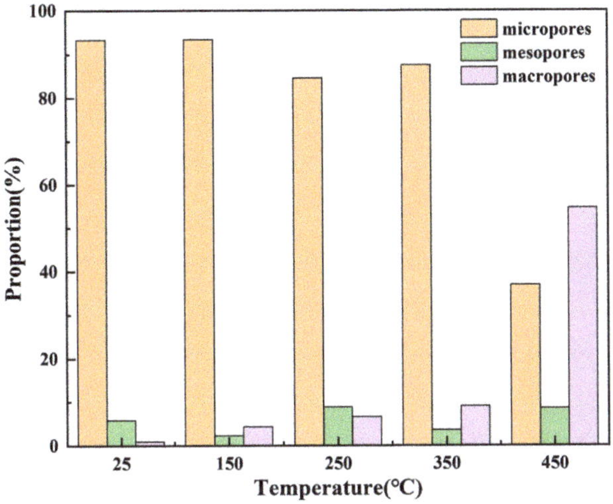

Figure 8. Histograms of pore distribution under different temperature.

Based on the Coates model, the T_2 cutoff value was selected according to the relevant data measured by NMR to calculate permeability, and the parameters related to pore fractures in lignite obtained from NMR testing are shown in Table 2 and Figure 9.

Table 2. The pore fracture-related parameters of lignite based on NMR testing.

Temperature (°C)	Effective Porosity (%)	Free Liquid Porosity (%)	Permeability (md)
25	8.73	2.18	1.37
150	16.94	1.89	5.38
250	23.82	4.27	29.80
350	23.07	3.23	13.08
450	30.22	13.01	856.99

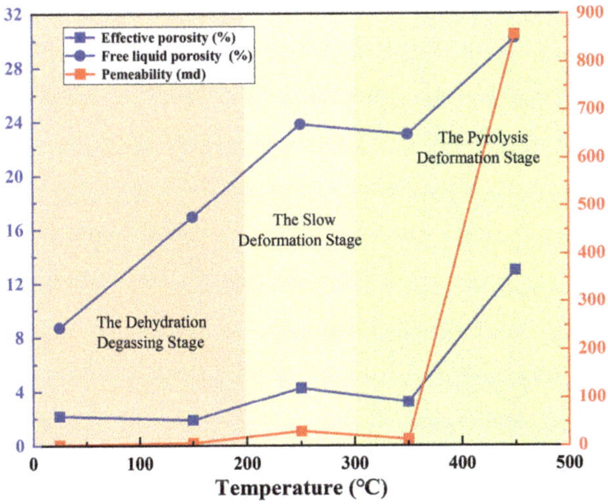

Figure 9. The effective porosity, free liquid porosity, and permeability of lignite under different temperature based on an NMR test.

Based on the analysis of the nuclear magnetic resonance test data, as the lignite temperature increased from 25 °C to 250 °C, its effective porosity rose from 8.73% to 23.82%, and permeability increased from 1.37 md to 29.80 md under dehydration and degassing effects. When the temperature reached 350 °C, the porosity and permeability decreased due to the softening of the lignite, and the production of tar, partial pores, and fractures became blocked, with the porosity being 13.07% and the permeability being 13.08 md. At 450 °C, due to lignite pyrolysis, the internal hard particles underwent thermal cracking, resulting in large fractures at the micrometer scale or above. The void ratio and permeability significantly increased, with the porosity reaching 30.22% and the permeability reaching 856.99 md.

3.3. Evolution in Porosity Fissure Structure of Lignite by X-CT [30]

A two-dimensional image can be insightful and convey information regarding the evolutionary patterns of individual pores and fractures; nevertheless, it is limited in its ability to depict the geographical distribution of lignite pores and fractures. Consequently, a 3D reconstruction of the CT picture was executed. Figure 10 illustrates the three-dimensional reconstructed grayscale computed tomography images of lignite subjected to various temperature treatments, together with images of the biggest pore clusters.

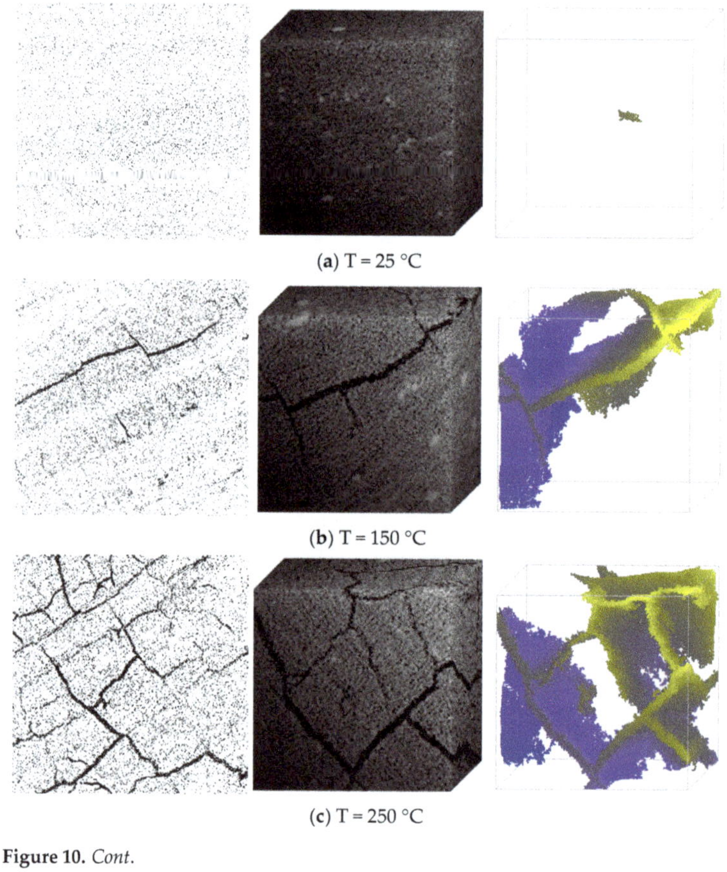

Figure 10. *Cont.*

(**d**) T = 350 °C

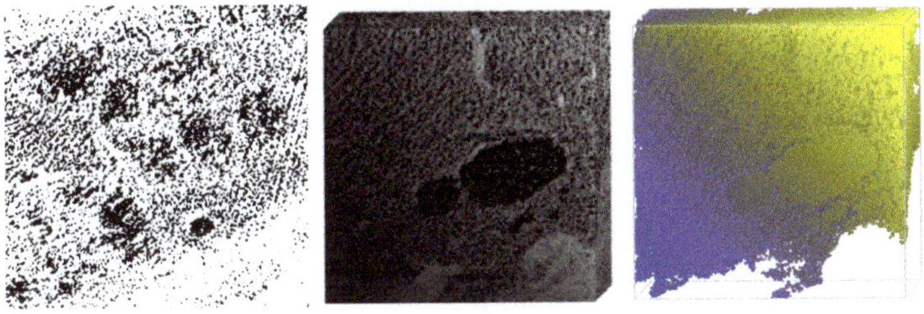

(**e**) T = 450 °C

Figure 10. Scanning binary image (**left**), 3D reconstructed grayscale CT image (**middle**), and maximum pore cluster image (**right**) of lignite under different temperature treatments.

3D reconstructed CT pictures help elucidate the development of pores and fractures in lignite under varying thermal treatments. Consequently, 3D reconstruction and data extraction were conducted on the binarized picture to quantitatively assess the development of the lignite pores and cracks. The primary structural parameters of the pores and fractures are presented in Table 3, while the variation in these properties at different temperatures is illustrated in Figure 11.

As the temperature rose, the porosity and the maximum cluster ratio initially increased, subsequently decreased, and then rose again. The total number of pore clusters initially rose and subsequently declined. At 25 °C, the maximum cluster ratio was 0%. Despite the comparatively high number of pore clusters at 25 °C, these clusters were predominantly tiny and poorly interconnected; hence, the percolation probability of lignite at this temperature was low. At 150 °C, the water and partially adsorbed gases (CO_2, CH_4, and N_2) of the lignite were released, new pores and fractures were formed, the connectivity of pore clusters increased, the porosity of lignite was 3.38%, and the maximum cluster ratio was 1.43%. At 250 °C, the predominant pore cluster, created by the interlinking of smaller pore clusters, nearly encompassed the entire area, the porosity of lignite was 6.54%, and the maximum cluster ratio was 3.41%. As the temperature escalated, at 350 °C, the lignite entered the initial stage of pyrolysis and softened, producing and releasing several volatile gases and tar, which blocked the pores, and the porosity decreased by 4.09%. The porosity of the lignite was merely 2.45%, and the maximum cluster ratio was 0.185%. At 450 °C, the semi-coke generated from lignite pyrolysis underwent further heating and conversion into coke, leading to the release of significant quantities of gas, the pores and fractures within the lignite were effectively interconnected, and the density of coke escalated while the volume diminished. The maximum cluster ratio was 12.57%, and the porosity was 16.02%.

Table 3. Structural properties of the pores and fractures in lignite, as depicted in the 3D reconstructed image.

Temperature (°C)	Effective Porosity (%)	Maximum Cluster Ratio (%)	Total Number of Pore Clusters
25	1.38	0	91,538
150	3.38	1.43	95,590
250	6.54	3.41	101,333
350	2.45	0.185	102,377
450	15.95	12.57	31,394

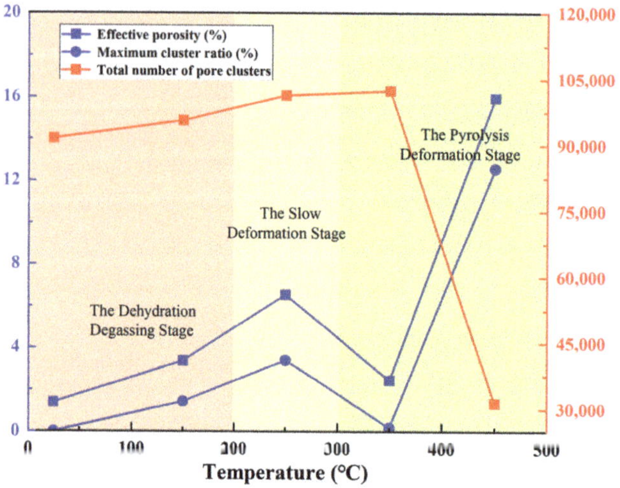

Figure 11. The effective porosity, maximum cluster ratio, and total number of pore clusters of lignite under different temperatures based on an X-CT test.

4. Discussion

These observations indicate that lignite, a complex organic rock, experiences substantial alterations in its internal structure due to variations in temperature and time, and its deformation under certain three-dimensional stress demonstrates a discernible pattern. The deformation law and mechanism of lignite under temperature and pressure are outlined as follows:

Within the temperature range of room temperature to 200 °C, lignite comprises a significant quantity of hard particles, specifically inorganic constituents. As temperature rises, the internal composition of lignite will also increase, and the thermal cracking of lignite is induced by the thermal expansion coefficient among various components, commonly known as the thermal cracking of lignite. At temperatures below 200 °C, the deformation behavior of lignite differs not only from that of other rocks but also from that of other hard coals [39,40]. Lignite exhibits swelling within certain temperature ranges. Nonetheless, in comparison to the internal moisture and partial gas release in lignite, the impact of coal expansion under triaxial pressure is inferior compared to that resulting from dehydration. In particular, as the temperature attains 100 °C, the dehydration of lignite accelerates, and its deformation exhibits no variability throughout the dehydration process. The water in the lignite transforms into vapor, leading to steam accumulation that generates pressure. This pressure forces some cracks in the lignite to penetrate, resulting in decreased pressure within the lignite and subsequent fluctuations and deformation of its structure.

Within the temperature range of 200 °C to 300 °C, lignite dehydration is largely accomplished, and owing to the comparatively low temperature, lignite does not see significant pyrolysis reactions. The carboxyl group starts to decompose with warmth, although the reaction's effect on the overall lignite structure is minimal, and the particle deformation in lignite results in its overall expansion. Consequently, the deformation of lignite is minimal, and its internal porosity nearly attains the maximum value, aligning with the permeability properties of lignite at elevated temperatures, specifically reaching a peak between the range of 250 °C to 300 °C.

The pyrolysis process of lignite within the temperature range of 300 °C to 650 °C can be categorized into two distinct stages. The initial stage primarily involves the pyrolysis reaction, occurring within a temperature range of 300 °C to 650 °C, whereas temperatures beyond 500 °C are classified as the second stage. These two stages correspond to the primary and secondary pyrolysis of lignite. The axial deformation of lignite indicates that secondary pyrolysis occurs at a far faster rate than primary pyrolysis. The lignite structure retains significant strength throughout the initial pyrolysis, while the second stage involves the polycondensation reaction of lignite pyrolysis. Lignite will inevitably soften, melt, flow, and expand across the temperature spectrum, ultimately solidifying as a colloidal entity, including gas, liquid, and solid phases. Lignite exhibits the greatest deformation and the most pronounced deformation within this temperature range, concurrently aligning with the significant fluctuations in lignite permeability seen in this range. This phase represents the critical temperature segment for lignite injection, as evidenced by the high-temperature region. The test results regarding the internal particle condition and permeability of lignite indicate that the skeleton retains significant support capacity, as illustrated in Figure 12.

Figure 12. Cross section and internal cracks of the specimen after triaxial testing.

Through high-temperature triaxial experiments using NMR and X-CT, the deformation and pore characteristics of lignite pyrolysis were studied. The deformation and pore development patterns of lignite under temperature and pressure effects were obtained. As temperature increases, the porosity of lignite rises, its mechanical strength decreases, and significant deformation occurs. High temperatures promote pore development in lignite. Between 250 °C and 650 °C, a large number of highly interconnected pore networks form. Investigating the deformation and pore structure characteristics of lignite is crucial for elucidating its mechanical and permeability features under varying temperature and pressure conditions.

5. Conclusions

This article studied the deformation and pore structure characteristics of lignite under triaxial stress from room temperature to 650 °C through high-temperature triaxial experiments using NMR and X-CT. The following conclusions were drawn:

(1) As temperature increases, the porosity of lignite rises, its mechanical strength decreases, and significant deformation occurs, and high temperatures promote pore development in lignite.
(2) The axial deformation of lignite pyrolysis is divided into three stages: dehydration and degassing at room temperature to ~200 °C, slow deformation between 200 °C and 300 °C, and pyrolysis deformation from 300 °C to 650 °C.
(3) Significant deformation occurs during both the dehydration degassing and pyrolysis deformation stages. Between 250 °C and 650 °C, a large number of highly interconnected pore networks form.

Author Contributions: F.Z.: investigation, conceptualization, data curation, formal analysis, writing—original draft, writing—review and editing. S.N.: conceptualization, writing—review and editing. J.H.: data curation. K.Z.: data curation. Z.Q.: writing—review and editing. All authors have read and agreed to the published version of the manuscript.

Funding: This research received no external funding.

Data Availability Statement: The original contributions presented in this study are included in the article. Further inquiries can be directed to the corresponding authors.

Conflicts of Interest: The author Shiwei Niu was employed by Xinwen Mining Group Co., Ltd. The authors declare that the research was conducted in the absence of any commercial or financial relationships that could be construed as potential conflicts of interest.

References

1. Li, C. *Advances in the Science of Victorian Brown Coal*; Elsevier: Amsterdam, The Netherlands, 2004.
2. Akbarzadeh, H.; Chalaturnyk, R.J. Structural changes in coal at elevated temperature pertinent to underground coal gasification: A review. *Int. J. Coal Geol.* **2014**, *131*, 126–146. [CrossRef]
3. Bhutto, A.W.; Bazmi, A.A.; Zahedi, G. Underground coal gasification: From fundamentals to applications. *Progress. Energyand Combust. Sci.* **2013**, *39*, 189–214. [CrossRef]
4. Li, C.-Z. Some recent advances in the understanding of the pyrolysis and gasification behaviour of Victorian brown coal. *Fuel* **2007**, *86*, 1664–1683. [CrossRef]
5. Mokhahlane, L.S.; Gomo, M.; Vermeulen, D. Temperature and electrical conductivity stratification in the underground coal gasification zone and surrounding aquifers at the Majuba pilot plant. *J. S. Afr. Inst. Min. Metall.* **2018**, *101*, 1053–1058. [CrossRef]
6. Shafirovich, E.; Varma, A. Underground coal gasification: A brief review of current status. *Ind. Eng. Chem. Res.* **2009**, *48*, 7865–7875. [CrossRef]
7. Li, B.; Yang, K.; Ren, C.; Li, J.; Xu, J. An adsorption-permeability model of coal with slippage effect under stress and temperature coupling condition. *J. Nat. Gas Sci. Eng.* **2019**, *71*, 102983. [CrossRef]
8. Zhao, Y.; Wan, Z.; Zhang, Y.; Qu Fang, X.G.; Wei, X.J.; Ma, W. Research and Development of 20MN servo controlled high temperature and high pressure rock mass triaxial testing machine. *Chin. J. Rock Mech. Eng.* **2008**, *27*, 1–8.
9. Perera, M.S.A.; Ranjith, P.G.; Choi, S.K.; Airey, D. Investigation of temperature effect on permeability of naturally fractured black coal for carbon dioxide movement: An experimental and numerical study. *Fuel* **2012**, *94*, 596–605. [CrossRef]
10. Liu, B.; Zhan, Z.; Cui, Z. Experimental study on compression and failure of coal under compression. *J. Coal Sci.* **1983**, *2*, 51–61. [CrossRef]
11. Shen, W.; Zhang, B. Testing of mechanical parameters of coal and rock of different coal ranks. *Chin. J. Rock Mech. Eng.* **2000**, *19*, 860–862.
12. Su, C.; Zhai, X.; Li, Y.; Li, S.; Liu, Z. Analysis of deformation and strength of coal samples under triaxial compression. *Chin. J. Rock Mech. Eng.* **2006**, *25* (Suppl. S1), 2963–2968.
13. Hou, Q.; Zhai, Y.; Han, Y.; Li, X. Discussion on the mechanism of deformation and gas production of coal. *Geol. Notif.* **2014**, *5*, 715–722.
14. Yang, Y. Basic experimental study on strength, deformation and microseismic characteristics of coal and rock. *Chin. J. Rock Mech. Eng.* **2006**, *25*, 1728.
15. Zhou, J.; Wang, G.; Shao, Z. Research on high temperature and high pressure experimental deformation of coal. *J. China Coal Soc.* **1994**, *19*, 324–332.
16. Ma, Z.; Mao, X.; Li, Y. Experimental study on the effect of temperature on coal mechanical properties. *Mine Press. Roof Manag.* **2005**, *22*, 46–48.

17. Bustin, R.M.; Ross, J.V.; Moffat, I. Vitrinite anisotropy under differential stress and high confining pressure and temperature: Preliminary observation. *Int. J. Coal Geol.* **1986**, *6*, 343–351. [CrossRef]
18. Jiang, B.; Qin, Y.; Jin, F. Experimental study on high temperature and high pressure of coal deformation. *J. China Coal Soc.* **1997**, *22*, 82–86.
19. Jiang, B.; Qin, Y.; Jin, F. Deformation characteristics of coal ultrastructure under high temperature and high pressure. *Geol. Sci.* **1998**, *33*, 18–25.
20. Liu, J.; Yang, G.; Ma, R. Macro- and microscopic mechanical behaviour of flow of coal samples experimentally deformed at high temperatures and pressures. *Chin. Sci. Bull.* **2005**, *50* (Suppl. S1), 56–63. [CrossRef]
21. Dong, F.; Feng, Z.; Yang, D.; Lu, X. Study on thermal deformation characteristics of Jimsar oil shale in Xinjiang under high temperature triaxial stress. *Coal Eng.* **2019**, *51*, 103–106.
22. Shkuratnik, V.L.; Filimonov, Y.L.; Kuchurin, S.V. Regularities of acoustic emission in coal samples under triaxial compression. *J. Ming Sci.* **2005**, *41*, 44–52. [CrossRef]
23. Shkuratnik, V.L.; Filimonov, Y.L.; Kuchurin, S.V. Experimental investigations into acoustic emission in coal samples under uniaxial loading. *J. Ming Sci.* **2004**, *40*, 458–464. [CrossRef]
24. Shkuratnik, V.L.; Kuchurin, S.V.; Vinnikov, V.A. Regularities of acoustic emission and thermoemission memory effect in coal specimens unde rvarying thermal conditions. *J. Ming Sci.* **2007**, *43*, 394–403. [CrossRef]
25. Voznesenskii, A.S.; Tavostin, M.N. Acoustic emission of coal in the postlimiting deformation state. *J. Ming Sci.* **2005**, *41*, 291–297. [CrossRef]
26. Nie, B.; He, X.; Wang, E.; Dou, L.; Liu, Z.; Sa, Z. Study on Electromagnetic Radiation and Acoustic Emission in Coal Shear Failure Process. *J. China Univ. Min. Technol.* **2002**, *31*, 65–67.
27. Wang, E.; He, X.; Liu, Z.; Li, Z. Study on the spectral characteristics of acoustic emission of coal body fracture. *J. China Coal Soc.* **2004**, *29*, 289–292.
28. Zuo, J.; Xie, H.; Zhou, H.; Peng, S. Experimental research on thermal cracking of sandstone under different temperatures. *Chin. J. Geophys.* **2007**, *50*, 1150–1155.
29. Wang, D.; Zeng, F.; Wei, J.; Zhang, H.; Wu, Y.; Wei, Q. Quantitative analysis of fracture dynamic evolution in coal subjected to uniaxial and triaxial compression loads based on industrial CT and fractal theory. *J. Pet. Sci. Eng.* **2021**, *196*, 108051. [CrossRef]
30. Liu, W.; Niu, S.; Tang, H. Structural characteristics of pores and fractures during lignite pyrolysis obtained from X-ray computed tomography. *J. Pet. Sci. Eng.* **2023**, *220*, 111150. [CrossRef]
31. Wang, D.; Tian, X.; Wei, J.; Zhang, H.; Yao, B.; Zhang, H.; Chen, C. Fracture evolution and nonlinear seepage characteristic of gas-bearing coal using Xray computed tomography and the lattice Boltzmann method. *J. Pet. Sci. Eng.* **2022**, *211*, 110144. [CrossRef]
32. Liu, W.; Niu, S.; Tang, H.; Zhou, K. Pore structure evolution during lignite pyrolysis based on nuclear magnetic resonance. *Case Stud. Therm. Eng.* **2021**, *26*, 101125. [CrossRef]
33. Asadi, I.; Ji, G.; Baghban, M.H. Evaluating the effective thermal conductivity of cement mortar through X-ray scanning. *Case Stud. Therm. Eng.* **2023**, *41*, 102686. [CrossRef]
34. Zhu, Q.; Wang, X.; Zuo, Y.; Pan, J.; Ju, Y.; Su, X.; Yu, K. Numerical simulation of matrix swelling and its effects on fracture structure and permeability for a high-rank coal based on X-ray micro-CT image processing techniques. *Energy Fuel* **2020**, *34*, 10801–10809. [CrossRef]
35. Cheng, M.; Fu, X.; Kang, J. Compressibility of different pore and fracture structures and its relationship with heterogeneity and minerals in low-rank coal reservoirs: An experimental study based on nuclear magnetic resonance and microCT. *Energy Fuel* **2020**, *34*, 10894–10903. [CrossRef]
36. Zhao, Y.; Lin, B.; Liu, T.; Zheng, Y.; Sun, Y.; Zhang, G.; Li, Q. Multifractal analysis of coal pore structure based on NMR experiment: A new method for predicting T2 cutoff value. *Fuel* **2021**, *283*, 119338. [CrossRef]
37. Zhang, J.; Hu, Y. Comparative evaluation of pore structure heterogeneity in low-permeability tight sandstones using different fractal models based on NMR technology: A case study of benxi formation in the central ordos basin. *Energy Fuel* **2020**, *34*, 13924–13942. [CrossRef]
38. Zhou, S.; Lin, B. *The Theory of Gas Flow and Storage in Coal Seams*; China Coal Industry Publishing House: Beijing, China, 1998.
39. Qu, F. *Experimental Study on Pyrolysis and Mechanical Properties of Coal In Situ*; University of Mining and Technology: Xuzhou, China, 2007.
40. Xie, K. *Domain Reactivity of Coal*; Science Press: Beijing, China, 2002.

processes

MDPI

Article

Research and Application of Gas Drainage Negative Pressure Regulation Method Considering Permeability Differences

Xiaoyu Cheng [1,*], Cheng Cheng [1], Hui Wang [2], Lu Xiao [1,3] and Xingying Ma [1]

1 China Coal Energy Research Institute Co., Ltd., Xi'an 710054, China
2 Shaanxi Qinan Coal Mine Safety Evaluation Services Limited, Xi'an 710054, China
3 Research Center of Fluid Machinery Engineering and Technology, Jiangsu University, Zhenjiang 212013, China
* Correspondence: chengxiaoyu@chinacoal.com

Abstract: This study investigates a dynamic regulation strategy for intelligent gas drainage negative pressure using a comprehensive approach involving numerical simulations, intelligent algorithms, and field experiments. In the numerical simulation component, a permeability evolution model was developed to characterize the area in front of the mining face. Simulations were performed under three negative pressure settings (13 kPa, 18 kPa, and 25 kPa) to investigate the relationships among drainage negative pressure, gas concentration, and flow rate. For the intelligent algorithm, a Long Short-Term Memory (LSTM) prediction model was developed to forecast drainage negative pressure. Based on the predictions, a dynamic regulation strategy for intelligent gas drainage negative pressure was formulated. For field validation, a 120-day in situ experiment was carried out. Intelligent control valves and monitoring instruments were deployed across various sections of the coal seam ahead of the mining face, validating the proposed regulation strategy. The results indicate that permeability is highest in the pressure-relief zone ahead of the mining face and lowest in the stress concentration zone. In the original stress zone, which is unaffected by mining disturbances, the permeability remains unchanged. Drainage negative pressure is positively correlated with gas flow rate, but negatively correlated with gas concentration. In the stress concentration zone, when drainage negative pressure reaches 25 kPa, permeability ceases to be the dominant factor influencing gas flow. At this stage, the pressure gradient between the gas in coal fractures and the drainage system becomes the primary driving force for gas flow. The intelligent dynamic regulation strategy for gas drainage, underpinned by the LSTM prediction model, demonstrated strong performance in field applications. Following intelligent regulation, gas concentrations in various regions showed significant improvement. The findings of this study actively contribute to the advancement of intelligent gas drainage technology.

Keywords: permeability; drainage negative pressure; drainage concentration; negative pressure dynamic regulation

1. Introduction

A considerable number of mines in China are characterized by low permeability, high gas content, and proneness to outbursts, posing serious risks such as gas over-limit incidents, outbursts, and explosions during mining operations [1–3]. Moreover, large-scale gas emissions contribute significantly to environmental pollution [4,5]. Gas drainage serves as a crucial measure for both gas disaster prevention and environmental management. Negative pressure drainage leverages the pressure gradient induced by suction to facilitate gas desorption from the coal seam and its subsequent flow into the drainage pipeline, effectively reducing gas content and pressure within the seam [6–11]. Coal seam permeability is also a key factor influencing gas migration. With the advancement of intelligent mining, smart gas drainage has emerged as an integral component [12–16]. Precise regulation of negative pressure plays a pivotal role in enabling intelligent gas drainage. Therefore, investigating the evolution of coal seam permeability and developing a dynamic negative pressure regulation strategy is of great theoretical and practical significance [17,18].

Cheng Yuanping et al. [19] developed a gas–solid coupled model of gas migration, incorporating matrix pseudo-steady-state diffusion, fracture seepage, permeability evolution, and coal deformation, to analyze the influence of diffusion and seepage on gas transport. Luo Zhiqiang et al. [20] proposed a multifactorial model incorporating intelligent algorithms to accurately forecast gas concentration. Wang Zhaofeng et al. [21] suggested that increasing negative pressure enhances drainage efficiency, though its effect on gas volume becomes marginal over extended drainage durations. Zhou Fubao et al. [22] found that higher negative pressures are associated with increased gas leakage from drainage boreholes. In summary, extensive theoretical research has been conducted on negative pressure regulation. However, current technologies still suffer from limitations, including imprecise control and the lack of real-time regulation capabilities. Xia Tongqiang et al. [23] established safety and efficiency criteria for gas drainage, proposed an adaptive control strategy, and developed a corresponding model based on particle swarm optimization. Liu Jun et al. [24] developed a theoretical model based on coupled coal seam seepage and borehole gas flow to quantify gas pressure variations under different influencing factors. Li Chuantian et al. [25] examined drainage effectiveness via gas pressure analysis and determined the pressure distribution around drainage boreholes. Zhou Aitao et al. [26] developed a three-dimensional gas distribution model for the goaf, assessing the effects of borehole spacing and negative pressure on drainage performance. Gao Han et al. [27] proposed a fully coupled gas–air flow model to evaluate the impacts of drainage duration, pressure levels, and leakage on gas concentration. Zhou Aitao et al. [28] proposed a graph-theory-based gas drainage network model, grounded in an air–gas mixed flow framework.

Furthermore, the influence of mining-induced disturbances on negative pressure regulation remains insufficiently studied. In response to these challenges, this study derives a permeability evolution model for the zone ahead of the mining face. The study also simulates the interrelationship between drainage pressure, gas flow rate, and concentration. Building on these results, an LSTM-based prediction algorithm was employed to formulate a dynamic regulation strategy for intelligent gas drainage. Finally, field experiments were conducted to validate the applicability of the proposed intelligent gas drainage regulation strategy.

2. Coal Permeability Evolution and Gas Flow Equation

2.1. Permeability and Porosity Evolution Equation of Coal Body Under Mining Disturbance

The deformation of the coal body within the seam is affected by various environmental factors, leading to corresponding changes in the pore structure of the coal matrix. During actual coal face mining, porosity and permeability exhibit dynamic variability. Therefore, deriving evolution equations for porosity and permeability is essential for developing a gas–solid coupling model of the mined coal seam [29].

The porosity evolution equation of the coal body, incorporating the initial porosity, is expressed as follows:

$$\phi = 1 - \frac{1-\phi_0}{1+\varepsilon_v}(1 + \frac{\varepsilon_s}{1-\phi_0} - K_Y \Delta p + \gamma \Delta T) \tag{1}$$

In the equation:

ϕ is the porosity of the coal body, %;

ϕ_0 is the initial porosity of the coal body, %;

ε_v is the total volumetric strain of the coal body, as the change in volume per unit volume of the object;

ε_s is the gas adsorption expansion strain of the coal body;

K_Y is the material volumetric compression coefficient, MPa^{-1}; $K_Y = 3(1-2v)/E$; its value is negative because compression leads to a reduction in volume;

E is the elastic modulus of the coal, GPa;

v is the Poisson's ratio of the coal;

Δp is the change in environmental gas pressure, MPa;

γ is the thermal expansion coefficient of the coal body, measured as the degree of thermal expansion of the coal body;

ΔT is the change in environmental absolute temperature, K.

Neglecting temperature fluctuations and gas adsorption-induced expansion, and assuming that the total surface area of coal particles per unit volume remains constant under stress-induced deformation, $\gamma(p) = 0$, the dynamic permeability evolution equation is derived as follows:

$$k = \frac{k_0}{1+\varepsilon_v}\left[1 + \frac{\varepsilon_v}{\phi_0} + \frac{(1-\phi_0)K_Y\Delta p}{\phi_0}\right]^3 \tag{2}$$

In the equation:

k_0 is the initial permeability of the coal body, m^2.

2.2. Gas–Air Flow Equation

A cubic microflow system is adopted as the representative elementary volume (REV) of the coal body seepage system for the formulation of the gas flow equation, as illustrated in Figure 1. Over a short time interval, Δt, and in accordance with the law of mass conservation, the net gas mass flux into and out of the elementary unit, combined with any internal mass sources, equals the total change in gas mass within the system.

$$\frac{\partial Q}{\partial t} = \left(\frac{c\rho_c\rho_0 ab}{(1+bp)^2} + \beta\phi\right)\frac{\partial p}{\partial t} + \beta p\frac{\partial \phi}{\partial t} \tag{3}$$

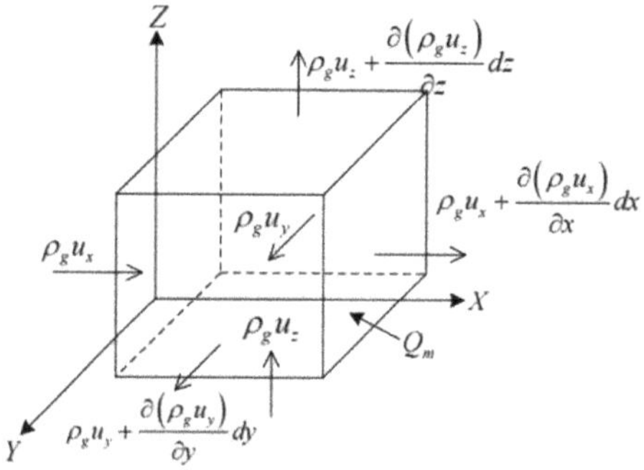

Figure 1. Gas mass conservation in the coal body. Q_m represents the source term, which is taken as 0 in this study; u_x, u_y, u_z represent the seepage velocities in the x, y, and z directions, m/s; $\rho_g u_x$, $\rho_g u_y$, $\rho_g u_z$ represent the seepage fluxes in the x, y, and z directions, in $kg/(m^2{\cdot}s)$.

In the equation:

a is the limiting adsorption capacity, m^3/t;

b is the gas adsorption constant, MPa^{-1};
p is the gas pressure, Pa;
ρ_c is the coal body density, kg/m^3;
ρ_0 is the gas density under standard atmospheric pressure, kg/m^3;
c is the correction factor, with a range of 0 to 1;
$\beta = \rho_n / p_n$;
ρ_n is the gas density at standard conditions, kg/m^3;
p_n is the gas pressure under standard conditions, Pa.
The final gas flow model equation is obtained as:

$$\left(\frac{c\rho_c\rho_0 ab}{(1+bp)^2} + \beta\phi\right)\frac{\partial p}{\partial t} + \beta p\frac{\partial \phi}{\partial t} - \nabla\left(\frac{k}{\mu}\beta p\nabla p\right) = 0 \tag{4}$$

In this study, it is assumed that the air present within the coal matrix originates solely from the diffusion of atmospheric air from the mining face and surrounding roadways. The governing equation for airflow is thus derived as follows:

$$\phi\frac{\partial p_1}{\partial t} + p_1\frac{\partial \phi}{\partial t} - \nabla\left(\frac{k}{\mu}\beta p\nabla p\right) = 0 \tag{5}$$

In the equation:
p_1 is the air pressure in the coal body fractures, Pa.

3. Permeability Evolution Simulation of Coal Body Ahead of the Mining Face

3.1. Establishment of the Geometric Model

As the coal body located farther from the mining face experiences minimal mining-induced disturbances, this study focuses on the region extending from 0 to 100 m in front of the mining face. The geometric model adopted in this study is illustrated in Figure 2. The model represents a rectangular domain measuring 100 m in length and 26 m in width, with a coal seam thickness of 6 m. A 5 m wide excavation zone is positioned at the leftmost boundary of the model to simulate mining activity.

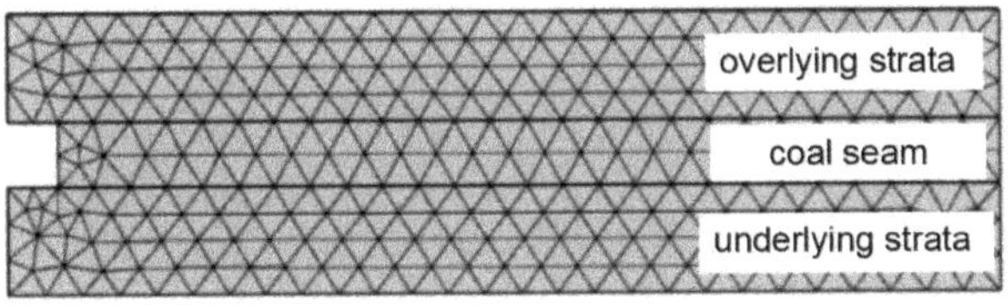

Figure 2. Geometric Model.

3.2. Establishment of the Physical Model

The finite element method is extensively employed to simulate optical, magnetic, and mechanical interactions [30,31]. In this study, COMSOL Multiphysics 6.2 was utilized to simulate the evolution of permeability in the region ahead of the mining face. The simulation incorporated two physics modules—the Solid Mechanics Module and the General Form Partial Differential Equation Module—to capture the coupled interactions among stress, strain, and gas flow within the coal seam. The key simulation parameters for the coal seam are presented in Table 1. The initial gas pressure within the coal seam was set at 1.5 MPa. A uniform load of 5 MPa was applied to the top boundary of the model to simulate the overburden stress, while displacements along the left, right, and bottom boundaries were constrained.

Table 1. Physical Parameters of Coal Seam.

Parameters	Value	Parameters	Value
Apparent density of coal, $\rho_c/\mathrm{kg/m^3}$	1300	Gas density, $\rho_0/\mathrm{kg/m^3}$	0.716
Elastic modulus of coal, E/GPa	2.63	Dynamic viscosity coefficient of gas, $u/\mathrm{Pa{\cdot}s}$	1.08×10^{-5}
Poisson's ratio of coal, v	0.32	Adsorption constant, $a/\mathrm{m^3/t}$	20
Initial porosity of the coal body, $\phi_0/\%$	5	Adsorption constant, $b/\mathrm{MPa^{-1}}$	3
Initial permeability of the coal body, $k_0/\mathrm{m^2}$	8.61×10^{-16}	Correction coefficient, c	1
Volumetric compressibility coefficient of the coal body, $K_Y/\mathrm{MPa^{-1}}$	410		

3.3. Analysis of Simulation Results

Following coal seam excavation, the distribution of vertical stress and corresponding permeability characteristics in the region ahead of the mining face are illustrated in Figure 3. As shown in Figure 3, the vertical stress ahead of the mining face can be categorized into three distinct zones: the pressure-relief zone, the stress concentration zone, and the original stress zone.

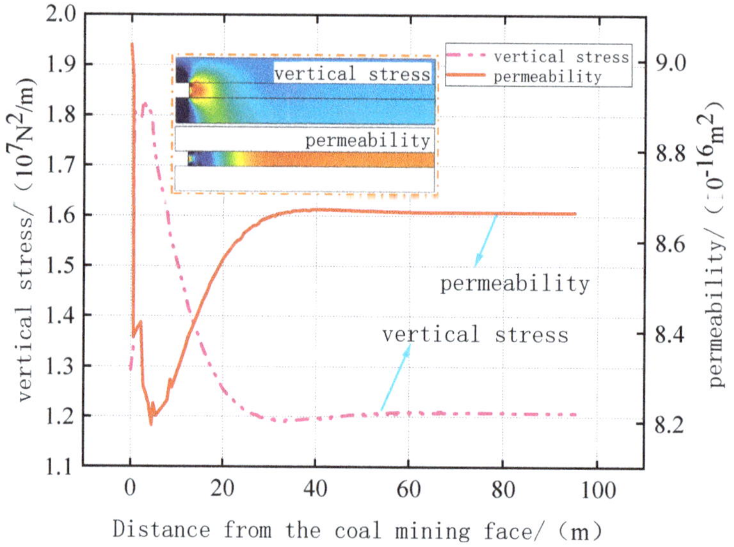

Figure 3. Vertical stress and permeability characteristic curve.

Pressure-relief zone: in this region, the coal seam is initially subjected to high vertical stress, leading to internal compression and fragmentation of the coal matrix. Consequently, vertical stress is alleviated, resulting in increased permeability.

Stress concentration zone: although the vertical stress in this zone remains relatively high, it is insufficient to induce coal failure. The elevated stress compresses fractures within the coal matrix, thereby decreasing permeability. With increasing distance from the mining face, vertical stress gradually diminishes, and permeability correspondingly increases.

Original stress zone: in this region, the vertical stress reflects the in situ stress present prior to mining activities. As the coal matrix remains undisturbed by excavation, its permeability is equivalent to the original in situ value.

4. Study of Gas Drainage Effect Under Different Negative Pressures and Dynamic Negative Pressure Regulation Strategy

To evaluate the influence of varying drainage negative pressures on gas extraction efficiency, three boreholes were strategically positioned within the pressure-relief, stress concentration, and original stress zones ahead of the mining face for simulation analysis. A schematic illustration of the borehole placement is presented in Figure 4.

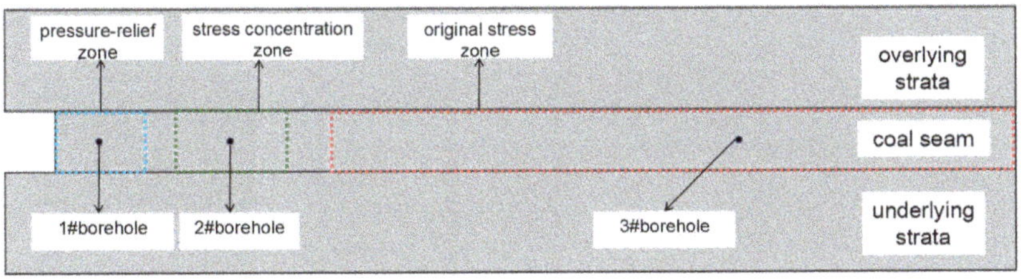

Figure 4. Schematic diagram of gas drainage drilling positions.

4.1. Study on Gas Drainage Effect Under Different Negative Pressures in the Pressure-Relief Zone

The variation trends of the pure gas flow rate and concentration in the borehole located within the pressure-relief zone over the drainage period are illustrated in Figures 5 and 6. As shown in Figure 5, the high permeability of the coal body in the initial phase of gas drainage significantly enhances gas flow. However, as the gas pressure rapidly declines, the pure gas flow rate correspondingly decreases. Higher negative pressure levels accelerate the decline in the pure gas flow rate. Specifically, within the first 0–20 days of drainage, the pure gas flow rate declined by 1.52 m^3/min, 1.59 m^3/min, and 1.88 m^3/min under negative pressures of 13 kPa, 18 kPa, and 25 kPa, respectively. During the mid-to-late stages of drainage, as gas extraction continues to reduce the gas content in the coal matrix, the pure flow rate becomes relatively low. Even with increased negative pressure, the enhancement in the pure gas flow rate remains marginal.

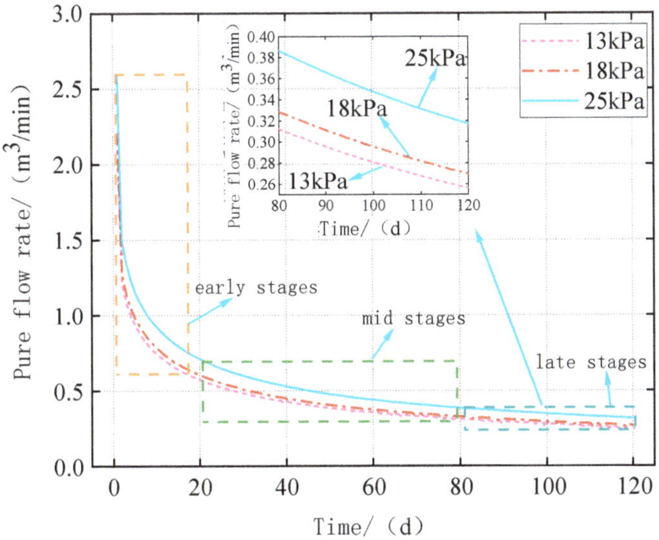

Figure 5. Pure flow rate of gas drainage in the pressure-relief zone.

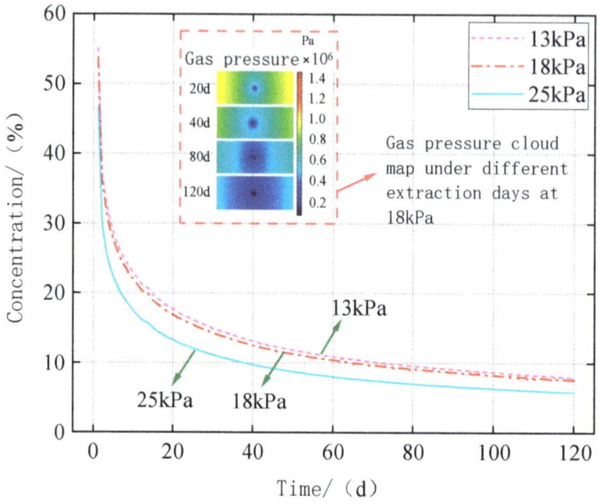

Figure 6. Gas concentration of gas drainage in the pressure-relief zone.

As shown in Figure 6, the high permeability in the pressure-relief zone leads to substantial air leakage from both the mining face and adjacent roadways. The extent of air leakage increases with higher drainage negative pressure. Consequently, during the initial phase of drainage, the gas concentration declines rapidly. In the mid-to-late drainage period, the gas concentration stabilizes at a relatively low level. At 80 days of drainage, gas concentrations under negative pressures of 13 kPa, 18 kPa, and 25 kPa were 9.6%, 9.2%, and 7.1%, respectively.

In the pressure-relief zone, to ensure safe and effective gas extraction, high negative pressure is recommended during the initial drainage phase to reduce the gas content in the coal matrix. As drainage progresses and the pure gas flow rate declines, reducing the negative pressure helps mitigate air leakage and enhance gas concentration. At this stage, gas extraction can be effectively conducted under reduced negative pressure.

4.2. Study on Gas Drainage Effect Under Different Negative Pressures in the Stress Concentration Zone

Figures 7 and 8 illustrate the variations in pure gas drainage flow rate and gas concentration in the borehole located within the stress concentration zone over time.

As shown in Figure 7, during the initial stage of drainage, low permeability causes a slower decline in pure gas flow rate compared to the pressure-relief zone. Under lower negative pressures, the decline in gas pressure is also more gradual. For example, within the first 0–20 days, the pure gas flow rate decreased by 0.35 m^3/min and 0.41 m^3/min under 13 kPa and 18 kPa, respectively. However, under a higher negative pressure of 25 kPa, the increased pressure differential within the coal matrix reduces the dominant role of permeability in governing gas flow. Instead, the pressure gradient between the drainage negative pressure and the gas pressure in coal fractures becomes the primary driver of gas flow. As a result, the gas pressure declines more rapidly under 25 kPa. In the mid-to-late drainage stages, since the gas content in the coal matrix remains higher than in the pressure-relief zone, increasing the negative pressure can moderately enhance the pure gas flow rate.

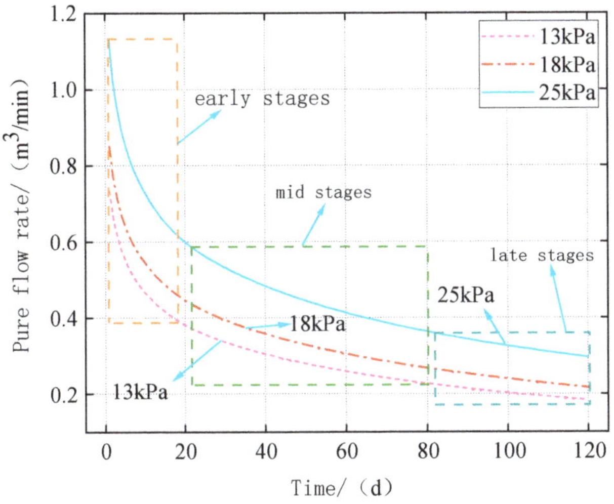

Figure 7. Pure flow rate of gas drainage in stress concentration zone.

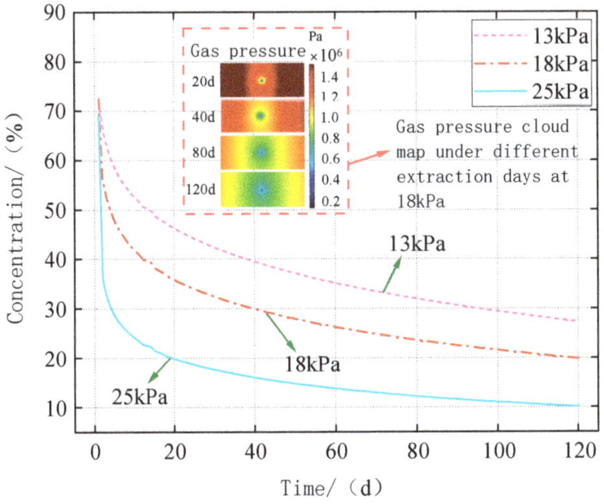

Figure 8. Gas concentration of gas drainage in stress concentration zone.

As shown in Figure 8, due to low permeability and limited air leakage under lower negative pressures, gas concentration decreases more slowly during the early drainage stage than in the pressure-relief zone. Furthermore, in the mid-to-late stages, gas concentration tends to stabilize at relatively higher levels. After 80 days, gas concentrations under 13 kPa and 18 kPa were 31.9% and 37.1%, respectively. However, under 25 kPa, although the pure gas flow rate was higher, the intensified air leakage due to the elevated negative pressure accelerated the decline in gas concentration. Within the first 20 days, the gas concentration dropped by 45.6%, reaching 12.2% by day 80. In the stress concentration zone, high negative pressure should be applied during the initial drainage stage to effectively reduce the gas content in the coal matrix. In the mid-to-late stages, once the gas content falls below the safety threshold and gas concentration declines, reducing the negative pressure can help improve gas concentration levels.

4.3. Study on the Effect of Different Negative Pressures on Gas Drainage in the Original Stress Zone

Figures 9 and 10 illustrate the variation in pure gas drainage flow rate and gas concentration in the borehole within the original stress zone over the drainage period.

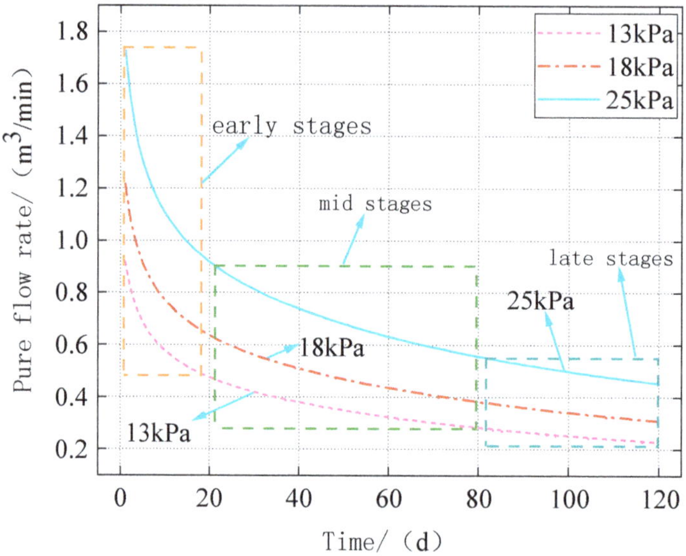

Figure 9. Pure flow rate of gas drainage in the original stress zone.

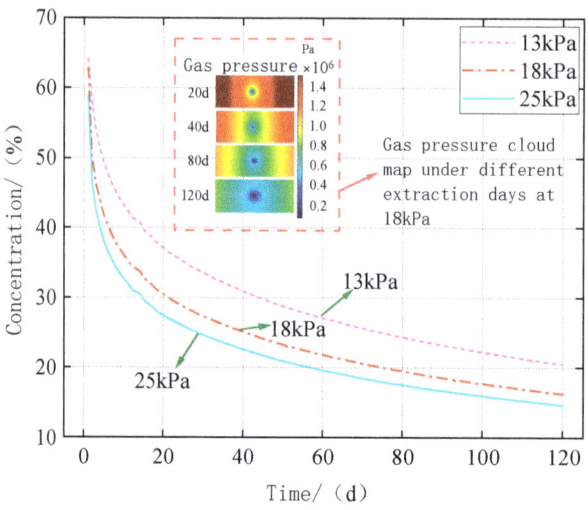

Figure 10. Gas concentration of gas drainage in the original stress zone.

As shown in Figure 9, since the original stress zone remains unaffected by mining disturbances, the permeability of the coal matrix retains its initial state. Gas flow primarily occurs through pre-existing fractures in the coal. Regarding the early stage of drainage, the large pressure differential between the internal gas pressure and the drainage negative pressure causes a rapid decline in gas pressure, leading to a sharp drop in the pure gas flow rate. The higher the applied negative pressure, the faster the decline. For instance, under negative pressures of 13 kPa, 18 kPa, and 25 kPa, the pure gas flow rates decreased by 0.45 m^3/min, 0.58 m^3/min, and 0.81 m^3/min, respectively, during the first 0–20 days. In the mid-to-late stage of drainage, as the pressure differential gradually decreases, the rate of decline in gas flow slows. However, due to the relatively stable gas content in the coal, the flow rate stabilizes at a relatively high level. Further increasing the negative pressure can enhance gas drainage efficiency during this phase.

According to Figure 10, although the original stress zone is unaffected by mining activity, air leakage from the roadway into the borehole still occurs. In the early stage, with permeability remaining at its initial level, the gas concentration decreases more rapidly than in the stress concentration zone. In the mid-to-late stage, as the gas flow rate decreases and air leakage intensifies, the gas concentration remains at a low level. Lowering the negative pressure reduces air leakage and improves gas concentration. For example, at 80 days of drainage, reducing the negative pressure by 5 kPa and 7 kPa resulted in gas concentration increases of 4.8% and 1.9%, respectively.

In the original stress zone, during the early stage, applying higher negative pressure is effective in reducing gas content in the coal. In the mid stage, moderately lowering the negative pressure helps maintain a higher gas concentration. In the late stage, when both gas pressure and gas content are significantly reduced, minimizing the negative pressure effectively improves gas drainage concentration.

4.4. Research on Intelligent Dynamic Regulation Strategy for Gas Drainage Negative Pressure

Based on the relationship between drainage negative pressure and gas drainage concentration obtained in the preceding analysis, this study proposes an intelligent dynamic control strategy for regulating drainage negative pressure. The Long Short-Term Memory (LSTM) prediction algorithm, a widely used model in big data forecasting, is adopted to predict optimal drainage negative pressure over time. Numerical simulation results, together with field-measured drainage data, are used to construct the training dataset for the LSTM model. After undergoing supervised training, the algorithm is capable of dynamically predicting the optimal negative pressure values to enhance drainage efficiency and gas concentration. On the basis of the prediction outcomes, an intelligent, adaptive regulation framework for drainage negative pressure is established. The complete process of intelligent dynamic control is illustrated in Figure 11.

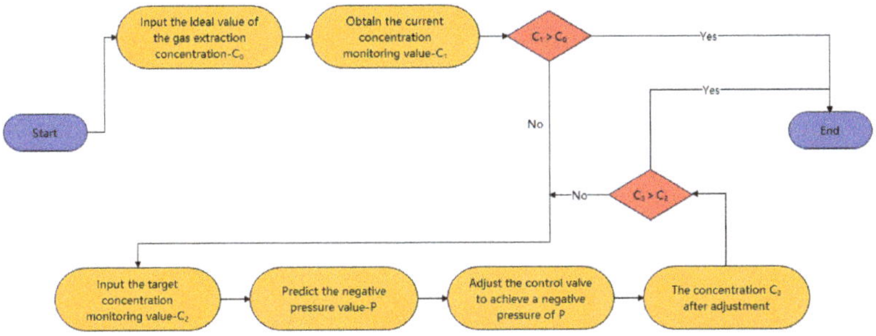

Figure 11. Flow chart of intelligent drainage negative pressure dynamic control.

The intelligent dynamic regulation process for gas drainage negative pressure is outlined as follows:

First, an ideal gas drainage concentration value (C_0) is set as the control target. The actual on-site monitored concentration (C_1) is then collected and compared with C_0. If $C_1 \geq C_0$, no adjustment is required. However, if $C_1 < C_0$, the system enters the concentration regulation mode.

A target concentration value (C_2) is then defined and input into the system. Using the LSTM prediction algorithm, the corresponding optimal drainage negative pressure is forecasted based on the given target concentration.

The system then regulates the drainage valve accordingly, adjusting the negative pressure to the predicted value. After regulation, the new concentration (C_3) is monitored. If $C_3 \geq C_2$, the regulation is considered successful, and the process ends. Otherwise, if $C_3 < C_2$, the regulation is deemed insufficient, and the cycle is repeated until the desired concentration is achieved.

During the drainage process, factors such as mining-induced disturbances and deterioration in borehole sealing may affect gas drainage performance. If multiple regulation attempts fail to meet the target concentration, the system will revise the target value (C_2) and initiate a new adjustment cycle accordingly.

5. Field Application

To validate the accuracy of the above research and assess the applicability of the proposed dynamic regulation strategy for drainage negative pressure, intelligent gas drainage regulation was implemented on-site. Control valves were installed in the pressure-relief zone, stress concentration zone, and original stress zone along the belt roadway of the working face.

A schematic diagram of the valve installation layout is presented in Figure 12, and the actual field installation is shown in Figures 13 and 14.

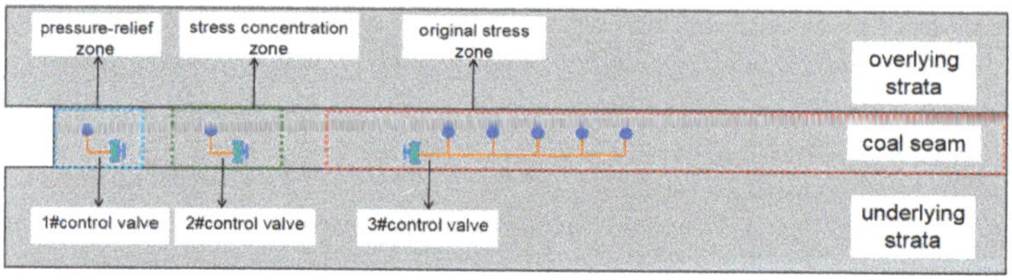

Figure 12. Schematic diagram of valve installation position.

Figure 13. Schematic diagram of single hole installation.

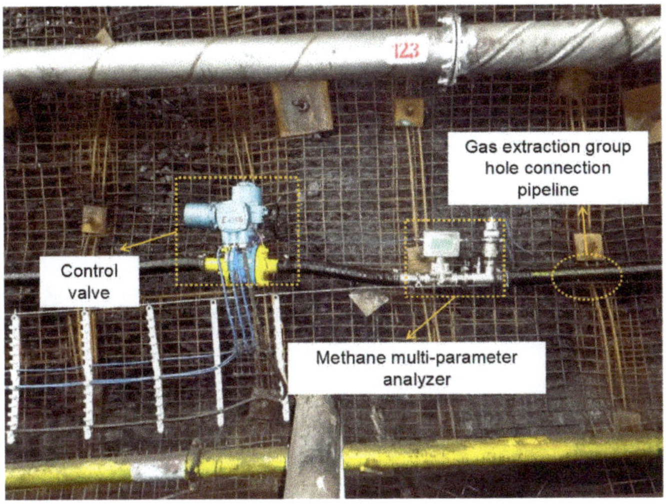

Figure 14. Schematic diagram of parallel holes installation.

Due to the relatively narrow strike widths in the pressure-relief and stress concentration zones, single-hole drainage systems controlled by valves were implemented in these areas. In contrast, the original stress zone, characterized by a wider strike, employed a group-hole drainage approach with valve-based control.

To accurately monitor variations in gas drainage negative pressure and concentration, a methane multi-parameter detector was installed upstream of each valve to measure both parameters in real time. All valves and detectors were integrated with an underground programmable logic controller (PLC) control cabinet. The PLC system transmitted real-time monitoring data to the surface gas drainage monitoring platform, which could also send control commands back to the PLC, thereby enabling remote surface-based regulation of underground valves.

During the trial, gas drainage concentrations were recorded every five days. The target drainage concentration value, C_0, was set at 20%. When measured concentrations fell below this threshold, dynamic regulation of the drainage negative pressure was triggered to restore concentration levels. Regulation was achieved by adjusting the valve opening—reducing the opening effectively lowered the drainage negative pressure.

Gas drainage concentrations were recorded at 5-day intervals throughout the 120-day field trial, with the results presented in Figure 15.

In the stress concentration zone, the gas drainage concentration first fell below the threshold of 20% at day 20. Utilizing the LSTM prediction algorithm to determine the appropriate negative pressure, the valve opening was adjusted to 72%. This adjustment resulted in a wellhead negative pressure of 20.3 kPa and an increase in gas drainage concentration to 32.1%. As drainage progressed, the concentration again dropped below 20% at day 80. A subsequent LSTM-based adjustment reduced the valve opening to 54%, achieving a wellhead negative pressure of 15.6 kPa and raising the concentration to 30.8%.

In the original stress zone, the gas concentration declined below 20% at day 60. Following LSTM prediction, the valve opening was adjusted to 67%, resulting in a concentration increase to 25.3%. At day 80, a further reduction in the valve opening size to 46% improved the concentration to 26.9%.

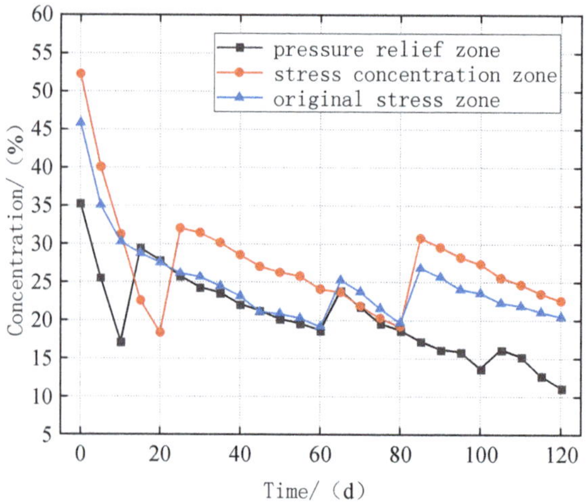

Figure 15. Experimental results.

In the pressure-relief zone, reducing the valve opening in the early stages of drainage significantly enhanced gas concentration. However, in the mid-to-late stages, increased air leakage from the working face and roadway weakened the regulation effect. Although further reductions in valve opening continued to improve gas concentration, the rate of improvement diminished and plateaued at relatively lower levels.

The results of the field trial closely matched the outcomes predicted by numerical simulation. Moreover, the application of LSTM-predicted negative pressure values effectively enhanced gas drainage concentrations. These findings validate the feasibility and practical applicability of the proposed intelligent dynamic regulation strategy for gas drainage negative pressure.

6. Discussion

With the advancement of electronic information and intelligent technologies, numerous traditional manual tasks in coal mining have been effectively automated. In recent years, intelligent mining has experienced rapid development, particularly in areas such as autonomous extraction and disaster prevention and control, where significant breakthroughs have been achieved. Although the adoption of intelligent technologies entails higher initial investments in terms of equipment and infrastructure compared to conventional methods, these technologies substantially enhance both production efficiency and operational safety.

However, as many of these technologies are still in the developmental phase, certain systems remain immature and require further refinement and validation. To accelerate their adoption, two critical aspects must be addressed. First, it is necessary to improve the reliability and robustness of intelligent mining technologies through continuous innovation, thereby expanding their applicability across diverse geological and operational conditions. Second, efforts should be directed toward reducing implementation and maintenance costs, making the technologies more accessible and economically viable for a broader range of mining operations.

By fostering a positive feedback loop between technological advancement and practical application, the coordinated progress of intelligent mining systems and coal mine production can be effectively promoted, ultimately leading to a safer, more efficient, and sustainable mining industry.

7. Conclusions

(1) Numerical simulation results reveal that permeability and drainage negative pressure are the two primary factors influencing gas flow within the coal seam. When the drainage negative pressure is below a certain threshold, gas flow is predominantly controlled by the permeability of the coal. However, once the negative pressure exceeds this threshold, it becomes the dominant factor governing gas migration.

(2) Based on the quantitative relationship between drainage negative pressure and gas concentration derived from simulation studies, an intelligent dynamic regulation strategy for gas drainage was developed using the LSTM prediction algorithm. This strategy offers a robust theoretical framework for real-time, intelligent adjustment of drainage parameters, enhancing the adaptability and responsiveness of gas control systems in underground coal mines.

(3) Engineering field trials confirmed the feasibility and practicality of the proposed intelligent regulation strategy. The system not only significantly improved gas drainage efficiency but also optimized resource utilization by preventing excessive negative pressure application. These results underscore the potential of intelligent technologies to advance gas drainage practices, thereby supporting the broader development of intelligent mining systems.

Author Contributions: Conceptualization, X.C.; methodology, C.C.; validation, X.C., H.W. and L.X.; formal analysis, C.C.; investigation, X.M.; resources, C.C.; data curation, X.M.; writing—original draft preparation, X.C.; writing—review and editing, H.W.; visualization, C.C.; supervision, L.X.; project administration, X.C.; funding acquisition, X.C. All authors have read and agreed to the published version of the manuscript.

Funding: This research is funded by the National Key Research and Development Program of China, grant number 2023YFF0615404.

Data Availability Statement: The data used to support the findings of this study are available from the corresponding author upon request.

Conflicts of Interest: Authors Xiaoyu Cheng, Cheng Cheng, Lu Xiao and Xingying Ma were employed by the China Coal Energy Research Institute Co., Ltd. Author Hui Wang was employed by the Shaanxi Qinan Coal Mine Safety Evaluation Services Limited. The remaining authors declare that the research was conducted in the absence of any commercial or financial relationships that could be construed as a potential conflict of interest. The China Coal Energy Research Institute Co., Ltd. and Shaanxi Qinan Coal Mine Safety Evaluation Services Limited had no role in the design of the study; in the collection, analyses, or interpretation of data; in the writing of the manuscript, or in the decision to publish the results.

References

1. Fan, C.; Li, S.; Luo, M.; Yang, Z.-H.; Zhang, H.-H.; Wang, S. Deep CBM extraction numerical simulation based on hydraulic-mechanical thermal coupled model. *J. China Coal Soc.* **2016**, *41*, 3076–3085.
2. Zhou, A.; Wang, K.; Wu, Z.; Li, S. Research on airflow catastrophic law induced by gas pressure in min. *J. China Univ. Min. Technol.* **2014**, *43*, 1011–1018.
3. Yao, Y.; Wang, Y.; Wu, Y. Research on measurement of extraction radius based on gas content of coal seam and drilling gas flow. *China Min. Mag.* **2020**, *29*, 116–120+128.
4. El Hachem, K.; Kang, M. Reducing oil and gas well leakage: A review of leakage drivers, methane detection and repair options. *Environ. Res. Infrastruct. Sustain.* **2023**, *3*, 012002. [CrossRef]
5. Uddin, J.; Dubey, R.; Balasubramaniam, V.S.; Kabel, J.; Khare, V.; Salimi, Z.; Sharma, S.; Zhang, D.; Yap, Y.K. Progress in electronic, energy, biomedical and environmental applications of boron nitride and MoS_2 nanostructures. *Micromachines* **2024**, *15*, 349. [CrossRef] [PubMed]
6. Wu, B.; Lei, B.; Hua, M.; Gua, H.; Zhao, Z. Parameters of gas tube extraction technology in the upper corner of working face. *J. Min. Saf. Eng.* **2014**, *31*, 315–321.

7. Zhang, W.; He, Z.; Kong, J.; Zhang, R.; Kang, J. Analysis of influencing factors on gas drainage effect of coal seam boreholes. *China Min. Mag.* **2023**, *32*, 98–102.
8. Liu, X. Numerical simulation and field application of sealing depth of gas extraction borehole in bedding. *China Min. Mag.* **2023**, *32*, 386–390.
9. Li, Q.; Xu, C.; Liu, J.; Fang, J.; Zhang, J.; Liu, F. Key technology and practice of directional drilling for gas drainage in all the mining time and space in underground coal mine. *J. China Coal Soc.* **2022**, *47*, 3108–3116.
10. Wang, L.; Liao, X.; Chu, P.; Zhang, X.; Liu, Q. Study on mechanism of permeability improvement for gas drainage by cross-seam cavitation borehole. *Coal Sci. Technol.* **2021**, *49*, 75–82.
11. Wang, D. Study on dynamic coupling law between drilling extraction radius and pre-pumping period in coal seam. *China Min. Mag.* **2020**, *29*, 144–148.
12. Zhang, H.; Qin, Y.; Su, W. Study on influence mechanism of negative pressure on gas drainage effect. *Saf. Coal Mines* **2019**, *50*, 23–26+31.
13. Lin, H.; Ji, P.; Kong, X.; Li, S.; Dou, G.; Li, K. Precise borehole placement model and engineering practice for pre-draining coal seam gas by drilling along seam. *China Coal Soc.* **2022**, *47*, 1220–1234.
14. Ji, C.; Cui, H. Key technology and practice of gas extraction in steep extra-thick coal seam. *Min. Saf. Environ. Prot.* **2021**, *48*, 99–103.
15. Xu, C.; Wang, J.; Du, C.; Zhou, A.; Wang, K. Simulation of attenuation mechanism of negative pressure along the hole length of long boreholes in gas drainage and its influencing factors. *J. Min. Saf. Eng.* **2021**, *38*, 419–428.
16. Zhang, C.; Gao, X.; Wang, S.; Ren, L. Experimental study and application of the influence of fracture development direction of coal seam on gas extraction effect. *Coal Eng.* **2020**, *52*, 96–100.
17. Lu, W. Research on application and development trend of intelligent technology in mining engineering. *China Min. Mag.* **2024**, *33*, 199–202.
18. He, S.; Wang, N.; Gao, B. Research on intelligent development and application of safety monitoring system based on KJ83X(A). *China Min. Mag.* **2021**, *30*, 149–152.
19. Cheng, Y.; Dong, J.; Li, W.; Chen, M.; Liu, K. Effect of negative pressure on coalbed methane extraction and application in the utilization of methane resource. *J. China Coal Soc.* **2017**, *42*, 1466–1474.
20. Luo, Z.; Zhai, H.; Tong, J. Incorporating multi-features and XGBoost algorithm for gas concentration prediction. *China Min. Mag.* **2024**, *33*, 359–363+370.
21. Wang, Z.; Tian, F.; Zhao, B.; Chen, X.J.; Yu, B.Z. The testing of gas drainage efficiency about feather-veined 1000 m length-drillhole. *J. China Coal Soc.* **2010**, *35*, 76–79.
22. Zhou, F.; Wang, X.; Xia, T. A model of safe drainage of coal seam gas. *J. China Coal Soc.* **2014**, *39*, 1659–1666.
23. Xia, T.; Lu, J.; Li, Z.; Duan, H.; Ren, H.; Zhang, Z.; Zhang, Y. Adaptive control strategy and model of gas-drainage parameters in coal seam. *Sustainability* **2022**, *14*, 9247. [CrossRef]
24. Liu, J.; Liu, Z.; Wei, Y.; Chen, X. Change characteristics of negative drainage pressure along the drill hole: Theoretical analyses and field tests. *ACS Omega* **2022**, *7*, 19948–19956. [CrossRef]
25. Li, C.; Sun, Y.; Wu, S.; Sun, X. Investigation on the gas drainage effectiveness from coal seams by parallel boreholes. *Sustainability* **2023**, *15*, 942. [CrossRef]
26. Zhou, A.; Xu, Z.; Wang, K.; Wang, Y.; An, J.; Shi, Z. Coal mine gas migration model establishment and gas extraction technology field application research. *Fuel* **2023**, *349*, 128650. [CrossRef]
27. Gao, H.; Du, F.; Cheng, X.; Zhang, J.; Zhou, A. Study on the influence mechanism of air leakage on gas extraction effect-a numerical case study of the coal mine site in Anhui. *Processes* **2023**, *11*, 2161. [CrossRef]
28. Zhou, A.; Du, C.; Wang, K.; Fan, X.; Wang, D.; Zhao, W.; Gao, H. Research on intelligent control theory and strategy of gas drainage pipe network based on graph theory. *Fuel* **2024**, *357*, 129867. [CrossRef]
29. Tao, Y.; Xu, J.; Cheng, M.; Li, S.C.; Peng, S.J. Theoretical analysis and experimental study on permeability of gas-bearing coal. *Chin. J. Rock Mech. Eng.* **2009**, *28*, 3363–3370.
30. Amirjani, A.; Abyaneh, P.Z.; Masouleh, P.A.; Sadrnezhaad, S.K. Masouleh finite and boundary element methods for simulating optical properties of plasmonic nanostructures. *Plasmonics* **2022**, *17*, 1095–1106. [CrossRef]
31. Zandi, E.; Alemrajabi, A.A.; Emami, M.D.; Hassanpour, M. Numerical study of gas leakage from a pipeline and its concentration evaluation based on modern and practical leak detection methods. *J. Loss Prev. Process Ind.* **2022**, *80*, 104890. [CrossRef]

processes

MDPI

Article

On the Fluid Behavior Response Characteristics During Early Stage of CBM Co-Production in Superimposed Pressure Systems: Insights from Experimental Analysis

Jiewei Ren [1], Qixian Li [2,*], Meichang Zhang [1,*], Jiang Xu [3], Yang Li [2] and Pengbin Yang [4]

1 College of Mining, Liaoning Technical University, Fuxin 123000, China
2 China Coal Research Institute, Beijing 100013, China
3 State Key Laboratory of Coal Mine Disaster Dynamics and Control, Chongqing University, Chongqing 400030, China
4 Shanxi Jinxing Energy Co., Ltd., Lyuliang 035300, China; yangpbxjw@163.com
* Correspondence: liqixian0219@163.com (Q.L.); zhangmeichang@stu.lntu.edu.cn (M.Z.)

Abstract: The fluid disturbance effect is a significant challenge in CBM (CBM) co-production within superimposed pressure systems in China. To address the unique CBM reservoir of superimposed pressure systems, a CBM co-production experimental apparatus for multi-pressure systems has been independently developed. To comprehensively understand fluid behavior during the early stage of CBM co-production, two sets of experiments were conducted using the self-developed physical simulation test device: one in single-production mode and the other in co-production mode. The dynamic response of reservoir fluids and gas production characteristics were analyzed, and the fluid disturbance mechanism under wellbore fluid confluences was explored. The method adopted in this study addresses the issues of traditional co-production equipment, such as the use of series-parallel core holders, small dimensions, limited monitoring capabilities, single loading methods, and the lack of consideration for wellbore co-production flow disturbance and fluid redistribution in superimposed pressure systems. The following results were obtained: ① A flow disturbance effect emerges when fluids from coal reservoirs with different pressure properties converge and mix in a main wellbore. The pressure inside the four horizontal wells simultaneously reaches 1.45 MPa at t = 0.03 min. ② Based on the fluid disturbance effect, the evolution process of wellbore pressure is categorized into two stages: the confluence disturbance stage and the confluence pressure drop stage. ③ This fluid disturbance effect exacerbates the disparities among coal reservoirs, facilitating fluid exchange between the main wellbore and coal reservoirs through branch wellbores. Under the co-production mode, the instantaneous gas production of the No. 1 coal reservoir reaches its maximum negative value at the moment of production, amounting to −3.85 L/min, indicating that a portion of the fluid from high-pressure coal reservoirs flows back into low-pressure coal reservoirs. ④ A dynamic characterization compatibility method is proposed based on the differences in fluid flow between the single and co-production modes during the early stage of CBM production. For example, at t = 0.1 min, the pressure compatibility coefficients of the No. 1–4 coal reservoirs are 0.72, 0.45, 0.34, and 0.33, respectively. The pressure compatibility and production compatibility coefficients exhibit rapid growth during the early stages, followed by a slight decrease during the middle and later stages. ⑤ The worst compatibility performances are observed during the early stage of CBM co-production, but these performances improve as the co-production time extends. ⑥ Optimizing superimposed pressure systems involves progressive co-production: dynamically introducing coal reservoirs, balancing reservoir pressure, minimizing fluid disturbance, and enhancing recovery efficiency.

 https://doi.org/10.3390/pr13041095

Keywords: superimposed pressure systems; CBM co-production; fluid disturbance; compatibility; physical simulation test

1. Introduction

Coal serves as the primary energy resource in China, accounting for 56.0% of its total national energy supply in 2021 [1]. The overuse of coal and oil resources has negatively impacted the country's ecosystems [2,3]. As public awareness of environmental conservation rises, coalbed methane (CBM) resources, regarded as efficient and clean unconventional natural gas resources, have been increasingly developed globally [4,5]. The CBM co-production method in areas with multiple coal reservoirs can convert uneconomic and thick marginal and thin coal reservoirs into valuable reservoirs [6,7].

In eastern Yunnan and western Guizhou, the most abundant resources of CBM are in southern China. The primary coal-bearing strata are the Changxing Formation and the Longtan Formation, with a total CBM resource volume of about 32.04 × 108 m^3, which accounts for about 10% of China's total national CBM resource volume [6,8,9]. In this area, the distances between coal reservoirs are relatively small, and the coal reservoirs are relatively closely spaced, exhibiting the distinctive characteristic of coal reservoir group development. Meanwhile, multiple sets of independent gas-bearing systems are developed vertically, characterized by superimposed pressure systems [10]. In the two pilot demonstration zones of coalbed methane development in eastern Yunnan, there are 8–16 minable coal reservoirs in the Enhong demonstration zone, with an accumulated thickness of 9.35–26.27 m. In the Laochang demonstration zone, there are 5–14 minable coal reservoirs with an accumulated thickness of 14–26 m and an average thickness of 20 m. The pilot demonstration area situated in western Guizhou is the Songhe block within the Tucheng syncline, encompassing 2–24 minable coal reservoirs that collectively possess an accumulated thickness ranging from 3.0 to 29.8 m [11]. In the eastern Guizhou region, CBM wells are primarily concentrated in the Tucheng Syncline and Yangmeishu Syncline in Panxian County. Among these, the Songhe exploration area has deployed clustered well groups GP-1 to GP-9 and the GC-1 CBM parameter well. Currently, the producing CBM wells in the Yangmeishu Syncline include three wells: DC-1, YMC-1, and QSC-1. Notably, the YMC-1 well exhibits high production characteristics, while the GP-3, GP-4, GP-5, GP-6, and GP-8 wells in the Songhe block are classified as medium-production wells, with the remaining wells being predominantly low-production. In contrast, CBM development in the eastern Yunnan region is mainly focused on the Laochang Basin, located in the southeastern part of Fuyuan County [12].

In contrast, the single-well production of CBM wells in eastern Yunnan has not achieved industrial gas flow breakthrough. The main reason for the above phenomenon is that many relatively thin coal seams exist in eastern Yunnan and western Guizhou. The frequent interaction of marine and continental coal measures deposits has resulted in the formation of superimposed pressure systems in the vertical direction, with a complex lithologic combination. The superimposed pressure systems provide superior gas sources and preservation conditions, which provide favorable conditions for CBM development [10–13]. On the other hand, it leads to the superposition of multiple fluid pressure systems to superimpose, making it challenging to regulate the redistribution and transmission of fluid energy among the systems involved in CBM co-production (as shown in Figure 1). It is easy to destroy the dynamic balance effect between systems due to poor compatibility, and the technology of the CBM co-production faces significant challenges. Based on the superimposed pressure systems, the fluid disturbance is common in the CBM co-production, limiting the maximum production capacity release. Hence, the investigation of the fluid flow mechanism underlying CBM co-production in superimposed pressure systems holds significant theoretical importance and practical worth, which is vital for guaranteeing the sustainable development and utilization of CBM resources.

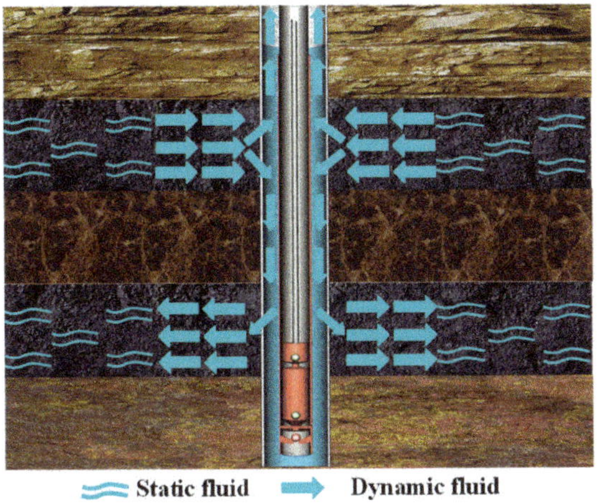

Figure 1. Schematic diagram of the occurrence process of fluid disturbance during CBM co-production.

To accurately understand the fluid effects in multi-gas reservoirs and effectively guide the efficient development of gas wells, a series of test devices have been designed for the development methods of multi-gas reservoirs. The disturbance of fluid in CBM co-production is still in the exploratory stage, and current physical simulation experiments have revealed common understandings [2,6,7,12,14–22]. In the physical simulation device for the co-production of a double-layer coalbed methane reservoir, the convergence point of the two gas flow paths in the wellbore simulation system creates a site for inter-layer interference, and the forward and reverse flow rates of the upper and lower layers are monitored through bidirectional flow meters [6]. The physical simulation device for co-production in superposed gas-bearing systems can simulate the gas and water flow processes during CBM co-production in superposed gas-bearing systems under high-temperature and high-pressure conditions, achieving separate measurements of gas and water flow rates entering and exiting the reservoir specimens [7]. The physical simulation device for combined desorption of double-layer coalbed methane reservoirs integrates the characteristics of multi-layer CBM reservoirs and CBM co-production features, enabling simultaneous simulation of the bottom-hole pressure in each co-production reservoir and each individual reservoir [14]. The physical simulation device for multi-layer co-production uses permeable materials between adjacent coal seams to better reflect actual formation conditions, allowing uniform interlayer crossflow and simulating the fluid flow processes within single layers and between multiple coal seams during CBM co-production [16]. The simulation experiment device for multi-reservoir production damage can replace traditional permeability measurements with constant pressure or constant flow rate methods, simulating the fluid flow state of multi-lithology reservoirs entering the wellbore simultaneously under different pressures [17,18]. In the simulation experiment device for CBM co-production of superposed gas-bearing systems, the coal reservoir simulation unit and the CBM extraction unit are the core components. This device can simulate the gas reservoir and CBM occurrence conditions of stacked gas-bearing systems under different well-type conditions, featuring advantages such as large-scale coal rock specimens, true triaxial stress loading, intelligent data monitoring, and multi-purpose integration [22]. The above research indicates that under the condition of multi-pressure system co-production, the borehole breakthrough disrupts the dynamic balance of fluid energy among various reservoirs, leading to fluid disturbance during co-production. Simultaneously, gases originating from multiple reservoirs interfere with each other as they converge into the wellbore from different pressure systems, affecting gas flow in the reservoirs, causing production damage, and reducing gas production capacity. Upon summarizing the existing physical simulation results, it has been found that multi-pressure system co-production is sensitive to reservoir physical properties and co-production patterns. Differences in interlayer pressure, permeability, effective stress, water saturation, and other factors may all induce fluid disturbance during co-production and production damage to the reservoirs. Strategies such as constant-pressure co-production, constant-production co-production, alternating and progressive production systems, and controlled production pressure differentials may be effective in reducing fluid disturbance and production damage during co-production.

Compared to physical simulation, numerical simulation boasts numerous advantages, such as repeatability and low cost. Currently, research on the dynamic changes of reservoir parameters in CBM co-production predominantly relies on numerical simulation and mathematical modeling. Xiong et al. established a dual-layer non-crossflow homogeneous gas reservoir model to analyze the influence of factors such as gas pressure and reservoir physical properties on co-production. They concluded that when the pressure coefficients are similar, differences in reservoir physical properties are the primary cause of interlayer interference [23]. Guided by the theory of superimposed pressure systems, Fu et al. proposed a progressive drainage strategy: first, the gas-bearing system with higher critical desorption pressure and gas production pressure is drained. When the pressure drops to the critical desorption and production pressures of another gas-bearing system, co-production of the two systems is initiated [24]. Jiang et al. used COMET 3 software to simulate the productivity of CBM co-production in the Bide-Santang Basin in western Guizhou, concluding that reservoir physical properties and hydrogeological conditions form the basis for co-production within and across gas-bearing systems, with permeability differential being the main factor affecting productivity contribution distribution, and the impact of production layer spacing reaching up to 70%, while the influence of reservoir pressure gradient is relatively limited [25]. Li et al. established a numerical model for the synergistic production of CBM and tight gas, systematically analyzing the dynamic changes in permeability and reservoir pressure during co-production, and suggested that single wellbore co-production of coal series gases also experiences interlayer interference due to pressure imbalance [26]. Wang et al. Simulated CBM co-production in the Huangnitang Syncline in Guizhou, with results showing that the effect of pressure-to-storage ratio and permeability on recovery rate is more than twice that of other factors [27]. Wang et al. constructed a geological-mathematical model for CBM co-production production layer groups, discussing the control mechanisms of permeability differential, reservoir fluid pressure difference, and production layer spacing on productivity contribution distribution and interlayer interference from an engineering adaptation perspective [28]. In summary, under CBM co-production conditions, the wellbore's connecting effect merges multiple gas-bearing systems into a single pressure system. When the attribute differences between the gas-bearing systems are significant, the dynamic equilibrium state of fluid energy among the systems is disrupted, leading to interlayer interference or incompatibility phenomena.

To address the aforementioned challenges, an independently developed physical simulation test system for CBM production in large-scale multi-field coupling superimposed pressure systems has been used to perform the physical simulations on CBM co-production in superimposed pressure systems. The fluid dynamic response characteristics of the reservoir and their gas-production characteristics during the early stages of the single production and co-production have been analyzed. Meanwhile, the formation mechanism of fluid behavior under well-hole fluid confluences has been explored, and the disturbance degrees of CBM co-production have been quantitatively evaluated.

2. Simulation Test Method

2.1. Simulation Test Device

The simulation test device is comprised of various modules, including a well-hole simulator, reservoir simulation, CBM production, multi-gas source, test control and data acquisition, real three-axis loading servo control, and coal-rock specimen formation (as shown in Figure 2). This device is a physical simulation test device for multi-field coupling CBM production with the functional advantages of large-scale coal-rock specimens, true triaxial stress loading, intelligent data monitoring, and integration of multi-purposes [29, 30]. Its primary features are ① Simulating the multi-pressure system composed of coal reservoirs and impermeable gas barriers. ② Simulating the real three-dimensional non-evenly distributed complex crustal stress state by controlling nine loading oil cylinders to independently apply stresses to different coal reservoirs. The large-scale physical simulation test device employed in this study effectively eliminates or mitigates the issues associated with the parallel core holder approach in constructing a co-production model, such as the monotony of reservoir samples, limited monitoring data methods, and uniform stress loading forms. This device enables the consideration of fluid pressure transmission characteristics in large-scale heterogeneous and multi-type reservoir samples under true three-dimensional and complex non-uniform geostresses, as well as between adjacent reservoirs.

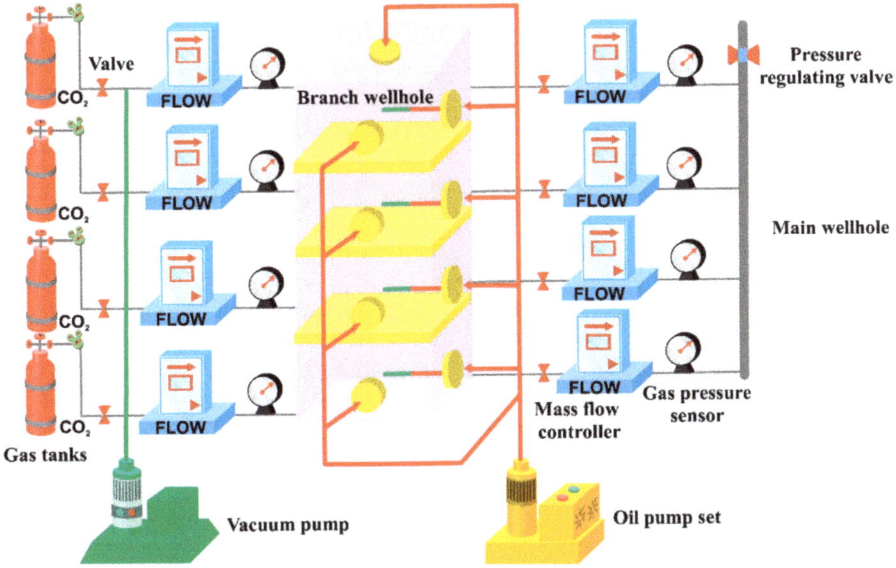

Figure 2. Schematic of the simulation test device.

It should be noted that the inhibitory effect of produced water on gas production in low-pressure coal reservoirs under different pressure systems cannot be ignored. However, since gas mass flow controllers can only monitor dry gases, the presence of moisture or other impurities in the gas can impair their measurement accuracy. Therefore, a gas-liquid separator must be installed at the inlet of the gas mass flow controller, as shown in Figure 3.

Without a gas-liquid separator, a spray-like mixture of gas and liquid will emerge from the outlet. The water produced in this study differs from that produced by on-site drainage, and due to experimental constraints, the role of water in different gas-bearing systems is neglected.

Figure 3. Schematic diagram of the gas-water separator.

2.2. Coal Reservoir Design and Material Selection

A superimposed pressure system is comprised of vertically stacked independent gas-bearing systems [31]. The material basis for its formation is low-permeability thick rock strata with high water and gas resistance between adjacent gas-bearing systems [32–34]. Thus, the gas and water within different gas-bearing systems are isolated, and the vertical gas and water exchanges between different gas-bearing systems are stopped. Given this characteristic of superimposed pressure systems, reproducing the test conditions constructed during the CBM production process in superimposed pressure systems with test methods must have four elements simultaneously: reservoir, CBM, impermeable gas barriers, and well-hole. To prevent inter-layer gas channeling resulting from fracture penetration, impermeable gas barriers are installed between adjacent coal reservoirs, along the four walls of the reservoir simulation module, and within each coal reservoir. Within the reservoir simulation module, there are four superimposed coal reservoirs, with each layer having a size of 212 mm × 380 mm × 390 mm, and the size of the whole sample is 1040 mm × 400 mm × 400 mm. Figure 4 shows the spatial layouts of the coal reservoir and impermeable gas barriers.

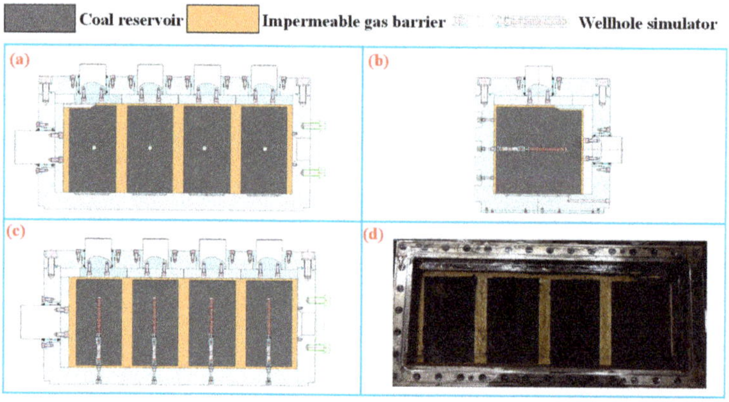

Figure 4. The spatial arrangement of coal reservoirs and impermeable gas barriers. (**a**) Front view, (**b**) left view, (**c**) vertical view, and (**d**) physical picture.

The materials used in the physical simulation test encompass coal reservoir materials, impermeable gas barrier materials, and cementing agents. In these experiments, briquette specimens are employed, utilizing raw coal sourced from the Jinjia coal mine in Guizhou. The test was conducted using crushed coal for two primary reasons. Firstly, acquiring large-scale raw coal is challenging, and its internal primary fractures are highly developed, potentially resulting in significant deviations in test outcomes. Conversely, briquette specimens can be regarded as an isotropic medium, offering more consistent results. Secondly, there are certain similarities between briquette specimens and raw coal in terms of physical mechanics and permeability [35–42].

The production flow chart of the briquette specimens is shown in Figure 5. After undergoing impurity removal, crushing, and screening processes, raw coal with four particle sizes of 0.250–0.425 mm, 0.180–0.250 mm, 0.150–0.180 mm, and 0.000–0.150 mm was obtained. To minimize the differences between the four groups of coal reservoirs, the raw material sources of the four independent coal reservoirs are consistent, and the conditions such as material proportioning, pulverized coal weight, molding pressure, and molding time are kept the same. Clay materials with dense texture, strong cementation, and viscosity were chosen for the impermeable gas barriers to realize their strong water and gas resistance characteristics. The impermeable gas barriers exhibit a wrapping effect on the coal reservoir.

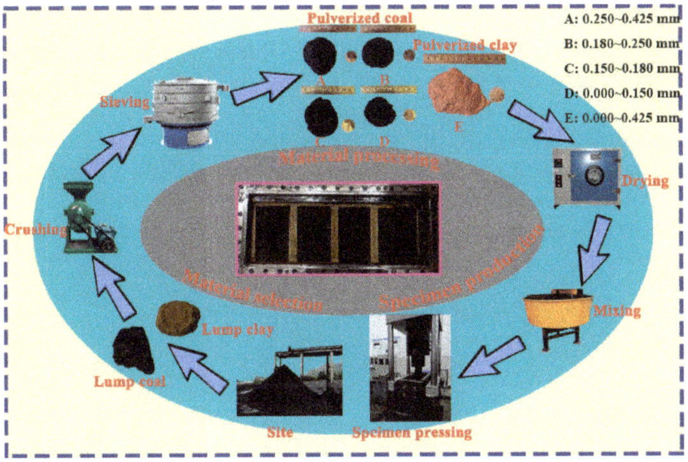

Figure 5. Production flow chart of the briquette specimens.

The optimized mix proportion scheme table of coal reservoir and impermeable gas barriers is detailed in the published papers [43]. Likewise, clay materials undergo the processes of impurity removal, crushing, and screening, with clay powder particles of 0.000–0.425 mm obtained. Cementing agents include the polyvinyl acetate emulsion adhesive and gypsum powder. Specifically, the polyvinyl acetate emulsion adhesive enhances specimen shaping, while the gypsum powder boosts specimen strength. The specific steps for specimen preparation are as follows: the material mixture was loaded into the reservoir simulation, and the specimen was molded under a cold forming pressure of 10.0 MPa for 1.0 h. After the first forming, the height of the specimen was 70 mm. The second batch of material mixture was loaded into the coal specimen box to form the second layer of the coal specimen (i.e., second forming). The extraction pipe and sensors are installed on the surface of the second layer. The third and fourth layers are formed correspondingly. The specimen height after the second, third, and fourth forming is 200 mm, 300 mm, and 390 mm, respectively. The fifth layer, an interlayer, is composed of clay materials, and the specimen height after the fifth forming is 400 mm. The gas pressure sensor model is

GB-Y-J6M, with a measurement range of −0.1 to 6 MPa and an accuracy of ±0.25% F.S. The temperature sensor selected is a platinum resistance thermometer, model Pt100, with a theoretical measurement range of −200 to 850 °C and an accuracy of ±0.15 °C.

2.3. Test Scheme and Steps

Two modes of single production and co-production have been designed in this scheme. The distinguishing feature between these two modes lies in the presence or absence of fluid confluences within the main well-hole. This distinction is further reflected in whether four independent horizontal wells are interconnected with the main well-hole. Four coal reservoirs in the superimposed pressure systems have been numbered, with the reservoirs along the Z axis being reversely numbered as 1, 2, 3, and 4. There is an independent horizontal well within each coal reservoir in the superimposed pressure systems. Therefore, the horizontal wells in the No. 1, 2, 3, and 4 coal reservoirs are numbered I, II, III, and IV. Multiple fluid pressure sensors are placed in different areas of the coal reservoir to obtain the changes in reservoir parameters in different positions during CBM production. The fluid pressure sensors model is GB-Y-J6M, with a measuring range of 0.1~6.0 MPa and a testing accuracy of ±0.25% F.S. The fluid pressure sensor is connected to the measuring point through a PU tube that can ignore the pressure loss. Figure 6a shows four independent cross sections, namely D1, D2, D3, and D4 (z = 141 mm, z = 395 mm, z = 645 mm, and z = 925 mm), and Figure 6b shows the specific positions of fluid pressure sensors in these cross sections. Different crustal stress values are set for different coal reservoirs. That is, loads of σH1, σH2, σH3, and σH4 are sequentially applied to No. 1–4 coal reservoirs along the *X*-axis direction, with loads of σh1, σh2, σh3, and σh4 applied along the *Y*-axis direction, and a load of σv applied along the Z-axis direction. Different initial reservoir pressures of PI, PII, PIII, and PIV are applied to coal reservoirs No. 1–4, respectively. The stress-loading process commenced, with a schematic representation of the stress application illustrated in Figure 7.

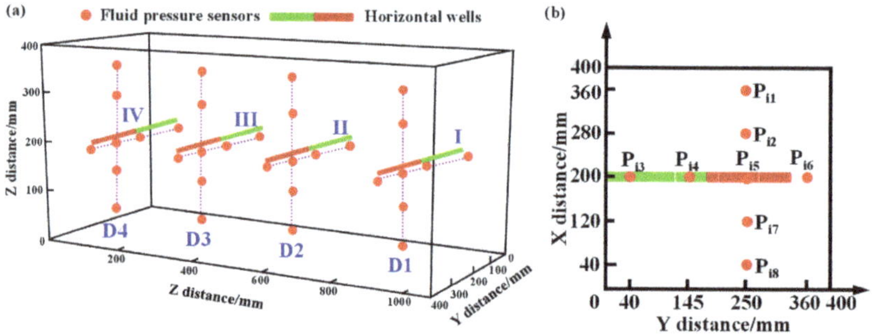

Figure 6. Layout of well-hole simulator and sensor. (**a**) Space layout, and (**b**) nomenclature of sensors.

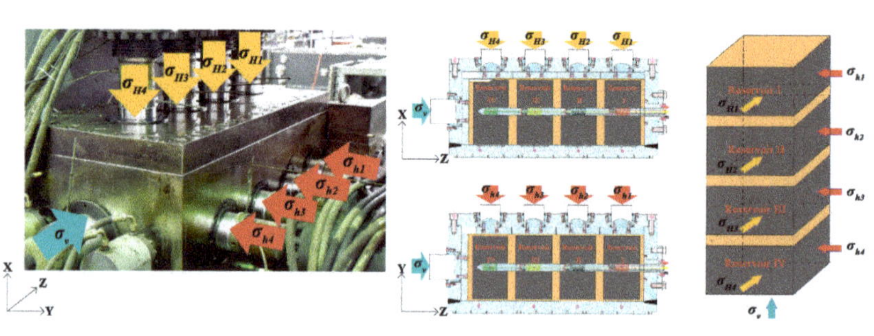

Figure 7. Stress loading scheme.

According to the analysis of the in-situ stress field characteristics in the eastern Yunnan and western Guizhou regions, it is evident that the three principal stresses and initial reservoir pressure exhibit a significant increasing trend with burial depth [44,45]. This results in a distinct vertical zonation of in-situ stress and initial reservoir pressure across different coal seams. Based on similarity theory analysis, the study determined a geometric similarity ratio of 2.31 and a bulk density similarity ratio of 1.04 between the prototype and the model. Consequently, the in-situ stress similarity ratio was calculated to be 2.4. In accordance with the fundamental principles of similarity theory, it is essential to strictly maintain consistency in the similarity ratios of in-situ stress and initial reservoir pressure to ensure the reliability of the model experiments. The study set the burial depth of the No. 1 coal reservoir at 350 m. Based on the vertical in-situ stress calculation formula and the empirical relationship for initial reservoir pressure in the eastern Yunnan and western Guizhou regions [44–46], the vertical in-situ stress of No. 1 coal reservoir was determined to be approximately 4.0 MPa, with an initial reservoir pressure of about 1.0 MPa. Considering that the ratio of minimum to maximum horizontal principal stress typically ranges between 0.4 and 0.8, this study selected the median value of 0.6 as the calculation parameter [47]. The lateral pressure coefficient (defined as the ratio of the average of the two horizontal principal stresses to the vertical principal stress) has a range of 0.39 to 1.95, and a value of 1.0 was adopted for this study [46]. According to the three-dimensional stress relationship theory, the maximum horizontal principal stress of the No. 1 coal reservoir was calculated to be 5.0 MPa, while the minimum horizontal principal stress was 3.0 MPa. Additionally, the reservoir pressure difference between adjacent coal reservoirs was set at 0.4 MPa. Based on the calculated results of the above parameters, the experimental scheme shown in Table 1 was ultimately determined.

Table 1. Test scheme.

No.	Crustal Stress (MPa)									Initial Reservoir Pressure (MPa)				Production Mode
	σ_{H1}	σ_{H2}	σ_{H3}	σ_{H4}	σ_v	σ_{h1}	σ_{h2}	σ_{h3}	σ_{h4}	P_I	P_{II}	P_{III}	P_{IV}	
1		5.0			4.0		3.0			1.0	1.4	1.8	2.2	Single production
2														Co-production

The test procedures follow (shown in Figure 8): ① Preliminary Preparation: Verify the proper functioning of the test system. Conduct air tightness detection to ensure the system is leak-free. Perform sensor calibration to guarantee accurate measurements. ② Vacuum Pumping: Initiate the evacuation of impurity gases from pipelines, cavities, and reservoirs to establish a controlled environment, reducing the reservoir pressure to −0.1 MPa. ③ Stress Loading: Following the vacuum pumping process, apply the predetermined stress to the specimen. In the X direction, σH1, σH2, σH3, and σH4 were set to 5.0 MPa; in the Z direction, σv was maintained at 4.0 MPa; and in the Y direction, σh1, σh2, σh3, and σh4 were 3.0 MPa. ④ Gas Injection and Adsorption: Inject gas incrementally in steps of 0.25 MPa until adsorption equilibrium is attained. Note: Adsorption equilibrium is determined by monitoring the pressure sensor. A stable pressure reading indicates that the system has reached equilibrium. ⑤ CBM Production: Upon completion of the preceding steps, commence the CBM production test. While the CBM co-production process typically spans 6.0 h, this study emphasizes the initial phase, specifically analyzing data collected within the first 10.0 min of the co-production process. This structured approach ensures clarity and enhances the readability of the experimental methodology.

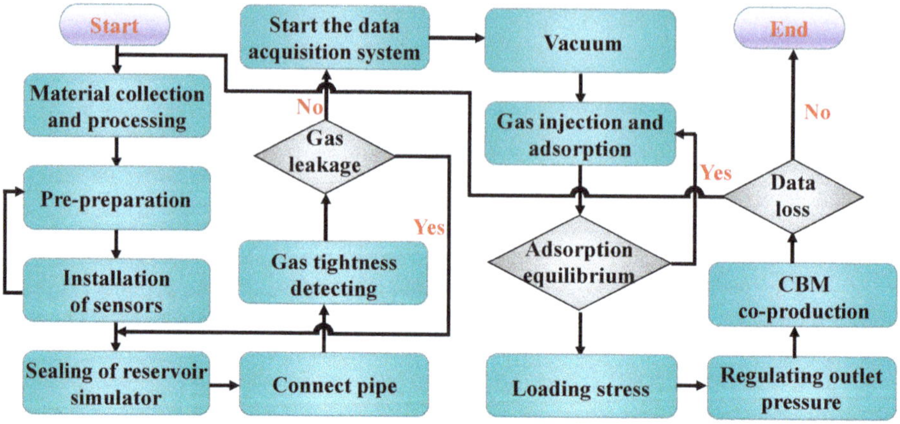

Figure 8. Physical simulation test flowchart.

2.4. Definition of Relevant Parameters

Using the single production mode as a benchmark, a pressure compatibility coefficient and a production compatibility coefficient are constructed from the perspective of dynamic variations of pressure and gas production during CBM production. Subsequently, the dynamic variation characteristics caused by fluid disturbance are explored, and the compatibility of CBM co-production is evaluated.

Based on the condition of the single production mode, the pressure compatibility coefficient is characterized by the ratio of the real-time difference in pressure between the co-production mode and single production mode to the maximum difference. The pressure compatibility coefficient is calculated as follows:

$$p_{i,r} = \frac{P_{i,C} - P_{i,S}}{\max(P_{i,C} - P_{i,S})} \quad (1)$$

Based on the condition of single production mode, the production compatibility coefficient is characterized by the ratio of the real-time difference in production between co-production mode and single production mode to the maximum difference. The production compatibility coefficient is calculated as follows:

$$q_{i,r} = \frac{q_{i,S} - q_{i,C}}{\max(q_{i,S} - q_{i,C})} \quad (2)$$

where p represents the pressure inside horizontal wells and q represents the instantaneous gas production volumes. The subscripts of i, C and i, S, i represents the number of the corresponding coal reservoir, with a value of 1, 2, 3, or 4 representing the No. 1, 2, 3, and 4 coal reservoirs, respectively. The C represents a single production mode, and the S represents a co-production mode.

3. Test Results and Analysis

3.1. Influences of Well-Hole Fluid Confluences on the Pressure Field

Figure 9 illustrates the dynamic evolution curves of pressure within four horizontal wells during the initial phase of CBM co-production. Under the co-production mode, the pressure in horizontal wells I and II demonstrates an increasing trend, while the pressure in horizontal wells III and IV exhibits a decreasing trend during the early stage of co-production. Notably, the pressure in all four horizontal wells simultaneously reaches

1.45 MPa at t = 0.03 min. Subsequently, the pressure curves for all four wells align, showing a consistent downward trend. Based on this observation, the pressure evolution in the horizontal wells can be divided into two stages: Stage 1, the confluence disturbance stage, and Stage 2, the pressure converging and drop stage.

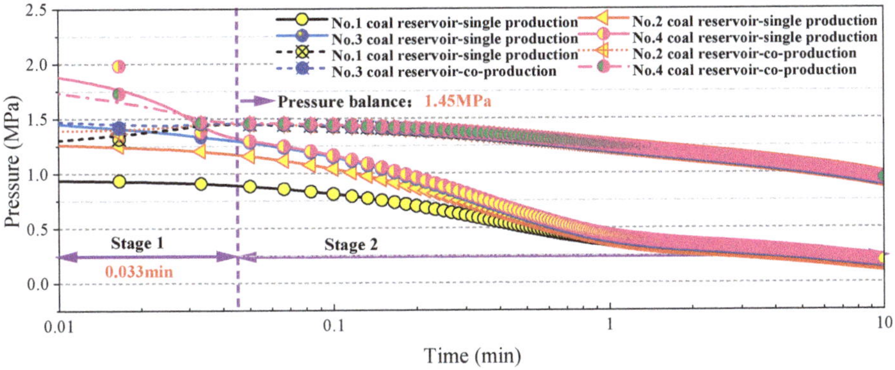

Figure 9. Dynamic evolution curves of pressure inside four horizontal wells during the early stage of CBM production under single-production mode and co-production mode.

Stage 1: The Confluence Disturbance Stage

This stage is attributed to the varying reservoir pressure properties of different coal reservoirs. During this stage, when fluids with different pressure properties converge in the main wellbore, the pressure in the horizontal wells corresponding to the low-pressure coal reservoirs does not decrease but instead rises due to the confluence effect. This phenomenon can be explained from a fluid dynamics perspective: when high-pressure fluids mix with low-pressure fluids in the main wellbore, the low-pressure fluids are pushed by the high-pressure fluids, leading to a temporary pressure increase. This pressure rise is particularly pronounced during the initial co-production period and differs significantly from the pressure evolution observed in single-production mode.

Stage 2: The Pressure Converging and Drop Stage

This stage follows the confluence disturbance stage, during which the pressures in the four horizontal wells reach equilibrium and begin to decline synchronously. In this stage, fluids with different pressure properties are fully mixed in the main wellbore, and the pressure gradient gradually becomes uniform, resulting in the four pressure decline curves becoming largely consistent. The downward trend in this stage reflects the uniform flow characteristics of the co-production fluids in the main wellbore.

In summary, under the co-production mode, fluids originating from coal reservoirs with varying pressure properties converge within the main wellbore, leading to the formation of confluences and subsequently inducing the co-production fluid disturbance effect. In contrast, this effect is absent under the single production mode.

To further investigate the differences in pressure curves between single production mode and CBM co-production modes, a pressure compatibility coefficient was proposed to quantify the degree of pressure difference between single production mode and CBM co-production modes. Specifically, based on the conditions of the single production mode, the pressure compatibility coefficient is characterized by the ratio of the real-time pressure difference between the co-production mode and the single production mode to the maximum pressure difference.

Figure 10 illustrates the evolution curves of the pressure compatibility coefficient during the early stage of CBM co-production. As shown in the figure, the pressure compatibility coefficients of No. 1–4 coal reservoirs all rapidly increase to 1.00 at the beginning of co-production, reaching this value around t = 1.4 min, and then gradually decline as co-production continues. During this stage, the pressure in the No. 1 coal reservoir exhibits significant fluid disturbance phenomena at the early stage of CBM co-production. Fluid exchange occurs between the main wellbore and each coal reservoir through the branch wellbores, meaning that a portion of the fluid from the high-pressure coal reservoir flows back into the low-pressure coal reservoir, creating pressure disturbances and inhibiting the productivity release of the low-pressure coal reservoir. The pressure compatibility coefficient curve of No. 1 is positioned at the top, followed by the curves of No. 2, 3, and 4 in descending order. For example, at t = 0.05 min, the pressure compatibility coefficients of coal reservoirs 1 to 4 are 0.72, 0.45, 0.34, and 0.33, respectively. This indicates that when coal reservoirs with significantly different pressure characteristics are co-produced, the coal reservoir with a higher initial reservoir pressure exhibits stronger resistance to pressure disturbances caused by fluid interference and demonstrates better compatibility. Conversely, coal reservoirs with lower initial reservoir pressures show weaker resistance to pressure disturbances and poorer compatibility. However, as co-production progresses, the pressure compatibility coefficients begin to decline from their peak values. After t = 10 min of co-production, the pressure compatibility coefficients of No. 1–4 coal reservoirs are 0.88, 0.89, 0.89, and 0.90, respectively. Numerically, the pressure compatibility coefficients of No. 1–4 coal reservoirs are relatively close, indicating that fluids from coal reservoirs with different pressure properties mix in the main wellbore, creating a fluid disturbance effect. As the pressures in the lateral wellbores stabilize, the impact of the fluid disturbance effect on coal reservoirs with similar pressure properties weakens, and the degree of influence becomes more uniform.

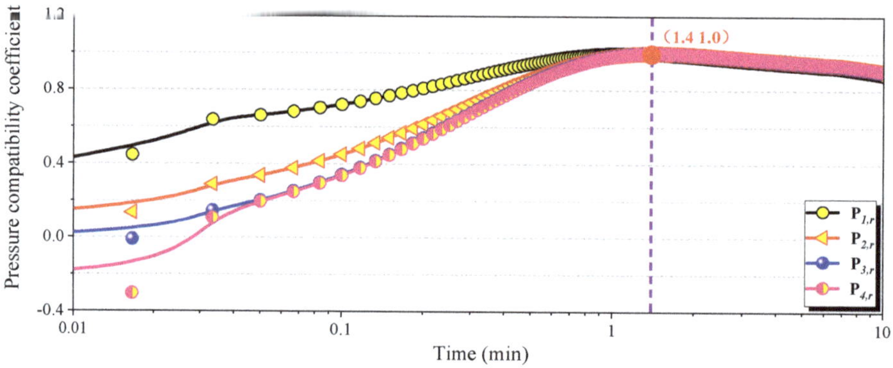

Figure 10. Dynamic evolution curves of the pressure compatibility coefficient during the early stage of CBM co-production.

Figures 11 and 12 show the cloud pictures of reservoir pressure at t = 0.08 min under CBM co-production and single-production modes, respectively. The figures show that under the co-production mode, there is a high-pressure area with a distribution shape of pear around the production section of the horizontal well in the No. 1 coal reservoir. Also, the pressure-rise trend develops towards the inner side of the coal reservoir. The reservoir pressure spatial distribution characteristics of the No. 2 coal reservoir are similar to those of the No. 1 coal reservoir. That is, their pressure rises. However, the pressure increase in the amplitude of the No. 2 coal reservoir is lower than that of the No. 1 coal reservoir. This observation suggests that pressure disturbances are more likely to emerge in low-pressure coal reservoirs and initially manifest in the near-well zones of these reservoirs. Under the single production mode, the pressure spatial distribution characteristics of the No. 1 and 2 coal reservoirs exhibit a distinct pattern. Specifically, a low-pressure area occurs around the production section of the horizontal well, and the pressure decrease trend develops towards the inner side of the coal reservoir. Their spatial pressure distributions differ from those under the co-production mode. The pressure spatial distribution characteristics of the No. 3 and 4 coal reservoirs under the single-production mode are similar to those under the co-production mode. Additionally, low-pressure areas are observed in the vicinity of the production sections of horizontal wells within these two coal reservoirs.

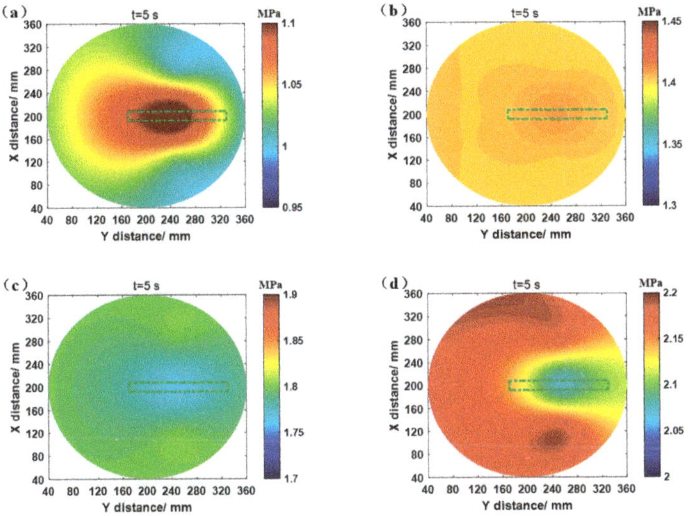

Figure 11. Cloud pictures of reservoir pressure under co-production mode. (**a**) No. 1 coal reservoir, (**b**) No. 2 coal reservoir, (**c**) No. 3 coal reservoir, and (**d**) No. 4 coal reservoir.

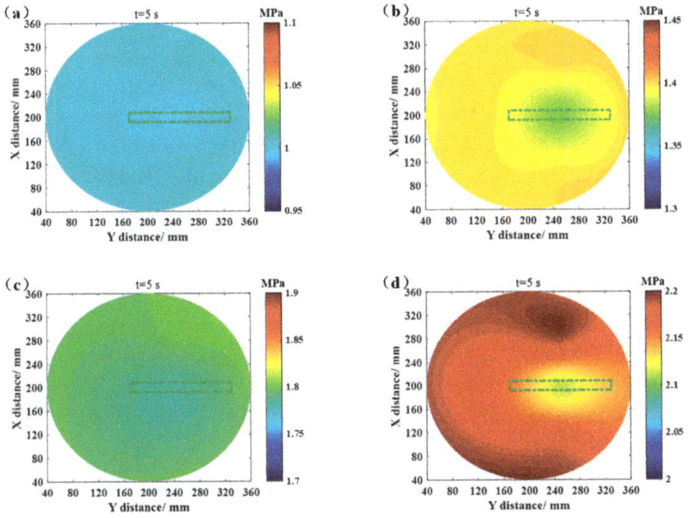

Figure 12. Cloud pictures of reservoir pressure under single production mode. (**a**) No. 1 coal reservoir, (**b**) No. 2 coal reservoir, (**c**) No. 3 coal reservoir, and (**d**) No. 4 coal reservoir.

In summary, during the single production mode, fluids flow continuously from individual coal reservoirs through independent branch shafts. Conversely, during the co-production mode, fluids exiting individual branch shafts converge within the main well hole, resulting in fluid interference between shafts and coal reservoirs. This interference extends into the coal reservoirs, leading to pressure disturbances. Its manifestation is that under the co-production mode, pressure drop rates will decrease, and even the pressure of low-pressure coal reservoirs will increase. According to CBM co-production cases in the Qinshui Basin, Hancheng area, and Laochang block in eastern Yunnan, strong interlayer interference occurs during multi-layer co-production due to phenomena such as fluid backflow, which aligns with the experimental results [48–51].

3.2. Influences of Well-Hole Fluid Confluences on the Flow Field

Figures 13 and 14 describe the flow field diagrams at t = 0.08 min under the CBM co-production and single production modes, respectively. Through observation of flow spatial distribution characteristics of fluids under the co-production mode, the flow vector arrows of fluids in the near-well zones of the No. 1 and 2 coal reservoirs all point to the deep sides of these reservoirs, showing a pattern of centrifugal flow. Also, the closer the areas are to the production sections of horizontal wells in these coal reservoirs, the faster the fluids flow. Under the single production mode, the flow vector arrows of fluids in the near-well zones of the No. 1 and 2 coal reservoirs all point to the production sections of horizontal wells, presenting a pattern of centripetal flow. It shows that under the co-production mode, the dynamic pressure equilibrium between well-holes and coal reservoirs is partially driven by the reverse injection of fluids from the high-pressure coal reservoirs to the low-pressure coal reservoirs. The fluids flow reversely towards the deep sides of the coal. Under both production modes, the flow field diagrams of the reservoirs create the centrifugal flow inside these reservoirs. e No. 3 and 4 coal reservoirs show dense equipotential lines in the near-well zones and sparse lines in the areas far from the wells. Their fluid flow vector arrows all point to the centers of well holes, presenting a pattern of centripetal flow. Although the pressure inside four horizontal wells reached equilibrium at t = 0.03 min with a pressure value of 1.45 MPa, the pressure at t = 0.08 min was higher than the original reservoir pressure of the No. 1 reservoir, indicating that the No. 1 reservoir was still affected by fluid disturbance, resulting in a reverse flow of fluid. The dynamic pressure balance between the coal reservoir and the wellbore drives the fluid from the coal reservoir with a high initial reservoir pressure to be injected in reverse into the coal reservoir with a low initial reservoir pressure, creating a fluid disturbance effect that results in reverse flow within the coal reservoir with a low initial pressure. As the production time extends, the fluid disturbance effect weakens its impact on the fluid within the coal reservoir with low initial pressure, and the reverse flow restores to a centripetal flow.

Figure 15 shows the evolution of fluid flow characteristics at the moment of t = 0.08 min under both production modes. The calculation principle of fluid flow characteristic parameters can be found in the published papers [43]. The figure shows that under the co-production mode, the evolution patterns of fluid flow characteristic parameters of the No. 1 and 2 coal reservoirs are similar, with negative relative flow velocities, deflection angles of about 180°, and absolute deviations of fluid flowing from the optimal flow trajectory trend line. The relative flow velocity of the No. 2 coal reservoir, measured at a distance of 140 mm from the center of the horizontal well's production section, returns to a positive value. This is accompanied by a sharp decrease in the deflection angle, which falls significantly below 90°, indicating a restoration of the fluid flow direction towards the optimal flow trajectory trend line. The relative flow velocities of the No. 3 and 4 coal reservoirs have positive values, with relatively small deflection angles. Under the single production mode, the relative flow velocities of fluids in the near-well zones of the No. 1–4 coal reservoirs all have positive values, with deflection angles below 90°. This observation indicates that fluids flow along the optimal flow trajectory trend line under the single production mode.

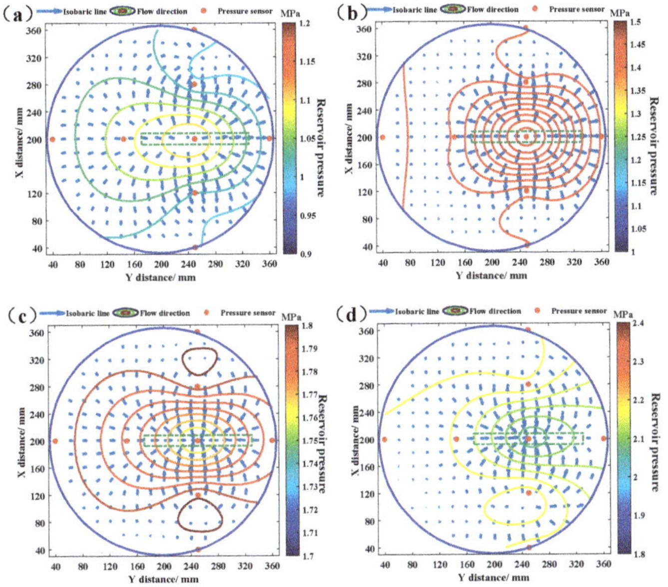

Figure 13. Flow field diagrams under co-production mode. (**a**) No. 1 coal reservoir, (**b**) No. 2 coal reservoir, (**c**) No. 3 coal reservoir, and (**d**) No. 4 coal reservoir.

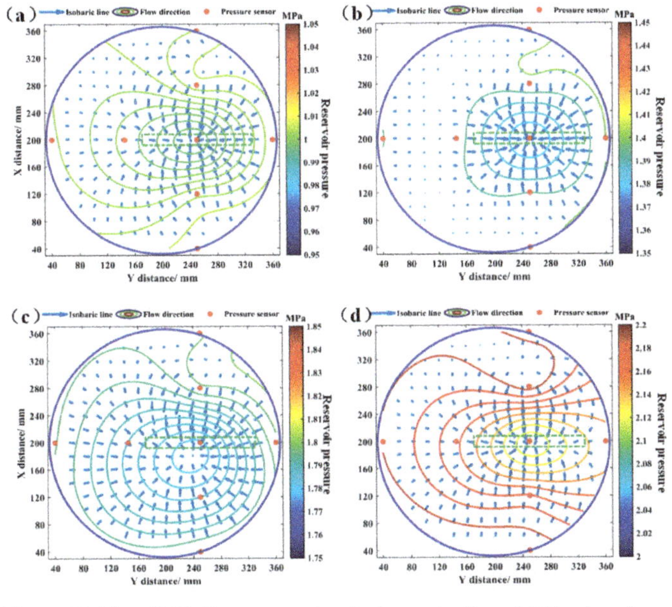

Figure 14. Flow field diagrams under single production. (**a**) No. 1 coal reservoir, (**b**) No. 2 coal reservoir, (**c**) No. 3 coal reservoir, and (**d**) No. 4 coal reservoir.

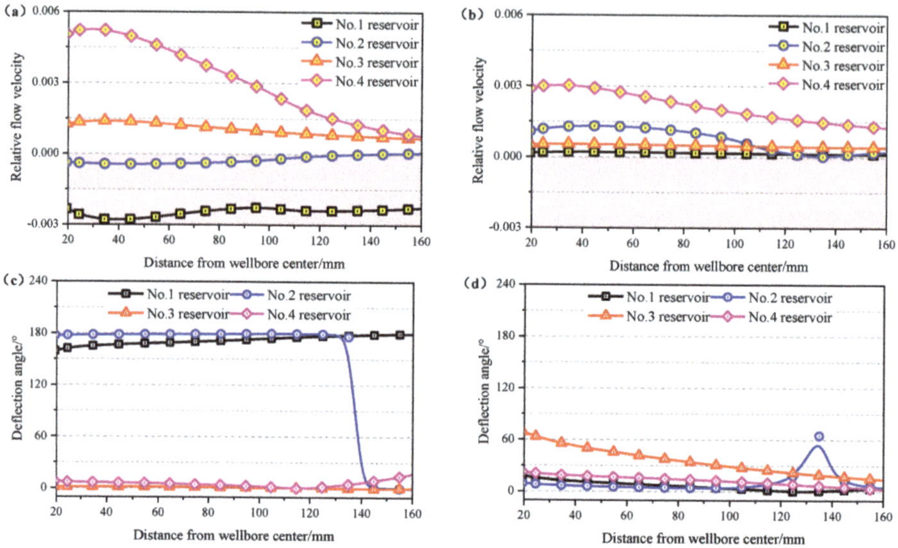

Figure 15. Fluid flow characteristics at the moment of t = 0.08 min under both production modes. (**a**) Relative flow velocity, co-production mode, (**b**) relative flow velocity, single production mode, (**c**) deflection angle, co-production mode, and (**d**) deflection angle, single production mode.

In summary, during the initial stage of CBM co-production, fluids from different coal reservoirs converge into well holes, resulting in interference. This interference, referred to as "well-hole fluid disturbance" in this study, has the potential to inversely impact the flow states of fluids within different coal reservoirs, ultimately altering their flow characteristics. Under the influence of a relatively strong fluid disturbance effect, high-pressure fluids flow into low-pressure coal reservoirs through centrifugal flow. Consequently, this alters the pressure distributions within the low-pressure coal reservoirs, leading to the formation of high-potential zones proximate to the well zones. Additionally, the influences of fluid disturbance on near-well zones are more significant than those in areas far from the wells. The relative reverse flow velocity of fluids exhibits a decreasing trend as the distance from the well hole increases.

3.3. *Influences of Well-Hole Fluid Confluences on Gas Production Characteristics*

Figure 16 shows the characteristic evolution curves of instantaneous gas production under the initial CBM co-production and single-production stages. It can be seen that under the single production mode, the evolution patterns of instantaneous gas production curves of four coal reservoirs are consistent. At the moment of production, their instantaneous gas production quickly reaches its peak values and then rapidly decreases, presenting a characteristic unimodal class exponential decline pattern. The No. 4, 3, 2, and 1 coal reservoirs, in descending order of instantaneous gas production amount, have a peak instantaneous gas production amount of 176.77 L/min, 43.96 L/min, 29.59 L/min, and 10.67 L/min, respectively.

Under the co-production mode, the instantaneous gas production of the No. 1 coal reservoir reaches its maximum negative value at the moment of production, amounting to −3.85 L/min. Then, it rises to 0.0 L/min at the moment of t = 3.6 min. The duration of negative gas production lasts for 3.6 min. The gas production of the No. 2 coal reservoir reaches its peak value at t = 0.53 min, which is 3.5 L/min. Before that moment, a suppression phase of gas production exists, then the production drops. The No. 3 and 4 coal reservoirs exhibit a pattern of instantaneous gas production reaching its peak value first and then decreasing quickly. This is comparable to that observed in the single production mode. The peak values of instantaneous gas production of the No. 3 and 4 coal reservoirs are 10.71 L/min and 21.92 L/min, respectively.

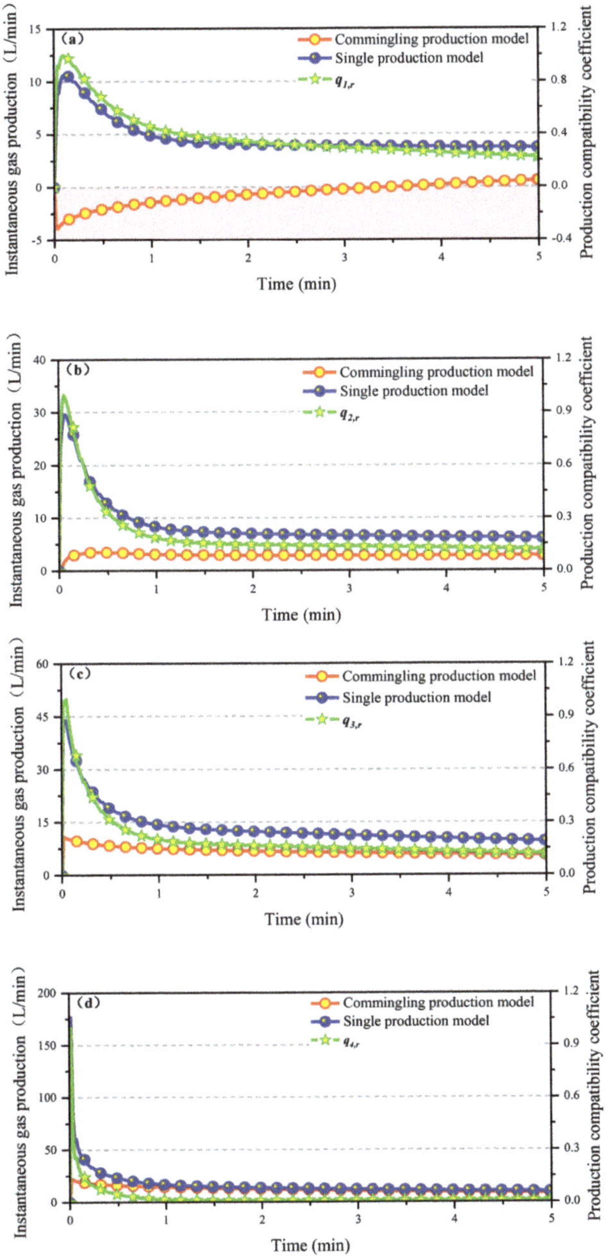

Figure 16. Dynamic characteristic evolution curves of instantaneous gas production during the early stage of CBM production under both production modes. (**a**) No. 1 coal reservoir, (**b**) No. 2 coal reservoir, (**c**) No. 3 coal reservoir, and (**d**) No. 4 coal reservoir.

Figure 17 shows the evolution curves of the productivity contribution rate during the early stage of CBM production under co-production and single-production modes. It can be seen that under the co-production mode, the productivity contribution rates of the No. 1 and 2 coal reservoirs exhibit an upward trend during the early stage. Conversely, the productivity contribution rates of the No. 3 and 4 coal reservoirs exhibit a downward trend. Under the single production mode, the productivity contribution rates of the No. 1–4 coal reservoirs at the moment of production are 3.1%, 9.3%, 11.5%, and 76.1%, respectively. Under the co-production mode, the productivity contribution rates of the No. 1–4 coal reservoirs at the moment of production are −13.0%, 3.2%, 36.0%, and 73.8%, respectively. Meanwhile, the productivity contribution of the No. 1 coal reservoir starts to rise from its initial rate of −13.0%. Then, it returns to a rate of 0.0% at the moment of t = 8.3 min.

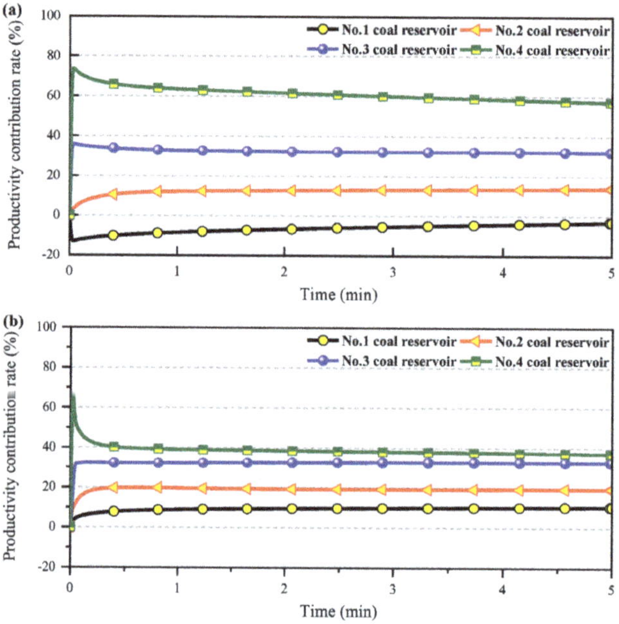

Figure 17. Dynamic characteristic evolution curves of productivity contribution rate during the early stage of CBM production under both production modes. (**a**) Co-production mode, and (**b**) single production mode.

In summary, the co-production mode accentuates productivity differences among coal reservoirs, amplifying the release of high-pressure coal reservoirs while suppressing low-pressure coal reservoirs. The disturbance during the early stage of co-production will force some fluids from high-pressure coal reservoirs to reverse flow into low-pressure coal reservoirs, thus inhibiting the gas production in those low-pressure coal reservoirs. Consequently, peak gas production may lag or even experience reverse flow. The pressure corresponding to individual coal reservoirs is different, so their instantaneous gas production caused by disturbance is also different. High-pressure coal reservoirs that encounter minimal disturbances gradually revert to single-production mode characteristics. However, their gas production capacity is suppressed for low-pressure coal reservoirs that undergo significant disturbances. Based on these findings, this study optimizes production strategies for superimposed pressure systems and proposes a progressive co-production schedule. The core of this schedule lies in the dynamic introduction of gas-bearing layers: initially, CBM from high-pressure coal reservoirs is extracted, and when the reservoir pressure decreases to a level comparable to that of low-pressure coal reservoirs, CBM from the low-pressure reservoirs is co-produced. The primary objective of the progressive co-production schedule is to achieve uniform reservoir pressure (i.e., similar fluid energy) across the superimposed pressure systems, thereby reducing fluid disturbance effects during CBM co-production and ultimately enhancing recovery efficiency.

4. Conclusions

The fluid disturbance effect is a prevalent challenge in coalbed methane (CBM) co-production in superimposed pressure systems in China. This study improved a large-scale experimental technique to evaluate the gas supply characteristics and visualize the fluid flow behavior based on CBM co-production experiments of superimposed pressure systems with four independent pressure systems. The formation mechanism of the fluid disturbance effect under well-hole fluid confluences was explored during the early stage of CBM production. Also, the coordination of CBM co-production has been evaluated through the utilization of the self-built pressure compatibility coefficient and production compatibility coefficient. The primary conclusions are drawn as follows:

a. Current laboratory research primarily focuses on seepage experiments involving series-parallel core samples, while large-scale physical simulation studies on CBM development in superimposed pressure systems are scarce. This study replicates the indoor simulation of superimposed pressure systems, which involve multiple pressure systems coexisting within a reservoir, providing a new approach for the safe and efficient extraction of superimposed pressure systems.
b. Based on the characteristics of superimposed pressure systems, the reconstructed coal simulates the coal reservoir and provides the occurrence place of CBM. The impermeable gas barrier of clay is used as the low-permeability thick rock strata to cut off the fluid connection between coal reservoirs. This work provides a new research approach for revealing fluid disturbance mechanisms and guiding CBM development.
c. The convergence and mixing of fluids from coal reservoirs exhibiting distinct pressure characteristics within the main well hole generate a fluid disturbance effect. The evolution process of well-hole pressure is categorized into two stages: the confluence disturbance stage and the confluence pressure drop stage. Due to well-hole fluid confluences, the well-hole pressure of branch well-holes in the corresponding low-pressure coal reservoirs increases in terms of reservoir pressure. This increase in pressure leads to fluid exchanges between the coal reservoirs and well-holes, manifesting in terms of production.
d. Based on the differences between the fluid disturbance effects of CBM co-production modes, a compatibility method of dynamic characterization has been put forward. The pressure compatibility and production compatibility coefficients exhibit rapid growth during the early stages, followed by a gradual decline during the middle and later stages. The worst compatibility is witnessed during the early stage of co-production, but the compatibility will improve the extension of co-production time.
e. Optimizing production strategies for superimposed pressure systems involves implementing a progressive co-production schedule. The core of this schedule lies in the dynamic introduction of gas-bearing layers: initially extracting CBM from high-pressure coal reservoirs and, once the reservoir pressure decreases to a level comparable to that of low-pressure reservoirs, co-producing CBM from the low-pressure reservoirs. The primary objective of this approach is to achieve uniform reservoir pressure (i.e., similar fluid energy) across the superimposed pressure systems, thereby minimizing fluid disturbance effects during CBM co-production and ultimately enhancing recovery efficiency.

f. This research is currently focused on the flow characteristics of superimposed pressure systems within the reservoir. Future studies should prioritize investigating the coupling effects of low porosity and permeability, gas-water two-phase seepage, the coexistence of multi-phase natural gas, and multi-type reservoirs on the dynamic evolution of the fluid behavior response characteristics induced by fluid interference. Additionally, it is essential to clarify the mechanisms and impacts of reservoir damage caused by the invasion of different fluid phases during CBM co-production and to reveal the coupled flow characteristics of interlayer crossflow and wellbore pipe flow.

Author Contributions: Software, Writing—original draft, J.R.; Conceptualization, Methodology, Software, Writing—original draft, Q.L.; Writing–Review and Editing, M.Z.; Validation, Resources, Supervision, J.X.; Review and Editing, Y.L. and P.Y. All authors have read and agreed to the published version of the manuscript.

Funding: This research was funded by the National Natural Science Foundation of China (No. 52304224).

Data Availability Statement: The data are available from the corresponding author on reasonable request.

Acknowledgments: We also would like to thank the anonymous reviewers for their valuable comments and suggestions that led to a substantially improved manuscript.

Conflicts of Interest: Authors Qixian Li and Yang Li were employed by the China Coal Research Institute, Author Pengbin Yang was employed by the Shanxi Jinxing Energy Co., Ltd. The remaining authors declare that the research was conducted in the absence of any commercial or financial relationships that could be construed as a potential conflict of interest.

References

1. National Bureau of Statistics of the People's Republic of China. Development of the People's Republic of China in 2021. In *Statistical Bulletin of National Economic and Social Development*; National Bureau of Statistics: Beijing, China, 2022.
2. Liu, H.; Su, G.; Okere, C.; Li, G.; Wang, X.; Cai, Y.; Wu, T.; Zheng, L. Working fluid-induced formation damage evaluation for commingled production of multi-layer natural gas reservoirs with flow rate method. *Energy* **2022**, *239*, 122107.
3. Pourhoseini, S.H. Enhancement of radiation characteristics and reduction of NOx emission in natural gas flame through silver-water nanofluid injection. *Energy* **2022**, *194*, 116900.
4. Quan, F.; Wei, C.; Ma, J.; Hao, S.; Song, Y. Modeling analysis of coalbed methane co-production interference: A case study in Eastern Yunnan Basin, China. *J. Nat. Gas Sci. Eng.* **2022**, *103*, 104631.
5. Huang, H.; Bi, C.; Sang, S.; Miao, Y.; Zhang, H. Signature of coproduced water quality for coalbed methane development. *J. Nat. Gas Sci. Eng.* **2017**, *47*, 34–46.
6. Wang, Z.W.; Qin, Y. Physical experiments of CBM co-production: A case study in Laochang district, Yunnan province, China. *Fuel* **2019**, *239*, 964–981.
7. Guo, C.; Qin, Y.; Sun, X.; Wang, S.; Xia, Y.; Ma, D.; Bian, H.; Shi, Q.; Chen, Y.; Bao, Y.; et al. Physical simulation and compatibility evaluation of multi-seam CBM co-production: Implications for the development of stacked CBM systems. *J. Petrol. Sci. Eng.* **2021**, *204*, 108702.
8. Li, S.; Tang, D.Z.; Pan, Z.J.; Xu, H.; Guo, L.L. Evaluation of coalbed methane potential of different reservoirs in western Guizhou and eastern Yunnan, China. *Fuel* **2015**, *139*, 257–267.
9. Lau, H.; Li, H.; Huang, S. Challenges and opportunities of coalbed methane development in China. *Energ. Fuel* **2017**, *31*, 4588–4602.
10. Qin, Y.; Moore, T.A.; Shen, J.; Yang, Z.B.; Shen, Y.L.; Wang, G. Resources and geology of coalbed methane in China: A review. *Int. Geol. Rev.* **2018**, *60*, 777–812.
11. Qin, Y.; Wu, J.G.; Zhang, Z.G.; Yi, T.S.; Yang, Z.B.; Jin, J.; Zhang, B. Analysis of geological conditions for coalbed methane co-production based on production characteristics in early stage of drainage. *J. China Coal Soc.* **2020**, *45*, 241–257.
12. Wang, Z.W. *Simulation and Geological-Mathematical Models of CBM Co-Production Interference: A Case of Longtan Formation in Western Guizhou and Eastern Yunnan, China*; China University of Mining and Technology: Xuzhou, China, 2021.

13. Qin, Y.; Wu, J.G.; Shen, J.; Yang, Z.B.; Shen, Y.L.; Zhang, B. Frontier research of geological technology for coal measure gas joint-mining. *J. China Coal Soc.* **2018**, *43*, 1504–1516.
14. Liang, B.; Shi, Y.; Sun, W.; Fang, S.; Jia, L.; Zhao, H. Experiment on influence of inter layer spacing on combined desorption of double-layer coalbed methane reservoir. *J. China Univ. Min. Techno.* **2020**, *49*, 54–61.
15. Guo, X. *Production Prediction of Multilayered CBM Reservoirs and Optimization of Production Strategy;* China University of Petroleum: Beijing, China, 2019.
16. Liu, L. *Experimental Study on Interlayered Crossflow of Multilayered CBM Reservoir in Eastern Yunnan and Western Guizhou;* China University of Petroleum: Beijing, China, 2019.
17. Zheng, L.; Li, X.; Su, G.; Zhao, W.; Kong, X.; Tao, X. Applicability of working fluid damage assessment methods for coalbed methane reservoirs. *Nat. Gas Ind.* **2018**, *38*, 34–45.
18. Zheng, L.; Tao, X.; Wei, P.; Wu, T.; Liu, H.; Cao, Z. Multi-reservoir production damage physical simulation system and its application in coal-measure gas production. *J. China Coal Soc.* **2021**, *46*, 2501–2509.
19. Wang, L.; He, Y.; Wang, Q.; Liu, M.; Jin, X. Improving tight gas recovery from multi-pressure system during commingled production: An experimental investigation. *Nat. Resour. Res.* **2021**, *30*, 3673–3694.
20. Liu, G.F.; Meng, Z.; Luo, D.Y.; Wang, J.N.; Gu, D.H.; Yang, D.Y. Experimental evaluation of interlayer interference during commingled production in a tight sandstone gas reservoir with multi-pressure systems. *Fuel* **2020**, *262*, 116557.
21. Wang, L.; Yang, S.L.; Liu, Y.C.; Xu, W.; Deng, H.; Meng, Z.; Han, W.; Qian, K. Experiments on gas supply capability of commingled production in a fracture-cavity carbonate gas reservoir. *Petrol. Explor. Dev.* **2017**, *44*, 824–833.
22. Xu, J.; Zhang, C.L.; Peng, S.J.; Jia, L.; Guo, S.C.; Li, Q.X. Multiple layers superposed CBM system commingled drainage schedule and its optimization. *J. China Coal Soc.* **2018**, *43*, 1677–1686.
23. Xiong, Y.; Zhang, L.; Yang, F.; Lu, Y.; Hu, S. Study on technical policy and threshold value of multiple-zone production with one well for multi-layered gas reservoir. *Nat. Gas Ind.* **2005**, *25*, 81–83.
24. Fu, X.; Ge, Y.; Liang, W.; Li, S. Pressure control and fluid effect of progressive drainage of multiple superposed CBM systems. *Nat. Gas Ind.* **2013**, *35*, 35–39.
25. Jiang, W.; Wu, C.; Wang, Q.; Liu, Y. Interlayer interference mechanism of multi-seam drainage in a CBM well: An example from Zhucang syncline. *Int. J. Min. Sci. Technol.* **2016**, *26*, 1101–1108.
26. Li, Y.; Meng, S.; Wu, P.; Wang, Z.; Yu, Z. Numerical simulation of coal measure gases co-production. *J. China Coal Soc.* **2018**, *43*, 1728–1737.
27. Wang, S.; Wang, Y.; Zhang, Z.; Li, Z.; Feng, Y. Analysis of influencing factors on productivity of coalbed methane of combined multi-layer drainage wells. *Coal Sci. Technol.* **2019**, *47*, 126–131.
28. Wang, Z.; Qin, Y.; Li, T.; Zhang, X. A numerical investigation of gas flow behavior in two-layered coal seams considering interlayer interference and heterogeneity. *Int. J. Min. Sci. Technol.* **2021**, *31*, 699–716.
29. Zhang, C.; Xu, J.; Yin, G.; Peng, S.; Li, Q.; Chen, Y. A novel large-scale multifunctional apparatus to study the disaster dynamics and gas flow mechanism in coal mines. *Rock Mech. Rock Eng.* **2019**, *52*, 2889–2898.
30. Zhang, C.; Xu, J.; Peng, S.; Li, Q.; Yan, F. Experimental study of drainage radius considering borehole interaction based on 3D monitoring of gas pressure in coal. *Fuel* **2019**, *239*, 955–963.
31. Qin, Y.; Xiong, M.H.; Yi, T.S.; Yang, Z.B.; Wu, C.F. On unattached multiple superposed coalbed-methane system: In a case of the Shuigonghe syncline, Zhijin-Nayong coalfield, Guizhou. *Geol. Rev.* **2008**, *54*, 65–70.
32. Shen, Y.L.; Qin, Y.; Guo, Y.H.; Yi, T.S.; Yuan, X.X.; Shao, Y.B. Characteristics and sedimentary control of a coalbed methane-bearing system in Lopingian (late Permian) coal-bearing strata of western Guizhou province. *J. Nat. Gas Sci. Eng.* **2016**, *33*, 8–17.
33. Shen, Y.L.; Qin, Y.; Wang, G.G.X.; Guo, H.Y.; Shen, J.; Gu, J.Y.; Xiao, Q.; Zhang, T.; Zhang, C.L.; Tong, G.C. Sedimentary control on the formation of a multi-superimposed gas system in the development of key layers in the sequence framework. *Mar. Petrol. Geol.* **2017**, *88*, 268–281.
34. Shen, Y.L.; Qin, Y.; Wang, G.G.X.; Xiao, Q.; Shen, J.; Jin, J.; Zhang, T.; Zong, Y.; Liu, J.B.; Zhang, Y.J.; et al. Sealing capacity of siderite-bearing strata: The effect of pore dimension on abundance and micromorphology type of siderite in the Lopingian (Late Permian) coal-bearing strata, western Guizhou province. *J. Petrol. Sci. Eng.* **2018**, *178*, 180–192.
35. Arenillas, A.; Pevida, C.; Rubiera, F.; Pis, J. Comparison between the reactivity of coal and the synthetic coal models. *Fuel* **2003**, *82*, 2001–2006.
36. Jessen, K.; Tang, G.; Kovscek, A. Laboratory and simulation investigation of enhanced CBM recovery by gas injection. *Transp. Porous. Med.* **2008**, *73*, 141–159.
37. Jasinge, D.; Ranjith, P.; Choi, S. Effects of effective stress changes on permeability of Latrobe Valley brown coal. *Fuel* **2011**, *90*, 1292–1300.
38. Dutka, B.; Kudasik, M.; Pokryszka, Z.; Skoczylas, N.; Topolnicki, J.; Wierzbicki, M. Balance of CO_2/CH_4 exchange sorption in a coal briquette. *Fuel Process Technol.* **2013**, *106*, 95–101.

39. Sobczyk, J. A comparison of the influence of adsorbed gases on gas stresses leading to coal and gas outburst. *Fuel* **2014**, *115*, 288–294.
40. Arenillas, A.; Pevida, C.; Rubiera, F.; Pis, J. Characterisation of model compounds and a synthetic coal by TG/MS/FTIR to represent the pyrolysis behavior of coal. *J. Anal. Appl. Pyrol.* **2004**, *71*, 747–763.
41. Sobczyk, J. The influence of sorption processes on gas stresses leading to the coal and gas outburst in the laboratory conditions. *Fuel* **2011**, *90*, 1018–1023.
42. Dutka, B.; Kudasik, M.; Topolnicki, J. Pore pressure changes accompanying exchange sorption of CO_2/CH_4 in a coal briquette. *Fuel Process Technol.* **2012**, *100*, 30–34.
43. Li, Q.; Xu, J.; Peng, S.; Yan, F.; Han, E.; Jiang, C. Physical simulation experiment on flow characteristics in a low-pressure reservoir under co-production. *J. China Coal Soc.* **2021**, *46*, 351–363.
44. Wen, Z.; Kang, Y.; Deng, Z.; Sun, L.; Li, G.; Wang, H. Characteristics and distribution of current in-situ stress at shallow-medium depth in coal-bearing basins in China. *Geol. Rev.* **2019**, *65*, 729–742.
45. Zhang, Z.; Qin, Y.; Yang, Z.; Jin, J.; Wu, C. Fluid energy characteristics and development potential of coalbed methane reservoirs with different synclines in Guizhou, China. *J. Nat. Gas Sci. Eng.* **2019**, *71*, 102981.
46. Chen, S.; Tang, D.; Tao, S.; Xu, H.; Li, S.; Zhao, J. Statistic analysis on macro distribution law of geostress field in coalbed methane reservoir. *Coal Sci. Technol.* **2018**, *6*, 57–63.
47. Zhou, B.; Xu, J.; Peng, S.; Yan, F.; Yang, W.; Cheng, L.; Yang, W. Dynamic response of coal seam and roadway during coal and gas outburst. *J. China Coal Soc.* **2020**, *45*, 1385–1397.
48. Zhang, Z. *CBM System and Drainage Optimization of Taiyuan Formation in Southern Qinshui Basin*; China University of Mining and Technology: Xuzhou, China, 2016.
49. Zhang, Z.; Qin, Y.; Fu, X. The favorable developing geological conditions for CBM multi-layer drainage in southern Qinshui basin. *J. China Univ. Min. Technol.* **2014**, *43*, 1019–1024.
50. Du, X.; Li, X.; Xu, B.; Ren, W.; Shao, C.; Hu, A.; Zhai, Y. Multi-layer production evaluation of CBM wells in Han-cheng area. *Coal Geol. Explor.* **2014**, *42*, 28–34.
51. Zhang, E.; Wu, C.; Dang, G. Interlayer interference analysis of the multi-layer drainage for multiple seams CBM in Laochang mining area of eastern Yunnan province. *J. Henan Polytech. Univ. Nat. Sci.* **2020**, *39*, 10–17.

processes

MDPI

Article

Study on the Influence of Drilling Parameters on the Mechanical Properties and Pressure Relief Effect of Coal Rock

Yujiang Zhang [1,2], Yexing Chen [1], Shuai Zhang [1,*], Guorui Feng [1,*], Yuguo Wang [3], Shule Li [1], Qian Wang [1], Bo Wang [1] and Liang Zhao [4]

1 College of Mining Engineering, Taiyuan University of Technology, Taiyuan 030024, China; ylczyj@yeah.net (Y.Z.); chenyexing0789@link.tyut.edu.cn (Y.C.); lishule0813@link.tyut.edu.cn (S.L.); wangqian0833@link.tyut.edu.cn (Q.W.); wangbo01@tyut.edu.cn (B.W.)

2 Postdoctoral Centre, Shandong Energy Group Co., Ltd., Jinan 250014, China

3 College of Safety and Emergency Management Engineering, Taiyuan University of Technology, Taiyuan 030024, China; wangyuguo1368@link.tyut.edu.cn

4 Yangcheng County Yangtai Group Industrial Co., Ltd., Jincheng 048100, China; zhaoliang202501@163.com

* Correspondence: zhangshuai01@tyut.edu.cn (S.Z.); fgr_tut@163.com (G.F.)

Abstract: Based on considering the stress state distribution and potential failure surface of the specimen during uniaxial compression, the drilling parameters (layout, spacing, position) are set. Thoroughly understanding the influence of different drilling parameters on the pressure relief effect is conducive to reducing the occurrence of coal mine rock burst accidents. Through laboratory tests and numerical simulation tests under different drilling parameters, the influence laws of mechanical parameters, failure characteristics, AE characteristic parameters and energy evolution of specimens under different drilling parameters were studied. The pressure relief effect under different drilling parameters was evaluated by using the pressure relief effect evaluation index (X), and the best combination of drilling parameters was obtained. The results show the following: (1) Compared with the intact specimen, the peak strength of the drilling specimen is significantly reduced, and the drilling layout has the greatest influence on the mechanical properties, followed by the drilling spacing and drilling position. (2) Different drilling layouts will form different weak-strength surfaces in the specimen, and lead the expansion and penetration of cracks, resulting in different failure modes of the specimen. The stress distribution inside the specimen will affect the stress concentration around the borehole, finally affect the damage degree of the specimen. (3) Drilling can not only effectively reduce the energy accumulation capacity, but also enhance the degree of energy dissipation. The AE ringing counts and energy of the triangular-drilling specimens are the least. The AE ringing counts and energy decrease first and then increase with the increase in drilling spacing, and are the smallest at three times the drilling diameter. The AE ringing counts and energy increase gradually with the upward movement of the drilling position. (4) The optimal combination of drilling parameters was obtained by the test, and it was triangular-layout drilling, drilling spacing three times the diameter, and the drilling position in the middle of the specimen, and the value of the pressure relief effect evaluation index (X) was 65.41. The research results can provide some reference for the selection and optimization of drilling pressure relief parameters in rock burst mines.

Keywords: drilling pressure relief; drilling parameters; mechanical properties; characteristics of destruction; acoustic emission characteristics; pressure relief effect

1. Introduction

With the increase in coal mining depth, dynamic disasters have gradually become the main disasters in deep mines in China [1–3], such as rock bursts and mine earthquakes. The regional prevention and control of rock bursts in coal mines is by means of manual intervention. By changing the mechanical properties of surrounding rock, the energy accumulation and stress concentration of surrounding rock can be reduced. This can reduce the impact tendency of surrounding rock. Among them, drilling pressure relief with a large diameter is the most commonly used means to prevent rock bursts in rock burst mines [4–6]. The basic principles for determining some drilling parameters are provided in the prevention method of drilling pressure relief [7,8]. However, the influence of drilling parameters on mechanical properties and failure characteristics of coal rock has not been fully explored, which affects the accurate evaluation of the pressure relief effect. Therefore, it is necessary to study the effects of different combinations of drilling parameters on strength, failure, coal–rock impact, energy dissipation and pressure relief effect evaluation.

Many scholars have carried out relevant research on the mechanism of drilling pressure relief. Wang et al. [9] analyzed the pressure relief mechanism of drilling in deep roadways. He proposed that the stress transfer effect and the surrounding rock deformation control effect can be used as the direct evaluation index of the pressure relief effect, and determined the key drilling parameters when the pressure is fully relieved. Zhao and Lin et al. [10,11] studied the failure evolution process of rocks with boreholes and concluded that the cracks around holes are characterized by vertical tensile cracks formed on the upper and lower sides of boreholes, and shear cracks or secondary cracks formed on the left and right sides of boreholes. Tensile, compressive and shear applications affect the final failure mode of brittle rock. Jia et al. [12] studied the crack propagation and evolution penetration process around the drilling and found that the release of energy and stress is the fundamental reason for the drilling pressure relief. Wang et al. [13] concluded that the radius of the plastic zone around the borehole is the key factor affecting the pressure relief effect, and when selecting the pressure relief parameters of the borehole, the first choice should be the joint pressure relief of several boreholes, followed by expanding the diameter of the borehole and finally increasing the depth of the borehole. Zhu et al. [14] analyzed typical rock burst accidents and believed that excessive pressure relief of coal would seriously weaken its own carrying capacity. This forces the abutment pressure to move back to the direction of the roadway, resulting in delayed rock burst disasters. He also put forward the prevention and control measures of near-field strong pressure relief and strong support, deep-control pressure relief and strong monitoring. Zhu et al. [15] studied the pressure relief index of the working face impact danger area. It is considered that drilling pressure relief can prevent rock bursts by reducing the impact energy index and energy dissipation index of coal and rock. Yao et al. [16] unified the pressure relief mechanism of boreholes with roadway support, studied the energy characteristics of the anchorage-strengthening zone and stress-reducing zone, and believed that boreholes should be placed between the anchorage end and the peak stress.

A large number of scholars have conducted in-depth research on drilling pressure relief parameters, such as the diameter, layout, depth and direction of pressure relief drilling. Reasonable setting of drilling parameters must meet the pressure relief effect in three aspects. First, it is to reduce the stress of surrounding rock, release the elastic energy of roadway surrounding the rock and achieve the purpose of reducing the deformation of surrounding rock and preventing dynamic disasters. The second is to improve the physical condition of surrounding rock and reduce the burst tendency of surrounding rock, so that the surrounding rock does not have the condition of rock burst. The third is to ensure the strength characteristics of the surrounding rock, and the surrounding rock structure has a certain bearing capacity to maintain the stability of the roadway [17–20]. Yin et al. [21,22] evaluated the pressure relief effect of drilling from the perspective of energy storage and dissipation. With the increase in drilling diameter and number, the energy pre-release index, energy dissipation-increasing index and impact tendency weakening index gradually increase, and the increase gradient gradually decreases. Ma et al. [23] obtained the influence law of different borehole diameters, borehole depths and borehole angles on the peak strength of coal through numerical simulation tests. Wu et al. [24] analyzed the weakening effect of drilling with different shapes on the mechanical properties of specimens and concluded that the stability order of different drilling shapes was round, arch, trapezoid, rectangular and square. Zhang et al. [25] studied the influence law of different pressure relief drilling layouts and numbers by the coal-like material test and numerical simulation, and found that the pressure relief effect of triangular-layout drilling was the best. And the higher the density of the drilling, the more developed the rock cracks around the drilling, and the better the pressure relief effect. Nian et al. [26] studied the effect of large-diameter drilling on the pressure relief and permeability improvement of coal seams by numerical simulation and found that the stress around the drilling presented a symmetrical distribution. With the increase in drilling diameter, the plastic zone, crack and gas permeability of coal also increase. Wang et al. [27] studied the effects of the drillings number and arrangement on the mechanics and impact tendency of specimens and believed that with the increase in the drillings number, the impact tendency would gradually weaken, and the pressure relief effect would be better. Zuo et al. [28] arranged pressure relief drillings at 180°, 120°, 90° and 60°, and the peak strength of the specimens was reduced by 14%, 27%, 20% and 22%, respectively. At the same time, the sensitivity of the drillings' diameter, arrangement angle and horizontal spacing was analyzed, and the influence of the drillings on compressive strength was gradually reduced. From the perspective of energy storage, Gong and He et al. [29,30] studied the influence of a different number of drillings and layout angles on the rock burst tendency of red sandstone. It was found that the peak-strength strain storage index decreases with the increase in the number of drillings. The peak-strength strain storage index is the lowest when the drilling angle and loading direction are 27°. Moreover, the angle of the drillings layout is more obvious to the burst tendency than the number of drillings. Lyu et al. [31,32] proposed a method of pressure relief by variable-diameter drilling and studied the effects of reaming diameter, reaming depth and reaming spacing on the pressure relief and energy dissipation of roadway surrounding rock. It was concluded that the position of the variable diameter should be within the range of the stress peak value, and the deep reaming should pass through the stress concentration area, so as to achieve the best pressure relief effect.

The existing research on the pressure relief effect of drilling on coal rock mainly focuses on drilling diameter and drilling layout. However, the selection and determination principles of other drilling parameters are not clear enough. The stress state distribution of the specimen and the relationship between the potential failure surface and drilling parameters in the uniaxial-compression process are ignored. And the relationship between the mechanical parameters of the specimen affected by drilling parameters, the failure mechanism and the evaluation of the pressure relief effect is unclear. Therefore, based on considering the stress state distribution and potential failure surface of the specimen during uniaxial compression, three kinds of drilling parameters, such as drilling layout, drilling spacing and drilling position, are studied in this paper, as shown in Figure 1. And the influence laws of mechanical parameters (peak strength, elastic modulus, peak strain), failure characteristics (crack distribution, failure mode, crack count), AE characteristics (ringing count, energy) and the energy evolution of specimens under different drilling parameters are obtained. The impact energy index, strength reduction index and energy dissipation index of the specimens under different drilling parameters are analyzed, and a new evaluation index (X) of the pressure relief effect is proposed according to these three indexes. The pressure relief effect under different drilling parameters is evaluated by using the value of X, and the best drilling parameter combination is obtained. The research results can serve as a valuable reference for selecting drilling parameters when implementing the drilling pressure relief with a large diameter in rock burst mines.

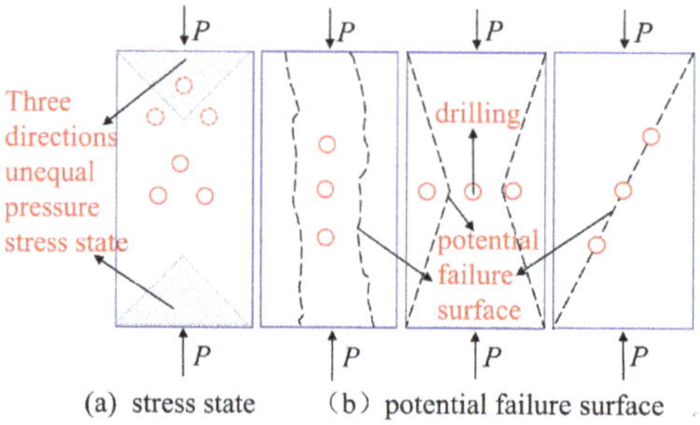

Figure 1. Schematic diagram of the stress state and failure mode of rock in uniaxial compression test.

2. Specimen Preparation and Test Scheme

2.1. Specimen Preparation

The internal joints and cracks of raw coal specimens are significantly developed, and their mechanical properties are unstable. In order to reduce the influence of the defects of the specimens themselves on the experimental results, many researchers choose to use physically similar materials to cast specimens for experimental research. They chose α-type high-strength anhydrous gypsum: water = 1:0.4 as the material to pour the cuboid specimen with side lengths of 100 mm $\times$ 100 mm $\times$ 200 mm [25]. The specimen preparation process is shown in Figure 2.

Figure 2. The specimens preparation process.

The impact tendency indexes of the standard specimen was tested, and the results are shown in Table 1. It can be determined that the gypsum specimen has a strong impact tendency, and its mechanical properties are similar to those of strong-impact tendency coal rock, which can be used in the research of drilling pressure relief parameters.

Table 1. Basic mechanical parameters and burst tendency indexes.

Specimen Number	Density/(g/cm^3)	R_c/MPa	D_T/ms	K_E	W_{ET}	Elastic Modulus/GPa
1	1.51	21.34	64.5	7.47	6.78	4.53
2	1.52	21.92	47.7	5.58	7.55	4.47
3	1.50	21.17	84.2	7.37	6.21	4.70
Average value	1.51	21.48	65.47	6.81	6.85	4.57

2.2. Test System

As shown in Figure 3, the test system consists of a microcomputer-controlled electro-hydraulic servo pressure test system, two high-speed cameras, and an acoustic emission acquisition and analysis system. The maximum static load that can be applied by this test system is 1000 kN. Two high-speed cameras were used to record the surface crack evolution process and fracture characteristics of the specimen. The camera resolution is 2448 × 2048 pixels, and the frame rate is up to 70 fps.

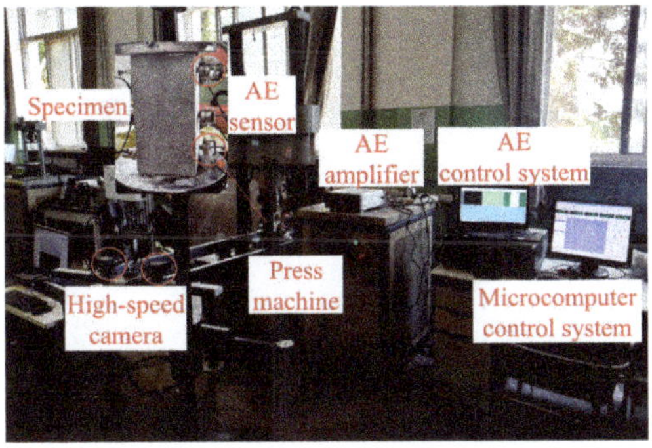

Figure 3. Test system and equipment.

The DS5-8B acoustic emission acquisition and analysis system was used to record the acoustic emission characteristics data of the specimen during the loading process. The acoustic emission system is composed of an acoustic emission sensor, amplifier, automatic data acquisition system and analysis software. The diameter of the sensor is 18 mm, and the resonance frequency is 100~400 KHz. The loading method of the test is displacement control, and the loading speed is controlled at 2×10^{-3} mm/s.

2.3. Test Scheme

According to the index of the drilling diameter-to-specimen width ratio (D/W) [33], the drilling diameter was determined to be 10 mm. The drilling parameters studied in this experiment include drilling layout, drilling spacing and drilling position.

Therefore, the three groups of test schemes designed are shown in Table 2. The first group determined the optimal drilling layout; the drilling spacing and drilling position were fixed; and the drilling layouts were changed to intact (I), horizontal (H), triangular (T), diagonal (D) and vertical (V). The second group determined the drilling spacing, selected the best layout in the first group and kept the drilling position unchanged. The drilling spacing was 20 mm, 30 mm, 40 mm and 50 mm, corresponding to 2 times, 3 times, 4 times and 5 times the drilling diameter. The third group determined the drilling position, selected the optimal layout and spacing determined by the first two groups; took the middle of the specimen as the starting point; changed the spacing between the centroid of the triangular layout of the drilling and the reference line in the middle of the specimen;

and set it at 0, 20, 40 and 60 mm. Therefore, specimen T30-0 is arranged in a triangular layout, the drilling spacing is 30 mm, and the drilling position is in the middle of the specimen. The drilling parameter settings and some specimens are shown in Figure 4. In order to ensure the accuracy of the test, three specimens were prepared for each drilling parameter combination.

Table 2. Experimental scheme.

Test Scheme	Drilling Layout	Drilling Spacing/mm	Drilling Position/mm
I	I, H, T, D, V	30	0
II	T	20, 30, 40, 50	0
III	T	30	0, 20, 40, 60

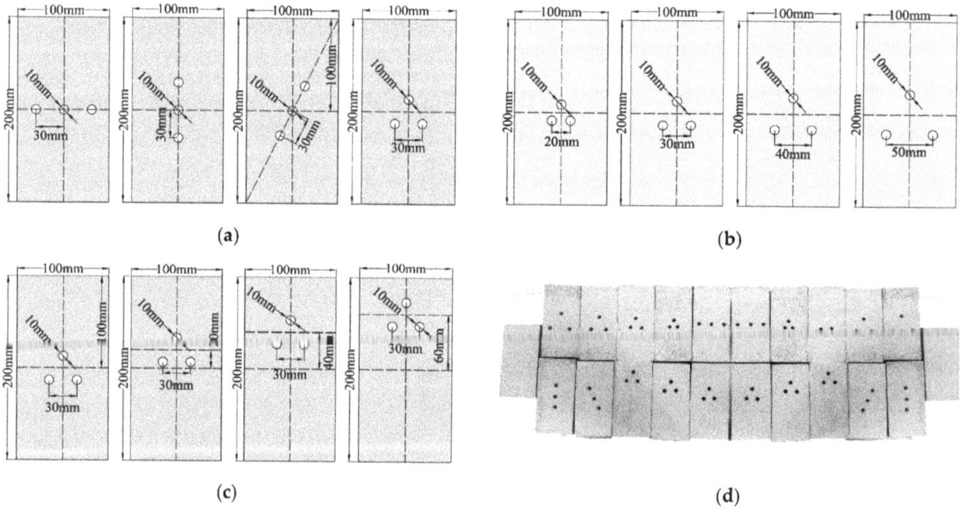

Figure 4. Drilling parameter setting diagrams: (**a**) Drilling layout; (**b**) Drilling spacing; (**c**) Drilling position; (**d**) Partial finished specimens.

3. Experimental Results Analysis

3.1. *Influence of Drilling Parameters on Mechanical Properties*

3.1.1. Influence of Drilling Layouts on Mechanical Properties

As shown in Figure 5a, the stress–strain curves of specimens with different drilling layouts were composed of four stages: the fracture compaction stage, the elastic deformation stage, the plastic deformation stage and the post-peak failure stage. It can be seen that the drilling layouts have an influence on the trend of the pre-peak curve and post-peak curve of the specimen. The pre-peak curve of the intact specimen and the vertical-layout drilling specimen increased rapidly. The pre-peak curves of the specimens with horizontal-layout, triangular-layout and diagonal-layout drilling become gentler successively. In the plastic deformation stage, the drilling specimens will demonstrate the phenomenon of stress reduction. The post-peak failure curves of both the intact specimen and the vertical-layout drilling specimen showed a rapid stress decline, showing the characteristics of brittle failure. However, the post-peak failure curves of horizontal-layout, triangular-layout and diagonal-layout drilling specimens changed from steep to gentle, and the triangular layout is the gentlest. It shows that different drilling layouts can effectively reduce the pre-peak energy accumulation capacity and post-peak energy release rate of specimens. And the triangular-layout drilling has the strongest ability to reduce energy accumulation.

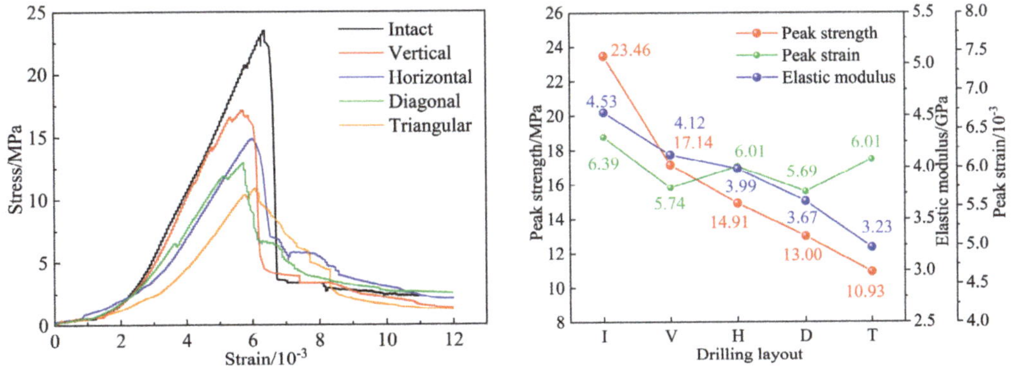

(a) (b)

Figure 5. Stress–strain curves (**a**) and mechanical parameters (**b**) of specimens with different drilling layouts.

As shown in Figure 5b, the drilling layout has a strong correlation with the peak strength and elastic modulus of the specimen, but has little effect on the peak strain of the specimen. The peak strength of the intact specimen is the highest, 23.46 MPa. The peak strength of drilling specimens with the vertical layout, horizontal layout, diagonal layout and triangular layout showed a decreasing trend, and were 17.14 MPa, 14.91 MPa, 13.00 MPa and 10.93 MPa, respectively. The elastic modulus of the intact specimen, vertical-layout, horizontal-layout, diagonal-layout and triangular-layout drilling specimen decreased from 4.53 GPa to 3.23 GPa, showing a decreasing tendency in sequence. The peak strength and elastic modulus are the smallest under triangular-layout drilling. This is due to the triangular-layout drilling having changed the stress distribution inside the specimen, and the stress concentration in the area around the drilling became more obvious, resulting in the initial initiation, development and penetration of cracks, and the formation of a large range of pressure relief areas in the middle of the specimen, resulting in the overall bearing capacity reduction [25].

3.1.2. Influence of Drilling Spacings on Mechanical Properties

As shown in Figure 6a, the stress–strain curves of specimens with different drilling spacings were composed of four stages: the fracture compaction stage, the elastic deformation stage, the plastic deformation stage and the post-peak failure stage. It can be seen that the curve coincidence degree of specimens with different drilling spacings is higher in the crack compression stage and the elastic deformation stage, but the curve difference between the plastic deformation stage and the post-peak failure stage is large. With the increase in drilling spacing, the curve of the post-peak failure stage first becomes gentle and then steep, indicating that the post-peak energy release rate of the specimen also decreases first and then increases. The residual strength of the specimens is roughly the same.

As shown in Figure 6b, there is a significant correlation between drilling spacing and peak intensity. When the drilling layout drilling is the same, the peak strength decreases first and then increases with the increase in drilling spacing. When the drilling spacing is three times the drilling diameter, the peak strength is the smallest (10.93 MPa). This may be caused by the largest holes group's linkage failure range of the specimen, whose drilling spacing is three times the drilling diameter. However, when the drilling spacing is more than three times the drilling diameter, the rock mass between the holes cannot be connected, resulting in the increase in the peak strength of the specimen. The change in the peak strain and elastic modulus is not obvious with the increase in drilling spacing.

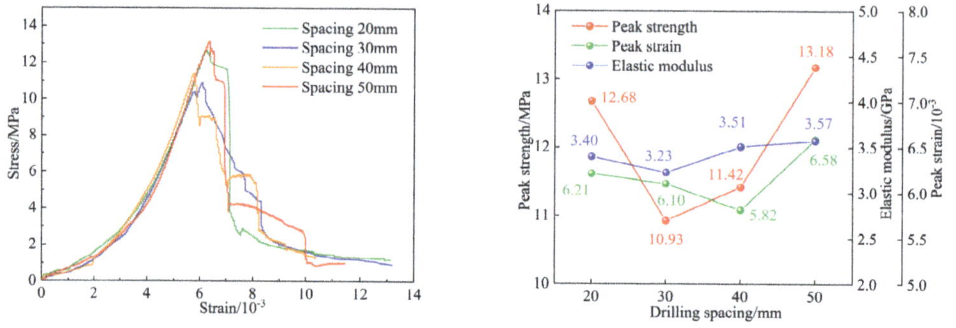

(a) (b)

Figure 6. Stress–strain curves (**a**) and mechanical parameters (**b**) of specimens with different drilling spacings.

3.1.3. Influence of Drilling Positions on Mechanical Properties

As shown in Figure 7a, the curve trend of specimens at different drilling positions under the fracture pressure tight stage and elastic deformation stage basically coincide. The change in drilling position has an obvious influence on the curve of the plastic deformation stage and post-peak failure stage. With the upward movement of the drilling position, the plastic deformation stage appears more and more late, and the curve of the plastic deformation stage is not obvious. With the upward movement of the drilling position, the trend of the post-peak curve becomes steeper obviously, indicating that the post-peak energy release rate of the sample is increasing. With the upward movement of the drilling position, the residual strength of the specimen increases gradually.

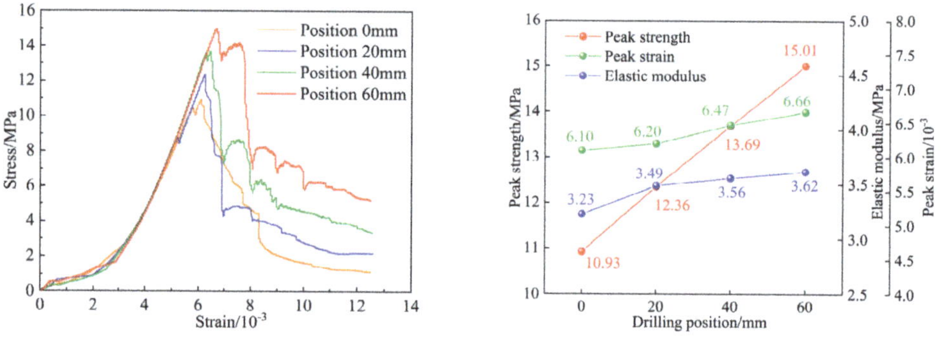

(a) (b)

Figure 7. Stress–strain curves (**a**) and mechanical parameters (**b**) of specimens with different drilling positions.

As shown in Figure 7b, there is a certain correlation between peak strength and drilling location. With the upward movement of the drilling position, the peak strength gradually increased to 10.93 MPa, 12.36 MPa, 13.69 MPa and 15.01 MPa, in that order. The peak strength is minimum when the drilling position is in the middle of the specimen, which may be caused by the maximum damage to the specimen by drilling in the middle position. The peak strain and elastic modulus are basically unchanged as the drilling position moves upward.

3.2. Influence of Drilling Parameters on Failure Characteristics of Specimens

3.2.1. Failure Characteristics Analysis of Specimens with Different Drilling Layouts

The shape of the specimens before and after the failure of the specimens with different drilling layouts is shown in Figure 8. Under axial compressive stress, the Poisson effect causes radial tensile stress in the intact specimen (Figure 8a), which results in the vertical distribution of surface failure cracks, showing a tensile failure mode caused by tensile stress. The surface of the specimen was spalling, a large amount of fragment was ejected, and the specimen was finally destroyed.

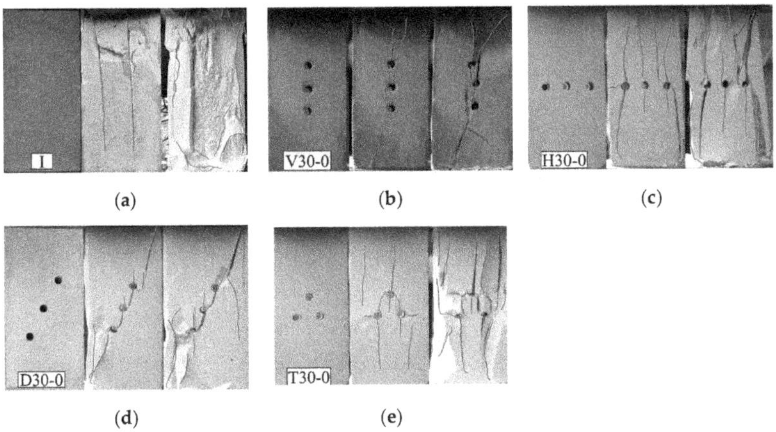

Figure 8. Surface failure diagrams of specimens with different drilling layouts: (**a**) Intact; (**b**) V30-0; (**c**) H30-0; (**d**) D30-0; (**e**) T30-0.

For drilling specimens, under uniaxial compression, obvious stress concentration occurs around the borehole [34]. Compressive stress is applied to the left and right sides of the drilling, and its magnitude is three times that of the applied stress. The tensile stress of the upper and lower ends is equal to the applied stress, but the tensile strength of the specimen is much smaller than the compressive strength. Therefore, different drilling specimens first produce tensile cracks at the upper and lower ends of the drilling, and then other cracks appear on the left and right sides [10,11].

Different drilling layouts will form different weak-strength surfaces in the specimen, which will guide the crack propagation and lead to different failure modes of the specimen. The vertical-layout drilling specimen (Figure 8b) formed a weak-strength surface on the central axis of the specimen, has a strong tensile stress zone along the axial direction, and the cracks are mainly tensile cracks. The tensile crack penetrated through the vertical drilling to form a vertical pressure relief zone, resulting in the macroscopic failure mode of the specimen being tensile failure. There was strong compression interaction between the holes of the specimen with horizontal-layout drilling (Figure 8c), and basically no cracks occurred on the left and right sides of the drilling, but only on the upper and lower ends of the drilling. Although the horizontal-layout drilling formed a horizontal weak-strength surface in the middle of the specimen, the linkage failure could not be formed between the drillings. The cracks at the upper and lower ends continued to develop and form macroscopic cracks, resulting in the macroscopic failure mode of the specimen being an X-conjugated inclined-plane shear failure. Diagonal-layout drilling specimens (Figure 8d) artificially form a weak-strength surface in the diagonal direction of the specimen, resulting in strong tensile interaction along the diagonal direction. The shear stress is the largest in the diagonal direction of the drillings, so the cracks eventually penetrate along the diagonal direction of the drillings, forming a diagonal pressure relief zone, and the final failure mode of the specimen is single-inclined-plane shear failure. The drilling pressure relief zones of the triangular-layout drilling specimen (Figure 8e) overlap each other, thus forming a weak-strength area in the middle of the specimen. With the further increase in axial stress, tensile cracks and shear cracks are fully developed in this area, and the drillings are connected with each other. In the middle of the specimen, a large range of drilling pressure relief zone is formed, resulting in the final failure mode of the specimen being tensile shear mixed failure.

3.2.2. Failure Characteristics Analysis of Specimens with Different Drilling Spacings

The surface failure of specimens with different drilling spacings is shown in Figure 9. The failure patterns of specimens with different drilling spacings can be divided into three categories: (1) Through-type failure: when the drilling spacing is small, it is easy to occur through-type failure; the specimens with a drilling spacing of 20 mm and 30 mm belong to this failure. The debris spalling on the surface of the specimen was obvious, and the secondary shear cracks developed and expanded around the boreholes, thus connecting the surrounding rock between the three boreholes. Finally, a large drilling pressure relief zone appeared in the middle of the specimen, resulting in a decrease in the bearing capacity of the specimen. (2) Independent-through-transitional-type failure: the failure mode of specimens with 40 mm drilling spacing belongs to this failure. The shear cracks between the boreholes were basically through, the tensile cracks in the lower two boreholes were not obvious, the debris spalling on the surface of the specimen was not obvious, and the middle area of the three boreholes was basically intact. (3) Independent-type failure: when the drilling spacing is large, it is easy to occur independent-type failure; the specimen with 50 mm drilling spacing belongs to this failure. Due to the large spacing between the boreholes, the interaction between the boreholes is weakened, and the stress state is gradually close to three independent boreholes. Although the specimen surface also showed the phenomenon of debris spalling, the cracks generated around the boreholes could not penetrate each other.

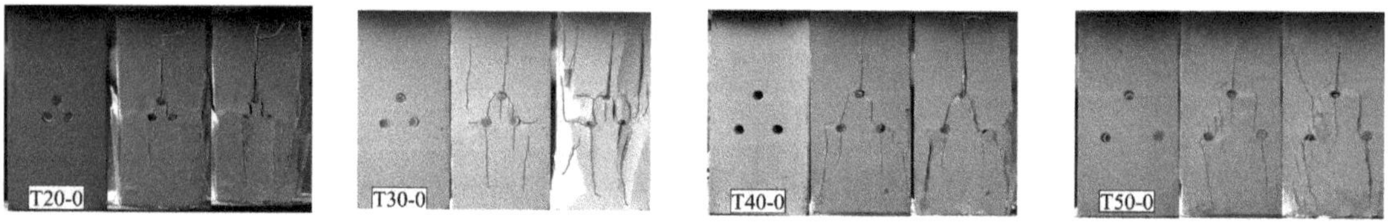

(a) (b) (c) (d)

Figure 9. Surface failure diagrams of specimens with different drilling spacing: (**a**) T20-0; (**b**) T30-0; (**c**) T40-0; (**d**) T50-0.

By comparing the four subfigures in Figure 9, it can be seen that the drilling spacing affects the interaction between boreholes, thus affecting the crack growth and failure mode of the specimen, and then affecting the failure degree and bearing capacity of the specimen.

3.2.3. Failure Characteristics Analysis of Specimens with Different Drilling Positions

The surface failure of specimens with different drilling positions is shown in Figure 10. The drilling position has little effect on the failure form of the specimen. The closer the drilling position is to the middle of the specimen, the greater the overall damage degree of the specimen and the lower the peak strength of the specimen. This may be due to the following reasons: (1) During the loading process, the stress distribution inside the specimen will affect the stress concentration around the drilling hole, thus affecting the failure pattern and strength characteristics of the specimen. When the drillings are placed in the upper part of the specimen, the load around the drilling is higher, the crack propagation is limited, and the peak strength of the specimen is increased. (2) The location of drillings also has a certain effect on crack propagation. When the drillings are arranged in the middle area of the specimen, the crack growth can affect the whole structure of the specimen to the greatest extent, and the damage degree of the specimen is the greatest.

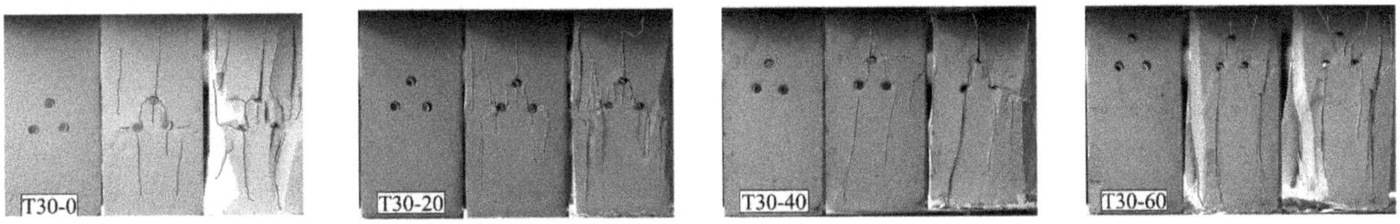

(**a**) Position 0 mm (**b**) Position 20 mm (**c**) Position 40 mm (**d**) Position 60 mm

Figure 10. Surface failure diagrams of specimens at different drilling positions: (**a**) T30-0; (**b**) T30-20; (**c**) T30-40; (**d**) T30-60.

3.3. Influence of Drilling Parameters on Acoustic Emission Characteristics of Specimens

3.3.1. Acoustic Emission Characteristics of Specimens with Different Drilling Layouts

AE signals are generated by crack initiation, expansion or merging of coal rock specimens, so the internal damage of specimens during loading can be evaluated based on the recorded acoustic emission signals. Two AE characteristic parameters, AE ringing count and AE energy, are selected in this paper. The former is used to characterize the frequency of microcrack activity in specimens, while the latter reflects the energy size and damage severity during the formation and expansion of microcracks in specimens.

According to Figure 11, the AE activities of specimens with different drilling layouts during uniaxial loading were studied. The AE characteristic parameters of the specimens with different drilling layouts have a strong correlation with the time–stress curve, and the peak value of the AE ringing count, AE energy and stress basically coincide. According to the characteristics of the stress–time curve, cumulative AE ringing count and cumulative AE energy distribution, acoustic emission activity can be divided into four stages: stable stage, rising stage, active stage and vanishing stage.

Stable stage (I): fracture pressure tight stage. The specimens were subjected to static load compaction, and the microcracks gradually closed. Due to the closure of natural defects, no significant AE activities were generated, and only a small amount of acoustic emission signals were recorded. The cumulative AE ringing count grows at a faster rate than the cumulative AE energy.

Rising stage (II): elastic deformation stage. The number of microcracks in the intact specimen increased steadily. Due to the small range and density of microcracks, only a small number of AE signals were recorded, and no macroscopic cracks occurred at this stage. For the drilling specimens at this stage, the tensile cracks around the drilling start and develop steadily along the loading direction, resulting in an increase in the number of AE signals compared with the number of intact specimens.

Active segment (III): plastic deformation stage. The internal damage of the specimen developed, resulting in a large number of cracks. The number of AE events increased significantly, and the AE ringing count and AE energy increased significantly.

Evanescent stage (IV): post-peak failure stage. For the intact specimen, a large number of AE signals are generated at the moment of failure, and basically no AE signals are generated at the post-peak failure stage. For different drilling specimens, the ratio of AE events in the post-peak failure stage to the total AE events is much higher than that of the intact specimens. This is because the drilling causes the failure process of the specimens to be more lasting, which weakens the failure strength, and the internal cracks continue to evolve in the post-peak failure stage. It can be well proved that different drilling layouts can reduce the impact inclination of specimens.

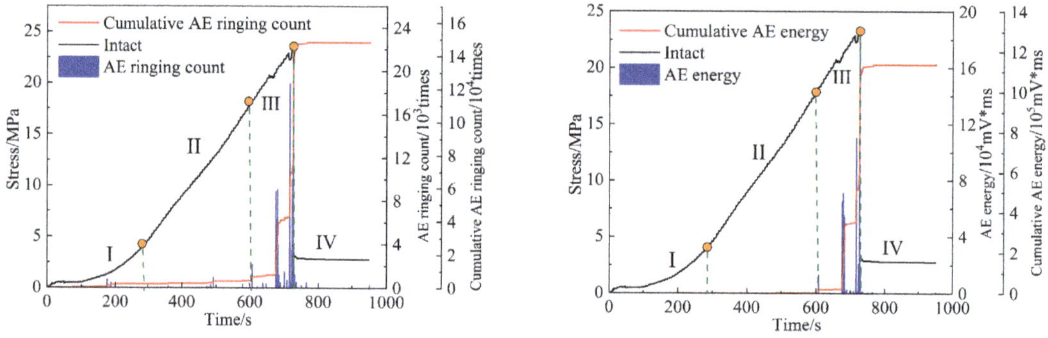

(**a**)

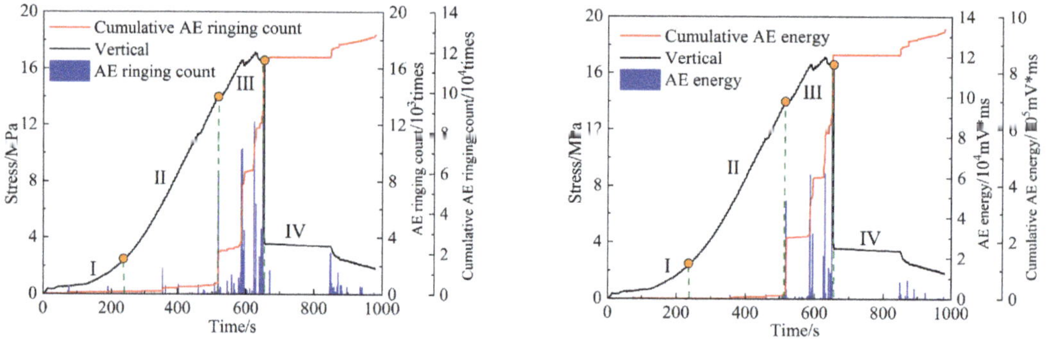

(**b**)

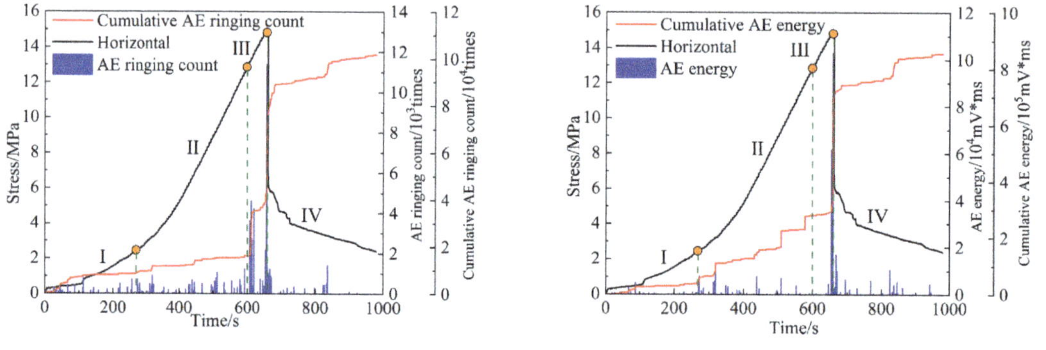

(**c**)

Figure 11. *Cont.*

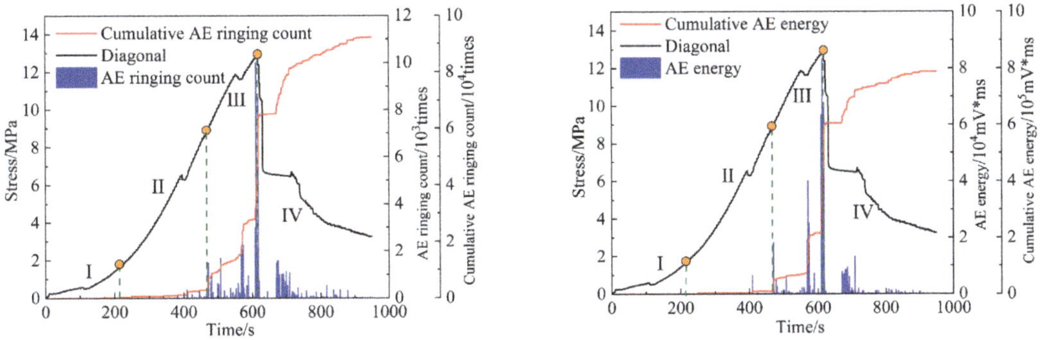

(**d**)

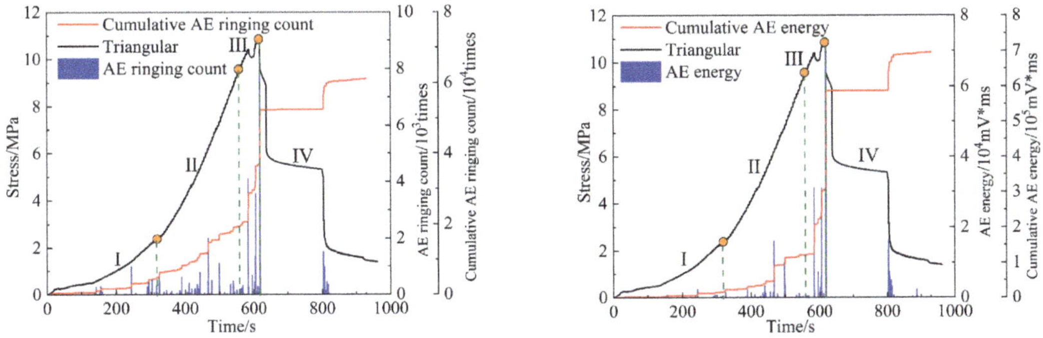

(**e**)

Figure 11. Evolution law of stress and AE characteristic parameters with time under different drilling layouts: (**a**) I; (**b**) V30-0; (**c**) H30-0; (**d**) D30-0; (**e**) T30-0.

Figure 12 shows the variation rules of the peak AE ringing count, peak AE energy, cumulative AE ringing count and cumulative AE energy under different drilling layouts, which are used to reflect the AE activity events during loading. The four indexes of peak AE ringing count, peak AE energy, cumulative AE ringing count and cumulative AE energy of intact specimens were the largest, which were 19.25×10^3 times, 17.94×10^4 mV*ms, 14.85×10^4 times and 11.41×10^5 mV*ms, respectively. The four acoustic emission characteristic parameters of the vertical-layout, horizontal-layout, diagonal-layout and triangular-layout specimens decreased successively, and the peak AE ringing count, peak AE energy, cumulative AE ringing count and cumulative AE energy of the triangular-layout specimens were the smallest. They were, respectively, 7.98×10^3 times, 7.10×10^5 mV*ms, 6.12×10^4 times and 6.95×10^5 mV*ms.

The AE activity events of the intact specimen were the most active, and the AE activity of the triangular specimen was the weakest. The reason may be that the presence of drilling weakens the peak strength and energy accumulation capacity of the specimen, resulting in the reduction of the total strain energy stored in the specimen. The release of strain energy stored in the specimen is not limited to the moment of failure, but begins at the initiation of cracks around the drilling and gradually releases after the specimen is damaged. The presence of drilling also causes the cracks to spread intensively during the loading process, reducing the number of microcracks and thus reducing the occurrence of acoustic emission events. In addition, the volume reduction caused by the excavation of holes in the specimen may also lead to a reduction in the number of cracks.

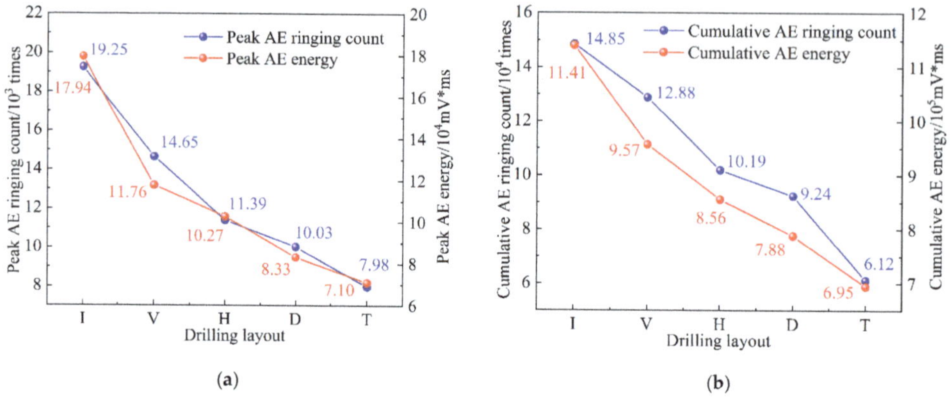

Figure 12. The variation law of characteristic values of AE parameters under different drilling layouts: (**a**) Peak ringing count and peak energy of AE; (**b**) Cumulative ringing count and cumulative energy of AE.

3.3.2. Acoustic Emission Characteristics of Specimens with Different Drilling Spacings

Figure 13 shows the changes in the peak AE ringing count, peak AE energy, cumulative AE ringing count and cumulative AE energy under different drilling spacings, which is used to reflect the activity of AE events in the loading process. With the increase in drilling spacing, the peak AE ringing count, peak AE energy, cumulative AE ringing count and cumulative AE energy decrease first and then increase. When the drilling spacing is three times the diameter, the four AE characteristic parameters are the smallest. The reason may be that the peak strength and pre-peak total strain energy of the specimen also decrease first and then increase with the increase in the spacing of the drillings, and the failure mode of the drillings gradually changes from through-type failure to independent-type failure, resulting in a decrease in the total strain energy stored in the specimen. However, when the spacing of the drillings is three times the diameter, the peak strength and pre-peak total strain energy of the specimen are the smallest, and the damage degree of the specimen is the largest.

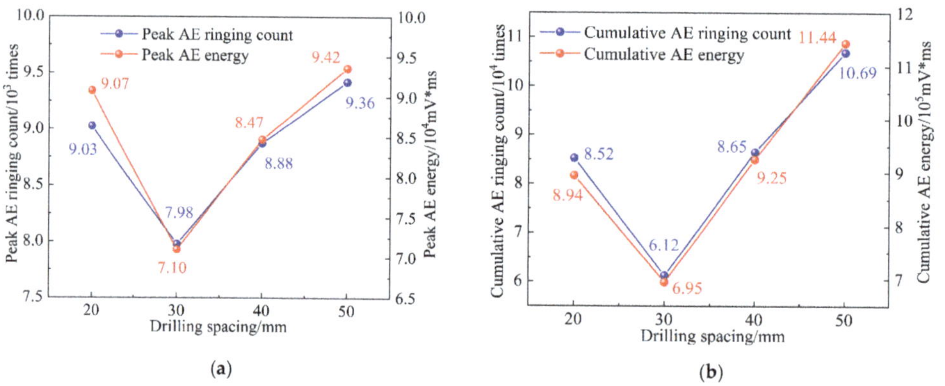

Figure 13. The variation law of characteristic values of AE parameters under different drilling spacings: (**a**) Peak ringing count and peak energy of AE; (**b**) Cumulative ringing count and cumulative energy of AE.

3.3.3. Acoustic Emission Characteristics of Specimens at Different Drilling Positions

Figure 14 shows the variation law of the peak AE ringing count, peak AE energy, cumulative AE ringing count and cumulative AE energy at different drilling positions, which is used to reflect the activity of AE events in the loading process. The four parameters of peak AE ringing count, peak AE energy, cumulative AE ringing count and cumulative AE energy increase gradually with the upward movement of the drilling position. When the drilling position is in the middle of the specimen, the four AE characteristic parameters are the smallest. The reason may be that the drilling layout in the middle of the specimen weakens the peak strength and energy accumulation capacity of the specimen to the greatest extent, resulting in the reduction of the total strain energy stored in the specimen, and the best pressure relief effect is achieved during the failure process.

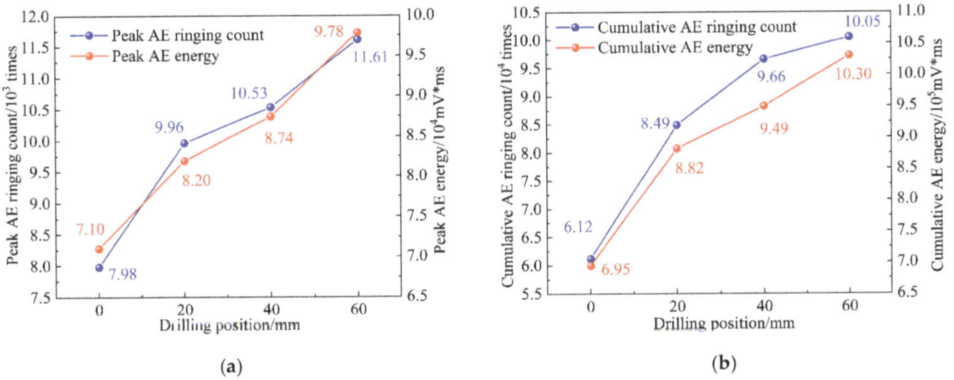

Figure 14. The variation laws of characteristic values of AE parameters under different drilling positions: (**a**) Peak ringing count and peak energy of AE; (**b**) Cumulative ringing count and cumulative energy of AE.

4. Microscopic Cracks and Energy Evolution Law

4.1. Numerical Model Establishment and Mesoscopic Parameter Calibration

PFC^{2D} 6.0 simulation software is capable of modeling the mechanical behavior of rock from a microscopic perspective, thereby revealing the underlying mechanisms of fracture propagation and energy evolution. Based on laboratory experiments, PFC software was employed to investigate the fracture propagation and energy evolution in samples with varying drilling parameters. A two-dimensional numerical model was constructed according to the dimensions of the experimental samples, with a simulated sample size of 100 mm × 200 mm. The model consisted of 14,132 randomly generated particles. Following the establishment of the model, the particles were bonded using a parallel-bond mode, and an initial equilibrium state was achieved. The boundary conditions of the model were consistent with those of the laboratory tests, where the lower boundary was fixed, a vertical downward velocity was applied to the upper boundary, and the loading rate matched that of the laboratory tests.

To calibrate the mesoscopic mechanical parameters of the model, a trial-and-error approach was utilized to establish the relationship between mesoscopic and macroscopic parameters. The calibrated mesoscopic parameters are presented in Table 3. The stress–strain curves obtained from the laboratory samples and the numerical simulation models are illustrated in Figure 15. The peak strength, peak strain and elastic modulus derived from the laboratory tests were 23.46 MPa, 6.39×10^{-3}, and 4.53 GPa, respectively. In contrast, the corresponding values from the numerical simulations were 23.59 MPa, 6.27×10^{-3}, and 3.91 GPa, respectively.

Table 3. Micromechanical parameters of the model.

Parameter	Value	Parameter	Value
Density (kg/m^3)	1510	Stiffness ratio	0.8
R_{min} (mm)	0.5	Parallel-bond modulus/GPa	1.7
R_{max} (mm)	0.75	Parallel-bond cohesion strength/MPa	12.0
Porosity	0.12	Parallel-bond tensile strength/MPa	5.5
Friction coefficient	0.5	Friction Angle/°	38

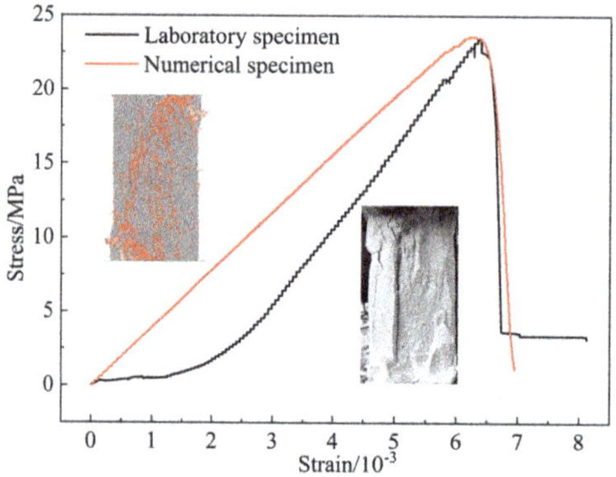

Figure 15. Stress–strain curves of laboratory specimen and numerical simulation model.

4.2. Influence of Different Drilling Parameters on Microscopic Crack Distribution

The microcrack distribution of the sample under different drilling parameters is shown in Figure 16. The microscopic crack distribution under the simulation is highly consistent with the macroscopic failure model shown in Figures 8–10. The crack distribution of the intact specimen is more scattered and random, the crack propagation of the drilled specimens is more orderly, and the crack distribution changes with the change in the drilling layout.

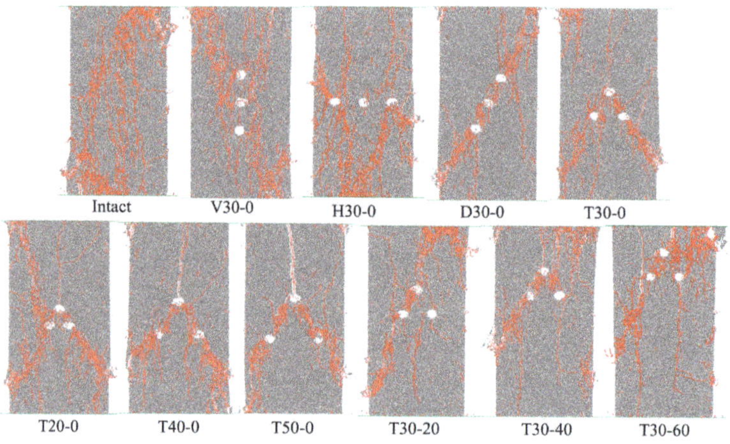

Figure 16. Microscopic crack distribution of specimen under numerical simulation.

In order to quantitatively study the relationship between the number of microscopic cracks and drilling parameters, the Fish language was used to define the total number of cracks in the PFC simulation software. The number of cracks under different drilling parameters is shown in Figure 17. With the change in drilling layout, the number of cracks decreased from 6734 in the intact specimen to 3513 in the triangular-drilling specimen. With the change in drilling spacing, the number of cracks decreases first and then increases, and is the smallest when the drilling spacing is three times the hole diameter. The number of cracks increases gradually as the drilling position moves up. These phenomena reveal the guiding effect of the drilling distribution on the expansion and penetration of cracks. The cracks gradually become orderly from disorder in the process of expansion.

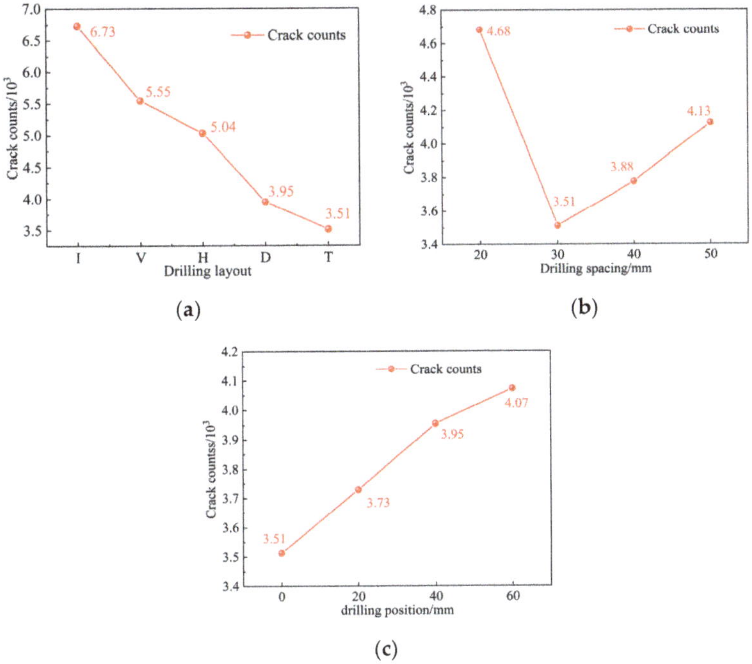

Figure 17. Number of cracks under different drilling parameters: (**a**) Drilling layout; (**b**) Drilling spacing; (**c**) Drilling position.

4.3. Influence of Different Drilling Parameters on Energy Evolution

In the process of rock deformation and failure, energy is continuously exchanged with the external environment. Assuming negligible heat exchange between the testing system and the surroundings, the total energy input into the system during rock failure is represented by boundary energy. During loading, this boundary energy is distributed among particle strain energy, cementation strain energy and dissipated energy (including slip energy and damping energy). The Fish language in PFC software is employed to monitor the various energy data during uniaxial compression [35], providing an accurate reflection of the energy evolution process in loading. Figure 18 illustrates the energy evolution curves for both the intact specimen and the T30-0 specimen.

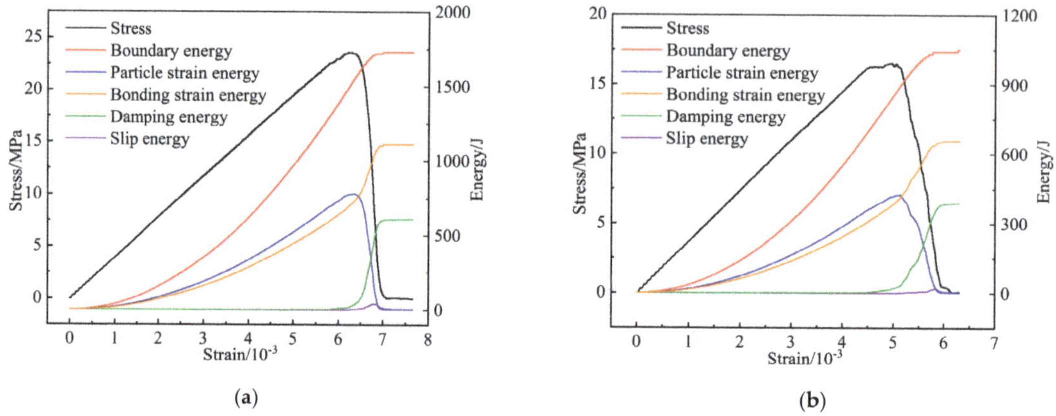

Figure 18. Energy evolution curve: (**a**) Intact specimen; (**b**) T30-0 specimen.

The energy evolution curve of the drilling specimen exhibits a shape that is largely consistent with that of the intact specimen. Prior to specimen failure, the boundary energy and strain energy increase as the strain increases. This occurs because, during the loading process, external boundary energy is initially converted into the strain energy of the specimen. When the specimen fails, dissipative energy and bonding strain energy rise rapidly, while particle strain energy decreases sharply due to sliding friction between particles and energy dissipation during the failure process. Table 4 presents the energy peaks of specimens under various drilling parameters. All forms of energy peaks in drilling specimens are lower than those in intact samples. Specifically, the boundary energy of the T30-0 sample is the smallest, and the ratio of dissipated energy to total energy is the highest, 38.99%. These findings indicate that drilling can not only effectively reduce the energy accumulation capacity, but also enhance the degree of energy dissipation, thereby achieving an excellent pressure relief effect.

Table 4. Various energy peaks under different drilling parameter combinations.

Specimen Shape	Peak Stress /MPa	Boundary Energy/J	Particle Strain Energy/J	Bonding Strain Energy/J	Damping Energy/J	Slip Energy/J
Intact	23.59	1730.02	778.32	1110.56	608.21	39.91
V30-0	19.41	1458.7	607.89	944.39	409.76	30.09
H30-0	19.08	1311.74	528.92	867.43	434.79	31.91
D30-0	17.18	1084.87	473.44	707.91	371.98	21.43
T30-0	16.5	1050.90	425.7	656.11	389.04	20.7
T20-0	17.28	1080.11	468.04	709.11	363.02	20.61
T40-0	16.71	1090.33	448.51	716.44	370.23	16.9
T50-0	18.34	1238.12	541.04	813.55	420.08	15.27
T30-20	17.15	1137.51	507.78	739.83	390.39	20.75
T30-40	18.29	1197.76	518.5	800.87	389.98	11.33
T30-60	18.83	1320.82	554.5	851.79	463.42	22.28

5. Analysis of Pressure Relief Effect and Discussion

5.1. Evaluation Method of Pressure Relief Effect

Burst tendency is an inherent attribute to determine whether rock burst can occur in coal, which is related to the mechanical properties of coal itself. The test process and classification standard of the burst tendency index of coal are provided [8]. The impact energy index (K_E) reflects the ability of rock to store elastic strain energy and is a direct criterion of impact risk. A large number of rock burst cases show that the impact energy index has a strong positive correlation with the probability of accident occurrence. In this paper, the reduction degree of the impact energy index (X_K) is used to evaluate the pressure relief effect.

$$K_E = \frac{S}{A_x} \tag{1}$$

$$X_K = \frac{\Delta K_E}{K_{E1}} = \frac{K_{E1} - K_{E2}}{K_{E1}} \tag{2}$$

where S is pre-peak accumulated strain energy, and A_x is the strain energy lost after the peak.

Zhu [15] first proposed the concept of the energy dissipation index method to evaluate the pressure relief effect. The energy dissipation index (X_S) refers to the ratio of the strain energy released during the unloading process to the strain energy accumulated before the peak in the whole stress–strain curve under uniaxial compression. The energy dissipation index (X_S) can reflect the proportion of energy consumed by plastic deformation in the process of pressure relief, reflecting the effectiveness of pressure relief measures. As shown in Figure 19, curve 1 is the stress–strain curve of coal rock before drilling pressure relief, and curve 2 is the stress–strain curve of coal rock after drilling pressure relief. The energy dissipation index is calculated by:

$$\Delta S = S_1 - S_2 = \int_0^{\varepsilon_1} \sigma_1 d\varepsilon - \int_0^{\varepsilon_2} \sigma_2 d\varepsilon \tag{3}$$

$$X_S = \frac{\Delta S}{S_1} = \frac{S_1 - S_2}{S_1} \tag{4}$$

where σ_1 is the peak strength before pressure relief; σ_2 is the peak strength after pressure relief; ε_1 is the peak strain before pressure relief; ε_2 is the peak strain after pressure relief; and ΔS is the dissipated strain energy. S_1 is the accumulated strain energy before the peak value of pressure relief, and S_2 is the residual strain energy before the peak after pressure relief.

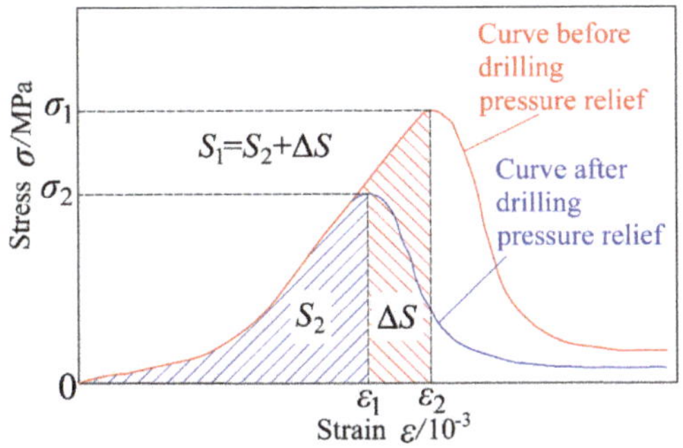

Figure 19. Calculation of energy dissipation index.

Zhang [25] proposed the strength reduction index to evaluate the pressure relief effect. The strength reduction index (X_σ) is the ratio of the peak strength of coal and rock mass after pressure relief under uniaxial compression to the peak strength before pressure relief. The strength reduction index reflects the attenuation degree of the peak strength of the sample and should be controlled within a reasonable range.

$$X_\sigma = \frac{\sigma_1 - \sigma_2}{\sigma_1} \tag{5}$$

Based on the above three indexes and considering the actual situation of drilling pressure relief in coal mines, applying different weights to the three indexes, the pressure relief effect evaluation index (X) is proposed. It can comprehensively consider the influence of drilling parameters on the energy storage capacity, energy dissipation capacity and peak strength of the specimen: the larger the X value, the better the pressure relief effect. The calculation formula of the pressure relief effect evaluation index (X) is as follows:

$$X = 45X_K + 30X_S + 25X_\sigma \tag{6}$$

5.2. Influence of Drilling Parameters on Pressure Relief Effect

5.2.1. Influence of Drilling Layouts on Pressure Relief Effect

The change in drilling layout will form different artificial weak planes inside the specimen, which will interact with the potential weak plane of the specimen itself. It affects the peak strength, crack propagation and energy dissipation capacity of the specimen and the final pressure relief effect. The calculation results of the pressure relief effect indexes of specimens with different layout modes are shown in Table 5, where $\Delta\sigma$ is the difference in peak strength before and after pressure relief. As shown in Figure 20, compared with the intact specimen, the strength reduction index of the vertical-drilling specimen, horizontal-drilling specimen, diagonal-drilling specimen and triangular-drilling specimen is 26.93%, 36.45%, 44.64% and 53.41%, respectively. The pre-peak accumulated strain energy of drilling specimens also decreased correspondingly, and the energy dissipation index was 33.17%, 41.29%, 47.39% and 58.01%, respectively. The impact energy index of the intact specimen, vertical specimen, horizontal specimen, diagonal specimen and triangular specimen decreased from 5.07 to 2.44, 1.33, 1.47 and 1.14, respectively. It can be seen that different drilling layouts can effectively reduce the burst tendency of the specimen.

Table 5. Evaluation index for the pressure relief effect of specimens with different drilling layouts.

Specimen Type	$\Delta\sigma$/MPa	X_σ/%	$S/10^{-3}$	$\Delta S/10^{-3}$	X_S/%	$A_x/10^{-3}$	K_E	X_K/%	X
Intact		-	54.62	-	-	10.77	5.07		
Vertical	6.32	26.93	36.50	18.12	33.17	14.98	2.44	51.88	39.18
Horizontal	8.55	36.45	32.07	22.55	41.29	24.14	1.33	73.77	54.45
Diagonal	10.47	44.64	28.73	25.89	47.40	19.58	1.47	71.01	57.45
Triangular	12.53	53.41	22.93	31.69	58.01	20.14	1.14	77.51	65.41

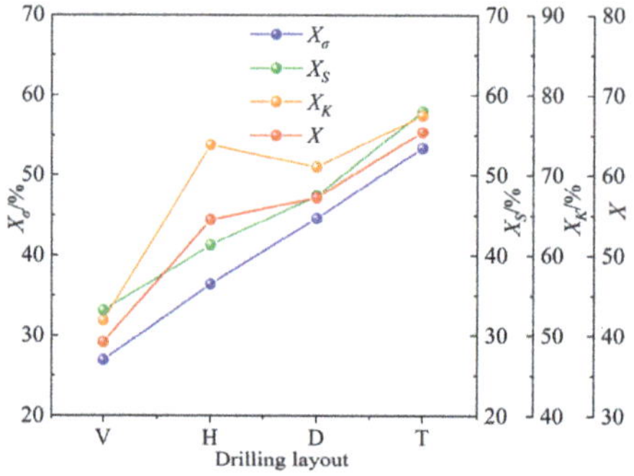

Figure 20. Relationship between different drilling layouts and the evaluation index of pressure relief effect.

Compared with the intact specimen, the pressure relief effect evaluation index (X) of the vertical-drilling specimen, horizontal-drilling specimen, diagonal-drilling specimen and triangular-drilling specimen is 39.18, 54.45 57.45 and 65.41, respectively. Therefore, the triangular-layout drilling has the best pressure relief effect.

5.2.2. Influence of Drilling Spacings on Pressure Relief Effect

When the drilling layout is determined, the different drilling spacings will affect the stress interaction and the crack propagation process between the drillings, thus affecting the final pressure relief effect. The calculation results of the pressure relief effect indexes of specimens with different spacing modes are shown in Table 6. As shown in Figure 21, compared with the intact specimen, with the increase in drilling spacing, the pressure relief effect evaluation index (X) was first increased and then decreased, and was 59.66, 65.41, 64.09 and 57.06, in sequence. This indicates that the pressure relief effect is best when the drilling spacing is three times the diameter.

Table 6. Evaluation index for the pressure relief effect of specimens with different drilling spacings.

Specimen Type	$\Delta\sigma$/MPa	X_σ/%	$S/10^{-3}$	$\Delta S/10^{-3}$	X_S/%	$A_x/10^{-3}$	K_E	X_K/%	X
Intact		-	54.62	-	-	10.77	5.07		
T20-0	10.78	45.95	26.93	27.69	50.70	20.26	1.33	73.77	59.66
T30-0	12.53	53.41	22.93	31.69	58.01	20.14	1.14	77.51	65.41
T40-0	12.03	51.31	21.52	33.10	60.60	16.70	1.29	74.56	64.09
T50-0	10.28	43.82	27.15	27.47	50.30	17.62	1.54	69.63	57.06

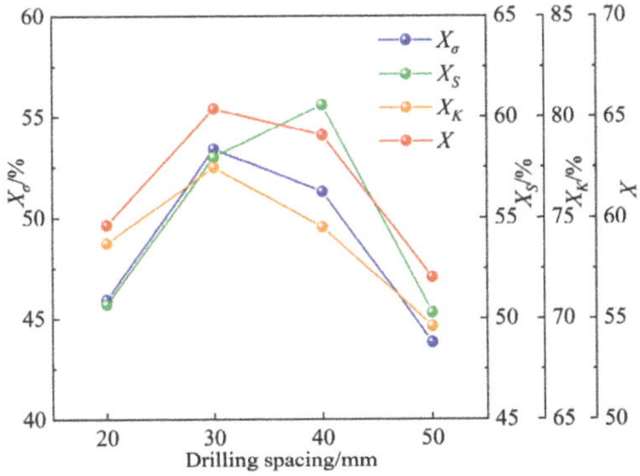

Figure 21. Relationship between different drilling spacings and the evaluation index of pressure relief effect.

5.2.3. Influence of Drilling Positions on Pressure Relief Effect

During the determination of drilling layout and spacing, because of the different stress states of specimens during uniaxial compression, the change in drilling position will also affect the pressure relief effect. The calculation results of the pressure relief effect indexes of specimens with different position modes are shown in Table 7. As shown in Figure 22, the strength reduction index and energy dissipation index decreased with the upward movement of the drilling position. The residual strength increases with the upward movement of the drilling position, and the post-peak strain energy is also increased, resulting in the impact energy index gradually decreasing with the upward movement of the drilling position. Compared with the intact specimen, the pressure relief effect evaluation index (X) decreased with the upward movement of the drilling position, and was 65.41, 60.67, 59.23 and 57.10, in sequence. It can be concluded that drilling in the middle of the specimen has the best pressure relief effect.

Table 7. Evaluation index for the pressure relief effect of specimens with different drilling positions.

Specimen Type	$\Delta\sigma$/MPa	X_σ/%	$S/10^{-3}$	$\Delta S/10^{-3}$	X_S/%	$A_x/10^{-3}$	K_E	X_K/%	X
Intact	-	-	54.62	-	-	10.77	5.07		
T30-0	12.53	53.41	22.93	31.69	58.01	20.14	1.14	77.51	65.41
T30-20	11.10	47.32	25.02	29.60	54.19	18.40	1.36	73.18	60.67
T30-40	9.77	41.66	28.34	26.28	48.11	24.40	1.16	77.12	59.23
T30-60	8.45	36.03	32.21	22.41	41.03	31.75	1.01	80.08	57.10

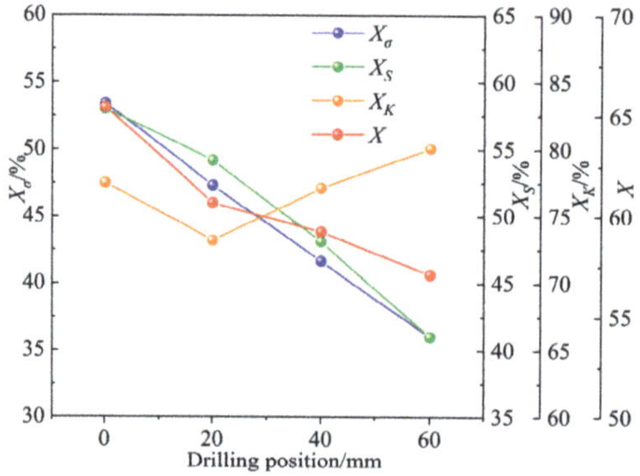

Figure 22. Relationship between different drilling positions and the evaluation index of pressure relief effect.

5.3. Discussion

(1) According to the value of the pressure relief effect evaluation index (X) under different drilling parameters, it is determined the influence degree of the drilling layout, drilling spacing and drilling position on the pressure relief effect. The X values of different drilling layouts ranged from 35.72–65.41, with a range of 26.93. The X value of drilling spacing ranged from 57.06–65.41, with a range of 8.31, and the X value of hole position ranged from 57.10–65.41, with a range of 8.35. Therefore, the drilling layout has the greatest influence on the pressure relief effect, and the change in drilling layout can be given priority in the actual working condition.

(2) In the process of carrying out the tests, the poured specimens were first placed in a constant-temperature curing room for maintenance, and the test conditions (temperature, homogeneity of specimens) were ensured as much as possible to reduce the influence of possible errors on the test results. Therefore, during the test, the results were valid at room temperature. The test results can serve as a valuable reference for selecting drilling parameters when implementing the drilling pressure relief with a large diameter in rock burst mines. However, from the laboratory test results to the application of drilling pressure relief under practical engineering conditions, it is necessary to consider many factors, such as geometric similarity and coal seam occurrence conditions. Therefore, in order to make the research results better meet the actual engineering conditions on the site, in the further research plan, the influence of temperature, gas pressure, original rock stress and disturbance stress on the drilling parameters, the mechanical properties of coal and the effectiveness of drilling pressure relief methods will be studied.

6. Conclusions

(1) The stress–strain curve shape of the specimen is affected by the drilling parameters, especially the drilling layout. Compared with the intact specimen, the peak strength of the drilling specimen is significantly reduced, and the drilling layout has the greatest influence on the mechanical properties, followed by the drilling spacing and drilling position. The peak strength of the specimen with a triangular layout is the smallest (10.93 MPa). The peak strength decreases first and then increases with the increase in drilling spacing, and the peak strength is the smallest when the drilling spacing is three times the diameter. The peak strength increases gradually as the drilling position moves upward.

(2) In the process of uniaxial compression, the stress distribution inside the specimen will affect the stress concentration around the drilling. Different drilling layouts will form different weak-strength surfaces in the specimen, and lead the expansion and penetration of cracks, resulting in different failure modes of the specimen. The failure modes of the specimens in vertical layout, horizontal layout, diagonal layout and triangular layout are tensile failure, x-conjugate shear failure, single-diagonal-plane shear failure and tensile shear mixed failure. According to the different drilling spacings, the failure modes can be divided into independent-type failure, independent-through-transitional-type failure and through-type failure. The drilling position will affect the crack propagation and finally affect the damage degree of the specimen.

(3) Drilling can not only effectively reduce the energy accumulation capacity, but also enhance the degree of energy dissipation. The AE ringing counts and energy of the triangular-drilling specimens are the least. The AE ringing counts and energy decrease first and then increase with the increase in the drilling spacing, and are the smallest at three times the drilling diameter. The AE ringing counts and energy increase gradually with the upward movement of the drilling position.

(4) Under different combinations of drilling parameters, the drilling layout being triangular, the drilling spacing being three times the hole diameter and the drilling position being in the middle of the specimen have the best pressure relief effects. Compared with the intact specimen, the strength reduction index (X_{σ}) is 53.41%, the energy dissipation index (X_S) is 58.01%, the reduction degree of impact energy index (X_K) is 77.51%, and the pressure relief effect evaluation index (X) is 65.41.

Author Contributions: Conceptualization, Y.Z. and S.Z.; methodology, Y.Z. and Y.C.; software, Y.W. and S.Z.; validation, S.L. and Y.W.; investigation, Y.C., Q.W., Y.W., S.L. and L.Z.; resources, G.F. and Y.Z.; writing—original draft preparation, Y.C.; writing—review and editing, Y.Z., S.Z. and Y.C.; project administration, Y.Z., S.Z. and G.F.; funding acquisition, Y.Z., S.Z. and B.W. All authors have read and agreed to the published version of the manuscript.

Funding: This research was funded by the National Natural Science Foundation of China (Nos. 52374100, 52304150, 52204107), China Postdoctoral Science Foundation-funded project (No. 2020M682215) and Fundamental Research Program of Shanxi Province (No. 202203021212252).

Data Availability Statement: The data presented in this study are available on request from the corresponding author.

Conflicts of Interest: Author Liang Zhao was employed by the Yangcheng County Yangtai Group Industrial Co., Ltd. Author Yujiang Zhang is part of the Postdoctoral Centre at Shandong Energy Group Co., Ltd. The remaining authors declare that the research was conducted in the absence of any commercial or financial relationships that could be construed as potential conflicts of interest.

References

1. Pan, Y.; Song, Y.; Liu, J. Pattern, Change and new situation of coal mine rock burst prevention and control in China. *Chin. J. Rock Mech. Eng.* **2023**, *42*, 2081–2095.

2. Dou, L.; Tian, X.; Cao, A.; Gong, S.; He, H.; He, J.; Cai, W.; Li, X. Present situation and problems of coal mine rock burst prevention and control in China. *J. China Coal Soc.* **2022**, *47*, 152–171.
3. Pan, J.; Liu, S.; Ma, W.; Xia, Y.; Wang, S.; Feng, M. Occurrence law and classification prevention of rock burst in coal mines of Shaanxi Province. *Coal Sci. Technol.* **2024**, *52*, 95–105.
4. Zhao, T.B.; Guo, W.Y.; Tan, Y.L.; Yin, Y.C.; Cai, L.S.; Pan, J.F. Case studies of rock bursts under complicated geological conditions during multi-seam mining at a depth of 800 m. *Rock Mech. Rock Eng.* **2018**, *51*, 1539–1564.
5. Gong, F.; He, Z.; Si, X. Experimental study on revealing the mechanism of rockburst prevention by drilling pressure relief: Status-of-the-art and prospects. *Geomat. Nat. Hazards Risk* **2022**, *13*, 2442–2470.
6. Pan, Y. Disturbance response instability theory of rock burst in coal mine. *Coal Sci. Technol.* **2018**, *43*, 2091–2098.
7. *GB/T 25217.10-2019*; Methods for Test, Monitoring and Prevention of Rock Burst—Part 10: Prevention Method of Drillhole Destressing on Coal-Seam. Standardization Administration of China: Beijing, China, 2019.
8. *GB/T 25217.2-2010*; Methods for Test, Monitoring and Prevention of Rock Burst—Part 2: Classification and Laboratory Test Method on Bursting Liability of Coal. Standardization Administration of China: Beijing, China, 2010.
9. Wang, M.; Wang, X.; Xiao, T. Borehole destressing mechanism and determination method of its key parameters in deep roadway. *Coal Sci. Technol.* **2017**, *42*, 1138–1145.
10. Zhao, X.; Zhang, H.; Zhu, W. Fracture evolution around pre-existing cylindrical cavities in brittle rocks under uniaxial compression. *Trans. Nonferr. Met. Soc. China* **2014**, *24*, 806–815. [CrossRef]
11. Lin, P.; Wong, R.H.; Tang, C.A. Experimental study of coalescence mechanisms and failure under uniaxial compression of granite containing multiple holes. *Int. J. Rock Mech. Min. Sci.* **2015**, *77*, 313–327.
12. Jia, C.; Jiang, Y.; Zhang, X.; Wang, D.; Ruan, H.; Wang, C. Laboratory and numerical experiments on pressure relief mechanism of large-diameter boreholes. *Chin. J. Geotech. Eng.* **2017**, *39*, 1115–1122.
13. Wang, P.; Jiang, Y.; Li, P.; Zhou, J.; Zhou, Z. Experimental analysis of pressure relief effect of surrounding rock in high-stress roadways under different drilling parameters. *Appl. Sci.* **2023**, *13*, 2511. [CrossRef]
14. Zhu, Q.; Yin, Y.; Liu, J.; Wen, J.; Shi, J. Mechanism of hysteretic rock burst and outburst induced by surrounding rock stress returning. *China Saf. Sci. J.* **2017**, *27*, 99–104.
15. Zhu, S.; Jiang, F.; Shi, X.; Sun, G.; Zhang, Z.; Cheng, X.; Zhang, H. Energy dissipation index method for determining rock burst prevention drilling parameters. *Rock Soil Mech.* **2015**, *36*, 2270–2276.
16. Yao, J.; Yin, Y.; Zhao, T.; Ren, W.; Qiu, Y.; Guo, W. Segmented enlarged-diameter borehole destressing mechanism and its influence on anchorage support system. *Energy Sci. Eng.* **2020**, *8*, 2831–2840. [CrossRef]
17. Ma, B.; Deng, Z.; Zhao, S.; Li, S. Analysis on mechanism and influencing factors of drilling pressure relief to prevent rock burst. *Coal Sci. Technol.* **2020**, *48*, 35–40.
18. Wang, Q.; Zhang, Y.; Cui, D.; Feng, G.; Du, Y.; Zhang, S.; Chen, H.; Zhang, Z. Effect of Indenter Angle on Bearing Performance of Coal and Accompanying Rocks Under Point Load and Characterization of Mechanical Parameters. *Energy Sci. Eng.* **2024**, *13*, 374–391.
19. Tai, L.; Li, C.; Hu, Y.; Yu, X.; Xu, Z.; Zhang, X.; Lu, S. Experimental study on the significance of pressure relief effect and crack extension law under uniaxial compression of rock-like materials containing drill holes. *Sci. Rep.* **2024**, *14*, 1104.
20. Zhang, Y.; Cui, B.; Wang, Y.; Zhang, S.; Feng, G.; Zhang, Z. Evolution Law of Shallow Water in Multi-Face Mining Based on Partition Characteristics of Catastrophe Theory. *Fractal Fract.* **2023**, *7*, 779. [CrossRef]
21. Yin, Y.; Chen, B.; Zhang, Y.; He, S.; Yao, C.; Liu, C. Experimental study and evaluation on the weakening of bursting liability of coal with boreholes. *Eng. Fail. Anal.* **2024**, *155*, 107754.
22. Yin, Y.; Chen, B.; Han, Y.; Qiu, Y.; Li, C.; Chen, Y. Evolution Law of Stress Field and Pressure Relief Effect of Coal Seam After Drilling. *Energy Sci. Eng.* **2024**, *12*, 5493–5503.
23. Ma, J.; Li, J.; Zhang, S.; Yang, S.; Dong, X.; Liu, D. Numerical simulation analysis of acoustic emission characteristics during the drilling pressure relief process in coal seams under different influencing factors. *Processes* **2023**, *11*, 3277. [CrossRef]
24. Wu, H.; Zhao, G.; Liang, W. Mechanical properties and fracture characteristics of pre-holed rocks subjected to uniaxial loading: A comparative analysis of five hole shapes. *Theor. Appl. Fract. Mech.* **2020**, *105*, 102433.
25. Zhang, S.; Li, Y.; Shen, B.; Sun, X.; Gao, L. Effective evaluation of pressure relief drilling for reducing rock bursts and its application in underground coal mines. *Int. J. Rock Mech. Min. Sci.* **2019**, *114*, 7–16.
26. Nian, J.; Zhao, B.; Zhang, W. Numerical simulation research on the pressure relief and permeability enhancement mechanism of large-diameter borehole in coal seam. *Geofluids* **2022**, *2022*, 2926213.
27. Wang, A.; Gao, Q.; Pan, Y.; Song, Y.; Li, L. Bursting liability and energy dissipation laws of prefabricated borehole coal samples. *J. China Coal Soc.* **2021**, *46*, 959–972.
28. Zuo, J.; Hu, S.; Zhou, X.; Zhang, C.; Guo, Y. Effective evaluation of pressure relief drilling layout for reducing rock bursts and sensitive factor analysis. *Arab. J. Geosci.* **2021**, *14*, 2643.

29. Gong, F.; He, Z.; Jiang, Q. Internal mechanism of reducing rockburst proneness of rock under high stress by real-time drilling pressure relief. *Rock Mech. Rock Eng.* **2022**, *55*, 5063–5081.
30. He, Z.; Gong, F.; Luo, S. Evaluation of the rock burst proneness of red sandstone with prefabricated boreholes: An experimental study from the energy storage perspective. *Geomat. Nat. Hazards Risk* **2021**, *12*, 2117–2154.
31. Lyu, J.; Qi, L.; Pan, Y.; Dai, L.; Tang, Z.; Wang, X. Study on the pressure relief energy dissipation law of variable-diameter boreholes in roadway surrounding rock under dynamic and static loads. *PLoS ONE* **2024**, *19*, e0306449.
32. Chen, M.; Zhang, Y.; Zang, C.; Zhang, G.; Li, Q.; Jiang, B. Experimental investigation on pressure relief mechanism of specimens with prefabricated reaming boreholes. *Rock Mech. Rock Eng.* **2023**, *56*, 2949–2966.
33. Qi, Y.; Jing, H.; Meng, B.; Dong, Z.; Liu, D. Experimental modelling on size effect of pressure relief hole. *J. Min. Saf. Eng.* **2018**, *35*, 538–544.
34. Zhang, T.; Jing, C.; Zhang, L.; Ji, X.; Pan, H. Strain localization characteristics of perforation failure of perforated specimens. *Coal Sci. Technol.* **2020**, *45*, 4087–4094.
35. Guo, J.; Feng, Z.; Li, X. Evolution of microcracks and energy of granite during shear test with PFC3D. *Coal Sci. Technol.* **2024**, *52*, 60–70.

processes

MDPI

Article

Molecular Simulation of Ultra-Microstructural Characteristics of Adsorption Pores in Terms of Coal and Gas Adsorption Properties

Pan Chen [1,2], Yanping Wang [1], Yanxia Zhao [3,*], Qi Wang [1], Zhihui Wen [1,4,5] and Ligang Tang [6,*]

1. College of Safety Science & Engineering, Henan Polytechnic University, Jiaozuo 454000, China; jmchenpan@163.com (P.C.); wyp@home.hpu.edu.cn (Y.W.); w19513251633@163.com (Q.W.); wenzhihui@hpu.edu.cn (Z.W.)
2. JiaoZuo Coal Industry (Group) Co., Ltd., Jiaozuo 454150, China
3. School of Mathematics & Information Science, Henan Polytechnic University, Jiaozuo 454000, China
4. State Key Laboratory Cultivation Base for Gas Geology and Gas Control, Henan Polytechnic University, Jiaozuo 454000, China
5. State Collaborative Innovation Center of Coal Work Safety & Clean-Efficiency Utilization, Jiaozuo 454000, China
6. School of Mine Safety, North China Institute of Science & Technology, Sanhe 065201, China

* Correspondence: zhaoyanxia001@126.com (Y.Z.); laotangx@126.com (L.T.)

Abstract: To investigate the ultra-microstructural characteristics and adsorption properties of coal pores, the pore structure of Dongsheng lignite and Chengzhuang anthracite in Qinshui Basin was characterized by the liquid nitrogen adsorption method. It was found that the SSA of micropores constituted more than 65% of the total SSA in both coal samples. The macromolecular model of coal and the N_2 molecular probe were used to obtain the ultrastructure parameters, and the gas adsorption behaviors of the two coals under different conditions were simulated by Grand Canonical Monte Carlo (GCMC) and Molecular Dynamics (MD). The results show that the pores of the lignite are mainly small pores, while the pores of the anthracite are mainly micropores. The specific surface area of the adsorption pores mainly constitutes micropores and ultra-micropores. The adsorption capacity of the CH_4 of anthracite is consistently higher than that of lignite. The CH_4 adsorption amount is positively correlated with the specific surface area and pore volume. This indicates that the gas adsorption capacity of coal is concentrated in micropores and ultra-micropores. The adsorption capacity increases with the increase in pressure and decreases with the increase in temperature. In the competitive adsorption of $CH_4/CO_2/H_2O$, the adsorption quantity is in the order of $H_2O > CO_2 > CH_4$. The research results provide a theoretical basis for coalbed methane exploitation and methane replacement.

Keywords: coal pore structure; molecular probe; ultra-micropore; coal macromolecular model; gas adsorption

1. Introduction

The coal industry, as the cornerstone of industrialization, will continue to dominate global energy production and consumption for a long time in the future [1–3]. Coal is an organic, porous, and heterogeneous material characterized by a vast system of pores and fractures, which forms an interconnected network structure [4,5]. This structure not only provides ample space for gas adsorption but also serves as a channel for gas migration [6,7]. Research on the ultra-microstructural characteristics and adsorption properties of coal pores holds significant theoretical importance for the development of coalbed methane and the methane displacement within coal seams.

As global research on coal pore structures deepens, more sophisticated testing methods have been applied to coal porosity measurements, each offering unique advantages in the structural characterization of pores. Common testing methods for studying pores in coal include scanning electron microscopy (SEM), nano-CT imaging, nuclear magnetic resonance (NMR), high-pressure mercury injection, low-temperature CO_2 adsorption, and low-temperature liquid N_2 adsorption [8–11]. The complexity of pore structures in coal can be quantitatively characterized based on pore volume (PV), specific surface area (SSA), porosity, pore shape factors, fractal geometry, and other features [12–14]. Liu et al. [15] used low-temperature liquid nitrogen adsorption to quantitatively characterize the microstructure and pore distribution of coal samples with different metamorphic degrees (anthracite, bituminous coal, and lignite). However, due to the inability of N_2 molecules to penetrate pores smaller than their own diameter, the low-temperature liquid nitrogen adsorption technique is limited to measuring only a fraction of micropores. Li et al. [16] employed a combination of high-pressure mercury injection, low-temperature CO_2 adsorption, and low-temperature liquid nitrogen adsorption to jointly characterize the full pore size range of four coal samples from the Lu'an mining area. Furthermore, Ren et al. [17] combined low-temperature CO_2 adsorption, low-temperature liquid N_2 adsorption, and NMR to systematically investigate the pore structure and fractal characterization of coal. Therefore, it is evident that the combination method for characterization serves as an effective means for comprehensively and quantitatively assessing pore characteristics.

It has been reported that the micropores and ultra-micropores of coal play a crucial role in CH_4 adsorption [18,19], and that oxygen-containing functional groups play a secondary role in gas adsorption [20]. Ren et al. [21] have demonstrated that during the surface acidification modification of coal using Fenton reagents, a significant amount of oxygen-containing functional groups can be generated on the inner surface of coal. However, the isothermal adsorption capacity of coal under different concentrations showed little difference, indicating that the impact of oxygen-containing functional groups on coal adsorption was relatively minor. Han et al. [22] constructed a 4 nm pore structure as a methane migration channel and then analyzed the differences in the methane adsorption characteristics of coal with different metamorphic degrees, temperatures, and pressures. Wang et al. [23] found that the total PVs and SSAs of different-rank coal samples all exhibited a U-shaped variation pattern and analyzed the influence of pore structure on CH_4 adsorption capacity. Jiang et al. [24] investigated the impact of pore structure on gas adsorption in low-rank coal, revealing a positive correlation between Langmuir volume and both micropore PV and SSA. Specifically, a greater micropore PV and SSA corresponded to a stronger gas adsorption capacity. However, due to experimental limitations, it is challenging to measure ultra-micropores in coal. Some researchers have employed molecular modeling and simulation techniques to determine ultra-micropore structures. Wang [25] simulated and calculated the pore parameters of Dongqu No. 2 coal with the aid of Materials Studio and measured the pore parameters in the aggregated structure model by utilizing molecules with different radii as probes. Wang et al. [26] employed van der Waals force probes to visualize coal pore space and quantitatively analyze pore structure parameters. Yan et al. [27] utilized molecular simulation to study the influences of temperature, pressure, and moisture content on CO_2 and CH_4 adsorption in a macromolecular model of coal from Chicheng Mine, as well as the underlying microscopic mechanisms. Combining physical experiments with molecular simulation offers remarkable advantages in the ultra-microstructural characterization of adsorption pores.

This study selected two coal samples of different metamorphic degrees, i.e., Dongsheng lignite and Chengzhuang anthracite, as research objects. The comprehensive quantitative characterization of their adsorption pore structures was realized by low-temperature nitrogen adsorption experiments. Then, macromolecular models of lignite and anthracite were constructed using Grand Canonical Monte Carlo (GCMC) and Molecular Dynamics (MD) methods. Meanwhile, N_2 molecules were employed as probes to simulate and calculate the SSAs and PVs of the coal samples, and thus their ultra-micropore structural characteristics were obtained. Moreover, the adsorption properties of the lignite and anthracite macromolecular models were simulated under various pressures, temperatures, and adsorbates. This study is expected to elucidate the ultra-micropore structural characteristics and adsorption properties of coal, thereby establishing a foundational basis for further exploration of the interaction mechanisms between gas molecules and coal. Additionally, it offers theoretical support for coalbed methane extraction and methane displacement within coal seams.

2. Experimental and Simulation Methodology

2.1. Molecular Model Construction

Two coal samples with different metamorphic degrees, i.e., low-rank Dongsheng lignite (hereafter referred to as DS-L) and high-rank Chengzhuang anthracite (hereafter referred to as CZ-A), were selected for the experiment. A previous study by our research team [28] has offered a comprehensive explanation of the specific methods and processes used to construct the molecular structure model of CZ-A. Correspondingly, the molecular structure model of DS-L was constructed in the same manner. The final initial 3D macromolecular models of coal are displayed in Figure 1, and the detailed parameters are listed in Table 1.

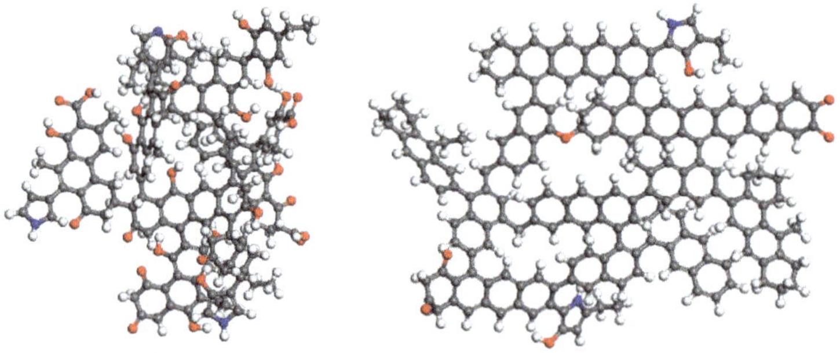

(**a**) (**b**)

Figure 1. (**a**) Initial 3D molecular model of DS-L. (**b**) Initial 3D molecular model of CZ-A.

Table 1. Detailed parameters of the molecular model.

Coal Sample	Element Formula	Ultimate Analysis			
		C	H	O	N
DS-L	$C_{153}H_{139}N_3O_{29}$	73.99	5.64	18.68	1.69
CZ-A	$C_{188}H_{138}N_2O_7$	89.00	5.48	4.41	1.10

The constructed model structures were optimized through geometric refinement and annealing dynamics to obtain globally energy-optimized molecular models. Subsequently, periodic boundary conditions were applied to the optimized macromolecular models for density simulations. By continuously varying the model density, molecular potential energy systems at different densities were obtained. The densities corresponding to the lowest energies in the DS-L and CZ-A molecular model systems were determined to be 1.30 g/cm^3 and 1.45 g/cm^3, respectively, representing the optimal densities for the molecular models. The corresponding periodic stable molecular models are depicted in Figure 2a,b.

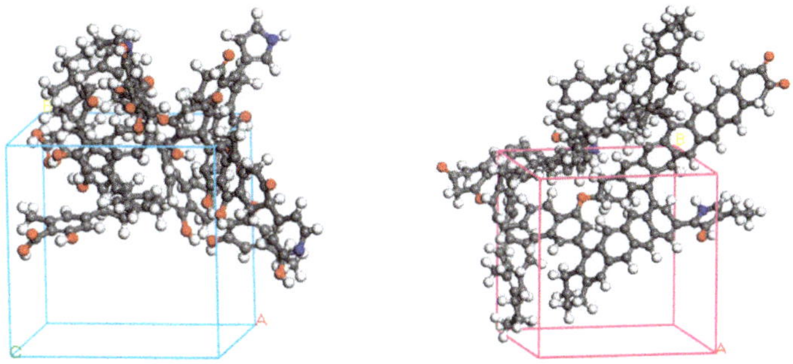
(**a**) DS-L (**b**) CZ-A

Figure 2. (**a**) Macromolecular model of DS-L under periodic boundary conditions. (**b**) Macromolecular model of CZ-A under periodic boundary conditions.

2.2. Liquid Nitrogen Adsorption Experiment

The pore structure characteristics of the experimental coal samples were determined through the liquid nitrogen adsorption experiment. The particle sizes of the experimental coal samples ranged from 0.18 to 0.25 mm. The measured contents included the SSA, PV, and pore structure characteristics of the coal samples. The experiment employed the V-sorb2800 SSA and pore size analyzer produced by Beijing Jinaipu Company (Beijing, China), which featured a pore size measurement range of 0.35–500 nm and an SSA lower limit of 0.01 m^2/g. The static volumetric method was employed, utilizing liquid nitrogen with a purity of ≥99.999% as the adsorbate at −196 °C (77.15 K) during the adsorption process.

2.3. Simulation of Coal Pore Structure Characteristics

The SSAs and PVs of the DS-L and CZ-A macromolecular models were simulated with the aid of Materials Studio. In the simulation, the Atom Volumes & Surface module within the Tools toolbar was utilized, and the coal macromolecular models were calculated based on the Connolly principle. In addition, the Atom Volumes & Surfaces tool was selected, and then a Connolly Surface was established, with the Connolly radius set to 1.84 Å (equivalent diameter of a nitrogen molecule). The remaining parameters were left at their default settings.

2.4. Optimization of the Model

The parameters were set as follows: Geometry optimization: The energy standard deviation was 0.00002 Kcal/mol; the atomic root mean square was 0.001 Kcal/mol/Å; the maximum iteration step was 50,000; the force field was COMPASS; the charge was Charge using QEq; the Coulomb force and van der Waals force were both atom-based; and the accuracy was ultra-fine. Annealing optimization: The number of cycles was 10; the initial temperature was 300 K; the maximum temperature was 600 K; the amplitude of each heating or cooling was 5 K; the ensemble was NVT ensemble; the time was 1.0 fs; the temperature control method was nose; the force field was Dreiding; the charge was charge using QEq; the Coulomb force and van der Waals force were both atom-based; and the accuracy was ultra-fine.

2.5. Adsorption Simulation

The specific parameter settings for the simulation were as follows: The task was set to adsorption isotherm, and the Monte Carlo method and the Grand Canonical ensemble were adopted. The simulation temperature was set to 303.15 K (30 °C), and the pressure ranged from 0 kPa to 10,000 kPa, with the gradient being 1000 kPa. The fugacity steps were set to 10, the equilibration steps to 100,000 (which meant that the first 100,000 steps of the simulation were used to reach equilibrium), and the production steps to 1,000,000 (which meant that the subsequent 1,000,000 steps were used to collect statistics on the studied adsorption properties). In addition, the COMPASS force field and Charge using QEq were selected. Both the electrostatic and van der Waals forces were set to atom-based, and the simulation precision was set to fine.

3. Results and Discussion

3.1. Pore Structure Characteristics of Coal

(1) PV distribution characteristics

Aiming at exploring the pore size distribution of coal samples with varying metamorphic degrees, the data obtained from the liquid nitrogen adsorption experiments were analyzed with the aid of the inherent Barrett–Joyner–Halenda (BJH) theoretical model in the software. This model, rooted in the Kelvin equation, is a robust model for determining pore size distribution. Based on the principle of capillary condensation, it can be expressed by Equation (1):

$$r_k = -\frac{2\sigma_1 V_{\mathrm{ml}}}{RT_{\mathrm{b}} ln(P/P_0)} \quad (1)$$

where r_k is the pore radius of the experimental coal sample, nm; σ_1 is the surface tension of the condensed phase, N/M; V_{ml} is the adsorption volume of the liquid under the action of liquid nitrogen, cm^3/g; R is the gas constant, with a value of 8.314 $J \cdot mol^{-1} \cdot K^{-1}$; T_{b} is the experimental temperature, at 77 K in this study; P is the pressure of the adsorbate, Pa; and P_0 is the saturation vapor pressure of nitrogen at −196 °C, Pa.

To investigate the PV distribution characteristics of coal samples with varying metamorphic degrees, it is necessary to classify the pore diameters. To achieve this goal, the decimal classification standard proposed by B.B.Hodot was adopted; that is, pores in the coal samples were divided into micropores (<10 nm), small pores (10–100 nm), mesopores (100–1000 nm), and macropores (>1000 nm). The PV distribution results obtained from the experiments are illustrated in Table 2.

Table 2. PV test results of coal samples.

Coal Sample	BJH PV (mL/g)			Total PV (mL/g)
	Micropores	Small Pores	Mesopores	
DS-L	0.003318	0.006256	0.00288	0.012454
CZ-A	0.009209	0.005768	0.004665	0.019642

As can be observed from Table 2, the total PV of DS-L is 0.012454 mL/g, whereas that of CZ-A is 0.019642 mL/g. For DS-L, the PV of small pores consistently exceeds that of micropores and macropores, indicating that small pores constitute the primary component of the total PV, followed by micropores and mesopores. In contrast, CZ-A is dominated by micropores, with small pores and mesopores being secondary contributors.

(2) Pore SSA distribution characteristics

The SSA was calculated utilizing the inherent Brunauer–Emmett–Teller (BET) calculation method in the analyzer, based on the different adsorption amounts of nitrogen for the coal samples with different metamorphic degrees under various pressures, as given by:

$$\frac{P/P_0}{V(1-P/P_0)} = \frac{1}{V_m C} + \frac{C-1}{V_m C}\left(\frac{P}{P_0}\right) \quad (2)$$

where P is the adsorbate pressure, Pa; P_0 is the saturation vapor pressure of nitrogen at −196 °C, Pa; V is the actual adsorption capacity on the coal sample surface, ml; V_m is the monolayer saturation adsorption capacity of nitrogen, ml; and C is a constant related to the adsorption capacity of the experimental coal sample.

The SSAs of the coal samples were determined through liquid nitrogen adsorption experiments, and the results are presented in Table 3. According to Table 3, the total SSA of DS-L is 4.839 m^2/g, whereas that of CZ-A is 5.362 m^2/g. Notably, for both coal samples, the SSA of micropores significantly exceeds that of small pores and mesopores, with the contribution ratio of micropores exceeding 65% in both instances. This underscores the pivotal role of micropores in contributing to the overall SSA of the coal samples. Moreover, as the metamorphic degree of the coal sample rises, the SSA of micropores exhibits an upward trend.

Table 3. SSA test results of coal samples.

Coal Sample	BET SSA (m^2/g)			Total SSA (m^2/g)
	Micropores	Small Pores	Mesopores	
DS-L	3.308	1.449	0.081	4.839
CZ-A	4.712	0.519	0.130	5.362

(3) Pore shape characteristics

The DE classification method was employed to analyze the experimental coal samples [29]. The varying metamorphic degrees of the coal samples resulted in differences in their liquid nitrogen adsorption isotherms, which in turn reflected distinct pore shape characteristics, as shown in Figure 3, where type A corresponds to cylindrical holes, type B relates to fissure holes, type C and type D correspond to wedge-shaped holes, and type E relates to ink-bottle-shaped holes in porous media.

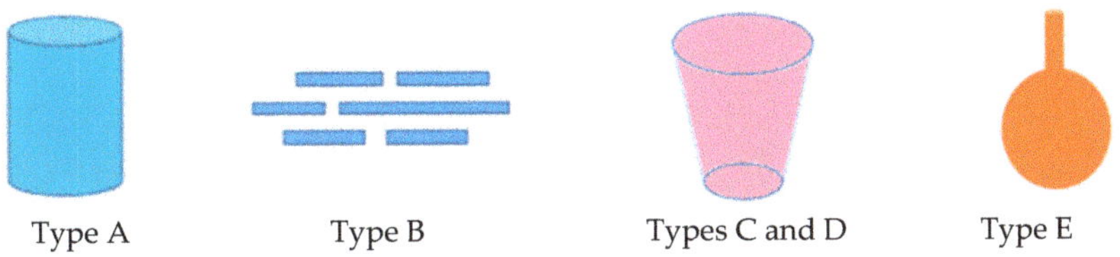

Figure 3. Pore shapes reflected by adsorption isotherms.

According to the experimental data, the liquid nitrogen adsorption isotherms for the coal samples are depicted in Figure 4.

As shown in Figure 4a, during adsorption, the liquid nitrogen adsorption and desorption isotherms of DS-L remain separated all the time, even at lower relative pressures. This observation suggests that the micropores in DS-L are well developed and dominated by cylindrical micropores with one end closed.

As shown in Figure 4b, during adsorption, the liquid nitrogen adsorption and desorption isotherms of CZ-A are also separated, with no overlap even at lower relative pressures. This finding reveals that micropores in CZ-A are also well developed and dominated by cylindrical micropores with one end closed. However, the desorption isotherm exhibits an inflection point at a relative pressure of approximately 0.5, indicative of the presence of ink-bottle-shaped micropores.

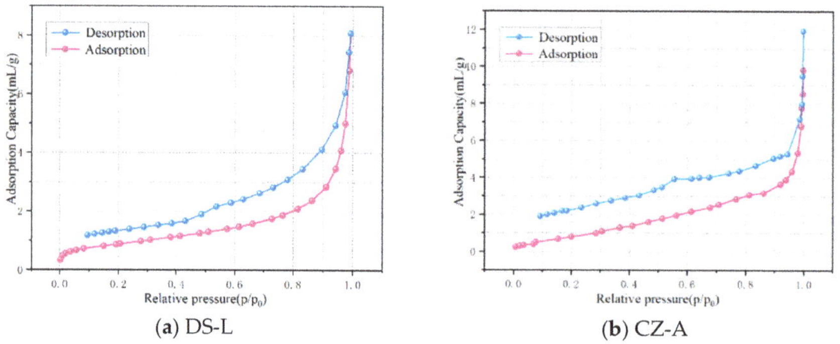

Figure 4. (**a**) Liquid nitrogen adsorption isotherms of DS-L. (**b**) Liquid nitrogen adsorption isotherms of CZ-A.

3.2. Simulation of Pore Structure Characteristics in Coal Macromolecular Models

Liquid nitrogen adsorption serves as a reliable method for the structural characterization of micropores. However, the limitations in testing methods pose challenges in acquiring the structural parameters for ultra-micropores. Consequently, this study adopts molecular simulation techniques to ascertain the pore structure parameters of ultra-micropores.

(1) SSA simulation

The SSAs of the DS-L and CZ-A macromolecular models were calculated based on the Connolly principle (Figure 5). The red area represents the surface area of the constructed molecular structure; the blue spheres represent the molecular probes used; the dashed lines depict the trajectories of the molecular probes; the arrows indicate the direction of probe movement; the gray outline represents the area enclosed by the molecular probes after completing one full circuit; and the area of the gray region corresponds to the surface area of the molecular structure.

$$Surface\ Area = \frac{Surface\ Area\ per\ cell}{Density \times Cell\ Volume} \times 10^4 \quad (3)$$

where $Surface\ Area$ denotes the Connolly SSA, m^2/g; $Surface\ Area\ per\ cell$ represents the surface area of the coal macromolecular model, $Å^2$; $Density$ stands for the density of the coal macromolecular model, g/cm^3; and $Cell\ Volume$ indicates the volume of the coal macromolecular model, $Å^3$.

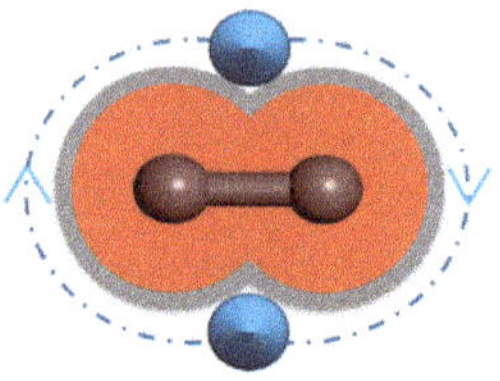

Figure 5. Diagram of the Connolly principle for the SSA calculation.

Firstly, the DS-L macromolecular model was simulated to obtain the molecular probe result under the Connolly surface (Figure 6a). By substituting the acquired result into Equation (3), the Connolly SSA of the DS-L macromolecular model was calculated to be 40.3707 m^2/g. Analogously, the Connolly SSA of the CZ-A macromolecular model was computed to be 45.5005 m^2/g, with the corresponding molecular probe result presented in Figure 6b.

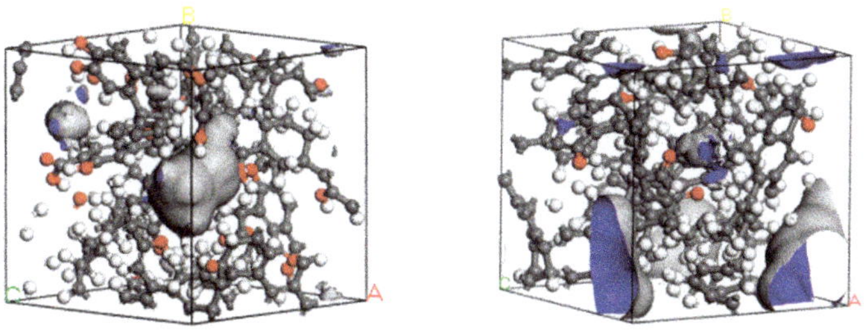

(**a**) DS-L (**b**) CZ-A

Figure 6. (**a**) Molecular probe result of DS-L. (**b**) Molecular probe result of CZ-A.

(2) PV simulations

Based on the Connolly principle, the PVs of the DS-L and CZ-A macromolecular models were formulated by:

$$Pore\ Volume = \frac{Cell\ Free\ Volume}{Density \times Cell\ Volume} \tag{4}$$

where *Pore Volume* denotes the Connolly PV, cm^3/g; *Cell Free Volume* represents the free volume of the coal macromolecular model, $Å^3$; *Density* stands for the density of the coal macromolecular model, g/cm^3; and *Cell Volume* indicates the volume of the coal macromolecular model, $Å^3$.

Initially, the DS-L macromolecular model was simulated, in which the parameters were set to be identical to those employed in the prior SSA simulation. Subsequently, the obtained parameters were substituted into Equation (4) to calculate the pore PV of DS-L, yielding a value of 0.022873 cm^3/g. Analogously, the PV of the CZ-A molecular model was computed to be 0.034971 cm^3/g. The comparison of the experimental and simulation results is shown in Table 4.

Table 4. Comparison of experimental and simulation results.

Coal Sample		Total SSA (Am^2/g)	Total PV (m^3/g)
DS-L	Experiment	4.839	0.012454
	Simulation	40.3707	0.022873
CZ-A	Experiment	5.362	0.019642
	Simulation	45.5005	0.034971

For both DS-L and CZ-A, the simulated SSA and PV results are numerically larger than the corresponding experimental values. In addition, the simulated SSA and PV results of CZ-A are both greater than those of DS-L, which is consistent with the experimental findings. This discrepancy arises because the liquid nitrogen adsorption method can hardly test ultra-micropores (<2 nm) in coal. The results provided by this method only represent the SSAs of a subset of micropores, as well as mesopores and macrospores. In contrast, during molecular simulation, the SSAs of various ultra-micropores in coal can be effectively measured by selecting molecular probes with an appropriate radius. This leads to a significant difference between the simulated and experimental results. This difference also highlights the limitations of current measurement techniques in accurately accessing ultra-micropore diameters under practical conditions.

In summary, combined with the results of liquid nitrogen adsorption experiments, it is found that micropores and ultra-micropores are extensively present in coal, and the structural characteristic parameters of the micropores and ultra-micropores of coal can be characterized by liquid nitrogen adsorption experiments and molecular simulation, and based on this, we can further carry out adsorption simulation studies on coal macromolecule models to analyze the adsorption characteristics of coal on different gases.

3.3. Adsorption Simulation on Coal Macromolecular Models

3.3.1. CH_4 Isothermal Adsorption Simulation

Based on the two macromolecular models of coal samples for CH_4 isothermal adsorption simulation, the number of CH_4 molecules adsorbed per unit cell at 303.15 K was obtained. The results are presented in Table 5. Additionally, the saturated adsorption configurations of macromolecules for DS-L and CZ-A after the completion of adsorption are illustrated in Figures 7a and 7b, respectively.

Table 5. CH_4 isothermal adsorption data (PCs/cell).

Pressure (MPa)	0	1	2	3	4	5	6	7	8	9	10
DS-L	0	1.76	2.85	2.98	3.09	3.18	3.25	3.31	3.35	3.38	3.4
CZ-A	0	2.28	3.98	4.36	4.67	4.89	5.08	5.19	5.28	5.35	5.37

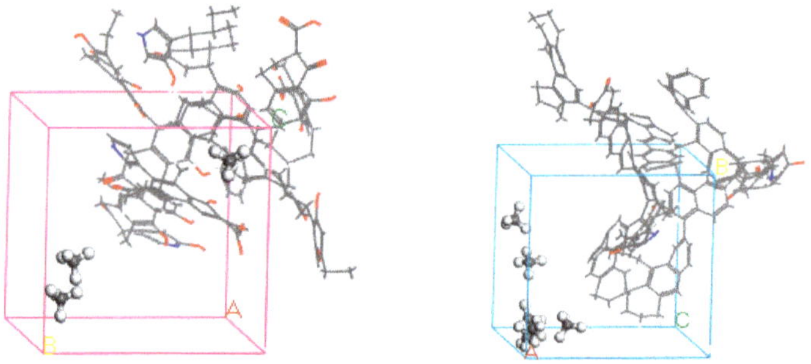

(**a**) DS-L (**b**) CZ-A

Figure 7. (**a**) Saturated adsorption configuration of macromolecules for DS-L. (**b**) Saturated adsorption configuration of macromolecules for CZ-A.

In the isothermal adsorption simulation of macromolecular models of coal, the results are reported as the number of methane molecules per unit cell. However, in actual methane adsorption experiments, the adsorption results are typically measured in units of cm^3/g. Therefore, to compare the simulation results with experimental results, a conversion is necessary, which can be given by Equation (5):

$$1\ \text{molecular/u.c} = 1/N_A \div M \times V_{mol}\text{cm}^3/g \quad (5)$$

where N_A is Avogadro's constant, with a value of 6.02×10^{23}; M is the mass of a single unit cell, g; and V_{mol} is the molar volume of a gas at standard temperature and pressure, with a value of 22.4×10^3 cm^3.

The isothermal adsorption data converted according to Equation (5) are shown in Table 6.

Table 6. CH_4 isothermal adsorption data (cm^3/g).

Pressure (MPa)	0	1	2	3	4	5	6	7	8	9	10
DS-L	0	15.84	25.65	26.82	27.81	28.62	29.25	29.79	30.15	30.42	30.6
CZ-A	0	20.43	35.66	39.06	41.84	43.81	45.52	46.50	47.31	47.94	48.12

As can be found in Table 6, at a constant temperature, the adsorption capacities of both the DS-L and CZ-A macromolecular models exhibit a gradual increase with the rise in pressure, indicating that pressure facilitates CH_4 adsorption. Moreover, within the entire pressure gradient range, the CH_4 adsorption capacity of the CZ-A macromolecular model is always greater than that of the corresponding DS-L model.

Based on the comprehensive analysis of the experimental and simulation results related to the pore structure characteristics, CZ-A exhibits a larger SSA and PV compared to DS-L. Additionally, the CH_4 adsorption capacity of coal is positively correlated with PV, SSA, and pore size. Notably, micropores and ultra-micropores are the primary contributors to the SSA and PV of coal, suggesting that the adsorption capacity of coal for gases lies within the range of micropores to ultra-micropores. Micropores, due to their smaller diameters, offer a larger space for gas molecule adsorption, thereby enhancing gas molecule adsorption. The experimental findings by Wang et al. [30] reveal that micropore filling accounts for 74% to 99% of the total gas adsorption amount, which further emphasizes the crucial role of micropores in augmenting the gas adsorption capacity. Prior research [31,32] demonstrates that an increase in SSA and PV leads to an improvement in methane adsorption capacity, and that the adsorption capacity strengthens with the rise in the metamorphic degree.

3.3.2. CH_4 Adsorption Simulation at Different Temperatures

With the aid of Materials Studio, simulations were conducted to assess the CH_4 adsorption capacity at temperatures of 303.15 K, 313.15 K, and 323.15 K, respectively. The Sorption module within the software was utilized for these simulations, with all parameter settings maintained to be consistent with those outlined in the preceding section. According to Equation (5), the simulation outcomes were then converted to derive the CH_4 isothermal adsorption data for both the DS-L and CZ-A macromolecular models at the aforementioned temperatures (Table 7).

Table 7. CH_4 isothermal adsorption data at different temperatures (cm^3/g).

Pressure (MPa)	DS-L			CZ-A		
	303.15 K	313.15 K	323.15 K	303.15 K	313.15 K	323.15 K
0	0	0	0	0	0	0
1	15.84	14.04	12.96	20.43	18.73	17.74
2	25.65	23.94	22.95	35.66	32.70	31.81
3	26.82	25.2	24.12	39.066	36.19	34.59
4	27.81	26.28	25.2	41.84	39.25	37.45
5	28.62	27	26.01	43.81	41.57	39.07
6	29.25	27.72	26.45	45.52	43.46	40.41
7	29.79	28.35	26.9	46.50	44.53	41.66
8	30.15	28.89	27.35	47.31	45.34	42.56
9	30.42	29.25	27.62	47.94	45.88	43.28
10	30.6	29.34	27.8	48.12	45.96	43.81

Based on the data in Table 7, Langmuir fitting was performed on the CH_4 adsorption simulation results for the DS-L and CZ-A macromolecular models at different temperatures (Figures 8 and 9).

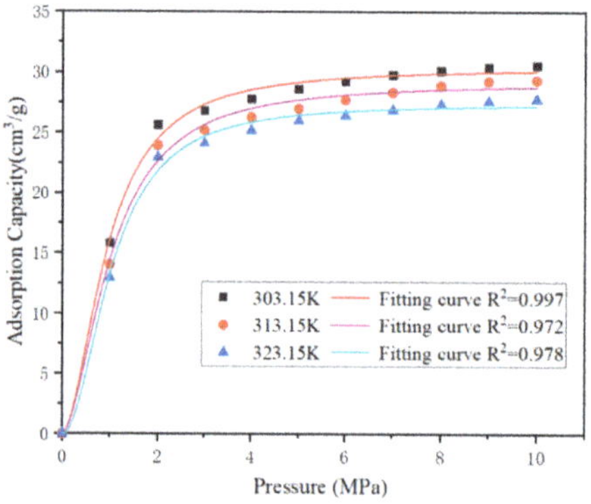

Figure 8. Isothermal adsorption curves of DS-L at different temperatures.

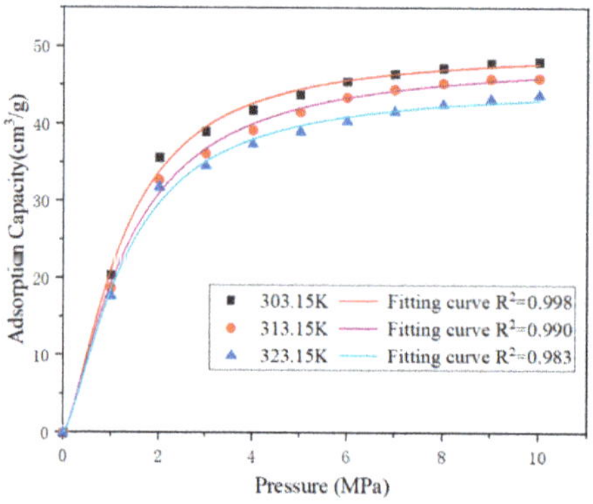

Figure 9. Isothermal adsorption curves of CZ-A at different temperatures.

Upon the analysis of Table 6 and Figures 8 and 9, it is evident that the CH_4 adsorption simulation results for both the DS-L and CZ-A macromolecular models across various temperatures adhere to the Langmuir equation. Specifically, during the initial stage of adsorption, the adsorption capacity increases rapidly with pressure. However, upon reaching a specific pressure threshold, the increase rate slows and subsequently stabilizes. Notably, with the increase in the simulation temperature, the maximum CH_4 adsorption capacity of both models has a declining trend, indicating that temperature has an inhibitory effect on CH_4 adsorption. This phenomenon is attributed to the fact that as the temperature rises, the kinetic energy of gas molecules on the coal surface is augmented, facilitating the escape of CH_4 molecules from the van der Waals forces, and consequently, reducing the CH_4 adsorption capacity. Liu and Xiao et al. [33,34] indicated that methane adsorption capacity in the coal sample decreased linearly with an increase in temperature, and the temperature effect on reducing the methane adsorption capacity was greater at low pressures. This is consistent with our research findings.

3.3.3. Isothermal Adsorption Simulation for Different Adsorbates

The analysis of the adsorption capacity of the lignite and anthracite macromolecular models for different adsorbates (CH_4, CO_2, H_2O) was conducted through adsorption simulations. Initially, the planar models of these three adsorbates were drawn in the Visualizer interface, followed by automatic hydrogen addition and clean processing for preliminary optimization. Subsequently, the Forcite module was employed to perform kinetic optimization and annealing optimization on the pre-optimized molecular models, with the optimization parameters set as described in Section 2.4. The molecular configurations of the three adsorbates after completing all optimizations are shown in Figure 10.

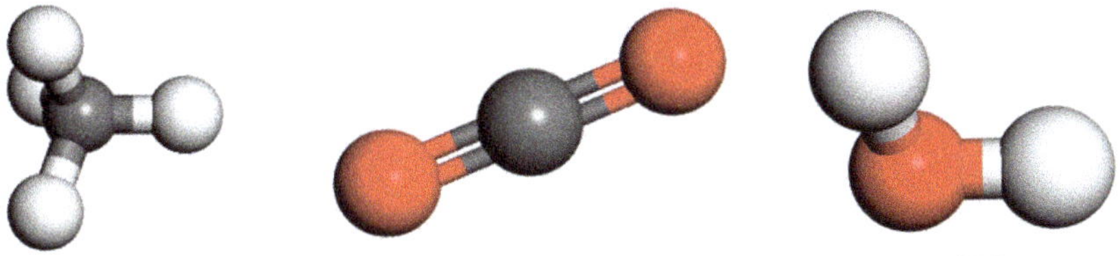

(**a**) CH_4 (**b**) CO_2 (**c**) H_2O

Figure 10. Molecular configurations of adsorbates.

In this simulation, the Sorption module of the software was used to investigate the adsorption behaviors of CO_2 and H_2O on the DS-L and CZ-A macromolecular models whose periodic boundary conditions had been constructed prior to the simulation. The simulation temperature was set to 303.15 K, while the other parameters were kept the same as those described in the previous section. The saturated adsorption configurations of the DS-L and CZ-A macromolecular models upon the completion of the simulation are presented in Figures 11 and 12, respectively.

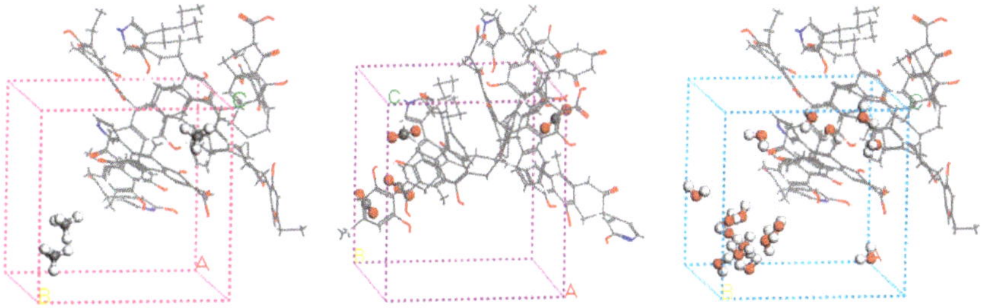

(**a**) CH_4 (**b**) CO_2 (**c**) H_2O

Figure 11. (**a**) Adsorption density field of the DS-L macromolecular model after adsorbing CH_4. (**b**) Adsorption density field of the DS-L macromolecular model after adsorbing CO_2. (**c**) Adsorption density field of the DS-L macromolecular model after adsorbing H_2O.

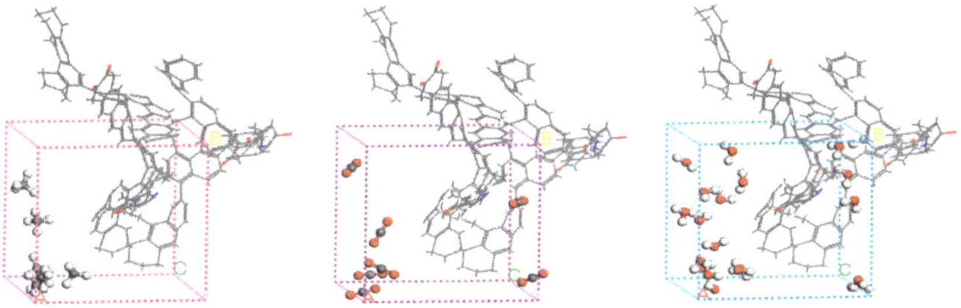

(**a**) CH_4 (**b**) CO_2 (**c**) H_2O

Figure 12. (**a**) Adsorption density field of the CZ-A macromolecular model after adsorbing CH_4. (**b**) Adsorption density field of the CZ-A macromolecular model after adsorbing CO_2. (**c**) Adsorption density field of the CZ-A macromolecular model after adsorbing H_2O.

Figure 11a depicts the adsorption configuration of the DS-L macromolecular model after adsorbing three CH_4 molecules, Figure 11b shows its configuration after adsorbing four CO_2 molecules, and Figure 11c displays its configuration after adsorbing fifteen H_2O molecules. Figure 12a–c provide the adsorption configurations of the CZ-A macromolecular model after adsorbing six CH_4 molecules, seven CO_2 molecules, and nineteen H_2O molecules, respectively.

Figures 13 and 14 show the density plots of the DS-L and CZ-A macromolecular models after adsorbing CH_4, CO_2, and H_2O, respectively. The colored patches indicate the potential positions of the adsorbates, and a denser distribution of these patches is indicative of a higher adsorption capacity.

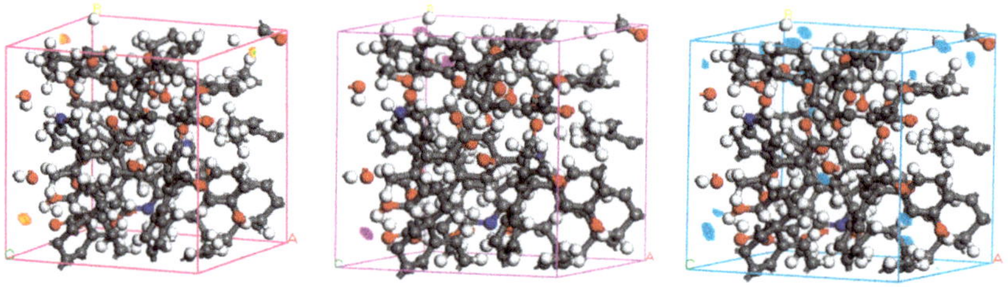

(a) CH_4 (b) CO_2 (c) H_2O

Figure 13. Adsorption density fields of the DS-L macromolecular model.

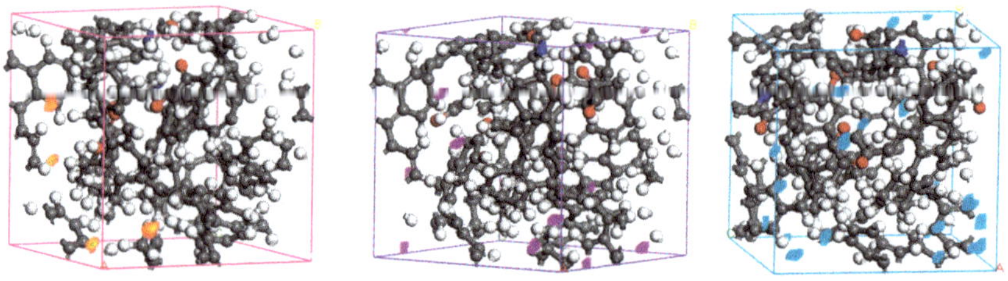

(a) CH_4 (b) CO_2 (c) H_2O

Figure 14. Adsorption density fields of the CZ-A macromolecular model.

According to the distribution of adsorbates within the CZ-A macromolecular model, it is evident that most adsorbates are adsorbed at the model boundaries. Among them, H_2O molecules exhibit the richest adsorption sites, while CH_4 molecules have the poorest. This observation can be attributed to two factors. The first factor is the Molecular Dynamic radii of the three adsorbates. Smaller radii allow for greater adsorption space within the micropores. The second factor is the properties of the adsorbates themselves. During adsorption, H_2O molecules form new adsorption sites due to capillary condensation, so they have the largest adsorption capacity. Consequently, competition for the same adsorption sites arises during multivariate adsorption processes.

Using Materials Studio, five simulations were conducted for each adsorbate under identical conditions, and their average adsorption capacities were calculated (Table 8).

Based on the data in Table 8, Langmuir fitting was performed on the simulation results obtained for the adsorption of these three different adsorbates on both the DS-L and CZ-A macromolecular models (Figure 15).

Table 8. Adsorption capacities of different adsorbates (cm^3/g).

Pressure (MPa)	DS-L CH_4	DS-L CO_2	DS-L H_2O	CZ-A CH_4	CZ-A CO_2	CZ-A H_2O
0	0	0	0	0	0	0
1	15.84	18.92	68.04	20.43	23.21	77.41
2	25.65	29.64	93.42	35.66	40.68	116.12
3	26.82	32.79	102.21	39.066	47.76	134.4
4	27.81	34.31	110.52	41.84	51.79	144.61
5	28.62	36.38	118.35	43.81	54.84	152.77
6	29.25	38.27	124.74	45.52	56.72	157.25
7	29.79	39.35	128.07	46.50	57.97	161.10
8	30.15	39.84	130.95	47.31	58.96	164.51
9	30.42	40.15	133.02	47.94	59.58	167.46
10	30.6	40.28	135	48.12	59.85	170.24

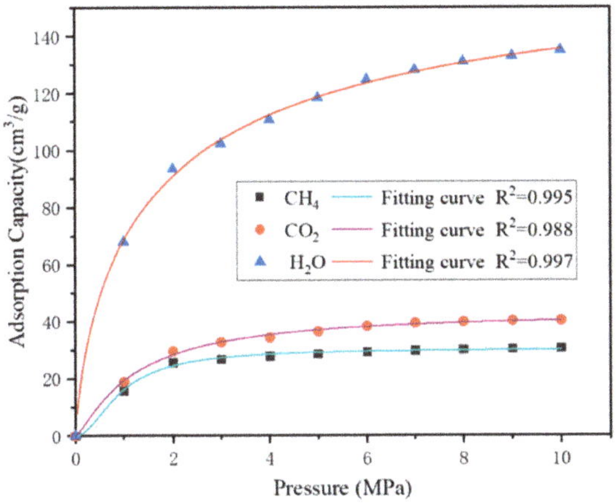

Figure 15. Isothermal adsorption curves of DS-L for different adsorbates.

As depicted in Figures 15 and 16, when the DS-L and CZ-A macromolecular models adsorb the three different adsorbates, the adsorption simulation results all conform to the Langmuir equation. Specifically, during the initial stage, the adsorption capacity increases significantly with the rise in pressure. However, the amplitude of the increase declines and tends to stabilize as the pressure reaches a certain value. Additionally, as the temperature rises, the adsorption capacity gradually declines. Under identical temperatures and pressures, the adsorption capacities of the DS-L and CZ-A macromolecular models for different adsorbates exhibit a consistent order: $H_2O > CO_2 > CH_4$. Moreover, the adsorption capacities of CZ-A surpass those of DS-L. These observations align with the research results reported by Jia [35]. That is, the competitive adsorption sequence of CO_2, CH_4, and H_2O follows the order of $H_2O > CO_2 > CH_4$. In the industrial applications of CO_2-enhanced coalbed methane recovery, the displacement efficiency often exhibits variability due to poorly understood competitive adsorption mechanisms, particularly the interference from H_2O. The simulations revealed a competitive adsorption sequence of $H_2O > CO_2 > CH_4$, indicating that priority should be given to controlling coal seam moisture to enhance CO_2 displacement efficiency. These findings can provide valuable theoretical insights for methane displacement in coal seams.

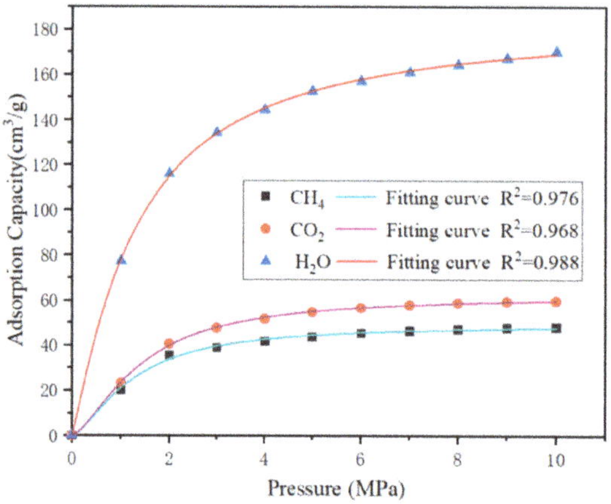

Figure 16. Isothermal adsorption curves of CZ-A for different adsorbates.

3.3.4. Energy Variations in Coal Before and After Adsorption of Different Adsorbates

To investigate the energy changes in the DS-L and CZ-A macromolecular models before and after the adsorption of CH_4, CO_2, and H_2O, isothermal adsorption simulations were conducted, and the energy change data were extracted accordingly. Tables 9 and 10 give the energy variations in the DS-L and CZ-A macromolecular models before and after the saturation adsorption of these three adsorbates, respectively.

Table 9. Energy changes in DS-L before and after adsorption of different adsorbates.

	Total Energy (Kcal/mol)	Valence Electron Energy (Kcal/mol)				Non-Bond Interaction Energy (Kcal/mol)		
		Bond	Angle	Torsion	Inversion	van der Waals	Electrostatic	H-Bond
Before adsorption	474.014	83.528	132.248	146.933	5.908	277.366	−152.935	−1.082
After CH_4 adsorption	443.325	83.527	132.136	146.932	5.907	241.572	−163.057	−1.094
After CO_2 adsorption	432.986	83.527	133.056	146.933	5.906	235.158	−172.356	−2.321
After H_2O adsorption	385.784	83.342	132.247	146.931	5.907	283.237	−246.236	−32.756

Table 10. Energy changes in CZ-A before and after adsorption of different adsorbates.

	Total Energy (Kcal/mol)	Valence Electron Energy (Kcal/mol)				Non-Bond Interaction Energy (Kcal/mol)		
		Bond	Angle	Torsion	Inversion	van der Waals	Electrostatic	H-Bond
Before adsorption	3595.511	1973.913	124.974	125.717	5.436	1233.536	−185.102	−3.327
After CH_4 adsorption	3157.256	1900.921	114.130	114.620	4.392	852.129	−225.942	−3.432
After CO_2 adsorption	2546.362	1836.315	120.074	113.646	5.931	901.308	−426.909	−5.366
After H_2O adsorption	1532.596	1364.367	132.368	120.575	6.354	1122.354	−1412.704	−235.653

As can be seen in Table 9, for DS-L, the total energy changes are 30.689 Kcal/mol, 41.028 Kcal/mol, and 88.23 Kcal/mol upon adsorbing the three adsorbates. The magnitude of the decrease follows the order of $H_2O > CO_2 > CH_4$. In terms of valence electron energy, minimal changes are observed; less than 1% on average. Regarding non-bond interaction energy, van der Waals energy changes more noticeably than electrostatic energy and hydrogen bonding energy for CH_4 and CO_2 adsorption, while electrostatic energy and hydrogen bonding energy change more significantly than van der Waals energy for H_2O adsorption. This suggests that CH_4 and CO_2 primarily adsorb on the models through the van der Waals forces, whereas H_2O predominantly adsorbs on the models through electrostatic and hydrogen bonding forces. Consequently, physical adsorption dominates in the adsorption of CH_4 and CO_2 by DS-L, whereas physical and chemical adsorption coexist in the adsorption of H_2O. As demonstrated in the study by Kang et al. [36], CO_2 exhibits not only physical adsorption but also chemical adsorption, while H_2O only demonstrates physical adsorption. Similarly for CZ-A, Table 10 reveals that physical adsorption is predominant in the adsorption of CH_4 and CO_2, and physical and chemical adsorption occur simultaneously in the adsorption of H_2O.

4. Conclusions

(1) The PV of DS-L predominantly comprises small pores, whereas the PV of CZ-A is chiefly characterized by micropores. The SSA of micropores constitutes more than 65% of the total SSA in both coal samples, suggesting that micropores are the primary contributors to the SSA of adsorption pores. Utilizing the constructed macromolecular models, the SSAs and PVs of the DS-L and CZ-A coal samples are simulated by employing N_2 molecule probes. The SSAs and PVs of the obtained coal samples are 40.3707 m^2/g and 45.5005 m^2/g, and 0.022873 cm^3/g and 0.034971 cm^3/g, respectively. The simulation outcomes for both coal samples exceed the experimental results, which demonstrates that molecular simulations represent an effective means of measuring pore structure parameters across various ultra-micropores in coal.

(2) The CH_4 adsorption amount of the CZ-A macromolecular model consistently surpasses that of the corresponding DS-L macromolecular model. Moreover, the CH_4 adsorption amount exhibits a positive correlation with the SSA and PV of coal.

(3) The maximum CH_4 adsorption amounts of both the DS-L and CZ-A macromolecular models diminish as the temperature increases, indicating a suppressive effect of temperature on CH_4 adsorption. When the two models adsorb CH_4, CO_2, and H_2O, the adsorption amounts follow the order of $H_2O > CO_2 > CH_4$. Analysis of the energy changes before and after adsorption reveals that the adsorption of CH_4 and CO_2 by DS-L and CZ-A is predominantly physical, whereas that of H_2O encompasses both physical and chemical adsorption.

Author Contributions: P.C.: Conceptualization, Funding acquisition, Methodology, Investigation, Writing—original draft, Writing—review and editing. Y.W.: Conceptualization, Methodology, Formal analysis, Investigation, Writing—original draft. Y.Z.: Methodology, Formal analysis, Investigation, Writing—review and editing. Q.W.: Investigation, Writing—original draft. Z.W.: Investigation, Writing—original draft. L.T.: Investigation, Writing—original draft. All authors have read and agreed to the published version of the manuscript.

Funding: This work was supported by the Program for National Natural Science Foundation of China (52274191), and the project was funded by China Postdoctoral Science Foundation (2021M700132) and the Basic Research Funds of Henan Polytechnic University (NSFRF220205).

Data Availability Statement: The raw data supporting the conclusions of this article will be made available by the authors on request.

Conflicts of Interest: Author Pan Chen was employed by the company JiaoZuo Coal Industry (Group) Co., Ltd.. The remaining authors declare that the research was conducted in the absence of any commercial or financial relationships that could be construed as a potential conflict of interest.

Nomenclature

r_k	Pore radius of the experimental coal sample	nm
σ_1	Surface tension of the condensed phase	N/m
V	Actual adsorption capacity on the coal sample surface	mL
V_m	Monolayer saturation adsorption capacity of nitrogen	mL

V_{m1}	Adsorption volume of the liquid under the action of liquid nitrogen	cm^3/g
R	Gas constant	$J \cdot mol^{-1} \cdot K^{-1}$
T_b	Experimental temperature	K
P	Pressure of the adsorbate	Pa
P_0	Saturation vapor pressure of nitrogen at −196 °C	Pa
Surface Area	Connolly SSA	m^2/g
Surface Area per cell	Surface area of the coal macromolecular model	$Å^2$
Density	Density of the coal macromolecular model	g/cm^3
Cell Volume	Volume of the coal macromolecular model	$Å^3$
Pore Volume	Connolly PV	cm^3/g
Cell Free Volume	Free volume of the coal macromolecular model	$Å^3$
Density	Density of the coal macromolecular model	g/cm^3
Cell Volume	Volume of the coal macromolecular model	$Å^3$
M	Mass of a single unit cell	g
V_{mol}	Molar volume of a gas at a standard temperature and pressure	cm^3

References

1. Ge, S.; Fan, J.; Liu, S.; Song, M.; Xian, Y.; Wang, B.; Teng, T. Low carbon modern coal-based energy technology system and development strategy. *J. China Coal Soc.* **2024**, *49*, 203–223.
2. Zhang, D.; Zhu, J.; Zhao, X.; Gao, Y.; Geng, M.; Chen, G.; Jiao, J.; Liu, S. Dynamic assessment of coalbed methane resources and availability in China. *J. China Coal Soc.* **2018**, *43*, 1598–1604.
3. Xu, F.; Hou, W.; Xiong, X.; Xu, B.; Wu, P.; Wang, H.; Feng, K.; Yun, J.; Li, S.; Zhang, L.; et al. The status and development strategy of coalbed methane industry in China. *Petroleum Explor. Dev.* **2023**, *50*, 765–783. [CrossRef]
4. Cheng, Y.; Hu, B. Main occurrence form of methane in coal: Micropore filling. *J. China Coal Soc.* **2021**, *46*, 2933–2948.
5. Zhou, J.; Guo, X.; Zhao, B.; Jin, Z.; Deng, C. Study on the influence of microscopic components and pore structure of coal-rock system on methane adsorption. *Min. Saf. Environ. Prot.* **2024**, *51*, 51–60.
6. Yi, M.; Cheng, Y.; Wang, Z.; Wang, C.; Hu, B.; He, X. Effect of particle size and adsorption equilibrium time on pore structure characterization in low pressure N_2 adsorption of coal: An experimental study. *Adv. Powder Technol.* **2020**, *31*, 4275–4281. [CrossRef]
7. Liu, H.; Mou, J.; Cheng, Y. Impact of pore structure on gas adsorption and diffusion dynamics for long-flame coal. *J. Nat. Gas Sci. Eng.* **2015**, *22*, 203–213. [CrossRef]
8. Zhang, C.; Jia, S.; Huang, X.; Shi, X.; Zhang, T.; Zhang, L.; Wang, F. Accurate characterization method of pores and various minerals in coal based on CT scanning. *Fuel* **2024**, *358*, 130128. [CrossRef]
9. Zhao, P.; He, B.; Zhang, B.; Liu, J. Porosity of gas shale: Is the NMR-based measurement reliable. *Pet. Sci.* **2022**, *19*, 509–517. [CrossRef]
10. Han, W.; Zhou, G.; Gao, D.; Zhang, Z.; Wei, Z.; Wang, H.; Yang, H. Experimental analysis of the pore structure and fractal characteristics of different metamorphic coal based on mercury intrusion-nitrogen adsorption porosimetry. *Powder Technol.* **2020**, *362*, 386–398. [CrossRef]
11. Dantas, S.; Struckhoff, K.C.; Thommes, M.; Neimark, A.V. Pore size characterization of micro-mesoporous carbons using CO_2 adsorption. *Carbon* **2021**, *173*, 842–848. [CrossRef]
12. Wen, H.; Liu, M.; Fan, S.; Wei, G.; Liu, Y.; Hao, J. Experimental study on variations of pore-fissure in coal-rock leached by liquid CO_2. *J. Xi'an Univ. Sci. Technol.* **2020**, *40*, 935–944.
13. Luo, M.; Glover, P.W.J.; Zhao, P.; Li, D. 3D digital rock modeling of the fractal properties of pore structures. *Mar. Pet. Geol.* **2020**, *122*, 104706. [CrossRef]
14. Xu, Z.; Lin, M.; Jiang, W.; Xu, L.; Ji, L.; Cao, G.; Hao, W. Pore-morphology-based pore structure characterization for various porous media. *Chem. Eng. Sci.* **2023**, *280*, 118982. [CrossRef]
15. Liu, S.; Sun, H.; Zhang, D.; Yang, K.; Li, X.; Wang, D.; Li, Y. Experimental study of effect of liquid nitrogen cold soaking on coal pore structure and fractal characteristics. *Energy* **2023**, *275*, 127470. [CrossRef]
16. Li, Y.; Zhang, Y.; Zhang, L.; Hou, J. Characterization on pore structure of tectonic coals based on the method of mercury intrusion, carbon dioxide adsorption and nitrogen adsorption. *J. China Coal Soc.* **2019**, *44*, 1188–1196.
17. Ren, P.; Xu, H.; Tang, D.; Li, Y.; Chen, Z.; Sun, C.; Zhang, F.; Chen, S.; Xin, F.; Cao, L. Pore structure and fractal characterization of main coal-bearing synclines in western Guizhou, China. *J. Nat. Gas Sci. Eng.* **2019**, *63*, 58–69. [CrossRef]

18. Li, S.; Zhou, Y.; Hu, B.; Qin, X.; Kong, X.; Bai, Y.; Zhang, J. Structural characteristics of adsorption pores in low-rank coals and their effects on methane adsorption performance. *Coal Geol. Explor.* **2023**, *51*, 127–136.
19. Zhang, S.; Tang, S.; Li, Z.; Liu, B.; Wang, R. Effect of pore structure on competitive sorption and diffusion of mixed methane and carbon dioxide in anthracite, South Qinshui Basin, China. *Int. J. Coal Geol.* **2022**, *253*, 103956. [CrossRef]
20. Dong, K.; Zhai, Z.; Guo, A. Effects of pore parameters and functional groups in coal on CO_2/CH_4 adsorption. *ACS Omega* **2021**, *48*, 32395–32407. [CrossRef]
21. Ren, X.; Wen, Z.; Song, H.; Liang, S. Experimental study on change of coal adsorption characteristics by using Fenton reagent. *Coal Technol.* **2016**, *35*, 173–175.
22. Han, Q.; Deng, C.; Jin, Z.; Gao, T. Molecular Simulation of the Adsorption Characteristics of Methane in Pores of Coal with Different Metamorphic Degrees. *Molecules* **2021**, *26*, 7217. [CrossRef]
23. Wang, Y.; Zhu, Y.; Liu, S.; Zhang, R. Pore characterization and its impact on methane adsorption capacity for organic-rich marine shales. *Fuel* **2016**, *181*, 227. [CrossRef]
24. Jiang, J.; Cheng, Y.; Zhang, S. Quantitative characterization of pore structure and gas adsorption and diffusion properties of low-rank coal. *J. China Coal Soc.* **2021**, *46*, 3221–3233.
25. Wang, Y. Characteristics of nano-pore structure of Dongqu No.2 coal and its interaction with $CH_4/CO_2/H_2O/N_2$. Master's Thesis, Taiyuan University of Technology, Taiyuan, China, 2022.
26. Wang, H.; Xiang, J.; Deng, X. Molecular Simulation of the Pore Structure and Adsorption Properties of Coal under Compression Stress. *Energy Fuels* **2023**, *37*, 1887–1895. [CrossRef]
27. Yan, J.; Jia, B.; Liu, B.; Zhang, J. Simulation Study on Molecular Adsorption of Coal in Chicheng Coal Mine. *Molecules* **2023**, *28*, 3302. [CrossRef]
28. Wen, Z.; Wang, Y.; Wei, J.; Wang, J.; Si, L.; Wang, Q. Study on Molecular Model Construction and Gas Adsorption Simulation of Anthracite in the Qinshui Basin. *Langmuir* **2023**, *39*, 12944–12955. [CrossRef]
29. Liu, X. Study on Microstructural Characteristics of Coal Surface and the Mechanism of Electromagnetic Radiation. Ph.D. Thesis, China University of Mining & Technology, Beijing, China, 2019.
30. Wang, Z.; Hao, C.; Wang, X.; Wang, G.; Ni, G.; Cheng, Y. Effects of micro-mesopore structure characteristics on methane adsorption capacity of medium rank coal. *Fuel* **2023**, *351*, 128910. [CrossRef]
31. Xiao, P.; Du, Y. Experiment on the influence of microstructure of tectonic coal on its adsorption characteristics. *J. Xi'an Univ. Sci. Technol.* **2021**, *41*, 237–245.
32. Xin, F.; Xu, H.; Tang, D.; Liu, D.; Cao, C. Problems in pore property testing of lignite: Analysis and correction. *Int. J. Coal Geol.* **2021**, *245*, 103829. [CrossRef]
33. Liu, Y.; Zhu, Y.; Liu, S.; Li, W.; Tang, X. Temperature effect on gas adsorption capacity in different sized pores of coal: Experiment and numerical modeling. *J. Pet. Sci. Eng.* **2018**, *165*, 821–830. [CrossRef]
34. Xiao, T.; Li, S.; Long, H.; Kong, X.; Bai, Y.; Qin, A. Experimental research on adsorption characteristics of N_2, CH_4, and CO_2 in coal under different temperatures and gas pressures. *Energy Sci. Eng.* **2023**, *11*, 637–653. [CrossRef]
35. Jia, J.; Xing, Y.; Li, B.; Jia, P.; Wu, Y.; Yang, Q.; Wang, D. Molecular simulations of multivariate competitive adsorption of CH_4, CO_2 and H_2O in gas-fat coal. *Colloids Surf. A* **2024**, *683*, 132917. [CrossRef]
36. Kang, G.; Zhang, B.; Kang, T.; Guo, J.; Zhao, G. Effect of Pressure and Temperature on CO_2/CH_4 Competitive Adsorption on Kaolinite by Monte Carlo Simulations. *Materials* **2020**, *13*, 2851. [CrossRef]

processes

MDPI

Article

Characterization of Damage and Infiltration Modeling of Coal-Slurry Consolidation Mechanics Under Loaded Conditions

Yaocai Tang [1,*], Peng Lu [2], Junxiang Zhang [3] and Wang Jian [2]

1 Yima Coal Industry Group Co., Ltd., Yima 472300, China
2 College of Safety Science and Engineering, Henan Polytechnic University, Jiaozuo 454003, China
3 School of Smarts Energy & Environment Engineering, Zhongyuan University of Technology, Zhengzhou 451191, China
* Correspondence: tangyaocai3653@163.com

Abstract: Coal seam gas drainage is a primary measure for mitigating coal and gas outburst hazards. Grouting sealing can form coal-slurry consolidated bodies, significantly improving the sealing quality of gas drainage boreholes and alleviating coal and gas outburst risks. Therefore, this study conducts triaxial loading and seepage experiments to analyze the mechanical failure characteristics and permeability variation of coal-slurry consolidated bodies under loading conditions following grouting sealing of gas drainage boreholes. Based on the "cube" model, a permeability model for the damaged coal-slurry consolidated body under loading conditions is established. The findings provide guidance for evaluating the leakage prevention performance of sealing materials in field engineering and optimizing the sealing efficiency of grouting materials. Future research may explore the damage and seepage evolution of coal-slurry consolidated bodies under various loading conditions and sealing material types.

Keywords: coal-slurry cementation; dual effective stress; damage; permeability modeling; triaxial seepage

1. Introduction

Coal remains China's primary energy source, but the depletion of shallow coal resources has led to increased mining depths. Deep coal seams exhibit complex geological conditions, characterized by low permeability, high in-situ stress, and high gas content [1–3]. These factors worsen mining conditions and increase the frequency of coal and gas outburst disasters [4–6]. Gas drainage in coal seams is the primary measure to mitigate such disasters, with the sealing quality of drainage boreholes directly affecting gas extraction efficiency. After grouting and sealing, the slurry fills fractures and bonds with the fractured coal and rock to form a high-strength, low-permeability, and stable consolidation body. However, under the combined effects of deep in-situ stress, mining-induced stress, and negative drainage pressure, fractures easily develop around the borehole, leading to gas leakage and seal failure. This results in a rapid decline in gas concentration and reduces drainage efficiency. Therefore, investigating the mechanical failure characteristics and permeability evolution of coal-slurry consolidation bodies under loading is crucial for optimizing sealing materials, enhancing borehole sealing quality, improving gas extraction efficiency, and ensuring safe coal mine operations.

Recent studies on grouting sealing have progressed significantly. Celik et al. [7] conducted experiments on permeability grouting for soil reinforcement, developed a flow model, and designed key slurry parameters. They analyzed the factors affecting differences between cylindrical and spherical flow models. Ren et al. [8] developed a fractal permeability model considering multiple influences of micro—pore fractures on the coal body. Zhu et al. [9] created a green grouting material and evaluated its anti-permeability properties using permeability tests. In coal-rock loading studies, many researchers simulated the seepage behavior of damaged coal under triaxial conditions. Du et al. [10] developed a true triaxial gas-solid coupling system to study deformation, failure, and permeability evolution of briquettes and raw coal. Liu et al. [11] used a true triaxial test system to analyze fracture failure modes and permeability variations in coal. Wang et al. [12–14] explored crack propagation and permeability evolution in gas-bearing coal under axial and confining stresses. In permeability modeling, Wu et al. [15] used CT scanning and a triaxial loading seepage system to test gas-containing coals and constructed a damage ontology model reflecting the deformation behavior of the coal body. Shi et al. [16] proposed a permeability model for elastic coal under uniaxial strain, incorporating pore pressure and gas adsorption. Xue Yi [17] refined Seidle's study by including damage effects in post-peak coal permeability models. Connell et al. [18] developed permeability models for coal under various rigid constraints. Bai et al. [19] established a cubic structural model for coal permeability under triaxial loading and unloading, validating it through experimental analysis.

Wang et al. [20] investigated the deformation and gas flow characteristics of coal under true triaxial stress conditions. They observed that during the entire stress-strain process of coal, the stress-strain curve initially increases and then decreases, while the permeability-strain curve initially decreases and then increases. A "stress-sensitive zone" exists regarding the influence of intermediate principal stress on the minimum permeability of coal. Lu et al. [21] measured the permeability of layered composite coal-rock under true triaxial stress conditions using a true triaxial apparatus and proposed a permeability model for layered composite coal-rock under such conditions. They found that the variation in the thickness of coal seams and sandstone layers significantly influences the evolution of composite coal-rock permeability, with permeability increasing as the ratio of sandstone to coal thickness increases. Liu [22] conducted a series of triaxial compression and unloading tests on cemented fractured specimens with varying coal particle contents. The results showed that as the coal particle content increases, both the peak strength and cohesion of the specimens decrease, while the internal friction angle in both tests increases. In the triaxial compression test, dissipated energy increased with coal particle content, while in the triaxial unloading test, dissipated energy initially increased and then decreased. Shen [23] investigated the mechanical properties, energy evolution, and brittleness of coal under various true triaxial stress states. They developed an energy dissipation evolution equation based on the logistic function, defining dissipated energy as a damage factor (D), and established a three-dimensional damage constitutive model for coal, considering residual strength. Cao [24] developed a method capable of reproducing in-situ stress and loading/unloading paths of coal under field conditions. They studied the damage and failure characteristics of coal samples under true triaxial loading and dynamic unloading conditions. As the level of the maximum principal stress ($\sigma 1$) increased, the peak number of acoustic emission (AE) events in coal samples initially increased and then decreased. With increasing triaxial load, the energy release zone shifted and expanded. Under higher triaxial load levels, the coal sample failed and was severely damaged, with dynamic unloading of the minimum principal stress ($\sigma 3$) leading to the ejection of coal fragments. Liang [25] studied the failure evolution process of coal under true triaxial cyclic loading. They analyzed the evolution of deformation, elastic modulus, Poisson's ratio, and peak strength, and characterized the volume strain of fractures, dissipated energy, and failure modes. The results indicated that, under true triaxial cyclic loading, internal damage and failure in coal were more severe, with shear failure being the predominant mode, complemented by tensile failure. Duan et al. [26] used a true triaxial system to investigate the impact of cyclic loading of each principal stress on the deformation, energy evolution, and damage characteristics of coal, revealing the mechanical evolution of coal under excavation disturbance in true triaxial conditions. They found that the total dissipated energy of coal increased exponentially with increasing amplitude of true triaxial layered cyclic loading. As the amplitude of the cyclic loading increased, the damage variable exhibited a pronounced S-shaped change, showing a trend of deceleration, acceleration, and then deceleration.

In conclusion, extensive research has been conducted on the mechanical properties and permeability models of coal-rock grout bodies, as well as the damage evolution of coal under triaxial loading conditions, yielding valuable results. However, most studies have focused on improving the mechanical properties of grout materials and developing theoretical permeability models for coal-rock structures. Research on the permeability characteristics of coal slurry grout bodies after grouting or their permeability evolution under loading conditions is limited. Unlike previous studies, this research integrates volumetric deformation and structural deformation, establishing a damage-permeability evolution model for coal slurry grout bodies. Triaxial creep-permeability experiments were conducted to investigate the mechanical and permeability changes under dynamic loading conditions. The results show that under stress loading, the permeability of coal slurry grout bodies exhibits a "V"-shaped trend before instability. After instability, the degree of permeability increase is higher with increasing initial axial and confining pressures. The average relative error between the experimental and theoretical permeability values is 8.7%, and the experimental results validate the theoretical model, providing guidance for optimizing grouting techniques and addressing the challenges of gas extraction in deep coal seams.

2. Model Building

2.1. Model Assumption

Gas flow in coal is influenced by the coupled effects of stress, seepage, and temperature fields. To analyze the seepage evolution of coal-slurry consolidated bodies under load, the study simplifies the physical model using a cubic structure. A theoretical permeability model is developed with the following assumptions:

(1) The consolidated body exhibits a dual-porosity structure, consisting of pores and fractures, and remains in an isothermal state during the seepage process.

(2) The gas within the consolidated body is considered an ideal gas, with its migration governed by Fick's law in the coal matrix and Darcy's law in the fractures.

(3) The deformation of the matrix elements within the consolidated body is isotropic, encompassing expansion and compression.

(4) The consolidated body undergoes two types of deformation mechanisms: bulk deformation and structural deformation.

2.2. Deformation of Coal-Slurry Cementation Under Effective Stresses

Before damage, the coal-slurry consolidated body remains in the elastic deformation phase, during which fracture aperture changes are reversible and no damage occurs. The study adopts a cubic model for analysis, as shown in Figure 1 [27].

The coal-slurry consolidated body, formed around grouted boreholes, undergoes strain under the combined effects of external stress and coal seam gas pressure. Based on the classical Terzaghi effective stress principle, the study establishes effective stress for the coal-slurry consolidated body using the dual effective stress principle for porous media [28]. The model considers two deformation mechanisms: bulk deformation and structural deformation, corresponding to bulk effective stress and structural effective stress, respectively.

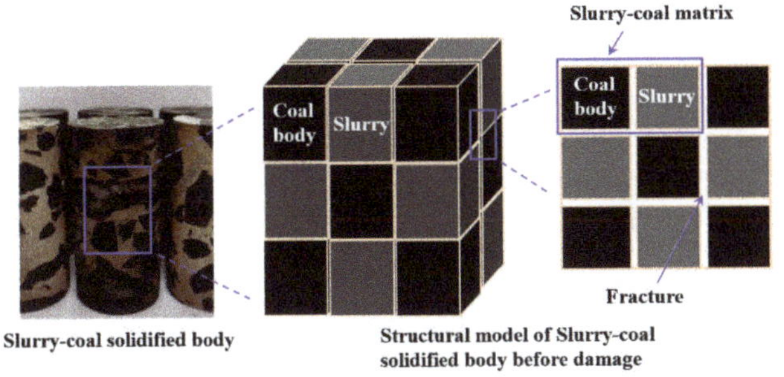

Figure 1. Physical model of coal-slurry consolidation body.

(1) Bulk Effective Stress [27]. Bulk deformation refers to the overall deformation of the coal-slurry matrix framework caused by bulk effective stress. In the coal-slurry porous medium, consider a cross-section OO^1 (Figure 2) with an area S (including pores and the matrix). The total stress σ applied on this section satisfies the equilibrium relationship expressed by the following equation:

$$\sigma S = p\varphi S + \sigma_a(1-\varphi)S \tag{1}$$

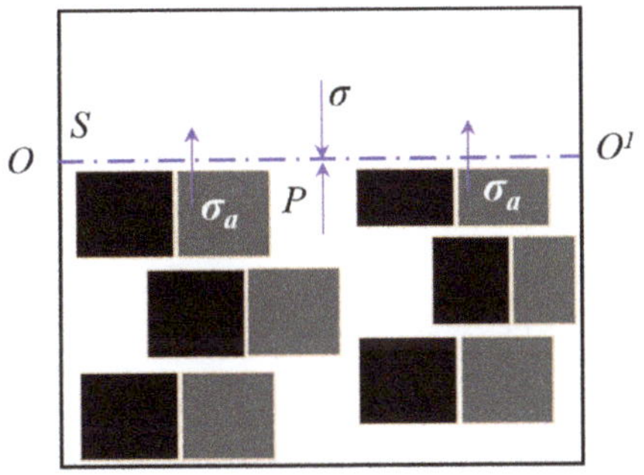

Figure 2. Relation established for the effective stress of the body.

The stress relation equation for coal-slurry porous media can be obtained from Equation (1):

$$\sigma = \varphi p + (1-\varphi)\sigma_a \tag{2}$$

Equating σ_a over the entire matrix cross-sectional area yields the effective stresses in the coal-slurry consolidation body proper:

$$\sigma_{\text{ef}}^{\text{B}} = \sigma_a(1-\varphi)S/S = (1-\varphi)\sigma_a \tag{3}$$

Equations (2) and (3) are obtained:

$$\sigma_{\text{ef}}^{\text{B}} = \sigma - \varphi p \tag{4}$$

From the relationship between effective stress and strain, it is obtained that:

$$\varepsilon_{\text{B}} = \frac{\sigma_{\text{ef}}^{\text{B}}}{E} \tag{5}$$

where, $\sigma_{\mathrm{ef}}^{\mathrm{B}}$: Effective stress of the body; σ: Total stress; p: Pore pressure (pore pressure between matrices); σ_a: Matrix average skeleton pressure; φ: Porosity of coal-slurry cementation body; ε_{B}: Intrinsic Stress; $\overline{E}$: Average modulus of elasticity of the sum of the modulus of elasticity of the coal body and the modulus of elasticity of the slurry body.

(2) Structural Effective Stress [27]. Structural deformation refers to changes in the spatial structure at the contact interface between the coal matrix and the slurry, including the contact interface between the slurry and the coal. Structural effective stress causes this deformation. In the coal-slurry porous medium, as shown in Figure 3, consider a curved surface AA^1 formed by connecting contact points. Contact points refer to the interface points between the matrix and the slurry or within the coal. The curved surface does not pass through the interior of the matrix. Let σ_{ci} represent the vertical component of the stress at the i-th contact point, and S_{ci} represent the vertical area corresponding to σ_{ci} (horizontal area). Based on the equilibrium of forces, the following equation is derived:

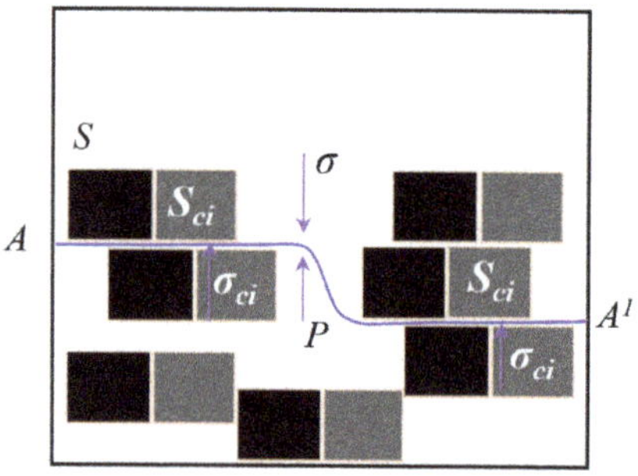

Figure 3. Structural effective stress establishment relationship.

Based on the balance of forces the following equation holds:

$$\sigma S = \sum \sigma_{ci} S_{ci} + \left(S - \sum S_{ci}\right) p \tag{6}$$

Equating $\sum \sigma_{ci}$ over the entire surface yields the following relation:

$$\sigma_{\mathrm{ef}}^{\mathrm{S}} = \sum \sigma_{ci} S_{ci} / S \tag{7}$$

$$\varphi_c = 1 - \sum S_{ci} / S \tag{8}$$

Associative Equations (6)–(8) give the structural effective stresses of the coal-slurry cementation:

$$\sigma_{\mathrm{ef}}^{\mathrm{S}} = \sigma - \varphi_c p \tag{9}$$

From the relationship between effective stress and strain, it can be obtained that:

$$\varepsilon_{\mathrm{S}} = \frac{\sigma_{\mathrm{ef}}^{\mathrm{S}}}{\overline{E}} \tag{10}$$

The total effective stress of the coal-slurry consolidation under load can be obtained by coupling Equations (4) and (9):

$$\sigma_t' = 2\sigma - p(\varphi + \varphi_c) \tag{11}$$

Associating (5) and (10) yields the total strain of the coal-slurry consolidation under load:

$$\varepsilon_t = \varepsilon_{\mathrm{B}} + \varepsilon_{\mathrm{S}} = \frac{\sigma_{\mathrm{ef}}^{\mathrm{B}} + \sigma_{\mathrm{ef}}^{\mathrm{S}}}{\overline{E}} \tag{12}$$

where, $\sigma_{\text{ef}}^{\text{S}}$: Structural effective stress; φ_c: Contact porosity; ε_{S}: Structural strain, where φ_c is between φ and 1, the value is determined according to the slurry-coal cementation body cementation and reinforcement condition, which is divided into two kinds: For loose and porous media with low degree of cementation, $\varphi_c \to 1$; For dense porous media with a high degree of cementation, $\varphi_c \to \varphi$.

After the formation of the coal-slurry consolidated body around the gas borehole, it undergoes deformation not only due to the applied stress but also from the deformation caused by the expansion of gas adsorbed by the consolidated body matrix. Since the strain induced by matrix adsorption is relatively small, it is assumed that the deformation of the coal-slurry consolidated body is primarily caused by the effective stress. Based on the coal permeability model considering effective matrix deformation [29], the permeability model for the coal-slurry consolidated body in the elastic phase can be derived.

$$\frac{k}{k_0} = \left\{ \begin{array}{l} 1 - \frac{\chi}{\varphi_0 K}[(\overline{\sigma} - \overline{\sigma}_0) - (p - p_0)] \\ -\frac{f_m}{\varphi_0}\left(\frac{\varepsilon_{\max} p}{p + p_L} - \frac{\varepsilon_{\max} p_0}{p_0 + p_L}\right) \end{array} \right\}^3 \tag{13}$$

where $\overline{\sigma} - \overline{\sigma}_0$ represents the average effective stress increment when loaded and $\frac{\varepsilon_{\max} p}{p+p_L} - \frac{\varepsilon_{\max} p_0}{p_0+p_L}$ represents the strain increment. Combining the principles of effective stress in the solid body and effective stress in the structure, i.e., Equations (11) and (12), it can be obtained:

$$\overline{\sigma} - \overline{\sigma}_0 = \sigma_t' - \sigma_{t0}' \tag{14}$$

$$\frac{\varepsilon_{\max} p}{p + p_L} - \frac{\varepsilon_{\max} p_0}{p_0 + p_L} = \frac{\sigma_{\text{ef}}^{\text{B}} + \sigma_{\text{ef}}^{\text{S}}}{\overline{E}} - \varepsilon_0 \tag{15}$$

Associative Equations (13)–(15) yield an elastic coal-slurry cementation permeability model based on the principles of principal effective stress and structural effective stress:

$$\frac{k}{k_0} = \left\{ \begin{array}{l} 1 - \frac{\chi}{\varphi_0 K}[(\sigma_t' - \sigma_{t0}') - (p - p_0)] \\ -\frac{f_m}{\varphi_0}\left(\frac{\sigma_{\text{ef}}^{\text{B}} + \sigma_{\text{ef}}^{\text{S}}}{\overline{E}} - \varepsilon_0\right) \end{array} \right\}^3 \tag{16}$$

where, k_0, k: represents the initial permeability and permeability of the cemented body, mD, respectively; χ: effective stress coefficient of the cemented body, $\chi = 1 - K/K_m$, $K = E/3(1-2\nu)$, K: overall bulk modulus of the cemented body, MPa; K_m: bulk modulus of the matrix of the cemented body, MPa; E: modulus of elasticity, MPa; ν: Poisson's ratio; φ_0: initial porosity, %; p_0: initial pore pressure, MPa; f_m: deformation influence coefficient of cemented body, 0~1; $\overline{E}$: average elastic modulus of coal body and slurry, MPa; ε_0: initial strain, %.

2.3. Modeling the Permeability of Damaged Coal-Slurry Consolidates

After damage, coal-slurry consolidation enters an elastic deformation stage. During this stage, the consolidation body experiences damage and cracking, leading to an increase in crack quantity, with irreversible changes in crack apertures. Based on the effective stress-permeability model principle [19], the cubic structural model shown in Figure 4 establishes a permeability model for damaged coal-slurry consolidation. This model also accounts for the combined effects of matrix effective stress and structural effective stress.

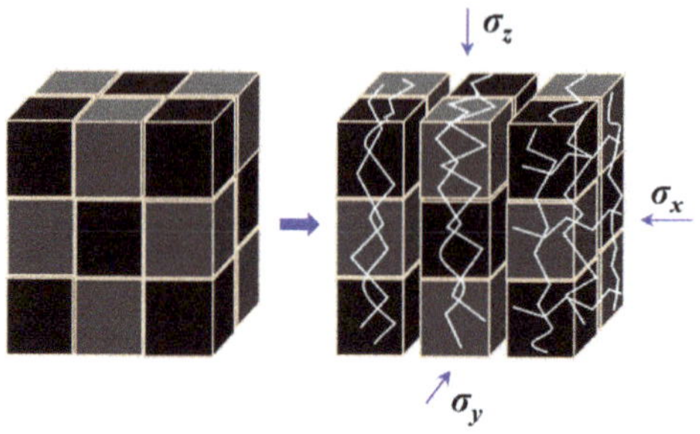

Figure 4. Structural model of damaged coal-slurry cementation body.

According to the classical Poiseuille's law, the flow of fluid through a single crack [30] is:

$$Q = \frac{b^3 l \Delta p}{12\mu L} \tag{17}$$

Since cracks develop in large numbers after damage to the solidified body and many cross-cracks appear, the flow rate of fluid through multiple cross-cracks is:

$$Q = \frac{b^3 l_n \Delta p}{12\mu L} \tag{18}$$

where, b: crack opening; l: crack length; l_n: sum of n cross-crack lengths; L: fluid flow distance; μ: hydrodynamic viscosity; Δp: pressure difference between the two ends of the crack.

Assuming that the fluid flow satisfies Darcy's law:

$$Q = \frac{Ak\Delta p}{\mu L} \tag{19}$$

Associative Equations (18) and (19) give the permeability of the solidus containing multiple cracks:

$$k = \frac{l_n b^3}{12A} \tag{20}$$

Assuming that the crack area per unit volume of the solidified body is S_1, combining Equation (20) yields:

$$k = \frac{l_n \varphi^3}{12AS_1^3} \tag{21}$$

The effective stress determines the change in permeability of the solidus, which can be obtained by biasing the effective force:

$$\frac{\partial k}{\partial \sigma_t'} = \frac{l_n \varphi^2 \partial \varphi}{4AS_1^3 \partial \sigma_t'} \tag{22}$$

Associative Formulas (21) and (22) are obtained:

$$\frac{\partial k}{\partial \sigma_t'} = \frac{3k\partial \varphi}{\varphi \partial \sigma_t'} \tag{23}$$

In Equation (23) let $d = \frac{1}{\varphi}\frac{\partial \varphi}{\partial \sigma_t'}$ denote the amount of variation of the solidus crack with effective stress, then Equation (23) can be expressed as:

$$\frac{\partial k}{\partial \sigma_t'} = 3dk \tag{24}$$

After the damage, the cracks in the solid body will be developed in large quantities, and this crack development brings about the damage destruction of the solid body in the macroscopic level. Therefore, in order to simplify the computational amount of $d = \frac{1}{\varphi}\frac{\partial \varphi}{\partial \sigma_t'}$, the degree of damage is introduced here to describe the crack development in the solidus according to the damage variables [31] defined based on the elastic modulus degradation:

$$D = 1 - \frac{\overline{E}}{E_0} \tag{25}$$

The damage variable D indicates the degree of development or damage of the cracks in the solidus; E_0 is the initial modulus of elasticity. It is also easy to see in Equation (25) that as the solid body is loaded, the cracks continue to develop, $\overline{E}$ continues to decrease, and the damage value D of the solid body continues to increase. D can be obtained by simplifying d with D:

$$d = \gamma D \tag{26}$$

In the formula, γ denotes the influence coefficient of effective stress change on crack development, $\gamma = 1 - K_p/K_m$, K_p is the volume modulus of the crack, and K_m is the matrix volume modulus of the cemented body, and the integral of the coupling Formulas (24) and (26) can get the model of the permeability of the damaged coal-slurry cemented body:

$$k = k_0 e^{3\gamma D(\sigma_t' - \sigma_{t0}')} \tag{27}$$

3. Coal-Slurry Cementation Seepage Test

3.1. Sample Preparation

The coal samples for the experiment were collected from the Jiulishan Mine, which belongs to the Jiaozuo Mining Company under the Henan Coal and Chemical Industry Group, located in Jiaozuo, Henan Province, China. The coal type is anthracite, and its industrial analysis results are shown in Table 1. To better simulate the fractured coal around actual boreholes, the coal samples were dried in the laboratory, then crushed into irregular pieces with dimensions smaller than 3 cm using a crusher. These crushed coal pieces served as the injected medium to replicate the grouting effect around fractured coal near gas drainage boreholes, as illustrated in Figure 5.

Table 1. Basic industrial parameters of coal samples.

Coal Sample	ρ/[g/cm^3]	ρ^t/[g/cm^3]	M_{ad}/[%]	V_{daf}/[%]	A_{ad}/[%]	φ/[%]
anthracite coal	1.56	1.49	4.12	11.18	9.72	4.49

The grouting material used in the experiment is a composite cement-based sealing material, primarily composed of P.O42.5 Portland cement, combined with accelerants, alkali activators, suspending agents, expanding agents, and reinforcing agents in specific proportions. The basic properties of the grouting material are listed in Table 2.

Table 2. Basic properties of grouting materials.

Performance Indicators	Compressive Strength/[MPa]	Mobility /[mm]	Initial Coagulation Time /[min]	Final Setting Time /[min]	Dilatation /[%]
Measured value	9.5	350	45	245	1.13

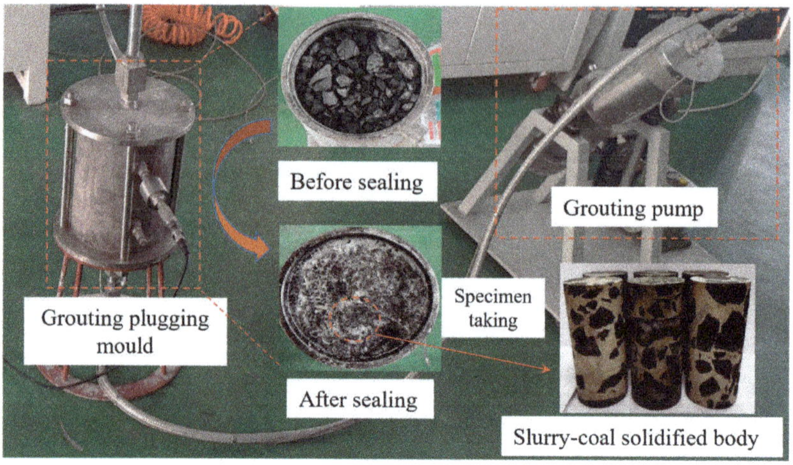

Figure 5. Coal-slurry consolidation system preparation process.

The grouting sealing simulation device, independently designed and constructed, injects grout into crushed coal blocks. The grout fully diffuses and fills the fractured coal, bonding and reinforcing it into a coal-slurry consolidation body. Cores are then extracted from the consolidated body and processed into standard cylindrical specimens with a diameter of 50 mm and a height of 100 mm. The preparation process for the coal-slurry consolidation body is shown in Figure 5.

3.2. Test Program and Procedures

This study uses the RLW-500G coal-rock triaxial creep-permeability test apparatus, independently developed by Henan Polytechnic University, located in Jiaozuo, Henan Province, China, to investigate the mechanical failure characteristics and permeability evolution of coal-slurry consolidation bodies under loading. The test apparatus consists of a system console, servo controller, triaxial pressure chamber, seepage control system, temperature control system, and pressure transmission device. It enables simultaneous monitoring of mechanical and seepage properties under temperature and stress loading conditions. The test apparatus is shown in Figure 6.

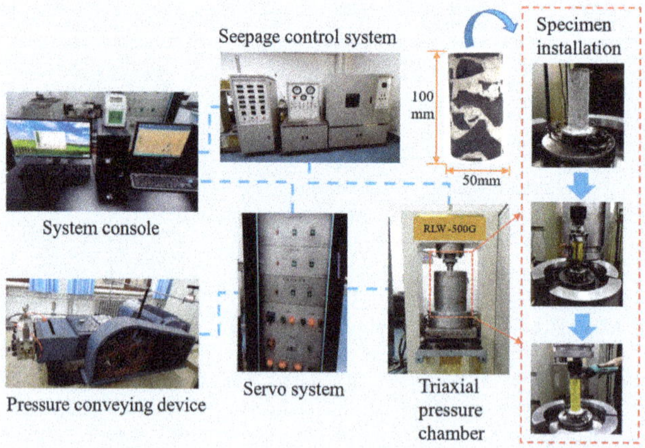

Figure 6. Schematic diagram of the test setup.

The coal-slurry grout bodies with dimensions of φ50 mm × 100 mm were prepared for the experiments. The flatness error of the sample's end faces was controlled to be no greater than 0.05 mm, and the end faces were parallel and perpendicular to the sample's axial line, with an error not exceeding 0.25°. During the experiments, the sample was first loaded to the preset axial and confining pressures using a servo control system. Axial pressure was then applied at a rate of 10 N/s to simulate the permeability variation under different loading conditions. The experimental plan is shown in Table 3, with no fewer than three samples used for each experimental condition to ensure the reliability of the results.

Table 3. Experimental program.

Specimen Number	Axial Pressure/[MPa]	Pressure on All Sides/[MPa]	Seepage Air Pressure/[MPa]
A_1	0.5	1	1
A_2	0.5	2	1
A_3	1	2	1

(1) Specimen Installation: Place the polished specimen on the installation base of the triaxial pressure chamber. Evenly apply silicone around the specimen, cover it with a heat-shrink tube, and shrink it with a hot air blower. Secure both ends with clamps and seal the specimen. After curing the silicone for 8 h, install strain sensors and connect the air pipes and circuits.

(2) Prestress Loading: Inject hydraulic oil into the triaxial chamber until filled. Use axial/confining pressure controllers to load the specimen to the test conditions. Stabilize the pressures via the servo controller.

(3) Nitrogen Injection: Connect the gas pipeline, open the seepage system's inlet and outlet valves, and adjust the pressure-reducing valve. Introduce nitrogen and monitor the flowmeter. Begin testing once the flow stabilizes.

(4) Loading Test: The system console controls the servo system to apply axial loading at 10 N/s until the specimen fails.

4. Analysis of Test Results

The experiment follows the test plan in Table 3 to obtain mechanical and seepage parameters of coal-slurry consolidation bodies under various conditions. The permeability of specimens is calculated using the following formula:

$$k = \frac{2QP_0\mu L}{\left(P^2 - P_0{}^2\right)A} \tag{28}$$

where, k is the permeability, mD; Q is the gas seepage flow rate, cm^3/s; μ is the gas dynamic viscosity coefficient, Pa-s; L is the length of the specimen, mm; A is the cross-sectional area of the specimen, cm^2; P is the pressure of the inlet end, MPa; P_0 is the pressure of the outlet end, MPa.

Changes in permeability reveal grouting sealing and fracture growth in coal-slurry bodies. Stress-strain and permeability curves under varying loads are shown in Figure 7.

Figure 7a illustrates the stress-strain and permeability variation curves of the coal-slurry consolidated specimen A_1 under initial axial stress of 0.5 MPa and confining pressure of 1 MPa. As shown in the figure, with the continuous loading of axial stress, the pore-fracture structure of the specimen is compressed under the applied load (OA stage), and the permeability decreases gradually from 9.66 mD to 6.35 mD. Under further compression of the axial stress (AB stage), the specimen does not exhibit a significant reduction in bearing capacity; instead, it gradually yields as the principal stress difference increases, reaching a peak strength of 10.1 MPa. During this stage, the cracks in the specimen develop and expand, and the permeability steadily increases to 25.8 mD, forming a 'V'-shaped variation trend with the permeability in the OA stage, which first decreases and then increases. At point B, the specimen reaches the maximum bearing capacity and undergoes an instantaneous instability failure, causing the permeability to rapidly increase from 25.8 mD to 35.45 mD, a 37.4% increase. After instability, internal damage in the specimen continues to develop, and during this stage, the permeability steadily increases to 36.35 mD, fluctuating until the specimen is completely destroyed.

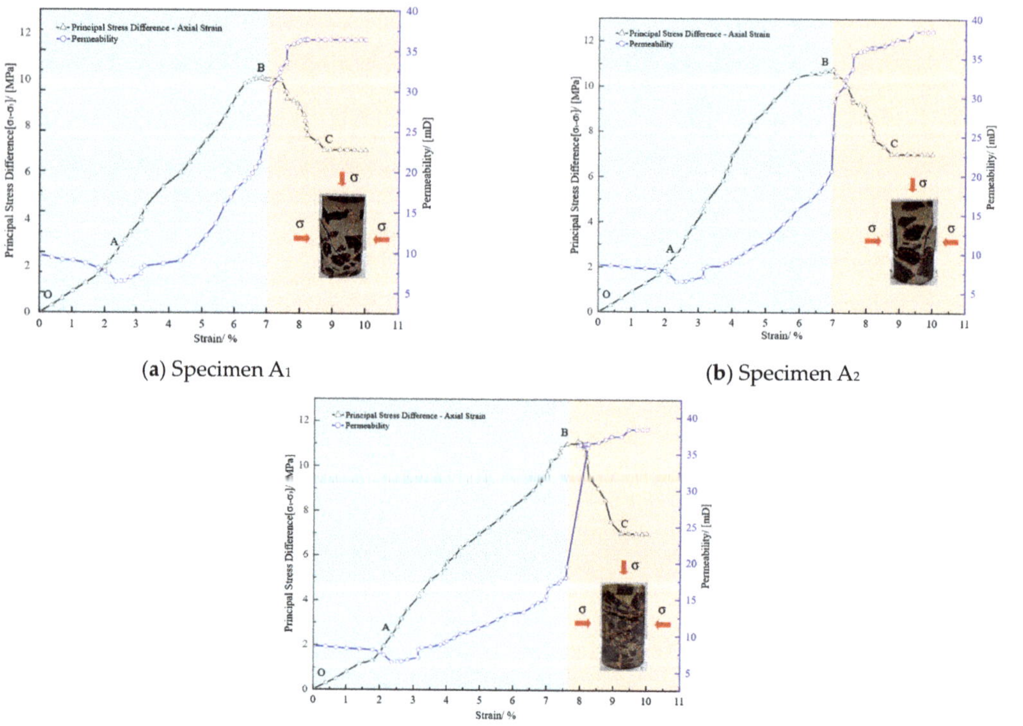

(**a**) Specimen A1

(**b**) Specimen A2

(**c**) Specimen A3

Figure 7. Variation of triaxial seepage test of coal-slurry cementation under different loadings.

Figure 7b illustrates the stress-strain and permeability variation curves of the coal-slurry consolidated specimen A_2 under initial axial stress of 0.5 MPa and confining pressure of 2 MPa. As shown in the figure, with the increase in principal stress difference, the permeability of the specimen decreases gradually from 8.56 mD to 6.34 mD (OA stage). When the principal stress difference gradually increases to 10.78 MPa (AB stage), the bearing capacity of the specimen increases, the slope of the stress-strain curve gradually decreases, and the volumetric strain of the specimen shifts from compression to expansion, with permeability steadily increasing to 20.55 mD. This forms a 'V'-shaped variation trend with the permeability in the OA stage. At the instability failure point B, the specimen undergoes an instantaneous failure, and the permeability rapidly increases from 20.55 mD to 32.78 mD, a 59.51% increase. Subsequently, internal damage continues to develop, and the permeability further increases to 35.85 mD.

Figure 7c illustrates the stress-strain and permeability variation curves of the coal-slurry consolidated specimen A_3 under initial axial stress of 1 MPa and confining pressure of 2 MPa. As shown in the figure, during the OA stage, the permeability of the specimen decreases gradually from 8.53 mD to 6.35 mD, maintaining a low level. During the AB stage, with the continued loading of axial stress, the specimen begins to yield, and the permeability steadily increases to 19.45 mD. Similar to specimens A_1 and A_2, the permeability during the OB stage exhibits a 'V'-shaped variation trend. When the axial stress increases to 11.01 MPa (point B), the specimen undergoes instability failure, and the permeability rapidly increases from 19.45 mD to 36.38 mD, a 87.04% increase. Subsequently, the permeability continues to increase to 38.39 mD.

The experimental study on the damage and seepage of coal-slurry consolidated specimens under different initial confining pressure and axial stress conditions indicates that the specimen is compressed under the applied load (OA stage), resulting in a decrease in permeability. With continued stress loading (AB stage), the specimen undergoes a transition from volumetric compression to expansion, internal damage begins to develop, cracks start to form and expand, and permeability gradually increases. The permeability exhibits a 'V'-shaped variation trend before instability failure (point B). When the specimen reaches its maximum bearing capacity, instability failure occurs instantaneously, and the bearing capacity sharply decreases. Micro-pores and cracks with poor internal connectivity begin to develop and connect over a wide area, causing permeability to increase rapidly and resulting in a seepage breakthrough phenomenon, as shown in Figure 8. The above analysis indicates that the permeability of the three specimens increased by 37.4%, 59.51%, and 87.04%, respectively, at the moment of instability failure, suggesting that the extent of permeability increase is greater with higher axial and confining pressures under different loading conditions. The analysis suggests that the cause of this phenomenon is that the specimen remains in a compressed state under the applied load, with continued axial stress loading leading to damage of the coal matrix and mineral particles in the slurry. However, due to the constant compression, the expansion of pores and cracks, as well as the development of new pore-crack structures, which could facilitate gas migration, are not fully expressed. When the coal body is suddenly decompressed, the specimen undergoes an instantaneous failure, and the bearing capacities of the three specimens decrease by 30.49%, 34.87%, and 36.23%, respectively. The internal damaged cracks open and become interconnected, allowing gas to rapidly desorb and seep, migrating from the connected pores and cracks, resulting in a continuous increase in permeability.

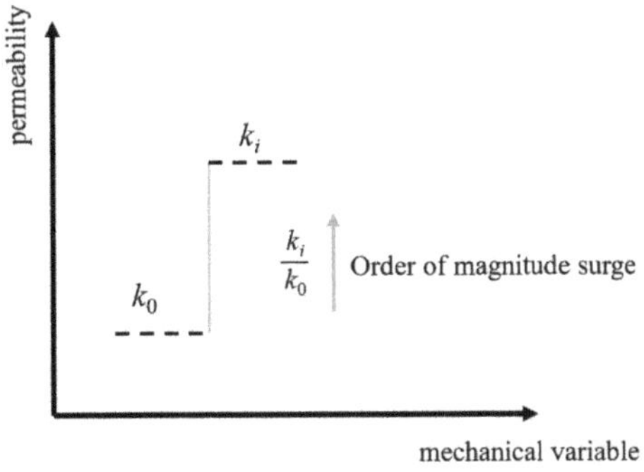

Figure 8. Generalized model for sudden changes in seepage.

5. Model Validation

5.1. Model Basic Parameters

To verify the rationality of the permeability model established in Section 2, theoretical analysis of the permeability is conducted using Equation (27), based on the triaxial seepage test results of coal-slurry consolidated specimens under different loading conditions in Section 4 and the quasi-Newton method. Table 4 lists the relevant basic parameter values of the coal-slurry consolidated specimen. The pore pressure P is taken as the triaxial seepage gas pressure of 1 MPa, and the initial pore pressure P_0 is set as atmospheric pressure. The contact porosity φ_c is assigned a fixed value of: $\varphi_o/2$.

Table 4. Values of basic parameters of coal-slurry consolidation body.

Specimen Number	$\overline{E}_0$/[MPa]	K_0/[MPa]	K_{m0}/[MPa]	K_{p0}/[MPa]	v	φ_o/[%]	φ_c/[%]	k_o/[mD]
A_1	2220	1002	960	875	0.19	0.18	0.09	9.66
A_2	2355	995	975	865	0.19	0.20	0.10	8.56
A_3	2100	1055	980	885	0.20	0.19	0.095	8.53

5.2. Validation of Experimental Versus Theoretical Values of Permeability of Coal-Slurry Consolidates

Figure 9 presents a comparison between the experimental and theoretical permeability values for the three specimens under different loading conditions, with some data shown in Table 5. Statistical methods are applied to analyze the mean, variance, and standard deviation of the theoretical and experimental permeability values. This results in the mean permeability variation curve, variance fluctuation curve, and error bar distribution. The formula for calculating the variance is as follows:

$$s^2 = \frac{\sum_{i=1}^{n} (x_i - \overline{x})^2}{n-1} \tag{29}$$

where, x_i is the ith sample data value; $\overline{x}$ is the sample mean, $\overline{x} = \frac{1}{n}\sum_{i=1}^{n} x_i$; n is the number of data points in the sample.

Table 5. Selected data related to permeability of coal-slurry cementation under different loadings.

Strain	Specimen A_1				Specimen A_2				Specimen A_3			
	Measured Value	Theoretical Value	Average Value	Variance	Measured Value	Theoretical Value	Average Value	Variance	Measured Value	Theoretical Value	Average Value	Variance
0.37485	9.3254	8.3254	8.8254	0.5	8.4522	7.4522	7.9522	0.5	8.4522	7.5565	8.00435	0.40114
2.12	7.3255	7.2255	7.2755	0.005	7.3255	5.8544	6.58995	1.08207	7.3255	6.1424	6.93395	0.30662
3.15	8.02	8.15	8.085	0.00845	8.0552	5.4552	6.7552	3.38	8.0552	5.4044	6.7298	3.51337
5.12	12.1	10	11.05	2.205	9.6345	7.6345	8.6345	2	9.6345	7.6544	8.64445	1.9604
6.45	19.2	15.8	17.5	5.78	14.6454	12.6454	13.6454	2	12.8454	11.6544	12.2499	0.70924
7	24.6	21.2	22.9	5.78	25.45587	18.5254	21.99064	20.01571	15.45587	16.3558	15.80584	0.24495
7.58	33.64027	30.2	31.92014	5.91773	33.64027	25.9402	29.79024	27.64554	17.94027	22.5544	20.24734	10.6451
8.16	36.3624	35.12	35.7412	0.77178	36.3624	34.9455	35.65395	1.0038	36.3624	38.6624	37.5124	2.645
9.23	36.3965	35.28	35.83825	0.62329	37.3965	36.8945	37.1455	0.126	37.3965	39.9864	38.69145	3.35379

As can be seen from the figure, the error bar distribution between the theoretical permeability values derived from the model established in this study and the experimentally measured permeability values of coal-slurry consolidated specimens under different loading conditions is relatively narrow, with no large gaps in the error bar distribution, indicating a high degree of agreement. The relative errors between the theoretical and experimental curves for specimens A_1, A_2, and A_3 are 9.5%, 8.6%, and 8.1%, respectively, with an average relative error of 8.7%, indicating that the theoretical permeability values calculated from the model are close to the experimental values and exhibit a strong agreement in terms of trend. From the variance fluctuation curves of the specimens in the figure, it can be observed that the theoretical and experimental permeability variations fluctuate significantly near the yield stage (around point B), while fluctuations are more stable in other stages. This indicates that the errors between the theoretical and experimental values primarily occur when the specimen reaches its maximum bearing capacity and undergoes instability failure. There are two main reasons for this: first, experimental errors. During the damage deformation stage of the coal-consolidated specimen, it is initially compressed under sustained stress, followed by volumetric expansion, resulting in an increase in experimental permeability; second, the permeability model established in this study is based on a cubic structure, assuming that fracture damage occurs as uniform, continuous plastic deformation without sudden changes. However, the actual slurry has a high degree of brittleness, causing the specimen to experience an instantaneous failure after reaching its maximum bearing capacity, leading to experimental values that exceed theoretical values.

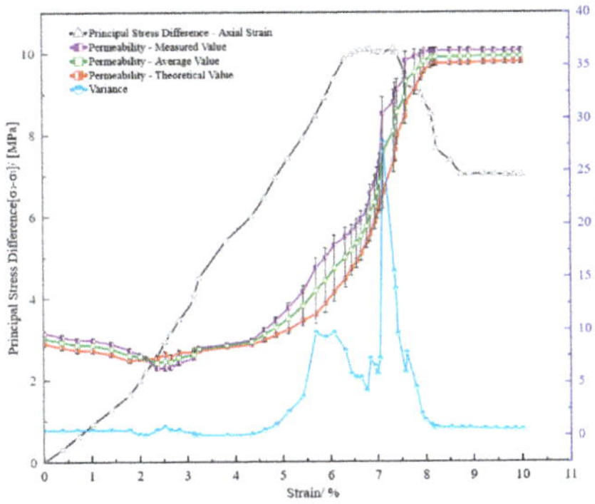

(**a**) Specimen A1

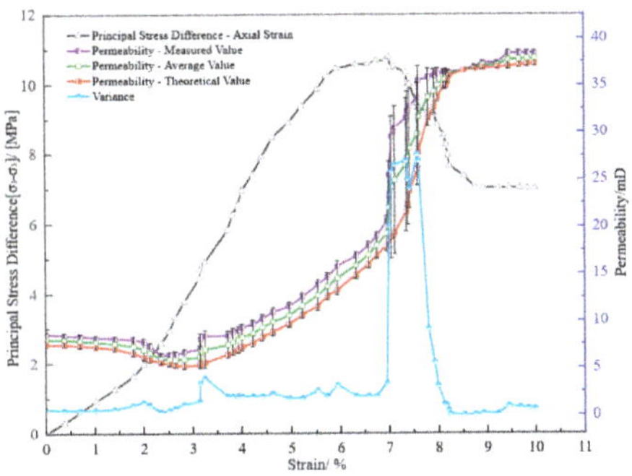

(**b**) Specimen A2

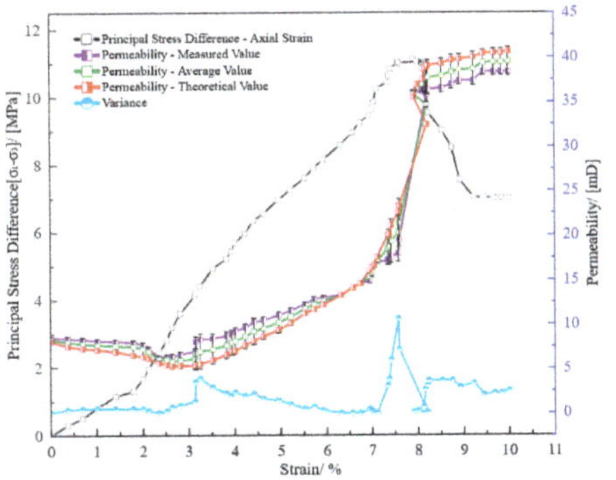

(**c**) Specimen A3

Figure 9. Comparison of experimental and theoretical values of permeability of coal-slurry consolidation under different loadings.

The above analysis indicates that the permeability model established in this study can accurately describe the dynamic evolution of permeability in coal-slurry consolidated specimens under loading conditions, and that analyzing the damage, rupture, and key gas leakage stages during loading reveals that optimizing and improving the plasticity of the slurry can enhance the impermeability of sealing materials under loading conditions, which is beneficial for gas extraction operations.

6. Conclusions

Grouting sealing can form coal-slurry consolidated bodies, significantly improving the sealing quality of gas drainage boreholes and mitigating coal and gas outburst hazards. Therefore, building on previous research, this study focuses on the coal-slurry consolidated bodies formed after grouting and sealing gas drainage boreholes. True triaxial loading and seepage experiments were conducted, and a permeability model of coal-slurry consolidated bodies under loading conditions was established. The findings provide guidance for analyzing the leakage prevention effectiveness of sealing materials in field engineering and optimizing the sealing performance of grouting materials. Future research can further explore the damage and seepage evolution of coal-slurry consolidated bodies under various loading conditions and sealing material types. The main conclusions are as follows:

(1) Based on the 'cubic' model, a permeability model for coal-slurry consolidated specimens under loading conditions was established, considering the intrinsic and structural deformations induced by consolidation body damage under triaxial stress.

(2) Triaxial stress loading permeability tests under different initial confining pressures and axial pressures show that the coal-slurry consolidated specimens remain in a compressed state under stress loading. In the compaction stage (OA phase), the micropores and fractures within the consolidated body are gradually compressed and closed, leading to a reduction in permeability. In the elastic-plastic stage (AB phase), damage and failure begin to develop within the specimen, and fractures gradually expand, causing an increase in permeability. The permeability exhibits a 'V' shaped trend prior to instability failure (point B). After continued stress loading, instability failure occurs at the maximum bearing capacity, and in the BC phase, the poorly connected micropores and fractures within the specimen begin to develop large-scale connections, causing a rapid increase in permeability and the occurrence of seepage mutation. A large number of damaged fractures rupture and interconnect, leading to a stabilization of permeability after reaching its maximum value. Under different loading conditions, the extent of permeability increase after instability failure of the coal-slurry consolidated specimens increases with higher initial axial and confining pressures.

(3) The comparison between experimental and theoretical permeability values under different loading conditions shows that the theoretical permeability values from the model are close to the experimental values, with an average relative error of 8.7%, indicating that the coal-slurry consolidated body damage permeability evolution model established in this study can accurately describe the dynamic changes in permeability under loading conditions. The research findings provide valuable guidance for analyzing the sealing effectiveness of sealing materials and optimizing their performance.

Author Contributions: Conceptualization, Y.T.; Methodology, P.L.; Software, J.Z.; Validation, J.Z.; Investigation, W.J.; Data curation, W.J.; Writing—original draft, Y.T.; Writing—review & editing, P.L. All authors have read and agreed to the published version of the manuscript.

Funding: This research was funded by the National Natural Science Foundation of China (52204264); Sponsored by Program for Science & Technology Innovation Talents in Universities of Henan Province (25HASTIT015); Training Plan for Young Backbone Teachers of undergraduate universities in Henan Province (2024GGJS092); Training Plan for Young Backbone Teachers of Zhongyuan University of Technology (2023XQG05); Subject Backbone Teacher Support Program of Zhongyuan University of Technology (GG202420); Key Scientific Research Project of Colleges and Universities in Henan Province (25A620001).

Data Availability Statement: The raw data supporting the conclusions of this article will be made available by the authors on request.

Acknowledgments: We would like to thank the Yima Coal Industry Group Co., Ltd., Henan Polytechnic University, and Zhongyuan University of Technology for their valuable suggestions and all the support provided for this research.

Conflicts of Interest: The author Tang Yao-Cai was employed by Yima Coal Industry Group Co., Ltd. The remaining authors declare that the research was conducted in the absence of any commercial or financial relationships that could be construed as a potential conflict of interest.

Nomenclature

σ	Total stress	MPa
σ_{ef}^{B}	Effective stress of the body	MPa
σ_a	Matrix average skeleton pressure	MPa
σ_{ef}^{S}	Structural effective stress	MPa
p	Pore pressure	MPa
φ	Porosity of coal-slurry cementation body	%
φ_0	Initial porosity	%
φ_c	Contact porosity	%
ε_B	Intrinsic Stress	%
ε_0	Initial strain	%
$\overline{E}$	Modulus of elasticity	MPa
k_0	Initial permeability	mD
K	Overall bulk modulus of the cemented body	MPa
K_m	Bulk modulus of the matrix of the cemented body	MPa
Q	Gas seepage flow rate	cm^3/s
μ	Gas dynamic viscosity coefficient	Pa·s
L	Length of the specimen	mm
A	Cross-sectional area of the specimen	cm^2

References

1. Li, S.; Gao, M.; Wu, B.; Xu, Y.; Li, Y.; Zeng, G. Dynamic compressive failure of coal at different burial depths. *Geomech. Geophys. Geo* **2023**, *9*, 53. [CrossRef]
2. Shi, Z.; Li, B.; Li, L.; Wang, N.; Zhang, J. Study on the directional extension law of hydraulic fractures induced by pre-cast slot under true-triaxial. *Theor. Appl. Fract. Mech.* **2024**, *133*, 104546. [CrossRef]
3. Li, B.; He, Y.; Shi, Z.; Jian, W.; Wang, N.; Zhang, Y. Mutual feedback and fracturing effect of hydraulic fractures in composite coal-rock reservoirs under different fracturing layer sequence conditions. *Int. J. Rock. Mech. Min. Sci.* **2024**, *184*, 105968. [CrossRef]
4. Fan, C.; Li, S.; Luo, M.; Du, W.; Yang, Z. Coal and gas outburst dynamic system. *Int. J. Min. Sci. Technol.* **2017**, *27*, 49–55. [CrossRef]
5. Chen, L.; Wang, E.; Ou, J.; Fu, J. Coal and gas outburst hazards and factors of the No. B-1 Coalbed, Henan, China. *Geosci. J.* **2018**, *22*, 171–182. [CrossRef]
6. Li, A.; Ding, X.; Yu, Z.; Wang, M.; Mu, Q.; Dai, Z.; Li, H.; Zhang, B.; Han, T. Prediction model of fracture depth and water inrush risk zoning in deep mining coal seam floor. *Environ. Earth Sci.* **2022**, *81*, 315. [CrossRef]
7. Celik, F. The observation of permeation grouting method as soil improvement technique with different grout flow models. *Geomech. Eng.* **2019**, *17*, 367–374. [CrossRef]
8. Ren, Y.; Wei, J.; Zhang, L.; Zhang, J.; Zhang, L. A Fractal Permeability Model for Gas Transport in the Dual-Porosity Media of the Coalbed Methane Reservoir. *Transp. Porous Med.* **2021**, *140*, 511–534. [CrossRef]
9. Zhu, X.; Zhang, Q.; Zhang, W.; Shao, J.; Wang, Z.; Wu, X. Experimental Study on the Basic Properties of a Green New Coal Mine Grouting Reinforcement Material. *ACS Omega* **2020**, *5*, 16722–16732. [CrossRef]
10. Du, W.; Zhang, Y.; Meng, X.; Zhang, X.; Li, W. Deformation and seepage characteristics of gas-containing coal under true triaxial stress. *Arab. J. Geosci.* **2018**, *11*, 190. [CrossRef]
11. Liu, Y.; Wang, E.; Jiang, C.; Zhang, D.; Li, M.; Yu, B.; Zhao, D. True triaxial experimental study of anisotropic mechanical behavior and permeability evolution of initially fractured coal. *Nat. Resour. Res.* **2023**, *32*, 567–585. [CrossRef]
12. Wang, K.; Du, F.; Zhang, X.; Wang, L.; Xin, C. Mechanical properties and permeability evolution in gas-bearing coal-rock combination body under triaxial conditions. *Environ. Earth Sci.* **2017**, *76*, 815. [CrossRef]
13. Wang, K.; Du, F. Experimental investigation on mechanical behavior and permeability evolution in coal-rock combined body under unloading conditions. *Arab. J. Geosci.* **2019**, *12*, 422. [CrossRef]
14. Wang, D.; Zhang, P.; Wei, J.; Yu, C. The seepage properties and permeability enhancement mechanism in coal under temperature shocks during unloading confining pressures. *J. Nat. Gas Sci. Eng.* **2020**, *77*, 103242. [CrossRef]
15. Wu, Y.; Wang, D.; Wei, J.; Yao, B.; Zhang, H.; Fu, J.; Zeng, F. Damage constitutive model of gas-bearing coal using industrial CT scanning technology. *J. Nat. Gas Sci. Eng.* **2022**, *101*, 104543. [CrossRef]
16. Shi, J.Q.; Durucan, S. Drawdown induced changes in permeability of coalbeds: A new interpretation of the reservoir response to primary recovery. *Transp. Porous Med.* **2004**, *56*, 1–16. [CrossRef]

17. Xue, Y.; Dang, F.; Cao, Z.; Du, F.; Ren, J.; Chang, X.; Gao, F. Deformation, permeability and acoustic emission characteristics of coal masses under mining-induced stress paths. *Energies* **2018**, *11*, 2233. [CrossRef]
18. Connell, L.D.; Lu, M.; Pan, Z. An analytical coal permeability model for tri-axial strain and stress conditions. *Int. J. Coal Geol.* **2010**, *84*, 103–114. [CrossRef]
19. Bai, X.; Wang, Y.; He, G.; Zhou, Z.; Wang, D.; Zhang, D. Research on a permeability model of coal damaged under triaxial loading and unloading. *Fuel* **2023**, *354*, 129375. [CrossRef]
20. Wang, G.; Liu, Z.; Wang, P.; Guo, Y.; Wang, W.; Huang, T.; Li, W. The effect of gas migration on the deformation and permeability of coal under the condition of true triaxial stress. *Arab. J. Geosci.* **2019**, *12*, 486. [CrossRef]
21. Lu, J.; Yin, G.; Deng, B.; Zhang, W.; Li, M.; Chai, X.; Liu, C.; Liu, Y. Permeability characteristics of layered composite coal-rock under true triaxial stress conditions. *J. Nat. Gas Sci. Eng.* **2019**, *66*, 60–76. [CrossRef]
22. Liu, B.; Zhu, L.; Liu, X.; Liu, Q.; Fan, Y.; Yao, W.; Deng, W. Energy evolution and damage deformation behavior of cemented broken coal specimen under triaxial compression condition. *Energy* **2024**, *310*, 133203. [CrossRef]
23. Shen, R.; Wang, X.; Li, H.; Gu, Z.; Liu, W. Brittleness characteristics and damage evolution of coal under true triaxial loading based on the energy principle. *Nat. Resour. Res.* **2024**, *33*, 421–434. [CrossRef]
24. Cao, A.; Wang, C.; Zhang, N.; Li, H.; Liu, Z.; Zhi, S. Experimental study on damage characteristics of coal samples under true triaxial loading and dynamic unloading. *Lithosphere* **2022**, *2022*, 5447973. [CrossRef]
25. Liang, Y.; Ran, Q.; Zou, Q.; Zhang, B.; Hong, Y. Experimental study of mechanical behaviors and failure characteristics of coal under true triaxial cyclic loading and unloading and stress rotation. *Nat. Resour. Res.* **2022**, *31*, 971–991. [CrossRef]
26. Duan, M.; Jiang, C.; Yin, W.; Yang, K.; Li, J.; Liu, Q. Experimental study on mechanical and damage characteristics of coal under true triaxial cyclic disturbance. *Eng. Geol.* **2021**, *295*, 106445. [CrossRef]
27. Lu, S.; Li, M.; Ma, Y.; Wang, S.; Zhao, W. Permeability changes in mining-damaged coal: A review of mathematical models. *J. Nat. Gas Sci. Eng.* **2022**, *106*, 104739. [CrossRef]
28. Liu, C.; Yu, B.; Zhao, H.; Hong, Z.; Tian, Z.; Zhang, D.; Liu, Y. Effective stress effect and slippage effect of gas migration in deep coal reservoirs. *Int. J. Rock. Mech. Min. Sci.* **2022**, *155*, 105142. [CrossRef]
29. Guo, P.; Cheng, Y.; Jin, K.; Li, W.; Tu, Q.; Liu, H. Impact of effective stress and matrix deformation on the coal fracture permeability. *Transp. Porous Med.* **2014**, *103*, 99–115. [CrossRef]
30. Kai, W.; Ang, L.; Zhou, A. Theoretical analysis of influencing factors on resistance in the process of gas migration in coal seams. *Int. J. Min. Sci. Technol.* **2017**, *27*, 315–319. [CrossRef]
31. Jiang, C.; Wang, L.; Ding, K.; Wang, S.; Ren, B.; Guo, J. Experimental study on mechanical and damage evolution characteristics of coal during true triaxial cyclic loading and unloading. *Materials* **2023**, *16*, 2384. [CrossRef]

processes

MDPI

Article

Stress Zoning Characteristics and Migration of Leaked Methane from Gas Wells Penetrating Protective Coal Pillars in Longwall Mining Areas

Jinhang Shen [1], Shun Liang [1,*], Yisong Hao [2], Zhi Ma [3], Weisheng He [3], Xu Liang [4], Shaoyou Xu [5] and Changheng Luo [5]

1 State Key Laboratory for Fine Exploration and Intelligent Development of Coal Resources, China University of Mining and Technology, Xuzhou 221116, China
2 Shaqu No. 2 Mine Huajin Coking Corporation Limited, Lvliang 033301, China
3 Ningxia Coal Industry Limited Liability Company Shuangma Coal Mine, China Energy Group, Yinchuan 750408, China
4 Shaanxi Xiaobaodang Mining Corporation Limited, Yulin 719302, China
5 No. 2 Gas Production Plant of Sinopec North China Oil and Gas Company, Ordos 017400, China
* Correspondence: 5756@cumt.edu.cn; Tel.: +86-151-6212-4780

Abstract: There are a large number of abandoned or casing-damaged oil/gas wells in the western mining areas of China. Under the influence of mining-induced stress, the methane leaked from the oil and gas wells will be transported through fracture within the coal pillar to the longwall working face, which will seriously threaten the safe mining of coal resources. There is no mandatory standard for the practice of coal pillars in coal and gas wells in coal/gas overlapping areas, and the problems of oversized coal pillars and waste of coal resources have occurred during the implementation. In this study, through finite element numerical simulation, three different sizes of protective coal pillars are modeled in the background of Shuangma Coal Mine. The impacts of different heights and widths of protective coal pillars on the evolution of stresses and the diffusion process of leaked methane are explored, and the spatial and temporal migration law of leaked methane under multi-field coupling is revealed. The results show that under mining-induced stress, the size of the protective coal pillar has a significant effect on the stress distribution and methane transport law. Compared with the 130 m coal pillar, the peak stress of the 150 m coal pillar decreased by 6.7%, and the peak stress of the 180 m coal pillar decreased by 9%. At 150 m and 180 m widths, the permeability difference between the two sides is only 1 mD, and the diffusion ranges are similar. From the stress distribution and gas diffusion law, it is shown that the effect achieved by 150 m and 180 m coal pillars is similar. This work is of great significance for the reasonable remaining protective coal pillars for oil/gas wells penetrating longwall mining areas, as well as the prevention and control of disasters caused by leaked methane from wells.

Keywords: abandoned oil/gas wells; mining stress; protective coal pillars; leaked methane

1. Introduction

As coal resources in eastern China gradually decreased, coal mining progressively shifted to the western regions of China. The engineering practices of coal–oil or coal–gas cross-exploitation are becoming more common, particularly in the Ordos, Junggar, and Tarim Basins [1–3]. These practices encounter many problems. Due to the early development of resources such as oil and gas, more and more production wells are in the late stage, leaving a large number of abandoned wells, which contain a large amount of natural gas, associated gas, as well as toxic and harmful gases such as carbon monoxide and methane [4,5]. These abandoned wells remaining from the early stage, due to the lack of casing or damage caused by aging, have led to the diffusion of toxic and harmful gases from the oil/gas wells into coal seams and even the atmosphere, seriously threatening coal mining safety and the surrounding ecosystem. This not only means a huge waste of resources but also brings many hazards to nearby coal seams mining [6–8]. Therefore, in these areas, maintaining the stability of wells in the longwall mining area is key to the safe mining adjacent coal seams [9–12].

Literature research reveals limited research on the stability of oil/gas wells affected by coal seam mining in common engineering practices, and no specific standards exist for remaining coal pillars in oil/gas wells [13]. Previous studies have focused on factors such as coal seam burial depth, mining-induced stress, coal pillar dimensions, and the relative position relationship between the well and the working face in the design of gas well protective pillars. As engineering practices increase, previous researchers have further pointed out that the horizontal and shear movements of strata caused by mining significantly impact the deformation of wells. The response modes of each stratum to horizontal and shear movements are different, resulting in the inconsistency of deformation between adjacent strata. This inconsistency in deformation can easily lead to shear deformation of the wellbore. Liang Shun et al. explored the distribution of axial stress, strain, and displacement in oil/gas wells under various factors, identifying the timing and layers where wells are most susceptible to destabilization. The goal is to focus on protecting these unstable areas and preventing instability and leakage in oil/gas wells during longwall mining [9,14,15]. However, most research has centered on the casing stability of wells, with limited focus on the safety of wells lacking casing or damaged due to aging.

In the coalbed methane field, scholars such as Gray, Bustin, and Cui [16–18] assume that the overburden stress of the coal seam is constant and the stress on the coal mass is uniform. By constructing different variables, they propose a classical permeability model, which is significant for studying CBM exploitation and methane migration from abandoned oil/gas wells. Subsequent scholars [19,20] studied the impact of oil and gas pipeline leaks on methane diffusion in shallow aquifers using methods such as multiphase, multi-component, and numerical models. They believed that factors such as geological conditions, phase behavior, and coal seam dip angle affect methane permeability, which in turn influences the diffusion range of methane. Krzysztof Skrzypkowski [21] used a number of experiments to show that cracked rocks are particularly susceptible to gas migration, which translates into a weakening of strength and deformation parameters.

In summary, previous studies on deformation, instability, and leakage in wells during the co-extraction of coal, oil, and gas resources have mostly focused on situations where the well casing is intact and there is no leakage. There is little research on the safety of oil/gas wells that have not been cased or have been damaged due to age. The study of methane migration often adopts the assumption of uniform stress, focusing on the impact of oil/gas pipeline leaks on methane diffusion in shallow aquifers. However, in the coal-oil–gas overlapping zone, the mining of the working face will cause the stress on the coal mass to be unbalanced, and previous studies have not considered the stress evolution caused by coal seam mining. Therefore, this study takes Shuangma Coal Mine's remaining protective coal pillars as the engineering background, considering the effects of mining-induced stress and casing damage on both the stability of wells and the design of protective coal pillars, and carries out the evolution of mining-induced stress and migration of leaked methane from gas wells penetrating protective coal pillars. The results of the study are of great significance for optimizing the size of protective coal pillars in oil/gas wells without casing or damaged due to age in longwall mining areas, improving the recovery rate of coal resources, revealing the mechanism of underground disasters caused by leaked methane from oil/gas wells, and realizing safe and efficient mining of coal, oil, and gas resources.

2. Characteristics of Stress Evolution Within Protective Coal Pillars for Gas Wells

The coal pillar is an enclosed, non-visible space, making it challenging to accurately monitor the evolution of the stress field within protective coal pillars of gas wells during mining due to technical and site limitations. Therefore, based on FLAC3D 6.0 simulation software and the actual working conditions of Shuangma Coal Mine, this study analyzes the stress evolution within coal pillars of varying widths and heights and its effects on the stability and deformation of gas wells, as well as their effects on the bearing and deformation of gas wells inside the coal pillar.

2.1. Engineering Geology Conditions

The Shuangma Coal Mine is located in the Ordos Basin, with an area of approximately 67.99 km^2 and an annual output of 4 Mt. It mines the 4 # coal seam and uses the caving method for roof management. There are small faults in the coal mine working face, but they do not affect normal mining operations. Overall, the geological structure of Shuangma Coal Mine is simple. More than 170 oil/gas wells have been explored in Shuangma Coal Mine, which has caused serious safety problems for the mining face. The main coal seam, the 4^{-1} seam, has a simple geological structure and a mining face length of 280 m. It includes one abandoned gas well without safety measures, where gas pressure ranges from 9 to 15 MPa. The average thickness of the coal seam is 4 m, and its burial depth is 290 m; vertical stress is 6.9 Mpa, and horizontal stress is 8.28 Mpa. As shown in Figure 1, the seam roof is primarily fine sandstone, while the floor is predominantly mudstone.

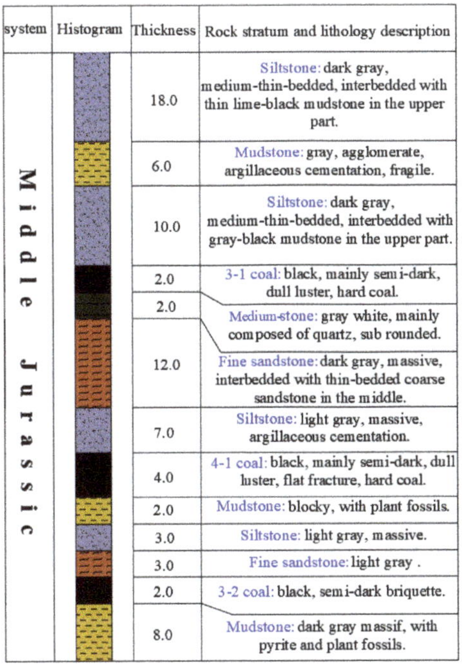

Figure 1. Histogram of the coal seam of Shuangma Coal Mine.

2.2. Distribution Characteristics of Stress Within Different Dimensional Protective Coal Pillars for Gas Wells

2.2.1. Simulation Scenario

Based on the geological conditions of Shuangma Coal Mine, and referencing coal mining methods used under buildings and protective coal pillar dimensions designed for oil/gas wells in similar mines, the protective coal pillar width is designed at 180 m in Shuangma; this method has a large margin coefficient, so three different pillar schemes were designed with a maximum boundary of 180 m, namely 130 m, 150 m, and 180 m. This study uses FLAC numerical simulation to create the geological and Mohr-Coulomb models to simulate rock formations. Physical and mechanical parameters are listed in Table 1. The lower boundary is constrained in the model, while the front-back and left-right boundaries restrict displacement in the Z, X, and Y directions, respectively; the upper surface remains free. The mining sequence follows site operations: the left working face is mined first, followed by the right working face. The specific model parameters are illustrated in Figure 2. Figure 3 presents the xz-plane section of the model. Figure 3a illustrates schemes with varying widths of coal pillars, where different advancement stages represent the retention of coal pillars with different widths. For instance, when the working face advances to position III, it signifies the retention of a coal pillar with a width of 180 m. Figure 3b depicts schemes with varying heights of coal pillars, where different mining heights represent the retention of coal pillars with different heights. For example, when the mining height is 1 m, the height of the retained coal pillar is 1 m.

Table 1. Physical and mechanical parameters of rock lithologies employed in numerical simulation.

Lithology	Density (Kg/m^3)	Bulk (GPa)	Shear (GPa)	Cohesion (MPa)	Tension (MPa)	Friction (°)
Medium-sandstone	2550	4.63	2.40	1.9	1.5	30.2
Mudstone	2430	3.5	1.85	1.2	0.8	25.1
Fine-sandstone	2620	5.32	2.98	1.7	1.3	28.0
Siltstone	2600	4.30	2.38	1.4	0.9	28
coal	1400	2.89	1.83	0.8	0.6	20.4

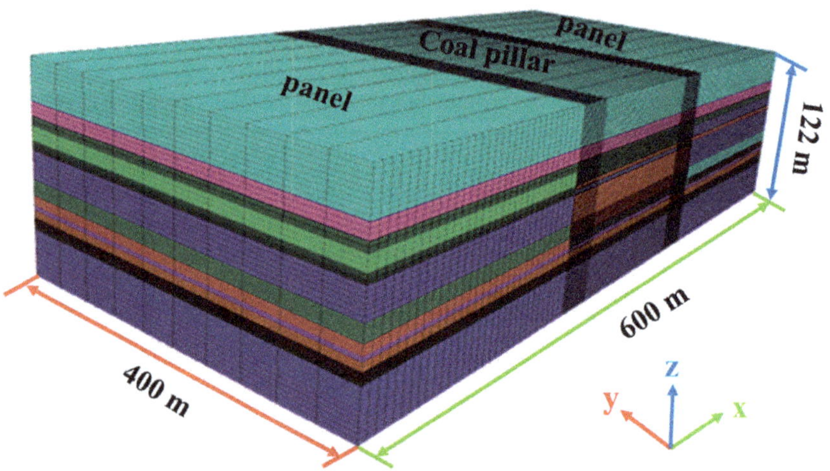

Figure 2. FLAC3D geological model.

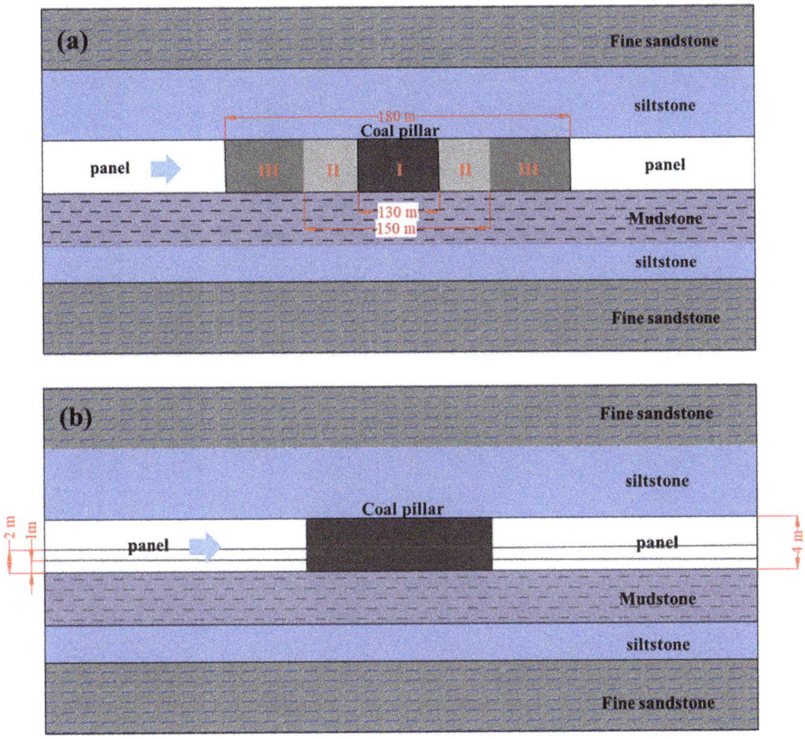

Figure 3. Schematic diagram of the simulation scheme ((**a**) schemes of varying widths (**b**) schemes of varying heights).

2.2.2. Simulation Results and Analysis

The vertical stress within the coal pillar is symmetrically distributed, forming a 'saddle' shape (Figure 4). As the working face advances, the roof forms a 'bottom-shaped' structure, and both sides of the pillar are subject to bearing pressure; the vertical stress increases, and the closer to the middle, the vertical stress decreases. When one side of the working face is mined, the coal pillar is subjected to mining-induced stress, leading to a redistribution of stress within the coal pillar ahead of the working face. Specifically, the vertical stress near the working face increases while the vertical stress farther from the working face decreases. The peak stress and influence range of protective coal pillars with different widths are different during the mining process of the working face.

Variations in mining height also lead to differences in the distribution characteristics of stress fields in the protective coal pillars of the working face, affecting the migration of leaked methane in gas wells. During coal seam mining, the stress concentration inside the protective pillar primarily appears in the triangular area where the coal wall contacts the roof and floor, and fractures in this area progressively develop and mature. As the 'second open-off cut through' and the working face continues to advance, the stress distribution in the protective pillar evolves from a "parabolic" to a "saddle" shape. Concurrently, as the height of the protective pillar increases, the vertical stress peak of the roof and floor in front of the coal also varies (Figure 4b).

Figure 5 shows that as mining progresses on the working face, stress on the left side of the coal pillar gradually increases until mining is completed, reaching the first peak stress at the protective coal pillar. The peak stresses for different widths of coal pillars are 26.48 MPa, 24.71 MPa, and 24.09 MPa, respectively. After the model reaches equilibrium, mining the right working face causes stress changes on the right side of the coal pillar. The peak values of the second stress on the right side of the pillar with different widths are 24.05 MPa, 22.76 MPa, and 22.31 MPa, respectively. Comparing the stress peaks reveals that the left side of the coal pillar has higher stress than the right. The stress distribution shows that the influence range of secondary mining is smaller than that of primary mining. For coal pillars of different widths, the range of the smoothed area is 30, 60, and 80 m (Figure 4). The larger the smooth area, the greater the protolith stress or elastic region within the coal pillar. This area has less stress on the coal pillar, and the oil well located here has higher stability. In addition, the stress on the roof sandstone is greatly affected, and the roof is prone to slip and failure. At this time, the protected area of the gas well is relatively small and prone to instability. When comparing the two coal pillar sizes of 180 m and 150 m, it is evident that the 180 m protective coal pillar provides a greater smoothing area, characterized by a smooth zone of 80 m, as illustrated in the accompanying diagram, with minimal differences in stress impact range and vertical stress peak value. The impact of mining around gas wells is smaller, resulting in fewer fractures and easier protection of wells. However, the increase in coal pillar width also means the waste of coal resources.

According to Figure 5, as the coal pillar width increases, the smooth areas between the first and second peak stress also increase, indicating that the elastic areas of the coal pillar gradually increase. At the same time, the peak vertical stress also decreases accordingly. After the completion of mining, compared to the width of the 130 m protective coal pillar, the first stress peak of the 150 m protective coal pillar width decreased by 6.7%, and the second stress peak decreased by 5.3%; the first peak stress of the 180 m pillar width decreased by 9%, and the second stress peak decreased by 7.2%. Through data comparison and analysis, the wider the protective coal pillar, the lower the peak stress. However, when the width of the pillar reaches 150 m and 180 m, the stress peaks are 24.71 MPa and 24.09 MPa, respectively, with a difference of 0.62 MPa—the peak stress and stress diffusion area tend to be similar. Therefore, in practical engineering, the width of the protective coal pillar can be slightly reduced to increase coal production. However, a too-small coal pillar width may threaten the safety of oil/gas wells. When choosing to protect the coal pillar width, it is necessary to consider factors such as coal resource utilization and the safety of oil/gas wells comprehensively.

As illustrated in Figure 6, during the mining process, the stress on the left side of the coal pillar experiences a gradual increase, ultimately reaching the first stress peak. The peak values corresponding to different heights of coal pillars are 27.37 MPa, 28.93 MPa, and 30.73 MPa. Subsequently, upon executing the second open-off cut, the stress distribution on the right side of the coal pillar changes; the peak values of the second stress on the right side of the pillar with different heights are 24.05 MPa, 22.76 MPa and 22.31 MPa. Under identical mining conditions, The higher the protective coal pillar, the greater the peak bearing pressure in front of it (Figures 6 and 7). This is because as the height increases, the roof's thickness decreases accordingly, and the decompression effect on the movement of the upper rock layer becomes smaller. The peak point of the bearing pressure shifts forward, and the range of influence increases. Under the same working face conditions, the peak vertical stress is directly proportional to the height of the protective coal pillar. When the pillar height is 2 m, the peak stress increases by 5.6% and 5.3%. At the height of 4 m, the peak stress rises by 12.2% and 11.2%. As the height increases, the distance between the two vertical stress peaks in the coal pillar shortens, indicating that taller mining heights make oil/gas wells more susceptible to destabilization and damage., and the leaked methane will quickly migrate to the working face. Therefore, appropriate protective measures should be taken. Within the same pillar width, the second peak vertical stress is lower than the first, indicating that the vertical stress borne by the coal pillar on the later mining side is smaller, and the range of stress influence is also smaller.

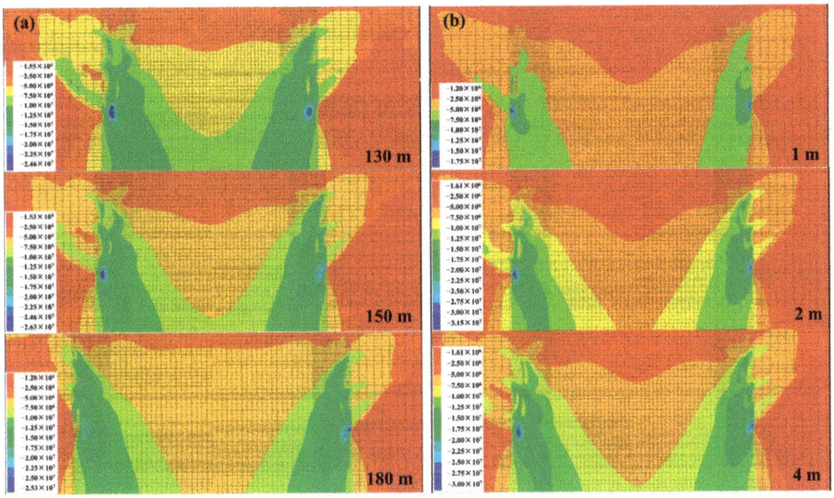

Figure 4. Cloud diagram of vertical stress distribution within protective coal pillars of gas wells of different dimensions ((**a**) different widths and (**b**) different heights).

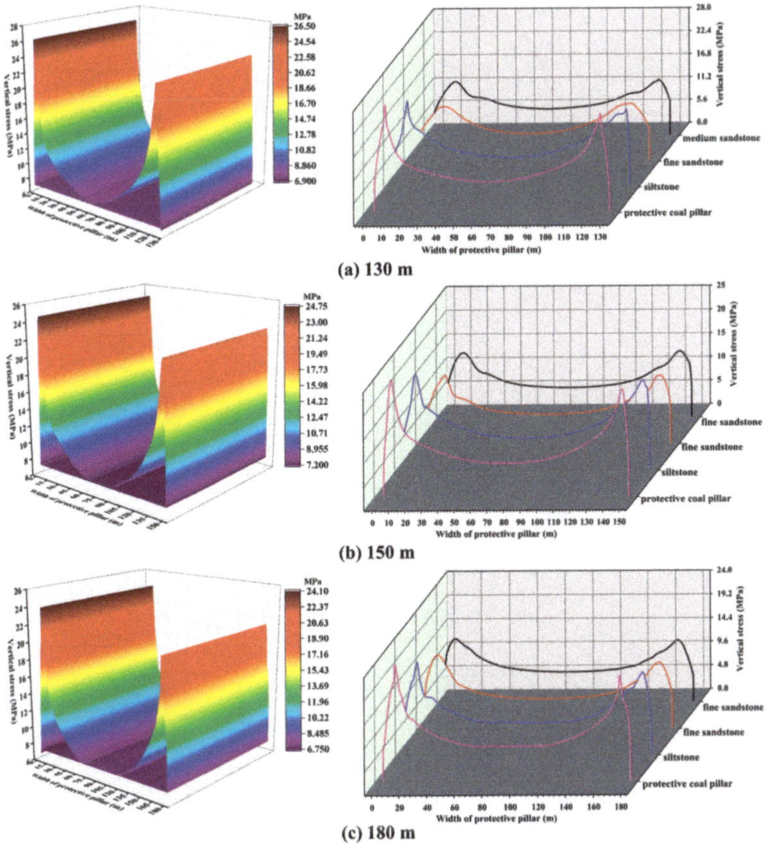

Figure 5. Vertical stress distribution in each rock layer of the protective coal pillar is achieved by gas wells of different widths.

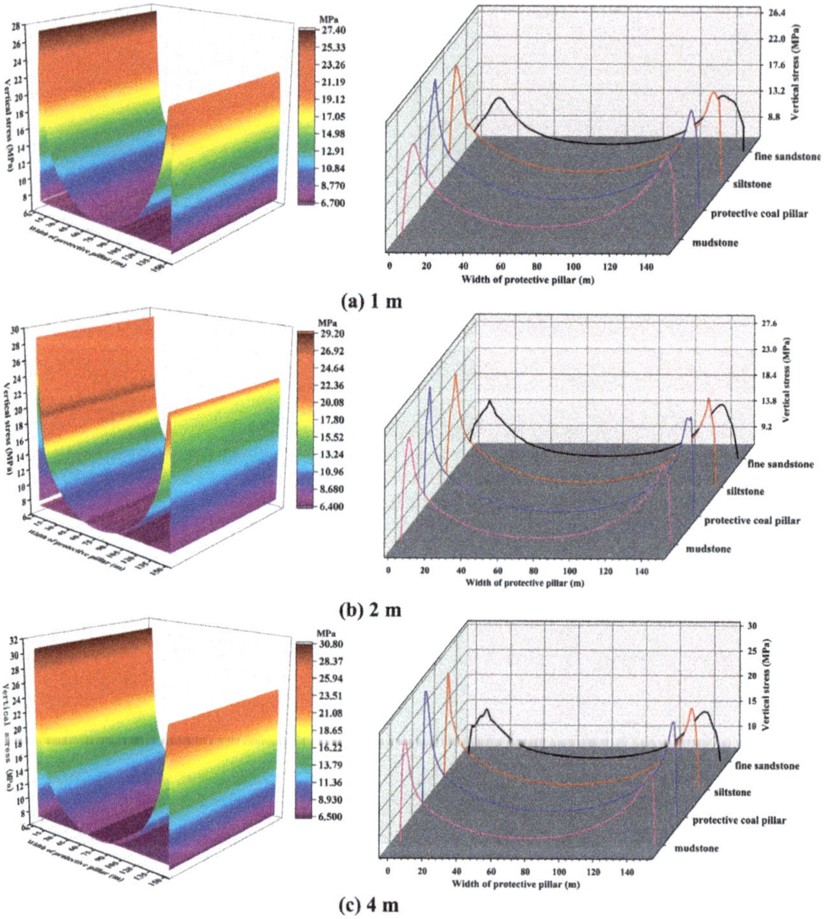

Figure 6. Vertical stress distribution in each rock layer of protective coal pillar by gas wells with different heights.

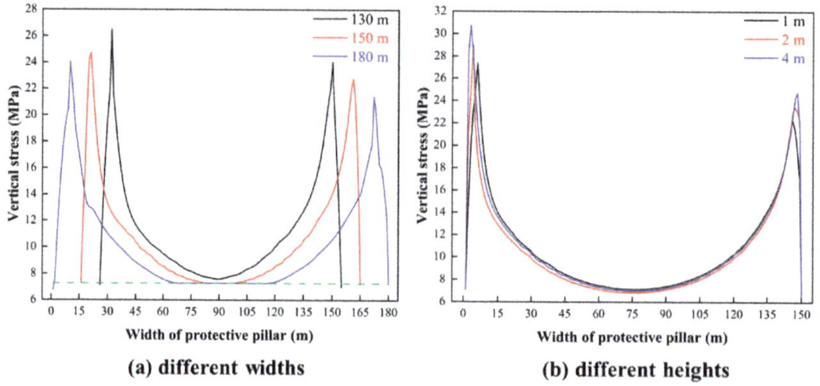

Figure 7. Vertical stress distribution line diagram inside the protective coal pillar of gas wells with different widths and heights.

3. Fracture Scale Division of Mining Disturbed Coal Mass and Control Equation of Methane Migration

During the mining process of the working face, the temporal and spatial evolution of the overburden fractures in the oil/gas wells protective pillars also occurs, resulting in changes in the internal structure of the coal seam and the formation of different scale fractures. The migration behavior of methane in different scale fractures of the coal mass is different, so the methane migration law in the mining fracture field of the coal mass in front of the working face is also changing. In view of this, this study first divides the fracture scale based on the Reynolds number of methane flow states at different fracture scales, establishes the relationship between permeability and each fracture scale, and then constructs the corresponding methane migration behavior control equations for each zone, thereby establishing a complete multi-scale fracture coal mass transport equation in the mining space.

The following assumptions were made for the equations based on coal stress field and methane migration characteristics [22].

(1) Coal is an anisotropic continuous porous medium with dual pores and a single permeability elastic continuous heterogeneous medium. The thermal expansion strain caused by temperature and gas adsorption was ignored in the study;
(2) The leaked methane is considered an ideal gas;
(3) Methane adsorption and desorption rates are very fast and negligible in the process of coal seam methane extraction;
(4) The process of diffusion of leaked methane conforms to Fick's law and is transported as seepage in the fracture;
(5) Methane migration processes are treated as isothermal processes.

3.1. Fracture Scale Classification of Mining Disturbed Coal Mass

According to the transport state of methane, multi-scale fractures within coal pillars can be divided into large-, medium-, and small-scale fractures. Multi-scale fracture classification needs to consider flow velocity and permeability and cannot be limited to scale fracture classification alone. Therefore, in this study, the Reynolds number of different fluid flow regimes combined with permeability are used to classify the scale of fracture.

According to Karman's criterion, the Reynolds number magnitude determines the fluid flow state. Therefore, this study uses the Reynolds number to quantify fracture size and identify the type of methane channel. The Reynolds criterion expression is:

$$\mathrm{Re} = \frac{\rho u r}{\mu} \tag{1}$$

where ρ is the density, u is the flow velocity, d is the diameter of the flow tube, and μ is the dynamic viscosity coefficient.

When the fluid flow state is laminar, its flow conforms to Darcy's law:

$$u = -\frac{k}{\mu}\nabla p \tag{2}$$

where u is the velocity of gas flow in the coal seam, m/s; ∇p is the fluid pressure gradient.

When the fluid flow state is turbulent, the power law can express the fluid velocity:

$$u = -\frac{k}{\mu}\left(\frac{\mathrm{d}p}{\mathrm{d}x}\right)^{s} \tag{3}$$

Relating the Reynolds number equation to the fracture of the methane channel, it can be obtained:

$$r = \mathrm{Re}\frac{\mu}{\rho u} \quad (4)$$

where r is the fracture scale. When the Reynolds number is less than 2320, the flow state of methane is linear laminar flow; when the Reynolds number is between 2320 and 4000, the flow state of methane is transitional flow; and when the Reynolds number is more than 4000, the flow state of methane is turbulent flow, which can be obtained by association of Equations (1)–(4) and Re the range of values:

Classified as small scale when the fracture width is $r \leq \frac{2320\mu^2}{\rho k} / \frac{dp}{dx}$, medium scale when the fracture width is $\frac{2320\mu^2}{\rho k} / \left(\frac{dp}{dx}\right)^s < r < \frac{4000\mu^2}{\rho k} / \left(\frac{dp}{dx}\right)^s$, and large scale when the fracture width is $r \geq \frac{4000\mu^2}{\rho k} / \left(\frac{dp}{dx}\right)^s$.

3.2. Control Equation for Methane Migration in the Mining-Disturbed Coal Mass

3.2.1. Governing Equations of Methane Migration Behavior in Mesoscale and Small-Scale Fractures

Regarding the flow of methane in medium scale (roof and floor fracture zones, advanced plastic coal mining fractures in the working face) and small-scale fractures within the mined coal rock mass. Essentially, it is an unstable gas flow process in the mining-induced stress field environment.

The tensor expression of the stress balance equation for an elastic porous medium system is:

$$\sigma_{ij,j} + f_i = 0 \quad (5)$$

where σ_{ij} is the component of the stress tensor and f_i is the volume force component.

The strain–displacement relationship is expressed as ε_{ij}, $u_{i,j}$ is the displacement component.

$$\varepsilon_{ij} = \frac{1}{2}(u_{i,j} + u_{j,i}) \quad (6)$$

In this study, it is assumed that the coal mass is an ideal elastic body, therefore, can be expressed by the generalized Hooke's law:

$$\varepsilon_{ij} = \frac{1}{2G}\sigma_{ij} - \left(\frac{1}{6G} - \frac{1}{9K}\right)\sigma_{kk}\delta_{ij} + \frac{\alpha_m p_m + \alpha_f p_f}{3K}\delta_{ij} + \frac{\varepsilon_s}{3}\delta_{ij} \quad (7)$$

where $G = E/2(1+v)$, $K = E/3(1-2v)$, $\alpha_m = 1 - K/Ks$, $\alpha_f = 1 - K(LmKn)$, $Ks = Es/3(1-2v)$, $\sigma_{kk} = \sigma_{11} + \sigma_{22} + \sigma_{33}$, $\varepsilon_s = \varepsilon_l p_m/(P_L + p_m)$, where G is the shear modulus of the coal, K is the bulk modulus of the coal, Ks is the elastic modulus of the coal skeleton, E is the elastic modulus of the coal, v is the Poisson's ratio of the coal, Kn is the fracture stiffness, Lm is the width of the matrix, δ_{ij} is the Kronecker's factor, α_m and α_f are the effective stress coefficients of the coal matrix and the fracture of the coal, p_m and p_f are the pressure of the matrix and the fracture pressure of the coal, ε_s is the volume strain of the matrix caused by gas adsorption/desorption, ε_l is the Langmuir-type strain coefficient, and P_L is the Langmuir pressure constant.

Associative Equations (5)–(7) can be obtained to control the stress field during the methane process in the coal seam:

$$Gu_{i,jj} + \frac{G}{1-2v}u_{j,ji} - \alpha_p p_{f,i} - K\varepsilon_{s,i} + f_i = 0 \quad (8)$$

In this paper, methane is treated as an ideal gas to simplify the calculation and research process. The ideal gas model can adequately explain the movement and transportation laws of methane within coal seams, and it also serves as a valuable reference for studying the pore structure and permeability characteristics of methane in coal-rock masses. By adhering to the ideal gas state equation, a more accurate description of the state and properties of methane in coal seams can be obtained.

$$pV = \frac{m}{M} R_0 T \tag{9}$$

The relationship between methane pressure and methane gas density can be derived as

$$\rho_g = \eta p \tag{10}$$

$$\eta = \frac{M}{RT} \tag{11}$$

Where T is the absolute temperature; R is the universal coefficient, M is the Mohr mass of the gas, kg/mol, and ρ_g is the density of methane at a methane pressure of p.

Methane in the coal seam has two states: adsorbed methane on the surface of coal rock pores and free methane in the form of free flow in fracture spaces. The content of free methane can be expressed as Q_1 [23], kg/m^3.

$$Q_1 = \eta p \phi \tag{12}$$

The formula for the adsorbed methane content per unit volume of coal surface can be obtained based on the Langmuir adsorption analytical equation [24]:

$$Q_2 = \frac{abp}{1 + bp} \rho_0 \rho_c \tag{13}$$

Q_2 is the adsorbed methane content, kg/m^3, a, b are adsorption constants, p is the adsorption equilibrium methane pressure, MPa, A is the ash content of coal, %, T_0 is the experimental temperature when the adsorption constants of coal were determined, W is the moisture of the coal seam, %, and the methane content can be expressed as:

$$Q = Q_1 + Q_2 = \eta p \phi + \frac{abp}{1 + bp} \rho_0 \rho_c \tag{14}$$

Conservation of mass in microunits in coal seams is obtained by considering the mass source term:

$$\nabla(\rho_g u) + \frac{\partial(Q)}{\partial t} = Q_m \tag{15}$$

Q is the methane content of the coal seam per unit volume of the coal seam, kg/m^3; u is the methane flow rate, m/s; ρ_g is the density at methane pressure P, kg/m^3; and Q_m is the mass source term, kg/m^3.

For small and medium-scale fracture, the association of Darcy's law and the continuity equation for methane yields.

$$\nabla(\rho_g u) + \frac{\partial(Q)}{\partial t} = Q_m \tag{16}$$

Combined with the methane equation of state $\rho_g = \eta p$, taking the mass source term Q_m = 0, the mathematical model of methane flow can be found.

$$\left(\frac{ab\rho_0\rho_c}{(1 + bp)^2} + \eta \phi \right) \frac{\partial p}{\partial t} + \eta p \frac{\partial \phi}{\partial t} - \nabla \left(\eta p \frac{k}{\mu} \frac{dp}{dx} \right) = 0 \tag{17}$$

Based on the theoretical model of matchstick permeability and the Kozeny-Carman equation, the permeability can be expressed as [25]

$$k = k_0 e^{-3c_f(\sigma_b - \sigma_{b0})} \tag{18}$$

where k is the permeability of coal mass, mD; ϕ is the fracture porosity; and c is the KC constant. Unfractured coal typically has a permeability in the range of 0.1 to 10 millidarcies (mD), while fractured coal can have permeabilities ranging from 10 to 1000 mD or more, depending on the extent and connectivity of the fractures.

The cubic law can obtain the single fracture flow rate of gas in coal mass as:

$$q = -\frac{r^3}{12\mu}\frac{\Delta p}{L} \tag{19}$$

where r is the width of coal fracture, m; μ is the dynamic viscosity of methane gas, Pa·s; Δp is the pressure difference of coal fracture, Pa; L is the length of coal fracture, m.

Coal seam methane is produced per unit of time:

$$q = S\frac{k}{\mu}\frac{\Delta p}{L} \tag{20}$$

where S is the cross-sectional area, according to the unit volume porosity $\varphi = bS$, denote the variation of coal porosity with effective stress by Λ, $\Lambda = \frac{1}{\varphi}\frac{\partial \varphi}{\partial \sigma'}$. The above equation can be rewritten as:

$$\frac{\partial k}{\partial \sigma'} = 3\Lambda k \tag{21}$$

The damage variable is invoked to define the fracture development and permeability changes within the coal mass under mining-induced stress, and the damage variable can be expressed in terms of the elastic modulus as [26]:

$$\frac{E}{E_0} = 1 - D \tag{22}$$

where E_0 is the modulus of elasticity of the elastic phase of the coal.

Λ can be represented by the damage variable D, γ which is the sensitivity impact factor.

$$\Lambda = \gamma D \tag{23}$$

Equation (22) can be written as:

$$\frac{\partial k}{\partial \sigma'} = 3\gamma D k \tag{24}$$

Below is the permeability modeling of the mesoscale fracture:

$$k = k_0 e^{3\gamma D(\sigma' - \sigma'_0)} \tag{25}$$

3.2.2. Governing Equations for Methane Migration Behavior in Large-Scale Fracture

In the goaf, cavities are formed after coal seam mining, which are filled with gangue and air. There are a large number of cavities and fractures between these gangues, and the gas flow state is between Darcy seepage and Navier–Stokes pipeline flow. To describe the rapid flow of fluids in this porous medium, the Brinkman equation can be used:

$$-\nabla \cdot \frac{\eta}{\varphi}\left(\nabla u_{\mathrm{br}} + (\nabla u_{\mathrm{br}})^{\mathrm{T}}\right) - \left(\frac{\eta}{\varphi}u_{\mathrm{br}} + \nabla p_{\mathrm{br}}\right) = 0 \tag{26}$$

where φ is the porosity; k is the permeability coefficient.

Since the roadways and working face ventilation can be considered pipe flow, the Navier–Stokes equations can describe the flow of gases in them.

$$-\nabla \cdot \eta \left(\nabla \boldsymbol{u}_{\mathrm{ns}} + (\nabla \boldsymbol{u}_{\mathrm{ns}})^{\mathrm{T}} \right) + \rho \boldsymbol{u}_{\mathrm{ns}} \cdot \nabla \boldsymbol{u}_{\mathrm{ns}} + \nabla p_{\mathrm{ns}} = 0 \tag{27}$$

$$\nabla \cdot \boldsymbol{u}_{\mathrm{ns}} = 0 \tag{28}$$

where η is the coefficient of viscosity and $\boldsymbol{u}$ is the velocity vector.

Methane migration obeys the hydrodynamic dispersion law due to the stochastic nature of the movement of methane molecules in the fluid. The study of convection–diffusion equations in porous media allows for a better understanding of the methane migration laws in mining:

$$\theta_s \frac{\partial c}{\partial t} + \nabla \cdot (-\theta_s D_L \nabla c + uc) = S_c \tag{29}$$

where θ_s is the volume of fluid per unit area per unit time,S_c is a measure of the mass of a substance transferred by diffusion per unit area over a unit time, c is the dissolved concentration, and D_L is the permeability tensor.

Before methane diffusion reaches equilibrium, the diffusion equation for the concentration of substance c with time t and position r is:

$$\frac{\partial c}{\partial t} = D\left(\frac{\partial^2 c}{\partial r^2} + \frac{2}{r}\frac{\partial c}{\partial r} \right) \tag{30}$$

A combination of these equations can be used to study the methane migration pattern in the goaf medium more accurately.

4. The Migration Law of Leaked Methane Within the Gas Well Protective Coal Pillar Under the Influence of Mining

Coal pillars are affected by stress during coal mining, resulting in varying degrees of damage within the coal mass and forming fractures of different scales. This increases the permeability of the coal mass by several orders of magnitude. There are significant differences in the migration behavior of methane within fractured coal masses at different scales. This chapter studies the influence of different factors on the migration and permeability evolution of leaked methane in coal pillars under the influence of mining, based on mathematical models of methane migration in fractures of different scales and taking Shuangma Coal Mine as the background.

4.1. Establishment of Numerical Model

An abandoned gas well in the working face is near the center of the working face (Figure 8), and this study simulates the gas pressure of 15 MPa for the oil and gas well. COMSOL Multiphysics was used to study the methane migration law of the gas wells in the coal pillar. A large amount of research has been conducted in the early stage to validate this model, and it has been confirmed that the model established in this paper is effective in simulating the entire process of methane seepage in coal seams [10] In order to better simulate the onsite work, this simulation models the overall range of the coal pillar. Model assignment parameters refer to previous research. The specific parameters are shown in Table 2 [27,28].

The model considers the length of the working face at the site, the location of the abandoned gas wells, and the extent and dimension of the protective coal pillar. A free tetrahedral mesh was applied, and three kinds of rectangular protective coal pillars of 130 × 130 × 4 m, 150 × 150 × 4 m and 180 × 180 × 4 m were constructed. The gas well, with a radius of 0.15 m, is situated at the center of the coal pillar. The radius of the gas well is consistent with that onsite, as shown in Figure 9.

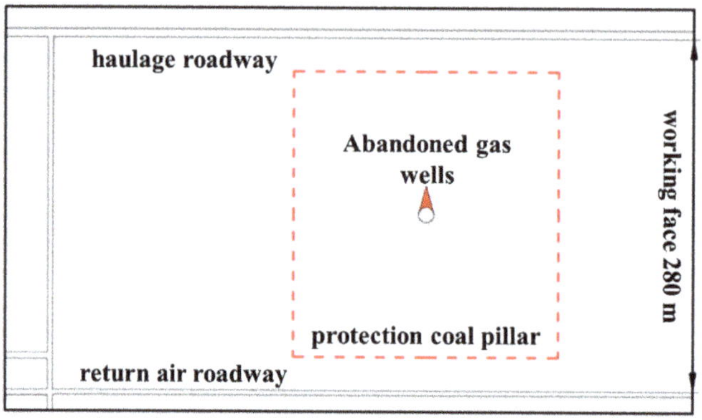

Figure 8. Location of the abandoned gas wells and range of protective coal pillars in Shuangma coal mine.

Table 2. COMSOL assignment parameters.

Parameter	Numerical Value	Parameter	Numerical Value
Initial fracture methane pressure, pfg0	0.1 [MPa]	Well pressure, pc	15 [MPa]
Langmuir volume parameters, VL	0.072 m^3/kg	Capillary weight of overburden, ρ_r	2550 kg/m^3
Langmuir pressure constant, P_L	3.034 MPa	Fracture stiffness, Kn	4.8 [GPa]
Coal Skeleton Density, ρ_c	1400 kg/m^3	Initial fracture porosity, φ_{f0}	0.021
Poisson's ratio of a coal, ν	0.3	Initial matrix porosity, φ_{m0}	0.045

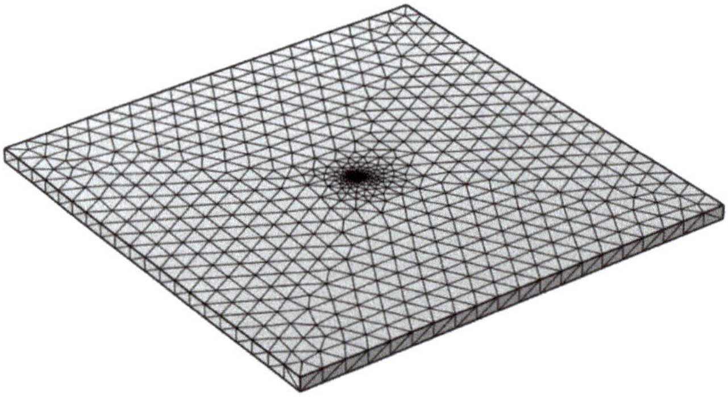

Figure 9. Model construction and meshing in COMSOL Multiphysics 6.2 (COMSOL Co., Ltd, Shanghai, China).

4.2. The Migration Law of Leaked Methane Within Protective Coal Pillars with Different Widths

When the protective coal pillar is 130 m wide, the diffusion time of leaked methane is within 0–80 days; during this period, methane has not yet diffused to the sides, resulting in higher pore pressure within the coal pillar and lower coal permeability. As the diffusion time reaches 100 days, the permeability of leaked methane in the pillar has reached equilibrium; between 100 and 120 days, external stress on both sides of the coal pillar increases, leading to a rise in pore pressure compared to levels before 80 days with permeability also increasing and reaching equilibrium. At this point, permeability on the sides of the pillar is higher than around the gas wells (Figure 10).

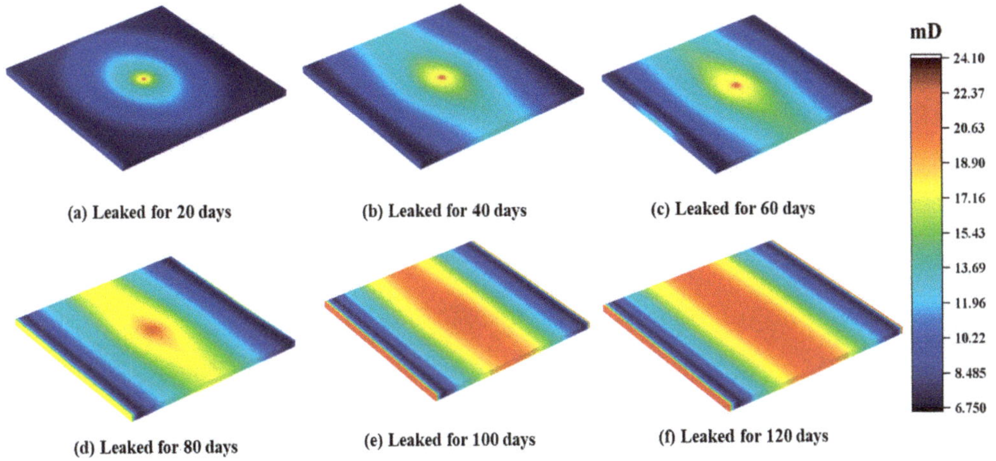

Figure 10. Spatiotemporal evolution cloud diagram of permeability of protective coal pillar with a width of 130 m.

The closer the protective coal pillar is to the oil/gas well, the greater the permeability change. However, in small- and medium-scale transition areas, the permeability change remains minimal as these areas maintain equilibrium for a long time. When the width of the pillar increases to 150 m and 180 m (Figures 11 and 12), the overall permeability evolution in the pillar is similar to that of the 130 m width. However, due to the wider width of the protective coal pillar and the wider elastic zone, it is more difficult for leaked methane to diffuse into mesoscale fractures. Therefore, when the width of the pillar is 150 m, the leaked methane reaches equilibrium within 160 days of diffusion, and when the width of the pillar is 180 m, the leaked methane reaches equilibrium within 180 days. As the width of the pillar increases, the permeability of leaked methane also increases. This is because the width of the protective coal pillar is different, and the external stress on both sides is different, resulting in a small difference in pore pressure and effective stress, leading to different permeability.

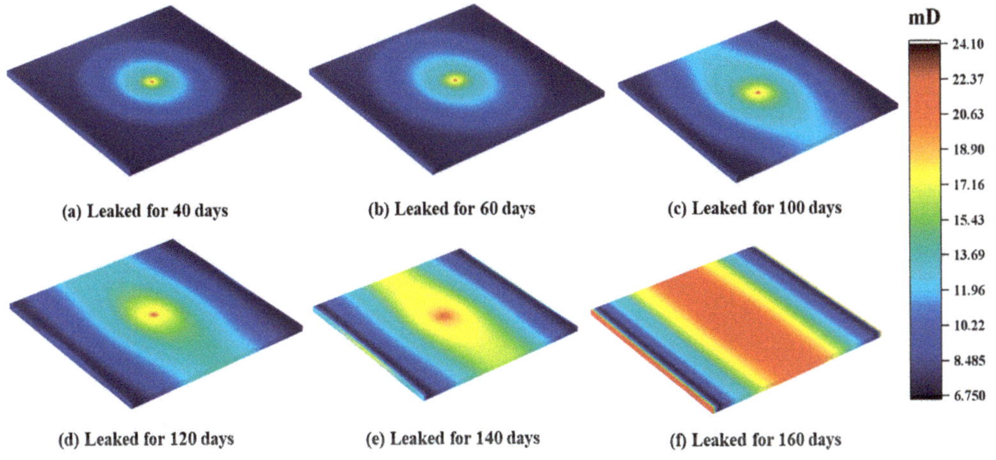

Figure 11. Spatiotemporal evolution cloud diagram of permeability of protective coal pillar with a width of 150 m.

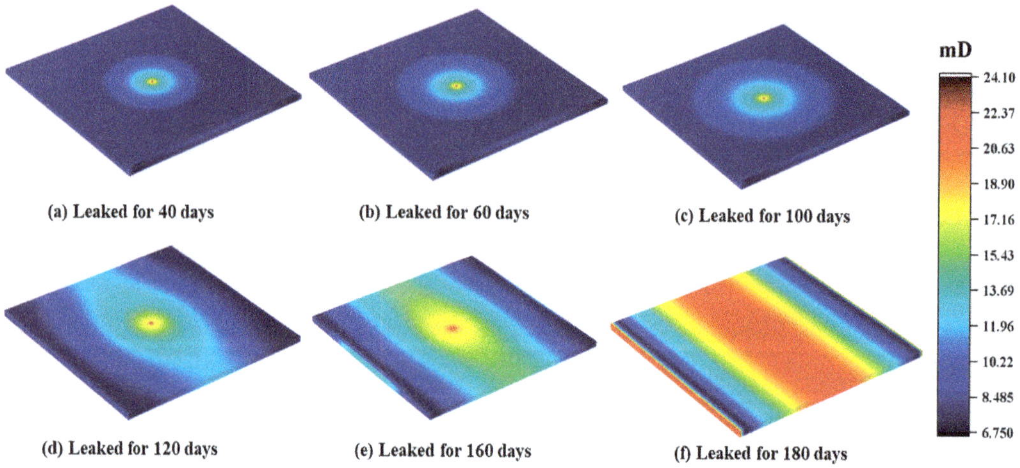

Figure 12. Spatiotemporal evolution cloud diagram of permeability of protective coal pillar with a width of 180 m.

The permeability evolution curve within the pillar exhibits a "W" shape (Figure 13). When the width of the protective coal pillar is 130 m, the permeability increases by 1.87 mD, 5.78 mD, and 10.52 mD at different time intervals during leaked methane from 0 to 80 days. When the methane leakage reaches 100 days, the permeability of the coal on both sides of the pillar slowly increases with the increase of leaked days, with the permeability increasing by 18.25 mD and 22.91 mD, respectively. This is because the leaked methane has reached equilibrium at this time, and the difference in pore pressure changes is small. With the increase of time, the difference between the permeability on both sides of the pillar and the permeability in the transition zone of small and medium scales also increases. At this point, a complete methane migration channel has been formed inside the pillar; At 120 days of leaked methane, the permeability on the left side of the pillar was 22.91 mD, and the permeability on the right side was 21.79 mD. The difference in peak permeability on both sides was due to the different external mining-induced stresses experienced. By studying the evolution law of permeability within the pillar, it can be concluded that under the influence of mining, the mesoscale fractures formed on both sides of the pillar due to the influence of mining-induced stress will lead to a sudden increase in the permeability of transition zone. The permeability is inversely proportional to the magnitude of mining-induced stress, which is mainly dominated by mining-induced stress.

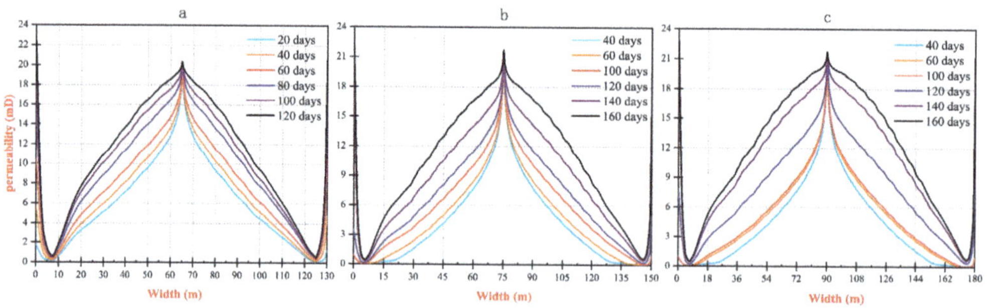

Figure 13. Line plots of the spatial and temporal evolution of permeability of protective coal pillars of different widths ((**a**) 130 m, (**b**) 150 m, and (**c**) 180 m).

From the transition zone to the surrounding area of the gas well, the evolution of permeability is mainly dominated by pore pressure. At this time, the leaked methane channel is a small-scale fracture, and the change in permeability is relatively small compared to the mesoscale. The maximum permeability around the gas well is mainly due to the high pore pressure. The speed at which coal is extracted can lead to different stress distributions within the coal seam and the surrounding strata. Faster exploitation rates can result in more rapid stress changes, potentially leading to earlier and more extensive fracturing. This increased fracturing can create additional pathways for methane to migrate, thus affecting the overall permeability of the coal mass. This can accelerate the migration of methane from the coal seam into the mining area, posing a greater risk of methane accumulation and potential gas outbreaks. In the initial stages of leakage, pore pressure around the gas wells is high, and the effective stress is low, resulting in maximum permeability. However, as leaked days increase, permeability on both sides of the pillar gradually rises. Due to continuous pressurization of the gas wells, leaked methane diffuses and transports through fractures of various scales. Ultimately, permeability on both sides of the pillar exceeds that around the gas wells. The increase in permeability near the working faces on both sides of the pillar indicates that the stress on both sides is smaller, and larger mesoscale fractures are generated, making it easier for leaked methane to diffuse. In addition, the permeability inside the pillar increases with the increase of the pillar width. After balancing, the peak permeability on both sides of the pillar is smaller when the pillar width is 130 m compared to 150 m and 180 m. This is the result of different mining-induced stresses. When the pillar widths are 150 m and 180 m, the permeability difference on both sides of the pillar is only 1 mD, indicating that these two widths achieve a similar effect.

4.3. The Migration Law of Leaked Methane from Protective Coal Pillars with Different Heights

Figures 14–16 show that as mining height increases, permeability on both sides of the protective coal pillars and around the gas wells decreases. This is because increased mining height results in higher mining-induced stress and lower effective stress, leading to reduced permeability. When the leakage days are 120 days, the permeability distribution in the 1 m-high pillar becomes stable and no longer changes. However, the permeability within pillar heights of 2 m and 4 m is still evolving, and the permeability on both sides of the pillar changes accordingly. This is mainly due to the high stress applied, which leads to slow diffusion of leaked methane.

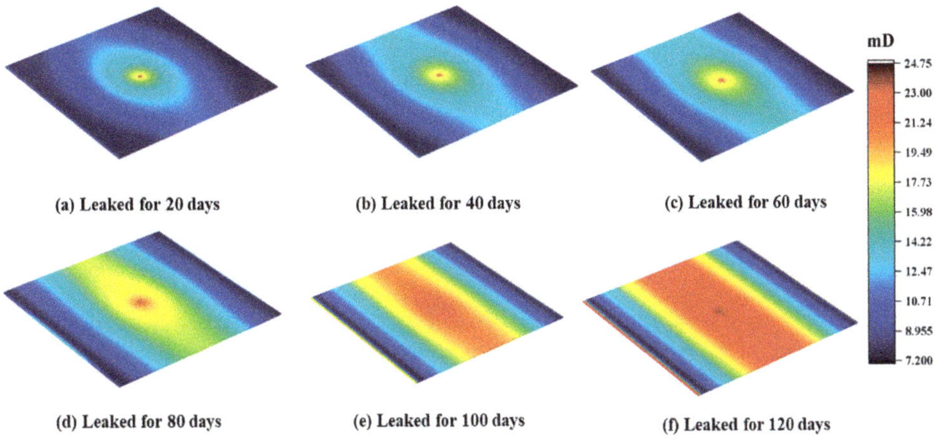

Figure 14. Spatiotemporal evolution cloud diagram of permeability at 1 m height of protective coal pillar.

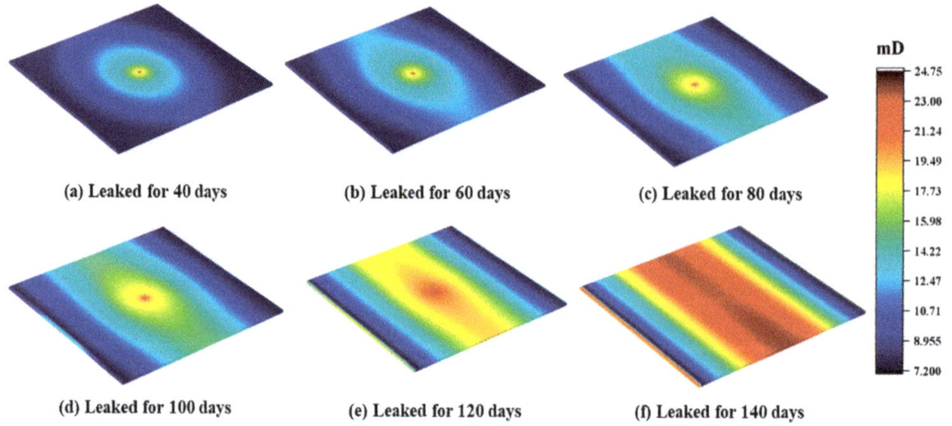

Figure 15. Spatiotemporal evolution cloud diagram of permeability at 2 m height of the protective coal pillar.

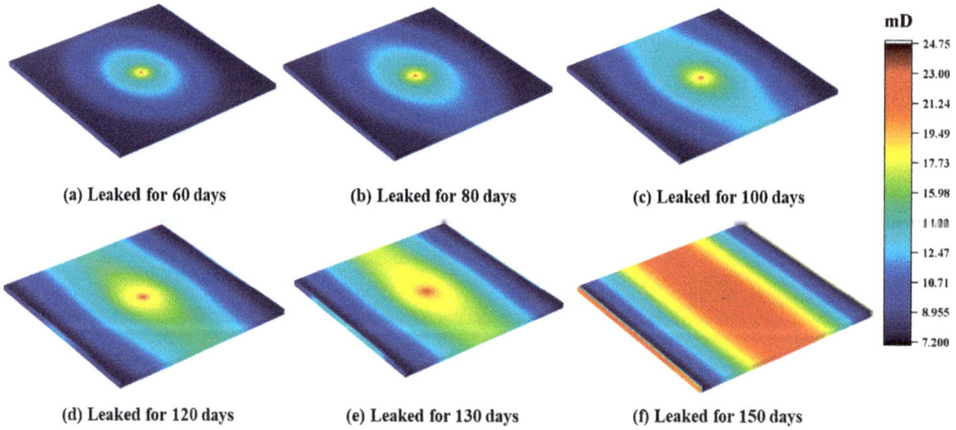

Figure 16. Spatiotemporal evolution cloud diagram of permeability at 4 m height of the protective coal pillar.

The overall permeability within the protective coal pillar at varying heights shows a trend of decreasing, increasing, then decreasing, and increasing again (Figure 17). This is because, under the influence of mining, both sides of the pillar are subjected to significant stress, leading to damage and plastic deformation of the coal mass, resulting in mesoscale fracture. When the supporting pressure decreases, the area between the plastic and elastic zones of the coal pillar is affected by stress, and fractures begin to close, resulting in lower permeability. After fully entering the elastic zone, the permeability is mainly affected by the stress around the gas well, resulting in high pore pressure and a further increase in permeability.

For instance, when the coal pillar height is 1 m and leaked methane reaches 120 days, a complete methane migration channel forms within the pillar. During this time, permeability on the left side drops sharply from 20.03 mD to 0.42 mD, which is a mesoscale fracture development stage. Afterward, the permeability around the gas well slowly increased, and the transition zone between small and medium scales slowly decreased, finally rising sharply to 22.02 mD. The different permeability on both sides of the pillar is caused by different peak stress.

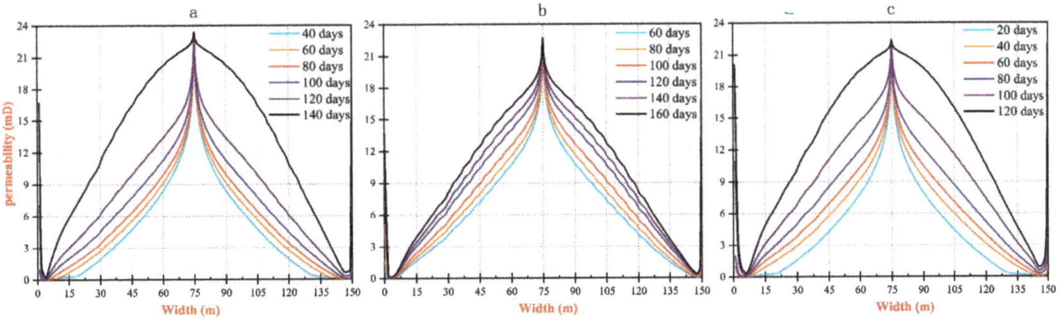

Figure 17. Line plots of the spatial and temporal evolution of permeability of protective coal pillars at different heights ((**a**) 1 m, (**b**) 2 m, and (**c**) 4 m).

Within protective coal pillars of varying heights, permeability trends are primarily influenced by mining-induced stress and the pressure from gas wells, while the magnitude of permeability values is mainly driven by mining-induced stress. As the pillar's height increases, the mining height and mining-induced stress also increase, expanding the damaged area. At this stage, the combined effects of mining-induced stress and gas well pressure lead to a decrease in permeability. When small-scale fracture transitions into medium-scale fracture, permeability values increase rapidly.

5. Discussion

This study is based on the background of the remaining oil/gas well-protective pillars in the Shuangma Coal Mine. The evolution characteristics and spatial distribution of mining-induced stress fields in pillars of different widths and heights were studied through numerical simulation. It was found that the stress inside the protective coal pillars of different sizes was asymmetrically distributed in a 'saddle' shape. As the pillar's height increased, the peak vertical stress of the roof and floor in front of the rib also increased. The higher the height, the greater the range of the caving zone and the height of fractured zone development, laying the foundation for subsequent research on the diffusion law of methane in the protective coal pillar. Similar patterns were observed by Manoj Khanal [29] and Manchao He [30] in their analyses of deep mining dynamics, mining-induced overburden deformation, and associated stress redistribution (Figure 18), validating the accuracy of the conclusions in this study.

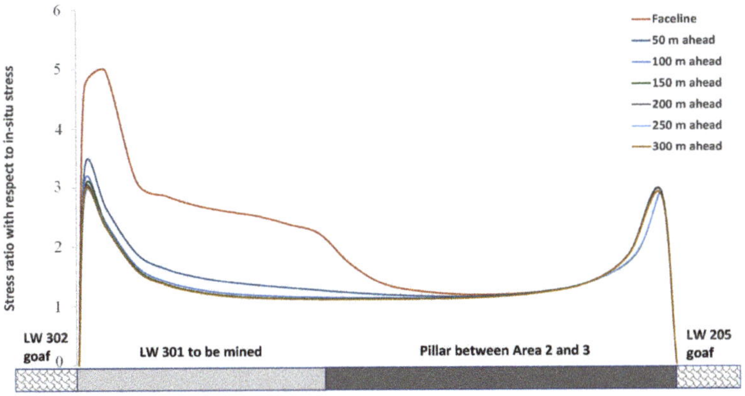

Figure 18. Vertical stress development ahead of longwall 301 [29].

Previous studies on the design of oil/gas wells protective pillar dimensions have primarily focused on intact wells with no leakage of toxic or harmful gases [3,11,12]. This study explores the evolution of stress and methane migration in protective coal pillars for oil/gas wells; from the perspective of leaked gas diffusion, it provides an idea for remaining coal pillars in under-protected oil/gas wells and abandoned oil/gas wells.

While the research provides valuable insights into the evolution of mining-induced stress and the migration of leaked methane in protective coal pillars for gas wells, it is essential to acknowledge and discuss the limitations of the study. The paper has modeled methane as an ideal gas, which simplifies the mathematical representation of fluid flow within the coal seam. This assumption is valid for many practical purposes, particularly when the pressure and temperature variations are not extreme. However, under conditions of high pressure and low temperature, deviations from ideal gas behavior may occur, which could affect the accuracy of flow predictions. Future studies could incorporate real gas behavior into the model to account for such deviations, especially in regions where the pressure and temperature conditions are known to be outside the ideal gas range.

In this study, simulations are based on specific geological and mining conditions at the Shuangma Coal Mine. The generalization of the findings to other coal mines with different geological structures and mining practices requires caution. Each mining site may have unique characteristics that could influence the stress distribution and methane migration pathways. Therefore, site-specific adjustments and validations may be necessary when applying findings to other locations.

6. Conclusions

(1) The stress of oil/gas well-protective coal pillars of different sizes is asymmetrically distributed in a saddle shape; vertical stress peaks on the initial mining side are higher than those on the later mining side, and the elastic region between these peaks expands with increasing pillar width. As the pillar's height increases, the bearing pressure's peak gradually shifts inward within the pillar, and the range of influence gradually increases. The stress in front of the pillar is more concentrated, and the vertical stress peak of the roof and floor in front of the rib is also increasing. The higher the height, the larger the collapse range of the overburden and the height of the fracture zone development, and the greater the damage to the stability of the oil/gas wells arranged in it.

(2) This study establishes the relationship between three different fracture widths and permeability, forming the methane migration equation for multi-scale fracture. Based on the matchstick theory model, a permeability model suitable for small fractures in the elastic stage was derived. The degree of damage was used to describe the development of fracture, and a permeability model suitable for mesoscale fractures was derived. For large-scale fractures in the goaf, the Brinkman equation was used to describe the rapid fluid flow in the pore medium.

(3) The overall evolution curve of methane permeability leakage in protective coal pillars with different widths and heights shows a "W" shape. In the initial stage of leaked methane migration, the permeability around the oil/gas well is the highest. As the number of leaked days increases, the permeability on both sides of the pillar gradually increases, and ultimately, the permeability on both sides of the pillar is greater than that around the oil/gas well. In addition, the permeability on the left side is greater than on the right side. Within the same period of time, the wider the pillar, the slower the diffusion of leaked methane and the lower the permeability. The permeability around the oil/gas well also decreases, and the pressure of leaked methane gradually

decreases. The slower the diffusion of methane, the higher the mining height, the greater the mining stress, and the lower the permeability.

(4) The simulation results show that when the protective coal pillar widths in the Shuangma Mine are 150 m and 180 m, the stress peak values and the stress smoothing areas are found to be similar, with a stress peak difference of 0.62 MPa and a stress smoothing area difference of 20 m. The permeability difference on both sides of the pillars is only 1 mD, indicating a similar effect for both coal pillar widths. Therefore, the site may consider reducing the coal pillar width to 150 m to increase coal production.

(5) The conclusions drawn in the paper are ideal and based on the geological conditions of the Shuangma Mine. Applying these conclusions to other sites will require site-specific adjustments and validation.

Author Contributions: Conceptualization: J.S., S.L. and Y.H.; methodology: J.S., S.L. and Y.H.; formal analysis and investigation: Z.M., W.H. and J.S.; writing—original draft preparation: J.S. and Y.H.; writing—review and editing: S.L., J.S., X.L. and S.X. funding acquisition: S.L.; supervision: J.S., S.L and C.L. All authors have read and agreed to the published version of the manuscript.

Funding: This work is supported by the National Natural Science Foundation of China (No. 52174139), the Postgraduate Research and Practice Innovation Program of Jiangsu Province (KYCX24_2880), and the Graduate Innovation Program of China University of Mining and Technology (2024WLJCRCZL019).

Data Availability Statement: The original contributions presented in this study are included in the article. Further inquiries can be directed to the corresponding author.

Conflicts of Interest: Author Yisong Hao was employed by Shaqu No. 2 Mine Huajin Coking Corporation Limited. Authors Zhi Ma and Weisheng He were employed by Ningxia Coal Industry Limited Liability Company Shuangma Coal Mine, China Energy Group. Author Xu Liang was employed by Shaanxi Xiaobaodang Mining Corporation Limited. Authors Shaoyou Xu and Changheng Luo were employed by No. 2 Gas Production Plant of Sinopec North China Oil and Gas Company. The remaining authors declare that the research was conducted in the absence of any commercial or financial relationships that could be construed as a potential conflict of interest.

References

1. Hao, D.Y.; Zhang, T.; Yang, Y.B.; Ren, J.M. Study on the disturbance characteristics and control strategies of coordinated exploitation of superimposed resources of coal and oil-type gas in Ordos Basin. *Energy Sci. Eng.* **2023**, *11*, 3964–3978. [CrossRef]
2. Liu, J.; Yang, T.; Wang, L.; Chen, X. Research progress in coal and gas co-mining modes in China. *Energy Sci. Eng.* **2020**, *8*, 3365–3376. [CrossRef]
3. Liang, S.; Han, H.; Elsworth, D.; Fu, X.; Yao, Q.; Kang, J.; Li, X.; Li, X. Evolution of Production and Transport Characteristics of Steeply-Dipping Ultra-Thick Coalbed Methane Reservoirs. *Energies* **2020**, *13*, 5081. [CrossRef]
4. Gianoutsos, N.J.; Haase, K.B.; Birdwell, J.E. Geologic sources and well integrity impact methane emissions from orphaned and abandoned oil and gas wells. *Sci. Total Environ.* **2024**, *912*, 169584. [CrossRef]
5. Wang, X.; Ma, H.; Qi, X.; Gao, K.; Li, S. Study on the Distribution Law of Coal Seam Gas and Hydrogen Sulfide Affected by Abandoned Oil Wells. *Energies* **2022**, *15*, 3373. [CrossRef]
6. Zhang, P.; Dougherty, H.; Su, D.; Trackemas, J.; Tulu, B. Influence of longwall mining on the stability of gas wells in chain pillars. *Int. J. Min. Sci. Technol.* **2020**, *30*, 3–9. [CrossRef]
7. Wang, X.; Zhang, Q.; Yuan, L. A coupled thermal-force-chemical-displacement multi-field model for underground coal gasification based on controlled retraction injection point technology and its thermal analysis. *Energy* **2024**, *293*, 130614. [CrossRef]
8. Raimi, D.; Krupnick, A.J.; Shah, J.; Thompson, A. Decommissioning Orphaned and Abandoned Oil and Gas Wells: New Estimates and Cost Drivers. *Environ. Sci. Technol.* **2021**, *55*, 10224–10230. [CrossRef] [PubMed]
9. Liang, S.; Elsworth, D.; Li, X.; Fu, X.; Sun, B.; Yao, Q. Key strata characteristics controlling the integrity of deep wells in longwall mining areas. *Int. J. Coal Geol.* **2017**, *172*, 31–42. [CrossRef]
10. Han, H.; Liang, S.; Liang, Y.; Fu, X.; Kang, J.; Yu, L.; Tang, C. The Role of Coal Mechanical Characteristics on Reservoir Permeability Evolution and Its Effects on CO2 Sequestration and Enhanced Coalbed Methane Recovery. *Geofluids* **2020**, *2020*, 8842309. [CrossRef]

11. Ajayi, K.M.; Schatzel, S.J. Transport model for shale gas well leakage through the surrounding fractured zones of a longwall mine. *Int. J. Min. Sci. Technol.* **2020**, *30*, 635–641. [CrossRef] [PubMed]
12. Su, D.W.H. Effects of longwall-induced stress and deformation on the stability and mechanical integrity of shale gas wells drilled through a longwall abutment pillar. *Int. J. Min. Sci. Technol.* **2017**, *27*, 115–120. [CrossRef]
13. Peng, S.S.; Cheng, J.; Du, F.; Xu, Y. Underground ground control monitoring and interpretation, and numerical modeling, and shield capacity design. *Int. J. Min. Sci. Technol.* **2019**, *29*, 79–85. [CrossRef]
14. Xu, Y.; Liang, S.; Liang, X.; Yang, B.; Shi, Z.; Wu, C.; Shen, J.; Yang, M.; Ma, Y.; Xu, P. Spatio-Temporal Evolution of Loading and Deformation of Surface Gas Pipelines for High-Intensity Coalbed Mining and Its Integrity Prediction Methodology. *Processes* **2024**, *12*, 213. [CrossRef]
15. Liang, S.; Elsworth, D.; Fu, X.; Li, X.; Yao, Q. Influence of Stratigraphic Conditions on the Deformation Characteristics of Oil/Gas Wells Piercing Longwall Pillars and Mining Optimization. *Energies* **2017**, *10*, 775. [CrossRef]
16. Cui, X.J.; Bustin, R.M. Volumetric strain associated with methane desorption and its impact on coalbed gas production from deep coal seams. *Aapg Bull.* **2005**, *89*, 1181–1202. [CrossRef]
17. Cui, X.; Bustin, R.M.; Chikatamarla, L. Adsorption-induced coal swelling and stress: Implications for methane production and acid gas sequestration into coal seams. *J. Geophys. Res. Solid Earth* **2007**, *112*, 1–16. [CrossRef]
18. Wang, J.G.; Liu, J.; Kabir, A. Combined effects of directional compaction, non-Darcy flow and anisotropic swelling on coal seam gas extraction. *Int. J. Coal Geol.* **2013**, *109*, 1–14. [CrossRef]
19. Rice, A.K.; McCray, J.E.; Singha, K. Methane Leakage from Hydrocarbon Wellbores into Overlying Groundwater: Numerical Investigation of the Multiphase Flow Processes Governing Migration. *Water Resour. Res.* **2018**, *54*, 2959–2975. [CrossRef]
20. Schout, G.; Griffioen, J.; Hassanizadeh, S.M.; de Lichtbuer, G.C.; Hartog, N. Occurrence and fate of methane leakage from cut and buried abandoned gas wells in the Netherlands. *Sci. Total Environ.* **2019**, *659*, 773–782. [CrossRef]
21. Skrzypkowski, K.; Zagórski, K.; Zagórska, A. Determination of the Extent of the Rock Destruction Zones around a Gasification Channel on the Basis of Strength Tests of Sandstone and Claystone Samples Heated at High Temperatures up to 1200 °C and Exposed to Water. *Energies* **2021**, *14*, 6464. [CrossRef]
22. Huo, B.; Jing, X.; He, A.; Fan, C. Hydraulic-Mechanical Coupling Model with Dual-Porosity Dual-Permeability for Anisotropy Coal Seams and Its Application in Mine Gas Extraction. *Adv. Civ. Eng.* **2019**, *2019*, 4534835. [CrossRef]
23. Ren, C.; Li, B.; Xu, J.; Wang, Z.; Li, J.; Zhang, Y.; Yu, J. Coal permeability evolution during damage process under different mining layouts. *Energy Sources Part A Recovery Util. Environ. Eff.* **2020**, 1–16. [CrossRef]
24. Li, Y.; Wang, Z.; Tang, S.; Elsworth, D. Re-evaluating adsorbed and free methane content in coal and its ad- and desorption processes analysis. *Chem. Eng. J.* **2022**, *428*, 131946. [CrossRef]
25. Gao, Q.; Liu, J.; Huang, Y.; Li, W.; Shi, R.; Leong, Y.; Elsworth, D. A critical review of coal permeability models. *Fuel* **2022**, *326*, 125124. [CrossRef]
26. Moortgat, J.; Schwartz, F.W.; Darrah, T.H. Numerical Modeling of Methane Leakage from a Faulty Natural Gas Well into Fractured Tight Formations. *Groundwater* **2018**, *56*, 163–175. [CrossRef] [PubMed]
27. Li, S.; Fan, C.; Han, J.; Luo, M.; Yang, Z.; Bi, H. A fully coupled thermal-hydraulic-mechanical model with two-phase flow for coalbed methane extraction. *J. Nat. Gas. Sci. Eng.* **2016**, *33*, 324–336. [CrossRef]
28. Zhu, W.C.; Liu, J.; Sheng, J.C.; Elsworth, D. Analysis of coupled gas flow and deformation process with desorption and Klinkenberg effects in coal seams. *Int. J. Rock. Mech. Min.* **2007**, *44*, 971–980. [CrossRef]
29. Khanal, M.; Qu, Q.; Zhu, Y.; Xie, J.; Zhu, W.; Hou, T.; Song, S. Characterization of Overburden Deformation and Subsidence Behavior in a Kilometer Deep Longwall Mine. *Minerals* **2022**, *12*, 543. [CrossRef]
30. He, M.; Wang, Q. Rock dynamics in deep mining. *Int. J. Min. Sci. Technol.* **2023**, *33*, 1065–1082. [CrossRef]

processes

MDPI

Article

Experimental Study on Spectral Response Characteristics and Mechanical State of Coal Based on Artificial Acoustic Signals

Jianguo Zhang [1,2], Wenlong Fu [1], Weilong Cui [3], Ji Ma [1,2,*] and Feng Du [3,4]

[1] State Key Laboratory of Coking Coal Resources Green Exploitation, China PingMei ShenMa Group, Pingdingshan 467000, China; zhangjg_z@126.com (J.Z.); fwlfuwenlong126@163.com (W.F.)
[2] School of Safety Science and Engineering, Henan Polytechnic University, Jiaozuo 454003, China
[3] School of Emergency Management and Safety Engineering, China University of Mining & Technology, Beijing 100083, China; cwlcuiweilong@163.com (W.C.); fdu@cumtb.edu.cn (F.D.)
[4] Key Laboratory of Xinjiang Coal Resources Green Mining, Ministry of Education, Xinjiang Institute of Engineering, Urumqi 830023, China
* Correspondence: showsma@hpu.edu.cn

Abstract: With the increase in coal mining depth, the stress and strain state of coal and rock mass affects the formation of dangerous zones of dynamic phenomena. In order to study the relationship between the frequency spectrum characteristics of artificial acoustic signals and the stress state of coal and gas pressure, a test device and system that can generate acoustic signals by mechanical vibration excitation are developed by using the design idea of the unit module. Firstly, the basic mechanical parameters of coal under uniaxial compression are analyzed. On this basis, we use the test device to study the qualitative and quantitative relationships between the relative stress coefficient K value of the coal body and the axial loading stress, whether it contains gas, and the mechanical vibration force. The test results show that when the gas-containing coal and the gas-free coal are subjected to the same external mechanical vibration knocking force to stimulate the artificial acoustic signal test, the relative stress coefficient K value increases first and then decreases with the increase in axial loading stress. The relationship between the relative stress coefficient K and the axial loading stress σ can be expressed in the form of exponential function $K = e^{-\frac{C}{\sigma}}$. When the axial loading stress and the external mechanical vibration force are both fixed values, the relative stress coefficient K value of the coal body with gas is smaller than that without gas. When the axial loading stress and gas-bearing pressure of the coal body are both fixed values, the relative stress coefficient K value decreases with the increase in the impact force of the external mechanical vibration. This experimental study can provide a reference for the identification and prediction of dynamic disasters based on artificial acoustic signals.

Keywords: mechanical vibration; artificial acoustic signal; coal stress state; relative stress coefficient

1. Introduction

With the development of the mine to the deep, the characteristics of strong adsorption and low permeability are more obvious. The coupling effect of high stress, high gas pressure, and low-permeability coal under mining disturbance is more obvious, which leads to an increase in coal and gas outburst accidents [1–4]. External vibration, such as coal cutter operation, roadheader excavation, artificial blasting construction, etc., induces frequent coal and gas outburst accidents. The dynamic change in ground stress and the characteristics of gas migration near the working face in the process of coal falling, blasting, and tunneling of shearers and coal cutters are very important for predicting coal and gas outburst and timely warning of the danger of dynamic phenomena.

In recent years, the detection and early warning technology of geological disasters, such as microseism, electromagnetic radiation, and acoustic emission, has developed rapidly [5–7]. Years of field application show that there is no single-value correspondence between the acoustic emission activity intensity of the coal body and the stress and strain state of the space near the working face, which leads to the low reliability of the seismic acoustic prediction method [8]. In the 1980s, Russia began to study the artificial acoustic prediction method for coal seam outburst risk. The artificial acoustic signal prediction method can realize the continuous online monitoring and early warning of the danger of various dynamic phenomena, geological damage, and stress change dynamics near the working face [9]. For the first time, Professor M.C. Анциферов [10] proposed a prediction method for outburst risk based on the number of acoustic emission pulses and their increase to evaluate the stress and strain state of coal seams. C.B. Мирер [11] was the first to study the artificial acoustic prediction method for outburst risk. He developed a method for predicting outburst hazard based on the analysis of amplitude-frequency characteristics of acoustic signals formed in the formation during the operation of process equipment. This lays the foundation for the artificial acoustic method to become the standard method of prominent risk prediction. Колчин et al. [12–15] solved a series of mine production problems, including the prediction of gas dynamic phenomena, the prediction of geological damage in front of the moving face, and the monitoring of the effect of anti-outburst measures. The developed method has been successfully applied in the prediction of mine dynamic phenomena. In 2015, the Russian company МНТЛ РИВАС developed the intelligent system of mine dynamic phenomenon prediction and rock mass status monitoring [16]. In China, in order to realize continuous automatic prediction of the stress and strain state of rock masses near working faces, Wei Fengqing et al. [17] conducted relevant research: Based on the roof deformation delay theory formed by the risk of dynamic phenomena and the elastic resonance spectrum characteristics of artificial acoustic signals, they developed CAKCM, an automated system for rock mass acoustic monitoring and prediction of dynamic phenomena, according to the results of many years of research and testing in Russia and the latest requirements of safety regulations. It has been tested and applied in a production mine.

The traditional contact discontinuous borehole index prediction method cannot meet the needs of complex geological conditions, complex dynamic phenomena, or high-yield, high-efficiency, and intelligent mines in many mining areas. The artificial acoustic signal prediction method can realize continuous, automatic, and quantitative prediction and early warning of the coal state and dynamic phenomena in the working face and in front of the working face. It can establish the functional relationship between the prediction index and the stress state of coal rock mass, and improve the reliability of the prediction results. However, the research on the spectral characteristics of artificial acoustic signals in China is still not perfect. How to use the amplitude-frequency characteristics of artificial acoustic signals to directly reflect the deformation and failure of coal and the gas content in coal has not been solved. The quantitative relationship between the amplitude-frequency characteristics of artificial acoustic signals and mechanical parameters such as the coal stress state and gas pressure is still not clear enough. Therefore, we need to carry out basic theoretical research based on artificial acoustic signals to reflect the stress state of coal and rock and the prediction of gas outburst risk. In view of the previous research experience, we have independently designed an experimental device and experimental method for mechanical vibration excitation of artificial acoustic signals. We have carried out qualitative and quantitative research on the spectral response characteristics of artificial acoustic signals generated by mechanical vibration excitation of coal and the stress and gas content of coal. We hope to provide a new direction and technical means for the identification of the front stress anomaly area and the risk identification and prediction of dynamic disasters such as coal outbursts in the mining process.

2. Artificial Acoustic Signal Test of Mechanical Vibration Excitation

2.1. Design Principle of Test Device

"Artificial acoustic signal" refers to the wide-band artificial acoustic vibration or artificial sound source excited by running mining equipment (such as shearers, cutting mechanisms of coal ploughs, drill bits, etc.) in the rock mass near the working face (by hammering the side wall of the roadway) to detect the rock mass. According to the amplitude-frequency characteristics of artificial acoustic signals, the stress state of rock masses near the working face can be analyzed. The risk of coal and gas outbursts in the mining face can be predicted, and the state of rock masses can be monitored (such as the geological damage in front of the working face and the stress state near the working face) [18]. The artificial acoustic signals of external vibration feedback carry plenty of important information, such as details about the rock stress, mechanical properties, structural changes, deformation and failure characteristics, and gas pressure. The artificially excited acoustic signal in coal rock is the superposition of inherent elastic vibration (hereinafter referred to as resonance). Resonance occurs in each rock layer from the free surface to the weakened contact surface [19]. Therefore, the acoustic signal excited in the layered coal-rock mass is the superposition of its own (resonant) elastic vibration in each individual layer. In the process of coal mining, the cutting mechanism of the running mining equipment is used as the source of acoustic vibration [20]. Taking the working face as an example, the acoustic vibration excites near the working face, with propagation to the deep coal body and the sound wave passing through the medium with variable parameters. The change in stress in the medium corresponds to the parameters of the stress diagram. Its length corresponds to the strike length of the increased mine pressure area. The propagation process of acoustic vibration is shown in Figure 1.

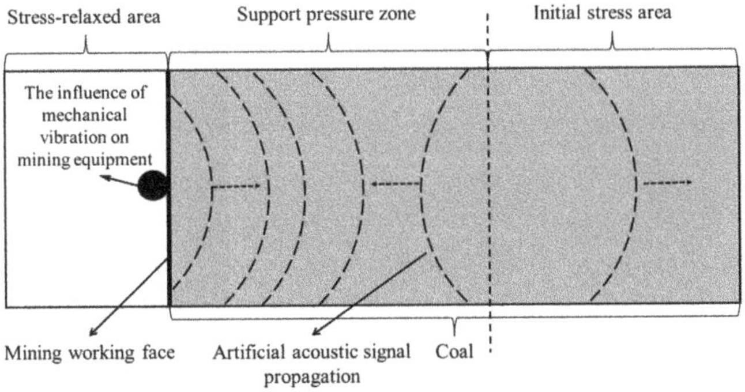

Figure 1. Propagation process of acoustic vibration in a coal seam.

Affected by external forces, resonance occurs in the coal seam. There is the following relationship between the thickness of the rock layer (h) that generates resonance in the coal seam and the resonance frequency (f_p):

$$f_P = \frac{v}{h} \tag{1}$$

where "v" is the phase velocity of the shear wave, m/s. The value of this speed is based on experience. The variable "h" is the thickness of the rock layer that generates resonance, m. The variable "h" can also be understood as the distance from the oscillation excitation point to the weakened mechanical (interlayer) contact surface in the rock mass.

According to the formula, the frequency of the natural resonance recorded during the monitoring process can determine the location of these weakened mechanical (interlayer) contact surfaces at the pulse action site of the rock mass relative to the coal seam. By analyzing the resonance frequency and amplitude, a conclusion can be drawn about the change in the stress and strain state of the rock mass in the section of the vibration excitation point. Considering that harmonic signals can be obtained only when the oscillating system is acted upon, an experimental device and a measurement method are designed for rock masses affected by mining operations on a small scale in the laboratory. In this method, the pulse effect of a small hammer on coal is used as the source of an artificial acoustic signal, and a seismometer is used to record it. Based on the elastic vibration pulse excitation in coal and rock masses, the seismometer converts the elastic vibration into an acoustic signal. On the spectrum of the recorded acoustic signal is the superposition of the elastic vibration, and the resonance frequency is distinguished by high amplitude. The obtained artificial acoustic signal spectrum reflects the stress–strain state of coal under loading stress and external mechanical vibration force.

2.2. Introduction of Experimental Device

Considering the complex disaster-pregnant environment of deep coal seams and the influence of mechanical vibration force in the process of coal seam mining, and based on the shortcomings of existing instruments and equipment, the design idea of the unitized module is adopted to develop a test device and system that can realize mechanical vibration force to stimulate the gas-bearing coal body to produce acoustic signals. The test device system includes inflatable and exhaust units, a mechanical vibration unit, an axial compression unit, a mechanical vibration force monitoring unit, and an artificial acoustic signal monitoring and collecting unit. Each unit module cooperates with the others to realize the coupling effect of gas pressure, a high-stress environment, and mechanical vibration force. The overall design is shown in Figure 2.

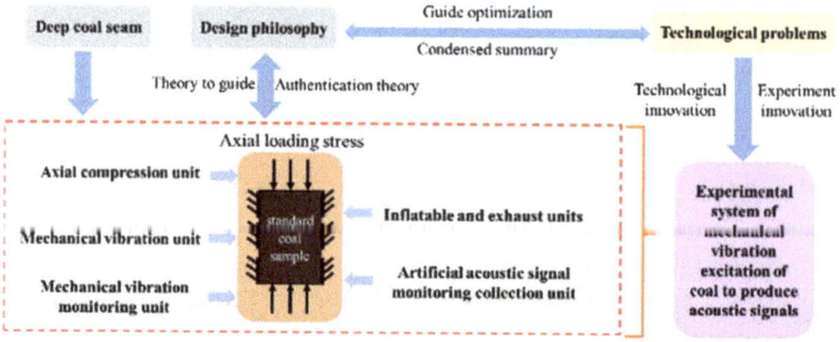

Figure 2. Test device and system design.

Considering that the harmonic signal can be obtained only when acting on the oscillating system, for the coal body affected by the mining operation, the pulse effect on the coal body is used as the source of the artificial acoustic signal, and the seismic geophone is used to record it. Based on the pulse excitation of elastic vibration in coal and rock masses, the seismic geophone converts the elastic vibration into an acoustic signal. Under laboratory conditions, using a portable artificial acoustic probe to collect the artificial acoustic signal excited by the mechanical vibration force of the coal body to the portable seismic geophone can be considered, and the spectrum characteristics and information parameters of the artificial acoustic signal can be analyzed. The overall structure of the experimental device for mechanical vibration excitation of artificial acoustic signals is shown in Figure 3.

Inflatable and exhaust units can realize coal with different gas pressures. Through the mechanical vibration unit and the mechanical vibration force monitoring unit, different mechanical vibration knocking forces can be realized. Axial compression with different degrees of pressure can be achieved by applying axial compression elements. The artificial acoustic signal monitoring and collection unit monitors and collects the artificial acoustic signal generated by the coal sample in the pendulum mechanical vibration excitation chamber. With the cooperation of each unit, the gas content, different stress and different mechanical vibration and knocking force can be set. The artificial acoustic signals generated by coal excitation under mechanical vibration are collected to study the dynamic response characteristics of coal under mechanical vibration. By analyzing and processing the data, the quantitative relationship between the amplitude-frequency characteristics of artificial acoustic signals and mechanical parameters such as coal stress state and gas pressure is established. The specific experimental device diagram is shown in Figure 4.

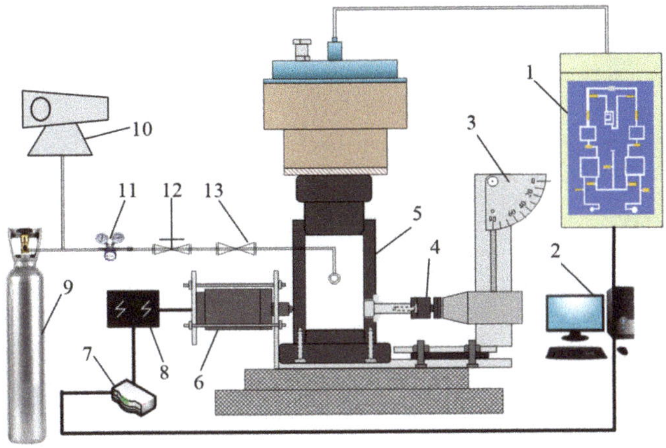

1-mechanical control system **2**-computer **3**-swinging hammer
4-pressure transducer **5**-processed cavity **6**-portable artificial acoustic probe
7-seismic detector **8**-power supply **9**-methane gas cylinder **10**-vacuum pump
11-pressure gauge **12**-reducing valve **13**-stop valve

Figure 3. Experimental device diagram.

Figure 4. Experimental device diagram.

2.3. Experimental Scheme

The coal samples used in the test were taken from No.8 Mine of Pingdingshan. In order to reduce the possible influence of the primary cracks of coal rock samples on the test results, the wave velocity and density of the specimens were measured before the test. The specimens with similar wave velocity and density were selected for the test. The coal sample selected in the test was a cylinder with a diameter × length of 50 mm × 100 mm. The average mass of the coal sample was 249.95 g, and the average density of the coal sample was 1.147 g/m^3. After measurement, the wave velocity of the selected coal sample was between 1930 m/s and 2013 m/s, and the texture was relatively uniform. The sample used in this test is shown in Figure 5.

Figure 5. Part of the coal samples required for the experiment.

The fixed pointer points to the dial at 15°, 20°, 25°, 30°, 35°, and 40°—six scales. Let the hammer strike the coal body freely, and ensure that the hammer strikes each time in each group of experiments are almost the same. The bottom swing of the small hammer strikes the middle and lower part of the coal body, and the instantaneous force of the knock can be recorded by the pressure sensor of the mechanical vibration force monitoring unit. Through the mechanical control system, the axial pressure controller is set to the axial pressure, loading rate, sample size, and other related parameters, so as to control the RMT-301 rock direct shear/triaxial compression composite testing machine to apply axial pressure to the indenter of the machining chamber. When each scale (percussion force) is tested, the control axial pressure is 10 kN, 15 kN, 20 kN, 25 kN, 30 kN, 35 kN, and 40 kN. When each axial pressure is set, the axial pressure is controlled to be stable, knocking once every 10 s, knocking about 6–10 times.

Firstly, the axial loading rate is set to 0.01 mm/s, and the mechanical parameters of the uniaxial compression of the coal are analyzed. Next, the experimental study of artificial acoustic signals stimulated by mechanical vibration is carried out under two conditions of no gas and 0.3 MPa gas pressure. Finally, combined with uniaxial compression parameters of the coal body, the mechanical state of the coal body under different axial loading stress and different mechanical vibration and knocking force is analyzed.

The control axial loading stress is divided into 5–30 MPa, with 2.5 MPa as a class, and the control axial loading stress does not change in each class. The hammer is located at 15°, 20°, 25°, 30°, 35°, and 40° of the dial, and freely falls to hit the coal body. Make sure that the small hammer strikes the coal body 6–10 times at each dial. The interval between each knock is 10 s. Table 1 shows the test of artificial acoustic signals stimulated by mechanical vibration in coal without gas. Table 2 shows the test of the artificial acoustic signal stimulated by mechanical vibration when the gas pressure in the coal is 0.3 MPa.

Table 1. Artificial acoustic signal test of gas-free coal body mechanical vibration excitation.

Number	Knock Scale/°	Axial Loading Stress/MPa	Interval Time/s	Number of Knocks
1	15	5 MPa, 7.5 MPa	10	6–10
2	20	10 MPa, 12.5 MPa		
3	25	15 MPa, 17.5 Mpa		
4	30	20 MPa, 22.5 Mpa		
5	35	25 MPa, 27.5 Mpa		
6	40	30 MPa		

Table 2. Artificial acoustic signal test of coal mechanical vibration excitation when gas pressure is 0.3 MPa.

Number	Knock Scale/°	Axial Loading Stress/MPa	Gas Pressure/MPa	Interval Time/s	Number of Knocks
1	15	5 MPa, 7.5 MPa	0.3	10	6–10
2	20	10 MPa, 12.5 MPa			
3	25	15 MPa, 17.5 MPa			
4	30	20 MPa, 22.5 MPa			
5	35	25 MPa, 27.5 MPa			
6	40	30 MPa			

3. The Spectrum Parameters of Artificial Acoustic Signals

3.1. Principle of Artificial Acoustic Signal Acquisition

The system converts the rock mass vibration (artificial acoustic signal) caused by the action of mining equipment into audio electrical signals. Based on the results of spectrum analysis, the prediction parameters based on the frequency and amplitude of the artificial acoustic signals were constructed [21].

By analyzing the frequency spectrum characteristics of the elastic vibration waves generated in coal and rock masses, the frequency spectrum parameters of artificial acoustic signals can be used for daily prediction of the dynamic phenomenon risk of mining faces in coal and gas outburst mines and rock burst mines, daily monitoring of the rock mass state of working faces, and monitoring of the safety and effectiveness of working face anti-outburst measures for relief drilling. According to the research results of C.B. Merpep in Russia [11], the artificial acoustic signal prediction method is based on the fact that the attenuation coefficient of sound waves decreases with the increase in spatial stress near the working face, and this reduction is more obvious at high frequency than at low frequency. Thus, the ratio of harmonic amplitudes of acoustic signals measured at high and low frequencies increases with increasing stress. The increase in the relative stress coefficient K corresponds to the increase in stress, potential energy accumulation, and dynamic phenomenon risk in the surrounding rock of the roadway.

In order to evaluate the relative stress dynamics of coal during the test, we calculated a third of the spectrum of the artificial acoustic signal. Based on this, the low-frequency component and high-frequency component of the spectrum corresponding to the sum of the amplitudes of the low-frequency and high-frequency parts of the spectrum were determined. Based on these, the relative stress coefficient K was calculated, as shown in Formula (2). The relative stress coefficient K value is equal to the ratio of the sum of the amplitudes of the high-frequency part of the spectrum to the sum of the amplitudes of the low-frequency part of the spectrum [22], as shown in Figure 6.

$$K = \frac{A_B}{A_H} \quad (2)$$

where "A_B" is the sum of the amplitudes of the high-frequency part of the acoustic signal spectrum, and "A_H" is the sum of the amplitude of the low-frequency part of the acoustic signal spectrum.

In the laboratory, the artificial acoustic signal probe needed to be installed on the surface of the coal body in order to be used. The metal gasket was in contact with the damping gasket on the back of the artificial acoustic signal probe and was not allowed to come into contact with the metal part of the geophone. Then, a series of external mechanical vibration shocks were carried out at a certain distance from the artificial acoustic signal probe. The acoustic pulses caused by the response of the coal body to the pulse action were recorded. In this case, in different areas of the coal body, it produced its own resonance oscillation. In the recorded acoustic signal spectrum, this is the superposition of the elastic vibration, with high amplitude to distinguish the resonance frequency. The obtained artificial acoustic signal spectrum reflects the stress and strain of the coal.

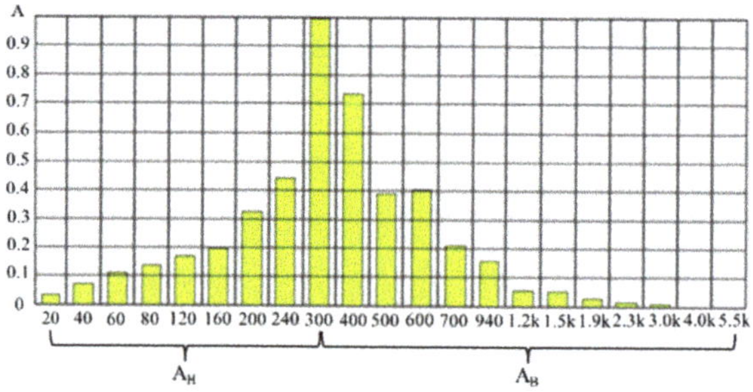

Figure 6. An example of one third of the frequency spectrum of an acoustic signal.

3.2. *Processing of Artificial Acoustic Signals*

In order to evaluate the resolution process of the artificial acoustic signal spectrum component of the relative stress dynamics of the coal body during the test, the determination of the relative stress coefficient K should follow the following order:

Firstly, a stable information spectrum of the acoustic signal is formed to exclude the "unqualified" spectrum. The acoustic detection method is used to collect enough information.

Secondly, a third of the spectrum is decomposed (the boundary of the spectrum is determined) so that the sum of the low-frequency amplitude on both sides of the spectrum is equal to the sum of the high-frequency amplitude.

Thirdly, after the above decomposition of the spectrum, the relative stress dynamics during the monitoring process can be evaluated by analyzing the relative stress coefficient K of the coal body during the test.

Before calculating the average frequency spectrum of the collected artificial acoustic signals, it is necessary to eliminate the frequency spectrum with poor quality. The artificial acoustic signal spectrum suitable for further calculation is a pulse with a clean profile, which arrives at the first order, and the amplitude of the single phase of the signal decreases gently along the pulse length. In the experiment, it is difficult to guarantee the effectiveness of the collected signal with only one knock or too few knocks. Tap 6–10 times to ensure that qualified spectrum signals can be screened for processing. The information spectrum of some artificial acoustic signals collected in the experiment is shown in Figure 7.

In the case of pulse excitation, the artificial acoustic signal can be recorded by oscillogram or linear spectrum, as shown in Figure 8. The fast Fourier transform was applied to the artificial acoustic signal excited by the coal body during the test, which can be expressed as the spectrum of the resonance frequency superposition. The resonance frequency is caused by the natural vibration generated in each layer [23]. An acceptable oscillographic image of an artificial acoustic signal has a stable sine wave form, which is represented by the truncation of the upper or lower part of the signal amplitude.

During the processing, the linear spectrum and the one-third multiple spectrum were scaled according to the amplitude maximum in the spectrum (normalization: that is, the ratio of each amplitude to the maximum value is calculated). This determined the "maximum shares" on both sides of the spectrum dividing line, which were then added together to obtain the sum of the low- and high-frequency parts of the spectrum. The calculation of one third of the spectrum of the acoustic signal reduced the number of "working" values of the discrete frequency amplitude, i.e., from 256 values in the linear spectrum to 22 values in the one third of the spectrum, thus simplifying the numerical processing of the artificial acoustic signal. Geoscan software is a data extraction and analysis software for portable artificial acoustic signal acquisition equipment. It was used to extract and screen the data collected from the experiment. The extraction and screening process of the spectrum information of the artificial acoustic signal is shown in Figure 9 below.

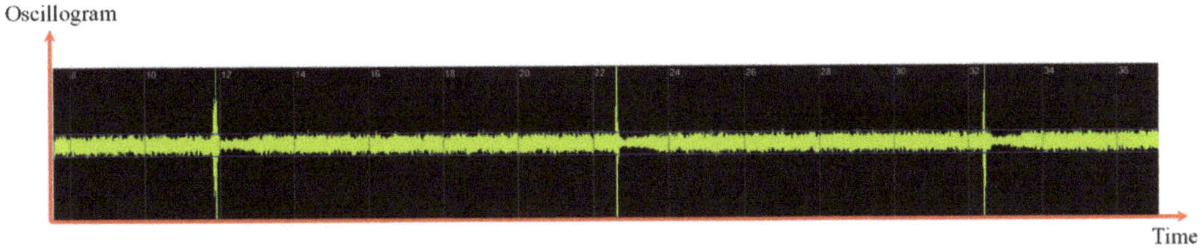

(**a**) The spectrum of artificial acoustic signals collected in the experiment.

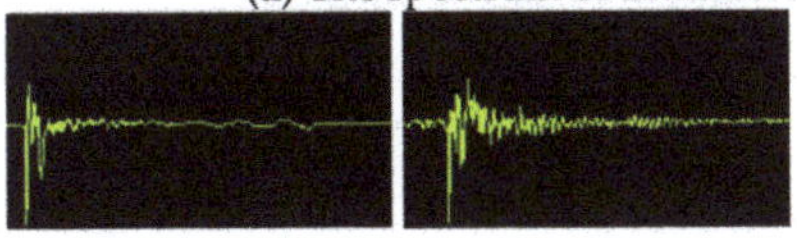

(**b**) Partially qualified spectrum available for analysis.

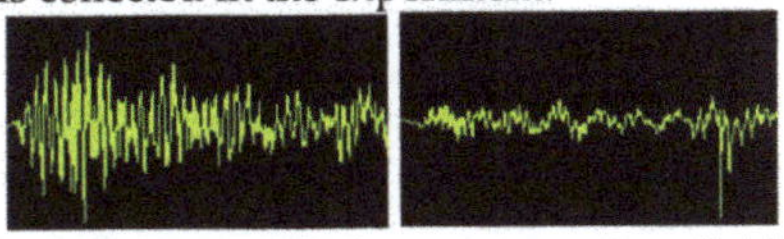

(**c**) Partially unqualified spectrum not available for analysis.

Figure 7. Artificial acoustic signal information spectrum example (part).

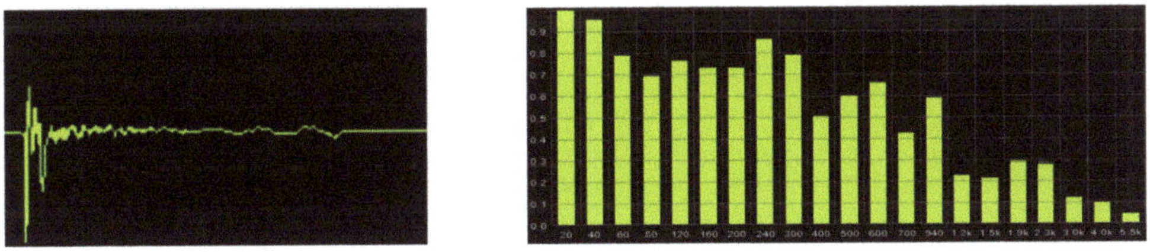

(**a**) Oscillogram (**b**) Spectrogram

Figure 8. Artificial acoustic signal oscillogram and spectrogram.

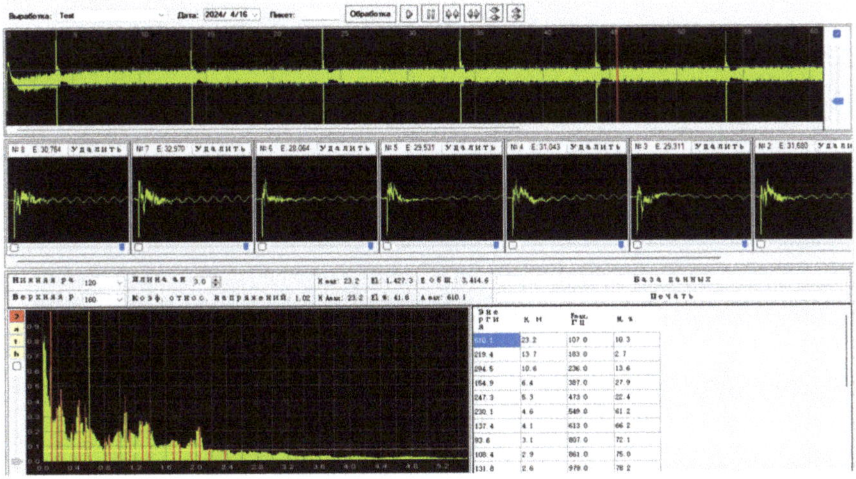

(**a**) Artificial acoustic signal extraction and screening.

Figure 9. *Cont.*

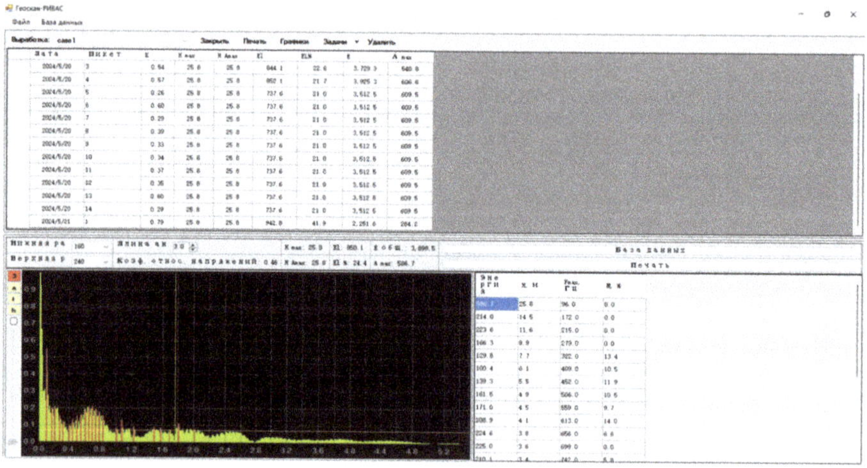

(b) Geoscan data processing interface.

Figure 9. Artificial acoustic signal oscillogram and spectrogram.

4. Experimental Results Analysis

4.1. Analysis of Mechanical Parameters of Coal Uniaxial Compression

The quality of the coal sample is 244.7 g. The coal sample height is 99.80 mm. The density of coal is 1.108 g/m^3. The stress-strain curve of the single coal under uniaxial load is shown in Figure 10. When the single coal is loaded uniaxially until it is destroyed, the stress-strain curve of the sample undergoes the crack compression compaction stage, elastic deformation stage, plastic deformation stage, yield stage and post-peak failure stage. The stress-strain curves of each stage show different variation characteristics. There is a significant stress drop after the peak stress. According to the test results, the uniaxial compressive strength of monomer coal is 17.846 MPa, the elastic modulus is 2.288 GPa, and the deformation modulus is 1.903 GPa.

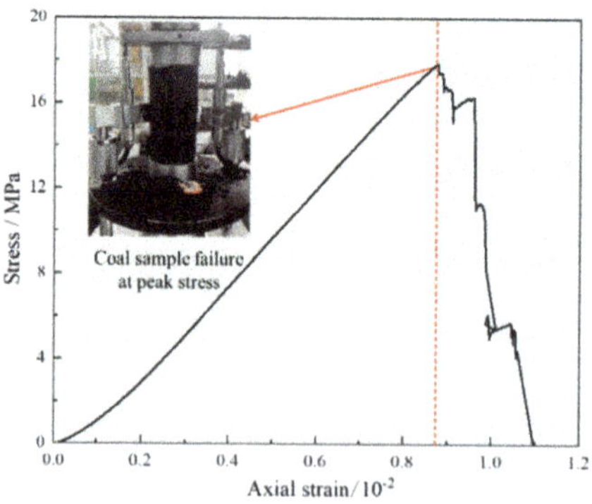

Figure 10. Artificial acoustic signal oscillogram and spectrogram.

When the loading rate is 0.01 mm/s, the failure mode of the single coal body is the tensile shear compound failure, which is dominated by a splitting crack. Among them, the middle and top parts of the sample are seriously damaged. With the increase in axial pressure, many small cracks appear in the sample. When the axial pressure is close to its peak value, several large, vertical cracks appear from both ends to the middle part. There are large cracks running through the upper and lower parts, which lead to the fracture of the coal body. The serious damage part of the sample has a local crushing fracture.

4.2. Analysis of Test Results of Artificial Acoustic Signal Excited by Mechanical Vibration

4.2.1. Gas-Bearing Situation and Influence of Axial Stress

When the hammer was located at the 20° free hem of the dial, the instantaneous knocking force measured by the mechanical sensor in contact with the small hammer in the mechanical vibration monitoring unit was 0.2 N. The small hammer was located at 20° of the dial to strike the coal body freely (that is, the external mechanical vibration knocking force was 0.2 N), and the axial stress was set to 10 kN, 15 kN, 20 kN, 25 kN, 30 kN, 35 kN, 40 kN, 45 kN, 50 kN, 55 kN, and 60 kN. Under these eleven different conditions, the spectrum information of the collected artificial acoustic signal was analyzed, and the relationship between the relative stress coefficient *K* and the axial loading stress under the condition of gas pressure or not was studied. For the case of no gas pressure and a gas pressure of 0.3 MPa, the relative stress coefficient *K* value of the coal body obtained by the test is shown in Tables 3 and 4. The data results are drawn as shown in Figure 11.

Table 3. Value of relative stress coefficient *K* without gas.

Axial Loading Stress/MPa	5	7.5	10	12.5	15	17.5	20	22.5	25	27.5	30
K	0.59	0.73	0.79	0.83	0.86	0.87	0.72	0.52	0.43	0.61	0.68

Table 4. The relative stress coefficient *K* value when the gas pressure is 0.3 MPa.

Axial Loading Stress/MPa	5	7.5	10	12.5	15	17.5	20	22.5	25	27.5	30
K	0.51	0.64	0.72	0.78	0.81	0.83	0.67	0.46	0.38	0.63	0.66

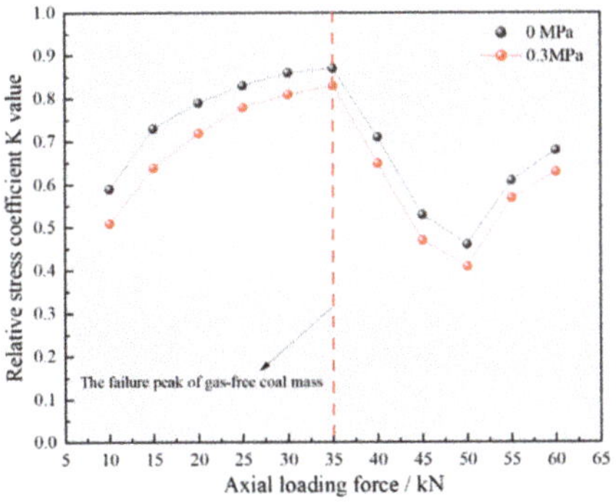

Figure 11. Variation of relative stress coefficient *K* value.

Through the analysis of the data, it was found that when the gas-containing coal body and the non-gas-containing coal body are stimulated by the artificial acoustic signal test under the external mechanical vibration impact force of 0.2 N, the relative stress coefficient *K* value increases first with the increase in the axial loading stress, then decreases, and then increases again. In the axial loading stress range of 10–35 kN, the relative stress coefficient *K* value is on the rise. In the early stage of loading, the growth rate of the *K* value is obvious, and the growth rate tends to be gentle in the later stage of loading. In the range of 35–50 kN of axial loading stress, the relative stress coefficient *K* value shows a downward trend, and the decline rate gradually slows down. When the axial loading stress is 50–60 kN, the relative stress coefficient *K* shows a rising trend again. However, under the same axial pressure, the relative stress coefficient *K* of gas-bearing coal is smaller than that of gas-free coal.

The reason is that the strength of the coal sample decreases with the increase in gas pressure. The cracks inside the coal are full of free gas. The free gas expands the volume of coal and reduces the density of coal. After the coal adsorbs the gas, the cohesive force between the coal particles is significantly reduced, which reduces the force and energy required when the coal is destroyed, and reduces the peak strength and peak strain. Chemical adsorption occurs after the coal adsorbs the gas, and these chemical reactions may also reduce the strength of the coal. The uniaxial compressive strength of the coal sample used in the test is 17.846 MPa, which is about 35 kN of the axial loading stress. This is the basic mechanical parameter measured by gas-free coal. When the coal gas pressure is 0 MPa and 0.3 MPa, there is no obvious damage to the coal body before reaching the peak strength measured by the gas-free coal body test. As the axial pressure continues to increase, the stress on the coal body increases, and the relative stress coefficient K increases. The strength of the coal sample decreases with the increase in gas pressure, and the relative stress coefficient K value of the coal containing gas is smaller than that of the coal without gas. After the peak strength, as the axial loading stress increases, the coal body is destroyed. The number of internal cracks increases and the coal body is broken, and the relative stress coefficient K shows a downward trend. With the continuous increase in axial stress, the surrounding of coal body is limited by the cavity, which is equivalent to the external confining pressure support. The confining pressure limits the development of coal body deformation and cracks to some extent. At this time, the radial deformation of the coal body is obviously limited. The process of the continuous application of axial pressure is the re-compaction of the coal body after peak failure. The compressive strength of the compacted coal body increases, and the relative stress coefficient K increases again.

Before the destruction of gas-free coal and gas-bearing coal, with the increase in axial loading stress, the relationship between the relative stress coefficient K value obtained by the artificial acoustic signal test of mechanical vibration excitation was fitted and analyzed. The results are shown in Figure 12. The fitting results show that both of them can be expressed in the form of exponential function, and the correlation coefficient is as high as 99%.

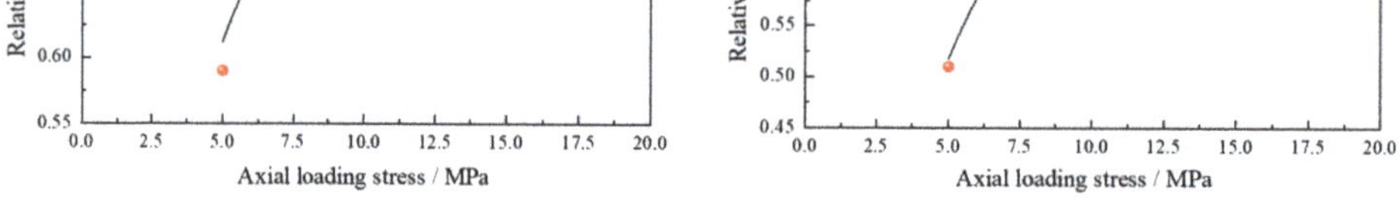

(**a**) The change relationship without gas pressure. (**b**) The change relationship when the gas pressure is 0.3 MPa.

Figure 12. Artificial acoustic signal oscillogram and spectrogram.

The results of the data fitting analysis show that the relationship between the relative stress coefficient K and the axial loading stress σ can be expressed by an exponential function.

$$K = e^{-\frac{C}{\sigma}} \tag{3}$$

where "C" is a constant term, which is determined by the combination of the coal sample properties, mechanical vibration force, mechanical vibration wave propagation path, axial loading stress, and gas pressure [24].

4.2.2. The Influence of Mechanical Vibration Force

The gas pressure and axial loading stress were set to be constant. The impact force of the mechanical vibration was changed to study the artificial acoustic signal generated by the external mechanical vibration excitation of gas-bearing coal. The small hammer was free to swing at different scale values, and the instantaneous knocking force could be measured by the mechanical sensor. The gas pressure was set to be 0.3 MPa. The axial loading stress was 20 kN. The external mechanical vibration knocking force was 0.15 N, 0.20 N, 0.25 N, 0.30 N, 0.35 N, and 0.40 N during the test. In other words, the hammers were placed at 15°, 20°, 25°, 30°, 35°, and 40° on the dial, respectively. Then the spectrum information of the artificial acoustic signal collected by the experiment was analyzed, and the relationship between the relative stress coefficient K and the external mechanical vibration force was summarized. The relative stress coefficient K value of the coal body obtained from the test is shown in Table 5. The data results are shown in Figure 13.

Table 5. Parameter table of relative stress coefficients under different mechanical vibration forces.

Mechanical vibration force/N	0.15	0.20	0.25	0.30	0.35	0.40
Relative stress coefficient K value	0.74	0.72	0.68	0.65	0.60	0.54

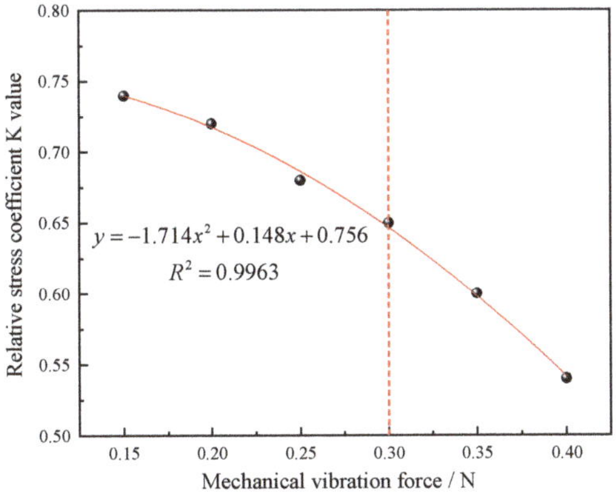

Figure 13. Relative stress coefficient K under different mechanical vibration forces.

By analyzing the data, it was found that there is a quadratic function variation trend between the mechanical vibration force and the relative stress coefficient K. When the gas pressure is fixed at 0.3 MPa and the axial loading stress is 20 kN, the relative stress coefficient K decreases with the increase in external mechanical vibration. Due to the instantaneous impact, the external mechanical vibration knocking force has an influence on the change in the internal structure of the coal body. Different mechanical vibration knocking forces have different degrees of cracks in the coal body. When the mechanical vibration knocking force is less than 0.2 N, due to the small knocking force, there are fewer cracks in the coal body. The attenuation degree of the acoustic signal from the signal source to the receiving position is small, and the difference in the relative stress coefficient K value is also small. When the mechanical vibration impact force is greater than 0.2 N, due to the limited area of the external mechanical vibration acting on the surface of the coal body, the greater the impact force, the more obvious the new cracks and fissures derived from the coal body. The more obvious the impact of the small hammer on the internal structure of the coal body, the greater the attenuation degree of the acoustic signal propagation process, resulting in a gradual increase in the difference of the relative stress coefficient K value measured by the test. When the mechanical vibration impact force is greater than 0.30 N, the relative stress coefficient K value decreases more obviously. When the external mechanical vibration knocking strength reaches a certain value, the impact force has a great influence on the internal structure of the coal body. There are many large cracks in the coal body, and even regional damage occurs in the knocked part due to the small force area. The above reasons lead to the serious attenuation of the signal source propagation process, so the relative stress coefficient K value decreases more obviously and the value is smaller.

5. Discussion

The artificial acoustic signal prediction method can realize the continuous online monitoring and early warning of various dynamic phenomenon risks, geological failures, and stress changes near the working face, and is one of the most promising development directions of dynamic mine phenomenon prediction. Now, a multifunctional intelligent monitoring and warning system based on amplitude-frequency characteristics of artificial acoustic signals has been developed and applied in Pingdingshan coal mine. With the help of shearers and roadboring machines in operation, the system can automatically determine the stress–strain state and relative stress dynamics of the roof by processing and analyzing the elastic wave spectrum characteristics of the roof, and issue a risk prediction conclusion, which can provide a scientific basis for the safety decision of mine dynamic phenomenon prevention and control. However, the small-scale laboratory theory based on the spectral characteristics of artificial acoustic signals needs to be studied urgently. The construction of an artificial acoustic signal monitoring test system of coal loaded with gas under mechanical vibration and the quantitative relationship between artificial acoustic signals and coal stress state and gas pressure are of great significance for guiding field engineering practice and verifying field monitoring rules.

In this study, based on artificial acoustic signals, a test device for artificial acoustic signals stimulated by mechanical vibration was independently designed to study the dynamic response characteristics of coal containing gas under mechanical vibration and collect artificial acoustic signals generated by coal excited under mechanical vibration. The stress–strain state of coal in different periods can be reflected by the relative stress coefficient K value in the basic parameters of artificial acoustic signals, so the qualitative and quantitative relationship between the relative stress coefficient K value and axial loading stress, gas content of coal, and different mechanical vibration forces is established. The experimental research can provide a theoretical basis for the prediction and early warning of dynamic disasters by artificial acoustic signals.

At present, the study on the frequency spectrum characteristics of artificial acoustic signals in China is still not perfect, and the quantitative relationship between artificial acoustic frequency spectrum signals and mechanical parameters such as the stress–strain state of coal and gas pressure is still not clear. As far as the relative stress coefficient K value is concerned, a preliminary qualitative and quantitative relationship has been obtained, but whether there is a clearer quantitative relationship between the relative stress coefficient K value and the gas pressure, gas migration, and axial loading stress of gas-containing coal remains to be studied further. Whether other parameters based on the artificial acoustic signal spectrum can be studied in a laboratory on a small scale, and what qualitative and quantitative relationship they have with the mechanical parameters of coal, remains to be studied further.

6. Conclusions

Considering the complex disaster-forming environment of deep coal seams and the influence of mechanical vibration force in the process of coal seam mining, an experimental device and system for generating acoustic signals from gas-bearing coal by mechanical vibration force are developed. Based on the analysis of mechanical parameters of coal under uniaxial compression, the effects of axial loading stress, gas content, and mechanical vibration force on the relative stress coefficient K of coal are tested. The main findings are as follows:

(1) During the uniaxial compression mechanical test of the coal sample, the stress–strain curve experienced the fracture compression compaction stage, the elastic deformation stage, the plastic deformation stage, the yield stage, and the post-peak failure stage. The uniaxial compressive strength of coal samples was 17.846 MPa and the elastic modulus was 2.288 Gpa. When the loading rate was 0.01 mm/s, the failure of single coal was mainly splitting failure and shear cracking.

(2) When the gas-bearing coal and gas-free coal were subjected to the same external mechanical vibration impact force excitation artificial acoustic signal test, the relative stress coefficient K value increased first, then decreased, and then increased with the increase in axial loading stress. Under the same axial pressure, the relative stress coefficient K value of coal containing gas was smaller than that of coal without gas.

(3) The relationship between the relative stress coefficient K value and the axial loading stress σ of the gas-containing coal body and the gas-free coal body can be expressed in the form of exponential function $K = e^{-\frac{C}{\sigma}}$. The constant term C value obtained by the test in the case of coal containing gas was larger than that without gas.

(4) When the axial loading stress of the coal body was the same as the gas pressure of the coal body, with the increase in external mechanical vibration, the relative stress coefficient K showed a downward trend. When the impact force of mechanical vibration was too large, it had a great influence on the damage to the coal body, and the relative stress coefficient K was smaller.

Author Contributions: Writing—original draft preparation, W.C. and J.M.; methodology, J.Z. and W.F.; validation, J.M. and F.D.; analysis, W.C., J.M. and J.Z.; supervision, W.F., J.M. and F.D. All authors have read and agreed to the published version of the manuscript.

Funding: This work was supported by the National Natural Science Foundation of China (52404119; 52374249), Natural Science Foundation of Henan (242300421451), Opening Foundation of State Key Laboratory of Coking Coal Resources Green Exploitation in China PingMei ShenMa Group (4104020211205z), Opening Foundation of Hubei Key Laboratory for Efficient Utilization and Agglomeration of Metallurgic Mineral Resources (2024zy001), and Open Research Fund of Key Laboratory of Xinjiang Coal Resources Green Mining, Ministry of Education (KLXGY-KB2421).

Data Availability Statement: Data are contained within the article.

Conflicts of Interest: The authors declare that this study received funding from State Key Laboratory of Coking Coal Resources Green Exploitation, China PingMei ShenMa Group. The funder had the following involvement with the study: Methodology; Writing-review&editing; Supervision.

References and Notes

1. Wang, K.; Du, F. Coal-gas compound dynamic disasters in China: A review. *Process Saf. Environ. Prot.* **2020**, *133*, 1–17.
2. Wang, Y.; Kong, D.; Wu, G.; Cheng, Z.; Song, G.; Shang, Y. Failure Mechanism and Movement Characteristics of Overlying Strata in Longwall Mining Face with Thick Aquifer. *Rock Mech. Rock Eng.* **2024**, *57*, 6787–6809.
3. Xiong, Y.; Kong, D.; Song, G. Research hotspots and development trends of green coal mining: Exploring the path to sustainable development of coal mines. *Resour. Policy* **2024**, *92*, 105039.
4. Shang, Y.; Wang, Y.; Zhang, L.; Wu, G.; Kong, D.; Zhang, P.; Jia, T. Impact of Lower Protective Coal Seam Mining on Overlying Strata: Deformation, Pressure Relief, and Permeability Enhancement. *Int. J. Geomech.* **2024**, *24*, 04024178. [CrossRef]
5. Li, J. Research on the Application of Acoustic Emission Intelligent Monitoring and Early Warning Technology for the Coal-rock Dynamic Disasters in Underground Mining Faces. *IOP Conf. Ser. Earth Environ. Sci.* **2020**, *526*, 012121. [CrossRef]
6. Di, Y.; Wang, E.; Li, Z.; Liu, X.; Huang, T.; Yao, J. Comprehensive early warning method of microseismic, acoustic emission, and electromagnetic radiation signals of rock burst based on deep learning. *Int. J. Rock Mech. Min. Sci.* **2023**, *170*, 105519. [CrossRef]
7. Liu, L.; Li, S.; Xiao, Y.; Chen, S.; Wang, Z.; Feng, G.; Wang, Y. An Inverted Heterogeneous Velocity Model for Microseismic Source Location in Deep Buried Tunnels. *Rock Mech. Rock Eng.* **2023**, *56*, 4855–4880. [CrossRef]
8. Zhou, Q.; Yin, X.; Li, K. Seismic rock physics modeling and shear wave velocity prediction method of coal measure strata. *Shiyou Diqiu Wuli Kantan/Oil Geophys. Prospect.* **2022**, *57*, 357–366.

9. Miller, C.B.; An, Z. Coal and gas outburst and other signs of gas dynamics. *Safety Coal Mines* **1989**, *4*, 29–33.
10. Shadrin, A.V.; Diyuk, Y.A.; Teleguz, A.S. Application of acoustic methods for rock mass hydrotreatment parameter control and efficiency assessment. *Mining Ind. J.* **2018**, *2*, 79–82. [CrossRef]
11. Мирер, С.В.; Хмара, О.И.; Масленщиков, Е.В. Оконтроле выбросоопасности забоев поспектральным характеристикам акустическихсигналов. *В*опросы предотвр*а*щения внез*а*пныхвыбросов. Н*а*учн. сообщ. ИГД им. *А.А.* Скочинского **1987**, 52–61.
12. Бобров, А.И.; Агафонов, А.В.; Колчин, Г.И. Контроль динамических процессов в призабойной части горного массива/Проблемы геодинамической безопасности. In Proceedings of the II международное рабочее совещание, СПб, Russia, 24–27 июня 1997; pp. 142–145.
13. Лунев, С.Г.; Колчин, Г.И. Акустический контроль выбросоопасности в очистном забое. J. Сб. науч. тр., Макеевка: МакНИИ, 2004: 97–107. Зингерман К М, Вершинин А В. К постановке задачи моделирования динамики пустотного пространства горного массива в нестационарном поле напряжений[C]//I МЕЖДУНАРОДНАЯ НАУЧНО-ТЕХНИЧЕСКАЯ КОН*ф*ЕРЕНЦИЯ «БЕЗОПАСНОСТЬ ТРУДА И Э*фф*ЕКТИВНОСТЬ ПРОИЗВОДСТВА ГОРНО-ДОБЫВАЮЩИХ ПРЕДПРИЯТИЙ С ПОДЗЕМНЫМ СПОСОБОМ РАЗРАБОТКИ». 2016: 38.
14. *ф*едеральные нормы и правила в областипромышленной безопасности “Инструкция попрогнозу динамических явлений и мониторингумассива горных пород при отработке угольныхместорождений. S. Утверждены приказом*ф*едеральной службы по экологическому,технологическому и атомному надзору от10.12.2020 г. №515: 114.
15. Брюханов, А.М.; Агафонов, А.В.; Рубинский, А.А.; Колчин, Г.И. Методы и средства прогноза и предотвращения газодинамических явлений в угольных шахтах. Расследование и предотвращение аварий на угольных шахтах. Том *3*. Д.: Вебер, 2007.
16. Копылов, К.Н.; Смирнов, О.В.; Кулик, А.И.; ПотаповП, В. Испытания автоматизированной системыакустического контроля состояния массива горныхпород. УГОЛЬ **2015**, *S7*(1072), 44–47, УДК: 622.861.
17. Zhang, J.; Wei, F.; Ivanovich, S.V. Monitoring system for rock mass state and dynamic phenomenon based on artificial acoustic signal САКСМ application. *Saf. Coal Mines* **2022**, *53*, 149–155.
18. Sun, L.; Lin, L.; Yao, X.; Zhang, Y.; Tao, Z.; Ling, P. Real-Time Recognition Method for Key Signals of Rock Fracture Acoustic Emissions Based on Deep Learning. *Sensors* **2023**, *23*, 8513. [CrossRef] [PubMed]
19. Брюханов, А.М.; Агафонов, А.В.; Рубинский, А.А.; Колчин, Г.И. Акустический контрольвыбросоопасности// Методы и средства прогнозаи предотвращения газодинамических явлений вугольных шахтах. М. Расследование ипредотвращение аварий на угольных шахтах. Том3.-Донецк. Вебер, 2007. 692.
20. Knyazkina, V.I.; Safronchuk, K.A.; Ivanov, S.L. About possibility of immediate evaluation of technical condition of mining equipment using signal value of acoustic emission friction. *IOP Conf. Ser. Mater. Sci. Eng.* **2019**, *560*, 012068. [CrossRef]
21. Копылов, К.Н.; Смирнов, О.В.; Кулик, А.И. Акустический контроль состояния массива и прогноз динамических явлений. Горный информ*а*ционно-*ана*литический бюллетень (н*а*учно-технический журн*ал)* **2015**, (S7), 574–583, УДК: 622.831.
22. Горбачев, А.С.; Шилов, В.И. МОНИТОРИНГ МАССИВА ГОРНЫХ ПОРОД ПО ПАРАМЕТРАМ ИСКУССТВЕННОГО АКУСТИЧЕСКОГО СИГНАЛА. ПРОГНОЗ ГЕОЛОГИЧЕСКИХ НАРУШЕНИЙ ПРИ ДОБЫЧЕ УГЛЯ. *В*естник н*а*учного центр*а* по безоп*а*сности р*а*бот в угольной промышленности **2022**, 6–16.
23. Dong, H.; Suo, J.; Zhu, Z.; Wang, H.; Ji, H. Enhanced Underwater Single Vector-Acoustic DOA Estimation via Linear Matched Stochastic Resonance Preprocessing. *Remote Sens.* **2024**, *16*, 1802. [CrossRef]
24. Шадрин, А.В.; Дегтярева, М.В. Акустический двухчастотный метод контроля напряженного состояния горного массива. *В*естник Н*а*учного центр*а* по безоп*а*сности р*а*бот в угольной промышленности **2013**, 55–59.

processes

MDPI

Article

Performance and Reliability of Thermoelectric Conversion Using a Crooked Thermosyphon to Enhance Heat Transfer from Coal Fires

Qingfeng Bao [1], Xiuting Guo [1], Bo Li [2], Wuyi Chen [2], Zhenping Wang [3] and Yang Xiao [3,*]

1 Ventilation Section, Shanxi Lu'an Huanbao Nengyuan Kaifa Co., Ltd., Changzhi 046204, China; baoqingfeng2003@163.com (Q.B.); guoxiuting2677@126.com (X.G.)
2 Changcun Coalmine, Shanxi Lu'an Huanbao Nengyuan Kaifa Co., Ltd., Changzhi 046204, China; libo19870039@163.com (B.L.); chenwuyi2024@163.com (W.C.)
3 School of Safety Science and Engineering, Xi'an University of Science and Technology, Xi'an 710054, China; wangzp@xust.edu.cn
* Correspondence: xiaoy@xust.edu.cn

Abstract: A large amount of energy can accumulate and be stored during underground coal fires. As thermal energy cannot be easily removed using the traditional technologies of fire prevention and extinguishment, there is a potential benefit to collecting and utilizing thermal energy from coal fires and converting it to electrical energy. Thus, this work proposes a thermoelectric generator as a solution to convert thermal energy from coal fires to electrical energy. To improve the thermal energy conversion efficiency, an experimental test system was established using a thermosyphon, an electric heating module, a cooling circulation module, a thermoelectric module, and a data acquisition module. Under the condition of ensuring the same input heat and cooling boundary conditions, the influence of three factors, namely the cooling method, the connection method, and the coverage rate of thermoelectric devices, on the performance of the coal fire waste heat conversion system was studied. The results show that, compared with air cooling, water cooling provides a greater temperature difference for the thermoelectric module, and the maximum temperature difference can reach 65.90 °C. Series connection between thermoelectric devices will generate a higher open-circuit voltage and output voltage. The maximum horizontal open-circuit voltage value can reach 3.34 V, and the maximum output voltage is 2.61 V. Compared with the coverage rates of thermoelectric devices of 15.0% and 30.0%, the output power under the coverage rate of 22.5% is the largest at 0.35 W, and its thermoelectric conversion efficiency is also the largest at 0.35%. The optimal combination of thermoelectric modules obtained from the research results can provide ideas for the application of in situ coal fire prevention and control.

Keywords: thermal energy; thermoelectric generator; energy conversion; heat transfer; thermosyphon

1. Introduction

In the context of the rapid development of global industrialization and socialization, heat energy utilization and conversion are a continuous and important concern [1–4]. In recent years, geothermal development has become a promising approach. Geothermal energy is a new clean energy source and is considered to be one of the largest renewable energy sources. However, it is slower than other power generation methods, owing to the low rate of return in the geothermal conversion process. On the other hand, coal fire disasters are extremely serious in China [5]. The principle of traditional fire prevention methods was to build a barrier, isolate oxygen, and eliminate heat energy. However, the huge heat energy accumulated and stored from a large range of underground coal fires could not be effectively transferred and eliminated through fire extinguishing technologies and could easily lead to the reignition of coal fires [6,7]. Unfortunately, this tremendous heat energy was also unwittingly wasted. According to statistics, globally, coal fires burn about 1.0 billion tons of coal resources every year, generating about 1000 GW of power energy. This is equal to the combined energy of the world's 500 nuclear power plants and 2.5 times as much as nearly 100 GW of hydroelectric power generation [8].

As efficient heat transfer equipment, thermosiphons have been applied in heat extraction from coal fires, owing to their simple structure, high heat transfer efficiency, and excellent isothermal performance [9]. The working medium, filling ratio, and geometric structure have been varied to improve the heat transfer performance within the thermosiphon. Sarkar et al. [10] found that regardless of the operating conditions, the working fluid enters the condenser in a superheated state and enters the evaporator in a subcooled state. Based on this, they derived expressions for predicting changes in temperature, pressure, specific volume, entropy, and enthalpy in each thermodynamic process, as well as expressions for estimating entropy generation. Machado et al. [11] estimated the thermal performance of thermosyphons through artificial neural networks and found that when the filling rate is 40%, the thermal performance is the best, and a slope of 45° shows better performance than a slope of 90°. Kim et al. [12] studied the relationship between the tilt angle, system charge, and heat transfer performance of a gravity thermosiphon. The optimal working conditions were a tilt angle of 30° and a system charge of 0.5 kg. Solomen et al. [13] found that the heat transmission coefficient of a thermosyphon's evaporation section increased by 45% when the inclination angle was 45°. Alammar et al. [14] performed a numerical simulation and found that in the positive temperature zone, the minimum thermal resistance of the thermosiphon appeared at a 90° inclination and a 65% filling factor. Khazaee et al. [15] obtained that when the angle of inclination was 40–60°, the evaporation section's heat transfer coefficient was the largest. In conclusion, changing the inclination angle could effectively improve the heat transfer performance of a thermosiphon. A thermosiphon's heat transfer capacity is closely related to the heat transfer process at the formation side, so strengthening the structural design of the thermosiphon and increasing the heat transfer coefficient were breakthrough points to realizing the effective use of heat energy from coal fires.

Based on the Seebeck effect, thermoelectric technology was used for the direct conversion of thermal to electrical energy, which could be regarded as a generator without any moving components [16–18]. As a result, this generator had higher reliability, low maintenance, and longer service life compared to conventional generators [19–22]. In addition, it was noiseless, environmentally friendly, and directly converted heat from the environment into electrical energy. Coal fires naturally play the role of thermal engines for semiconductor power generation because of their huge thermal energy. Moreover, thermoelectric conversion efficiency was affected by many factors during its continuous improvement. Under the TE module for TE materials, Hu et al. [23] developed a three-dimensional numerical model and verified it using an experimental device. Shi et al. [24] found a way to obtain higher heights, larger cross-sectional areas, and even smaller TEs under matched load resistances, as well as to build a re-sized T-module under ideal conditions of constant heat input. Hans et al. [25] established a parallel connection to the multi-stage thermoelectric device skewers using the control volume method, which analyzed the output characteristics of the coal fire waste heat conversion system from three different connection modes (all in parallel, all in series, and mixed), and proposed a selection indicator of a string parallel circuit: load resistance, load required voltage, and current. Zhao et al. [26] applied different thermoelectric devices to the coal fire waste heat conversion system and explored the influence of the coverage rate of thermoelectric devices on the heat transfer capacity and the thermosiphon's thermoelectric conversion efficiency. Li et al. [27] studied the tube-fin heat exchangers in a TEG as the heat and fluid flow of hot side channels. Nugraha et al. [28] utilized experimental and computational studies to research the influence of the heating rate of thermoelectric materials on the hot side on the electrical output. They discovered that when the heating rate was within the range of 0.94 to 5.67 K/s, the voltage, current, and power of the thermoelectric generator increased with the increased temperature.

The aforementioned achievements helped researchers to select and design high-performance thermoelectric generators. However, underground coal fires are complicated events with diverse causes and various constraints. Therefore, this work will consider coal fire prevention and thermoelectric resource utilization and will investigate the conversion performance of thermoelectric generators. A thermosiphon with a curved well structure was designed to remove thermal energy. On this basis, a waste heat conversion system was created by adding thermoelectric modules in order to explore the effect of different combinations. Furthermore, the optimization of the combined thermoelectric modules aimed to improve the thermal energy utilization and conversion capability of the system, inhibiting coal fire spread and achieving resource utilization.

2. Theoretical Consideration

The thermoelectric generator (TEG) is mainly composed of four parts, namely the heat pipe serving as a heat source, the copper block enhancing the heat conduction ability, the thermoelectric device for thermoelectric conversion, and the water-cooled heat exchanger. The heat exchanger on the hot side of the TEG has the function of absorbing heat from the outside and transferring it to the thermoelectric module. When there is a temperature difference between the hot and cold sides of the TEG, the combination of p-type and n-type side semiconductor pairs has the function of generating electricity. The schematic diagram of the TEG system is shown in Figure 1. The thermoelectric devices of TEG components are connected in series with the thermally conductive copper block. Heat is transferred to the thermally conductive copper block through the heat pipe. The thermally conductive copper block is connected to the hot side of the thermoelectric device and serves as a heat exchanger. The cold side is connected to the water-cooled heat exchanger and serves as a radiator. If the hot side of the TEG absorbs a large amount of heat, the electron energy on the hot side inside the thermoelectric device will be higher than that on the cold side. This will cause the electrons on the hot side of the TEG to enter the cold side at a higher speed than that with which the electrons on the cold side enter the hot side [4]. At this time, the hot side end of the thermoelectric device will carry a positive charge and the cold side end will carry a negative charge. Equation (1) shows the relationship between the induced voltage difference and the temperature difference.

$$\alpha = -\frac{\Delta V}{\Delta T} \tag{1}$$

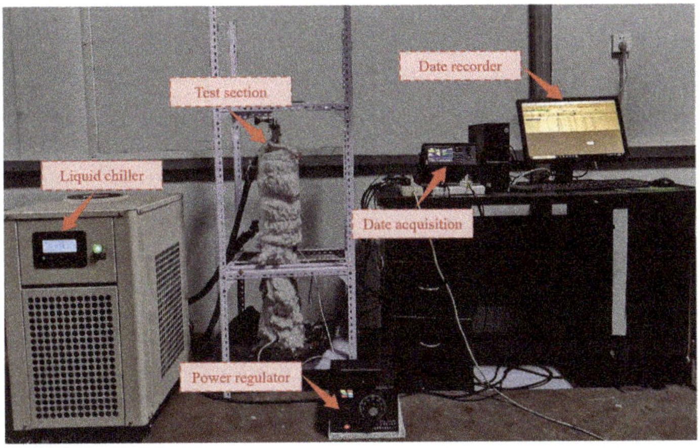

Figure 1. Physical picture of thermoelectric performance test system of coal fire waste heat conversion system.

Among them, α refers to the Seebeck coefficient of the semiconductor material and ΔV and ΔT represent the potential difference and temperature difference in the TEG, respectively. Equation (2) can be used to calculate the voltage generated by an open circuit when no current is flowing.

$$V_K = \alpha \Delta T_{TEG} = \alpha(T_h - T_c) \tag{2}$$

T_c and T_h represent the temperatures of the cold side and the hot side of the TEG, respectively. V_K represents the voltage generated by an open circuit when no current is flowing, and ΔT_{TEG} represents the temperature difference between the two sides of the TEG.

At present, the equivalent thermal conductivity (effective thermal conductivity, equivalent-value thermal conductivity, or comparable thermal conductivity) is used to evaluate the heat transfer performance of a thermosyphon. Specifically, the thermosyphon is considered to be equivalent to a solid-metal pipe with the same external dimensions as the thermosyphon. Although it reflects the quality of the heat transfer performance of the thermosyphon to a certain extent, it is greatly affected by geometric factors, which weakens the influence of the boiling heat transfer coefficient of the evaporation section and the condensation heat transfer coefficient of the condensation section on the heat transfer performance. Considering the convective heat transfer process of the working medium inside the thermosyphon, the equivalent convective heat transfer coefficient, as a new calculation method, is used to quantitatively evaluate the heat transfer ability of the thermosyphon. Equation (3) is as follows:

$$\alpha_{eff} = \frac{2\lambda Q_{in}(L_e + L_c)}{2\pi\lambda d_i(T_{w,h,o} - T_{w,c,o}) - Q_{in} d_i (L_e + L_c)\ln(d_o/d_i)} \tag{3}$$

In the formula, α_{eff} represents the heat transfer performance of the thermosyphon; Q_{in} is the input heat, in watts (W); L_e and L_c are the lengths of the evaporation section and the condensation section of the thermosyphon, respectively, in meters (m); d_i and d_o are the inner and outer diameters of the thermosyphon, respectively, in meters (m); λ is the thermal conductivity coefficient of carbon steel as the material, 48 W/(m·K); and $T_{w,h,o}$ and $T_{w,c,o}$ are the wall temperatures of the evaporation section and the condensation section, respectively, in degrees Celsius (°C).

3. Materials and Methods

3.1. Experimental System

The experimental test system comprised five elements: the thermosiphon, the electric heating device, the water-cooled cycle device, the thermoelectric device, and the data acquisition device. The physical picture is shown in Figure 1. Recent studies have shown that thermal resistance is mainly controlled at the formation side in thermal energy devices such as thermosiphons, and inclined or curved structures revealed better heat transfer performance than vertical structures. When the angle was 30°, the pipette's heat transfer performance could be improved by increasing the evaporation section's boiling heat transfer coefficient [8]. Therefore, the angle of the evaporation section to the coalbed was optimized and a thermosiphon with an evaporation section inclination angle of 30° was designed in this work. In the experiment, carbon steel, which is commonly used in normal-temperature and medium-temperature thermosyphons, was selected as the material for processing thermosiphon, and the processed thermosiphon was further cleaned. The carbon steel has an outer diameter of 22 mm, an inner diameter of 19 mm, and a height of 0.8 m. Conventional distilled water was selected as the internal working medium of the thermosiphon with a 30% filling ratio.

The electric heating device was composed of flexible heating wire (HTC-060, Omega, Norwalk, CT, USA) and a power regulator (STG-1000, Xishuo, Beijing, China). The wire's diameter, total length, and power were 4.80 mm, 1.80 m, and 125 W, respectively. A power regulator with an auto-regulator range of 0–300 V was used with the heating wire.

The water-cooled cycle device mainly includes a chiller (DIC006ASS-LA2, Ruitou, Xi'an, China) and a water-cooled heat exchanger. The chiller can provide constant heat transfer boundary conditions for the thermoelectric system by setting the cooling flow rate and water temperature. The size and thickness of the thermoelectric device (TEP-1263-2.4, Nanchang, China) were 30 × 30 mm and 2.4 mm, with 126 pairs of P-N sections internally. The thermoelectric conversion performance parameters of the thermoelectric devices under a constant cold surface temperature of 30 °C are shown in Table 1. To enhance heat conduction, thermal copper blocks were chosen as the material. It has an inner diameter of 23 mm, and the external dimensions of the thermal copper blocks are length × width × height: 30 × 30 × 30 mm, as shown in Figure 2.

Table 1. Operating performance parameters of thermoelectric device.

Model	Open-Circuit Voltage (V)	Matched Load Value (Ω)	Matching Load Value (Ω)	Nominal Power (W)	Heat Flow Through the Module (W)
TEP1-1263-2.4	10.8	5.4	5.4	5.4	96

Figure 2. Physical picture of the sleeve-type heat-conducting copper block.

The heating system is composed of a flexible heating wire and a power regulator. In the experiment, the HTC-060 flexible heating wire manufactured by OMEGA was chosen. This wire has a diameter of 4.8 mm, a total length of 1.8 m, and a power of 125 W. During the experiment, the heating wire was evenly and densely wound around the evaporation section of the thermosyphon to obtain a stable and uniform heat-flux density. Given that the power in the experimental design was relatively low, the power regulator needed to reduce the voltage for the purpose of adjusting the power of the heating wire.

Therefore, in combination with the selected range of input power values, in the experiment, a power regulator of model STG-1000W with auto-coupling voltage regulation from 0 to 300 V manufactured by Xishuo Technology Co., Ltd. (Suzhou, China) was selected for use in conjunction with the heating wire. A physical picture of the heating wire and the power regulator is shown in Figure 3.

The data acquisition device mainly consists of a multi-function data conversion module, a data acquisition switch unit, a PC computer, special processing software, and thermocouples. The data acquisition device can directly measure the thermocouples, resistance, AD/DC voltage, and other parameters. The thermocouples used mainly included ETAGK30K high-temperature thermocouples (Aiyong, Suzhou, China) (up to 480 °C, accuracy 0.3 °C) and ETA1006 T thermocouples (Aiyong, Suzhou, China) (up to 150 °C, accuracy 0.1 °C).

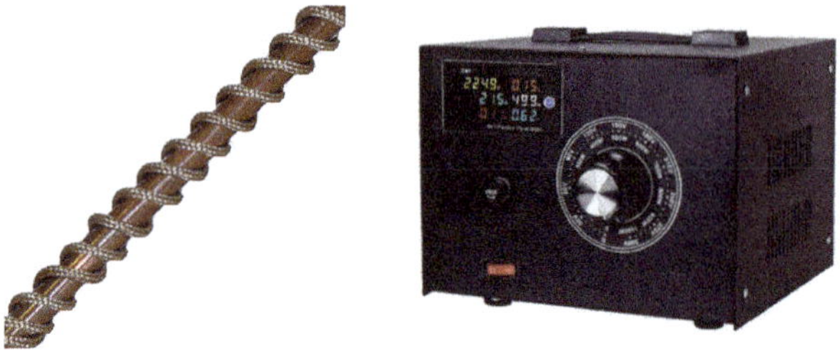

(**a**) (**b**)

Figure 3. The physical pictures of the heating wire and the power regulator. (**a**) The heating wire. (**b**) The power regulator.

The data acquisition device is shown in Figure 4, andit will be used in conjunction with the Bench Vue Data Acquisition (DAQ) software (Keysight DAQ970A). In the experiment, the data acquisition unit is used in conjunction with the corresponding module. It can directly measure parameters such as thermocouples, resistances, AC/DC voltages, etc. The characteristics are as follows:

(1) The scanning rate supports up to 450 readings per second at most.
(2) Each channel can be set with Ax + B to calibrate thermocouples and support over-temperature alarms.
(3) The configured Bench Vue DAQ data analysis software supports a graphical network interface, which is convenient for the real-time display and analysis of measured values.
(4) With a built-in signal conditioning function, it minimizes the noise from external wiring entering the system and avoids subjective speculation in error analysis.

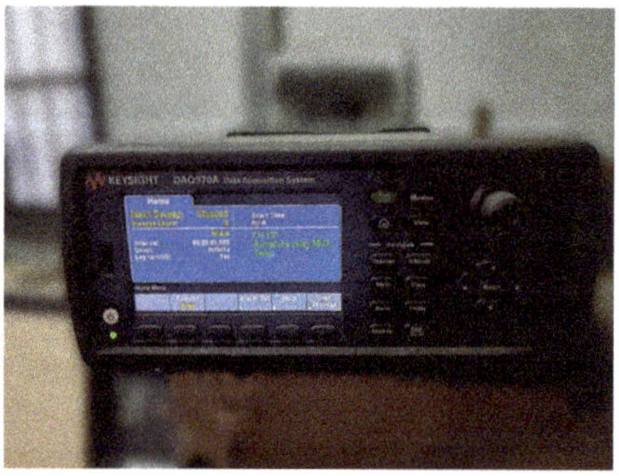

Figure 4. The data acquisition device Keysight DAQ970A.

3.2. Design Scheme

The experiments were carried out by arranging a thermoelectric module in the condensing section based on the determined tilted structure of the thermosiphon's evaporating section to ensure the same conditions of heat input. The thermoelectric conversion performance of the system was investigated by examining the effects of three factors, namely the thermoelectric device coverage (15.0%, 22.5%, and 30.0%), connection mode (series and parallel), and cold-end heat dissipation mode (air-cooled and water-cooled), on the heat transfer process. Figure 5 shows the schematic diagram of the covering principle of thermoelectric devices. The optimum operating parameters for each factor were determined to obtain the optimum thermoelectric conversion efficiency of the coal fire waste heat conversion system at constant heat flow, where the thermoelectric device coverage was the area of the thermoelectric unit covered to the overall area of the condensing section.

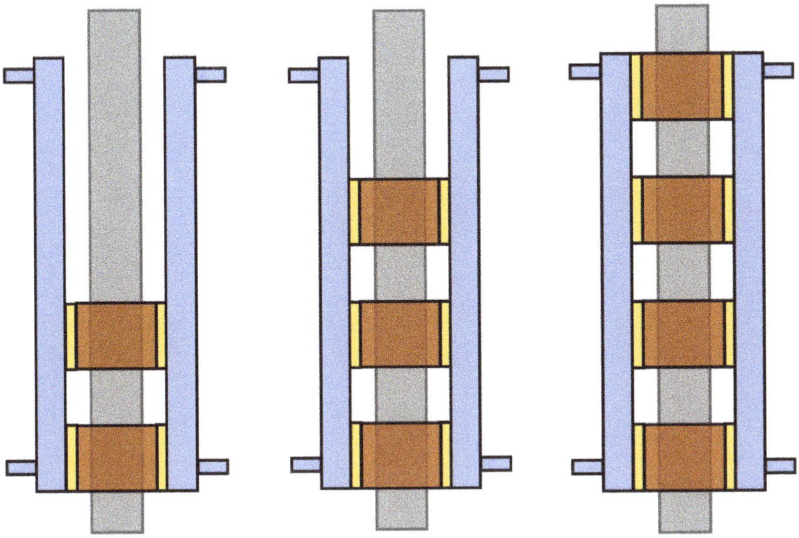

Figure 5. Schematic diagram of the thermoelectric device coverages. (**a**) 15.0% coverage rate. (**b**) 22.5% coverage rate. (**c**) 30.0% coverage rate.

3.3. Experimental Conditions

Taking the experimental test system with a thermoelectric device coverage rate of 22.5% under the water-cooled mode as an example, as shown in Figure 6, the input power was set to 100 W and a thermosiphon with an inclination angle β = 30° was chosen as heat extraction device. The pipeline and heating section of the cooling circulation system were wrapped in thermal cotton to reduce heat loss. Based on the measuring point of the external wall surface of the heat exchanger, the heating was increased in the four-way outlet, obtaining a fixed difference in temperature at the cold and hot ends of the thermoelectric devices at different levels of the thermosiphon. T-type thermocouples were arranged at the hot and cold ends of each of the thermoelectric devices and fixed with high-temperature thermal conductive tape to measure the temperature change at the cold end. The operation process was as follows:

- The cooling circulation system. The circulation water temperature was set to 10 °C (with the ambient temperature as a reference), and the flow rate was 8.3 L/min. Subsequently, the water temperature was stabilized.
- The heating system. The input power was adjusted to be stable near 100 W, and the thermosiphon was heated for a period until the coal fire waste heat conversion system ran stably and the temperature of the relevant measuring point did not fluctuate greatly.
- The data acquisition system. During the stable operation of the coal fire waste heat conversion system, the horizontal open-circuit voltage and total open-circuit voltage of the thermoelectric devices were measured and recorded by a multimeter. In addition, the load coil was added, forming a loop to measure the output voltage and current.

It can be concluded from the test that during the experiment, a thermosiphon with an inclination angle β = 30° should be chosen as the heat extraction device, the cooling circulation pipeline and heating section need to be wrapped with thermal insulation cotton, and the cold and hot sections of each thermoelectric device should be arranged with T-type thermocouples to measure temperature changes. The difference between air-cooled and water-cooled modes was that in the air-cooled mode, there was natural air convection, and the condensing section and thermoelectric module carried out heat exchange with the surrounding environment of the experiment without increasing the cooling circulation system. The measuring points are distributed on the outer wall of the thermosiphon at the four-way water inlet and outlet points.

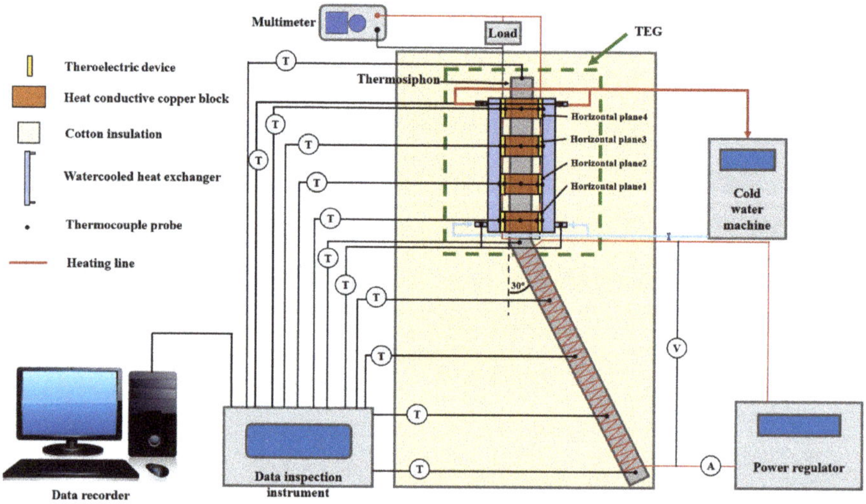

Figure 6. Schematic diagram of thermoelectric performance test system of coal fire waste heat conversion system.

3.4. Data Processing and Uncertainty

Under the premise of a single material length scale and module, the thermoelectric device's power performance was consistent with the heat source's hot end temperature and the cold end's heat dissipation mode. The temperature difference between the hot and cold ends of the thermoelectric devices can be calculated as follows:

$$\Delta T = \overline{T_{TEM,h}} - \overline{T_{TEM,c}} \quad (4)$$

where ΔT is the temperature difference between the hot and cold ends of the thermoelectric devices. $T_{TEM,h}$ and $T_{TEM,C}$ are the average temperatures of the hot and cold ends for the coal fire waste heat conversion system under stable operation, respectively, in °C.

The nominal power calculation formula is as follows:

$$P = \frac{U^2}{R} \quad (5)$$

where P is the nominal power, W. U is the output voltage, V. R indicates the load resistance, Ω.

The formula for the thermoelectric conversion efficiency of the coal fire waste heat conversion system is as follows:

$$\eta = \frac{P}{Q_{in}} \quad (6)$$

where η is the thermoelectric conversion efficiency, %. Q_{in} is the input power, W.

4. Results and Discussion

4.1. Compatibility Analysis of Thermosiphon and Thermoelectric Module

The thermoelectric conversion performance is mainly affected by the heat source temperature and cold end temperature. When the boundary conditions of heat transfer at the cold end are determined, the fit between the thermosiphon and the thermoelectric module determines the heat transmission. The heat of the thermosiphon absorption transfer in the coal fire waste heat conversion system was taken as the thermoelectric device's converted electrical energy. The other part is the heat loss generated during the transfer, including the heat loss of the heat transfer medium (conductive copper block) and the loss to the environment. Figure 7 describes the variation in the wall temperature of the condensing section of the thermosiphon and the thermoelectric device's hot end temperature with the thermoelectric device's coverage rate under air-cooled and water-cooled modes. The compatibility of the thermosiphon and the thermoelectric module was explored by comparison.

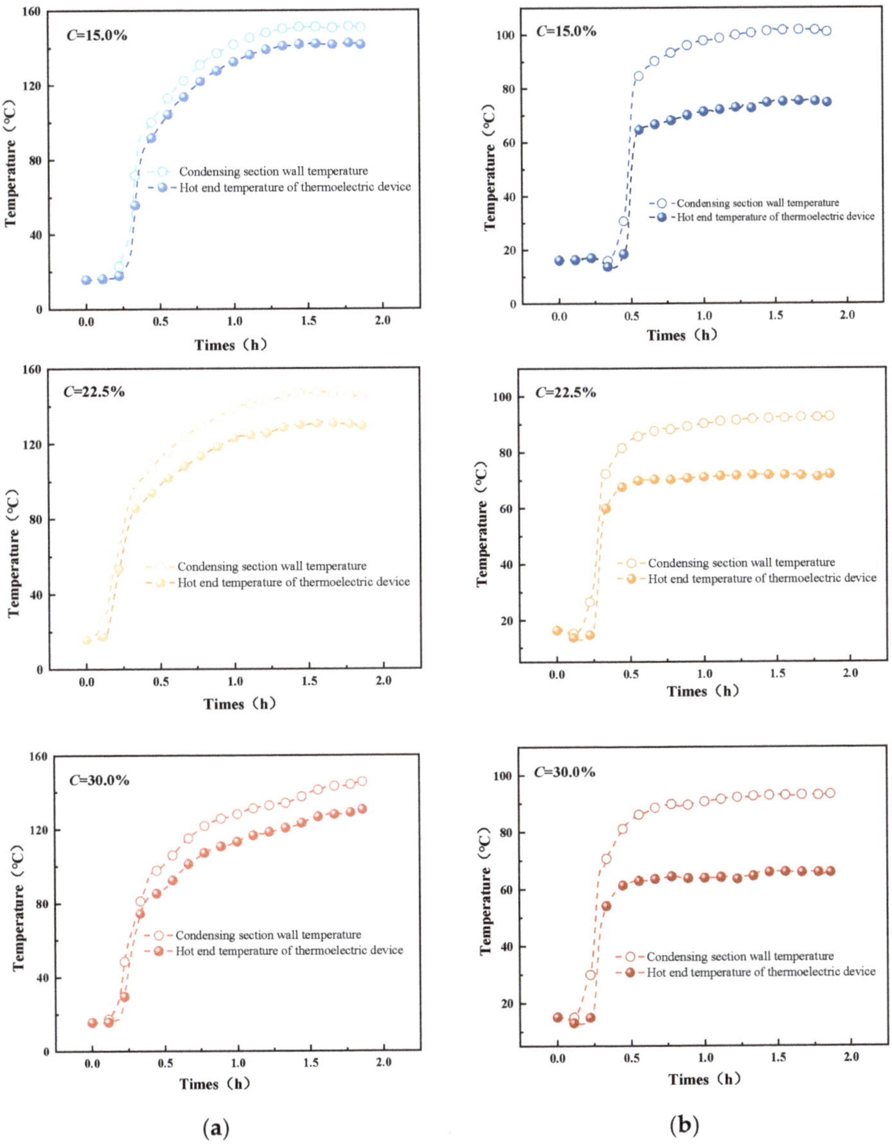

Figure 7. Variation in wall temperature in the condensing section and the thermoelectric device's hot end temperature. (**a**) Air-cooled mode and (**b**) water-cooled mode.

As Figure 7a shows, under air-cooled conditions, there is a similarity between the thermoelectric device hot end temperature trend and the condensation section temperature trend. The heating system was initiated, but due to the short heating time of the thermosiphon, it failed to make any changes. When the internal working medium of the thermosiphon was heat-evaporated, the system started to work immediately. The sharp rise in temperature in the condensing section also caused the hot end temperature of the thermoelectric device to rise rapidly. After a period of time, the heat input and output of the system were in balance, and the temperatures of the condensing section and the thermoelectric device also tended to stabilize. When the thermoelectric device had a coverage of 15.0%, the condensing section's wall temperature was similar to the temperature of the hot end of the thermoelectric device, but the latter was slightly lower. When the coverage of the thermoelectric devices increased to 22.5%, the gap between the two increased, and the thermoelectric devices' temperature decreased compared with previous values; this is mainly due to the increased number of thermoelectric devices, which caused the volume of the heat transfer medium to increase and the heat loss to also increase. As the number of thermoelectric devices continued to increase, the gap between the two decreased, and since the coverage of the thermoelectric devices increased, the heat exchange area was reduced from the condensate section to the external environment, thus delivering more heat to the thermoelectric devices. Therefore, the temperature of the hot end of the thermoelectric device did not change based on the amount of heat transfer medium. The wall temperature of the condensing section presented very similar changes under the three coverage rates. It could be seen that, compared with 15.0% coverage, 22.5% coverage had a higher utilization rate, although the heat loss increased, while 30% coverage was allocated by the same amount of heat due to the number of media; therefore, its utilization rate was also less than that of 22.5%.

As presented in Figure 7b, the water-cooled mode had a more pronounced impact on the coal fire waste heat conversion system than the air-cooled mode. With the increase in the coverage rate of thermoelectric devices, the air-cooled temperature difference gradually rises, with the maximum temperature difference increasing from 5 °C to 20 °C and the wall temperature of the condensation section reaching a maximum close to 160 °C. Under water-cooled conditions, however, as the coverage rate of thermoelectric devices increases, the maximum temperature difference does not change much, and the wall temperature of the condensation section reaches a maximum close to 100 °C. The temperature difference between the condensing section and the hot end of the thermoelectric device continued to grow over time as they were both trending in the same direction. The thermoelectric device's hot end temperature dropped initially and then abruptly rose during the early stages. The thermal conductivity of liquid water was roughly 0.59 W/m/K, which was significantly higher than that of air (0.02 W/m/K) at standard pressure and temperature [29]. The water-cooled heat exchanger's cooling force was greater than that of the thermoelectric device at first, causing a temporary fall in the temperature of the thermoelectric device's hot end. The latter, which was consistent with that of the air-cooled mode, had a rapid increase to a stable change. Under the water-cooled mode, the temperatures of both the condensing section and the thermoelectric element were lower than under the air-cooled mode, and the temperature difference between the two significantly increased. The reason for this phenomenon was that, on the one hand, the water-cooled heat exchanger had a significant impact on the system and more heat was absorbed by the internally circulating water after the thermal copper block and the thermoelectric device, which accelerated heat loss in the condensing section. On the other hand, the thermoelectric device made direct contact with the heat exchanger. Due to the low temperature of the hot end, the condensing section and the thermoelectric device's hot end had an increased temperature difference. The temperature difference between the condensing section and the thermoelectric device's hot end when the coverage rate was 22.5% was relatively minor when compared to the temperature difference between the thermoelectric device coverage rates of 15.0% and 30.0%. Compared with the 22.5% air-cooled rate, the trend flattened earlier and the temperature differential grew because of the difference in the cooling method. The increase in heat dissipation caused by the increased volume of the heat transfer medium led the condensation section temperature and the thermoelectric device hot end temperature to range between 15% and 30% coverage. Simultaneously, the distance between the two lines gradually increased with the number of thermoelectric devices.

In summary, under both air-cooled and water-cooled methods, there is a huge difference in the temperature difference between the wall temperature of the condensation section and the hot end of the thermoelectric device. The temperature difference in the air-cooled mode is concentrated within 0–30 °C, and that in the water-cooled mode is concentrated within 0–40 °C. Meanwhile, under the air-cooled method, the maximum wall temperature of the condensation section reaches nearly 160 °C, while under the water-cooled method, the maximum temperature is only around 100 °C. This indicates that when water is used as a medium, it will store a portion of heat by itself, and the effect of using the water-cooled method on the coal fire waste heat conversion system is more significant.

4.2. Temperature Difference Between Cold and Hot Ends in Thermoelectric Modules

Considering the non-uniformity of the temperature distribution on the wall surface of the condensing section, the thermocouple devices' cold and hot end temperature differences at different levels were further compared and discussed in detail. The two thermoelectric devices were arranged on the same horizontal plane. The position close to the evaporation section was called horizontal plane 1 (including the two thermoelectric semiconductor cooling plates). The other positions were named horizontal plane 2, horizontal plane 3, and horizontal plane 4 in the vertical direction, and each horizontal plane was equally spaced. Figure 8a,b present the temperature difference between the thermoelectric devices at different levels under air-cooled and water-cooled conditions, respectively. The trend in the different levels of temperature differences between thermoelectric-cooled pieces was observed under air-cooled and water-cooled modes. Although the coverage of the thermoelectric devices was different in the air-cooled mode, the temperature difference between thermoelectric devices increased sharply at first, then decreased rapidly, and finally tended to be stable.

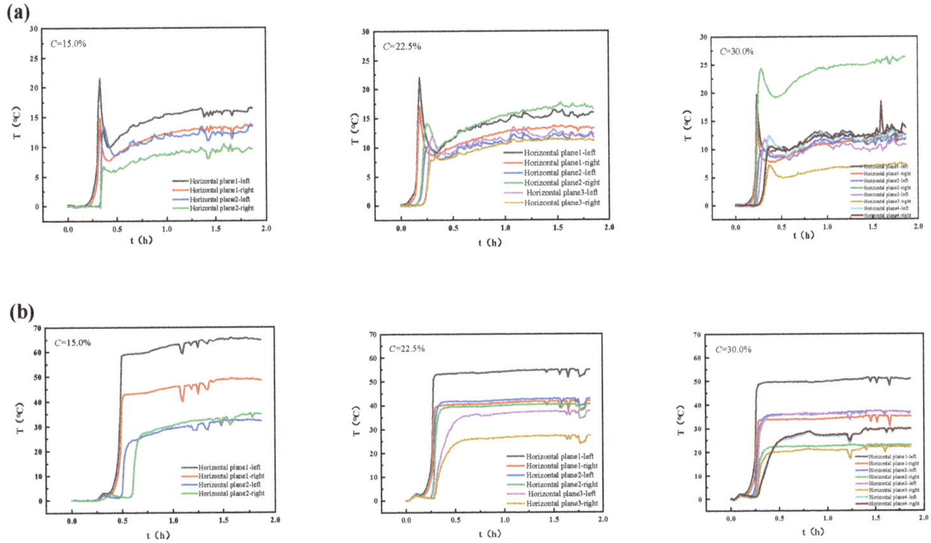

Figure 8. Temperature difference diagram of thermoelectric module with (**a**) air-cooled and (**b**) water-cooled modes.

Due to the complexity of the exothermic process of the condensation of the working medium in the thermosiphon, the thermoelectric components had different temperature differences at the same level. When the air-cooled mode was selected and the thermocouple coverage was 15.0%, the thermocouple temperature difference fluctuated in the range of 5–15 °C. When the thermocouple coverage increased to 22.5%, the thermocouple temperature difference at all levels was significant and fluctuated in the range of 10–17 °C. With the same input power and convective heat transfer, the increase in temperature difference indicated that more heat released from the condensing section was transferred to the thermoelectric device for thermoelectric conversion and less heat was dissipated to the environment. The overall temperature difference showed a decreasing trend as the coverage of the thermoelectric device increased. This was because the increased coverage of the thermoelectric module in the condensing section distributed the same heat more uniformly to the individual thermoelectric devices, which led to a reduction in the hot ends' temperature, resulting in a decrease in the hot and cold ends' temperature difference. The size of the difference in temperature was almost determined by the distance of the thermosiphon evaporation section: the closer the distance, the greater the difference in temperature, regardless of the coverage rate. In the scenario where horizontal plane 1 was greater than horizontal plane 2 and horizontal plane 3, temperature differences almost always existed. This was mainly caused by the higher temperature of the inner wall of the condensing section at that location. Nonetheless, the temperature difference between the left and right thermal electric devices was irregular. The temperature difference at level 2 was maximized at a coverage of 30%, which was due to the uneven heat distribution.

According to Figure 8b, when the water-cooled mode is selected, regardless of the thermoelectric device coverage rate, the temperature difference in the level 1 thermoelectric device is always greater than or equal to that of other levels. When the thermoelectric device coverage rate is 15%, the left–right temperature difference in level 1 is approximately 20 °C, and that of level 2 is approximately 3 °C. When the thermoelectric device coverage rate is 22.5%, the left–right temperature difference in level 1 is approximately 15 °C, while that of level 2 is still approximately 3 °C and that of level 3 is approximately 15 °C. When the thermoelectric device coverage rate is 30%, the left–right temperature difference in level 1 remains approximately 15 °C, while that of level 2 also increases to 15 °C, the left–right temperature difference in level 3 remains unchanged at 15 °C, and there is almost no temperature difference at level 4. When the temperature coverage was 22.5%, the overall difference in the temperature of the thermoelectric cooler increased, the maximum difference in temperature decreased, and the difference in temperature of the thermoelectric cooler was concentrated within 30–50 °C. With a further increase in the coverage of the thermoelectric device, more heat was dispersed toward the thermal device's hot end, the thermoelectric device's temperature difference was significant, and the fluctuation range of the change was large. Due to the lower coverage heat transfer efficiency, the rapid increase in the difference in temperature was delayed for 15% coverage of horizontal plane 2. As the coverage continued to increase, the maximum temperature difference progressively decreased, and the temperature differences at each level stage became closer and more likely to overlap.

The trends in thermoelectric equipment were different when choosing between air-cooled and water-cooled modes of heat dissipation. This was attributed to the superiority of forced convection heat transfer, which accelerated the heat loss from the cold end of the thermal power plant. In addition, the air-cooled mode was more unstable than the water-cooled mode.

The temperature profile became more and more compact as the coverage increased, and the ends of the air-cooled level 1 tended to be far apart. In addition, the overlap of cooling tended to be higher than that under the water-cooled mode at the same coverage. According to the principal analysis, the thermoelectric conversion relied on the cold and hot ends' temperature difference for power generation, and the larger the temperature difference was, the more it improved the power generation performance of the system.

Therefore, compared with the air-cooled mode, the water-cooled mode can enable the system to achieve better working conditions and output performance.

4.3. Horizontal Open-Circuit Voltage and Output Performance

Based on the good matching degree between the above systems and the advantages of the water-cooled mode, the effects of the coverage of thermoelectric devices on the thermoelectric module under the water-cooled mode were further explored. The open-circuit voltage of a battery is calculated as the difference between the electrode potential of the positive terminal and the electrode potential of the negative terminal when the battery is disconnected (when no current is flowing through the terminals). Therefore, the open-circuit voltage can directly reflect the ability of thermoelectric conversion. Just as the connection mode of the circuit elements would affect the current pass method, thus impacting the performance of the circuit elements, the connection mode between the thermoelectric devices affected the thermoelectric module's power generation performance. Therefore, based on the cooling mode and the coverage rate of the thermoelectric device, series and parallel connection methods were added, and then the influence of these three factors on the thermoelectric conversion performance of the coal fire waste heat conversion system was also analyzed. Figure 9 depicts the open-circuit voltage at each level in different thermoelectric device coverages. Although the thermoelectric device had different positions, the open-circuit voltage at different levels showed the same change trend with the change in connection mode. With the increase in coverage at all levels, both series and parallel voltages gradually decreased. Under the same cooling conditions, when the thermoelectric devices were connected in series with each other, the open-circuit voltage was greater than when they were connected in parallel. For the same coverage, the increase multiple in each horizontal plane was similar, the open-circuit voltage at horizontal plane 1 was higher than that at the other horizontal planes, and the open-circuit voltage gradually decreased as the distance from the evaporation section increased. When the thermoelectric device coverage was 15.0%, the maximum horizontal open-circuit voltage could reach 2.34 V. When the coverage of the thermoelectric devices increased to 30.0%, the open-circuit voltage at level 2 was slightly lower than that at other levels. This was mainly due to the uneven heat distribution on the wall of the condensing section, which led to differences in heat distribution at different levels. As the coverage of the thermoelectric devices increased, the heat of the thermoelectric device allocated to different locations tended to the average value, leading to an overall decrease in the horizontal open-circuit voltage.

Based on the above comparison and the analysis of each open-circuit voltage under different connection modes, the experiment added a load to form a complete loop on this basis, taking the voltage and nominal power as the visual data. Then, the influence of the connection mode and load value selection on the thermoelectric module's output performance was investigated, and the connection mode and the best matching load value were optimized. Based on the empirical values in previous research [21], 3, 7, 11, 15, and 20 Ω were chosen as the experimental values to measure the horizontal output voltage and the nominal power of different resistance values. To make the obtained rules more general and convincing, the horizontal output voltage and nominal power under the coverage of three kinds of thermoelectric devices were measured and calculated, as presented in Figures 5 and 6, respectively. By observing the value of the output voltage and nominal power, it could be judged that the loop's output voltage and nominal power in series were better than those in parallel connection, and the maximum output voltage's corresponding load value was different from the maximum nominal power's corresponding load value. As shown in Figure 10, the load resistance and its output voltage were approximately linearly related when connected in series and in parallel. The temperature of the voltage value at both ends of the load increased linearly with the increase in the load value when it was farther from the evaporation section. When farther away from the evaporating section, the voltage value at both ends of the load increases linearly with the increase in the load value. However, when closer to the evaporating section, when the load resistance increased, the output voltage also appeared to increase slowly. In addition, with more levels, the output voltage difference between the levels would decrease relatively.

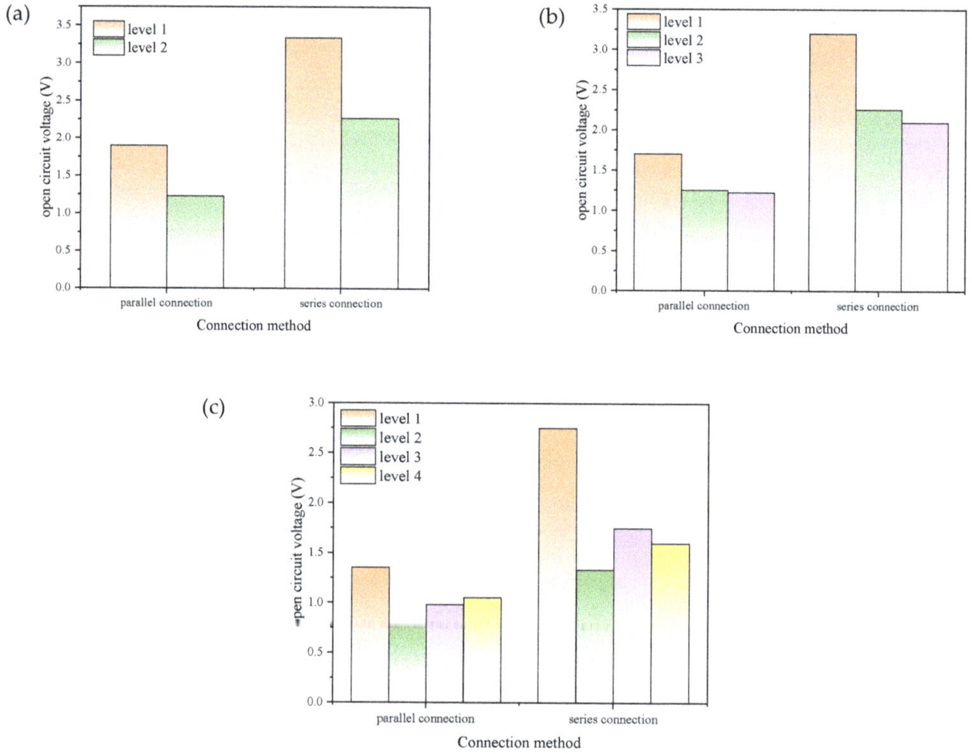

Figure 9. Open-circuit voltage under different thermoelectric device coverage rates. (**a**) 15.0% coverage rate. (**b**) 22.5% coverage rate. (**c**) 30.0% coverage rate.

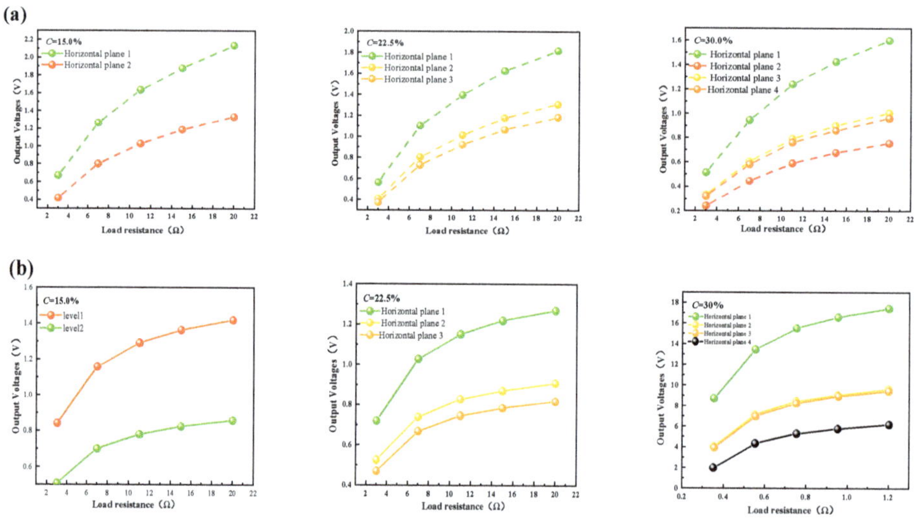

Figure 10. Output voltage at different levels under (**a**) series connection and (**b**) parallel connection.

The trends in the nominal power under the two connections were very different, as illustrated in Figure 11. As the load resistance value increased, the series connection nominal power presented an approximate parabolic variation. As the resistance value increased, the power increased at the beginning. In the parallel connection, the nominal power varied linearly with the increase in load resistance, and the load resistance corresponding to the maximum nominal power was 3 Ω. According to the maximum power transfer theorem, the maximum nominal power could be obtained when the value of the load resistance was equal to the value of the internal resistance of the DC circuit. In addition, it was estimated that the value of a single thermal device's internal resistance was 5.5 Ω.

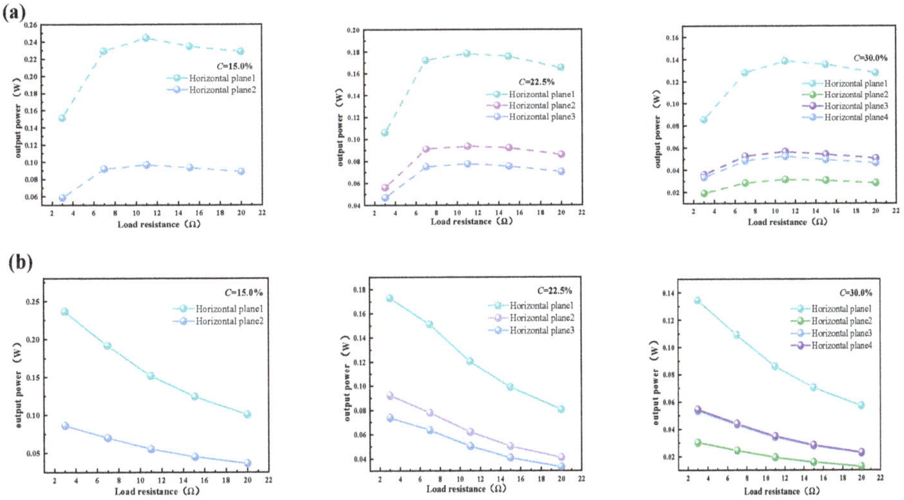

Figure 11. Nominal power at different levels under (**a**) series connection and (**b**) parallel connection.

4.4. Heat Transfer and Power Generation Performance

The thermoelectric conversion performance of the coal fire waste heat conversion system was closely related to the heat transfer process between the various components of the system. Figure 12 plots the temperature distribution and trends in each part of the system during stable operation. As the coverage of thermoelectric devices increased, the temperature in different regions of the system also changed. The horizontal coordinates E_{vap}, cond, TEM_h, TEM_c, W_{in}, and W_{out} represent the evaporation section, the condensing section, the thermoelectric module's hot end, the thermoelectric module's cold end, and the water-cooled heat exchanger's inlet and outlet in the system, respectively. For the thermosiphon and the thermoelectric module's hot end, the temperature decreased as the thermoelectric device's coverage increased. However, the trend in the temperature of the thermoelectric module's cold end was the opposite, and the increased coverage under the same cooling conditions added to the heat transfer area for the heat exchanger and thermoelectric module, thus increasing the heat exchanger's work force. Considering the heat exchanger's higher cooling water flow rate, the water-cooled heat exchanger's inlet and output temperatures varied less.

Based on the system's temperature distribution in different areas, the heat transmission thermal resistances within the system can be further calculated to evaluate the coal fire waste heat conversion system's heat transfer performance. The thermal resistance between the thermosiphon, the thermoelectric module, and the cold end of the thermoelectric module and the cooling water was calculated as follows [21]:

$$R_{hp} = \frac{T_{hp} - T_{tem,h}}{Q_{in}} R_{tem} = \frac{T_{tem,h} - T_{tem,c}}{Q_{in}} R_w = \frac{T_{tem,c} - \frac{T_{w,in}+T_{w,out}}{2}}{Q_{in}} \quad (7)$$

where R_{hp} is the thermal resistance of the thermosiphon tube, R_{tem} is the thermoelectric module's thermal resistance, and R_w is the thermal resistance between the cold end of the thermoelectric module and the heat exchanger, K/W; T_{hp} is the wall temperature of the thermosiphon tube, $T_{tem,h}$ and $T_{tem,c}$ are the temperatures of the thermoelectric module's hot and cold ends, and $T_{w,in}$ and $T_{w,out}$ are the inlet and outlet water temperatures of the heat exchanger, °C.

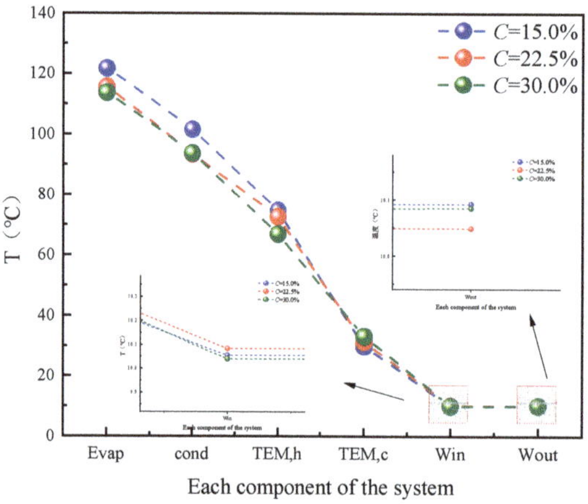

Figure 12. Temperature distribution of components of coal fire waste heat conversion system.

From Figure 13, it can be seen that when the thermoelectric components' coverage increases, the coal fire waste heat conversion system's total heat transfer thermal resistance increases, and the maximum heat transfer thermal resistance of the system is 1.64 K/W. Comparing the thermal resistance of the three parts of the coal fire waste heat conversion system, the thermosiphon had the smallest thermal resistance, followed by the thermoelectric module's cold end temperature and the heat exchanger, and the thermoelectric module had the largest thermal resistance. The heat transfer thermal resistance of the three parts of the system (R_{hp}, R_{tem}, and R_w) increased with the increase in coverage. The thermoelectric device's coverage had little effect on the thermosiphon's thermal resistance, which remained almost unchanged for the three coverage rates. The increase in coverage caused the thermoelectric module's capacity and thermal resistance to increase accordingly. Meanwhile, the heat transfer area between the water-cooled heat exchanger and the thermoelectric module increased, and their contact increased thermal resistance. The temperature difference between the evaporation and condensation sections also had great influence on the thermoelectric module, and the thermosiphon's thermal resistance could not be ignored. For the thermoelectric module, the coverage rate had a greater influence on the thermoelectric module's thermal resistance, the thermal resistance of the cold end of the module, and the heat exchanger's thermal resistance. In the selection of the coverage rate, researchers should consider the combined influence of heat transfer thermal resistance and system output performance.

Figure 14 describes the open-circuit voltage value of the coal fire waste heat conversion system under the coverage of three types of thermoelectric devices. When the coverage rate increased from 15.0% to 22.5%, the system's open-circuit voltage significantly increased, and the growth rate was 24%. It was found that the system's open-circuit voltage with 22.5% coverage and 30.0% coverage was approximately equal and did not increase significantly because of the increase in coverage. The results showed that the thermoelectric conversion performance of the coal fire waste heat conversion system did not increase with the increase in the coverage of the thermoelectric devices; the coverage had a peak value, and if this was too large, it caused resource waste.

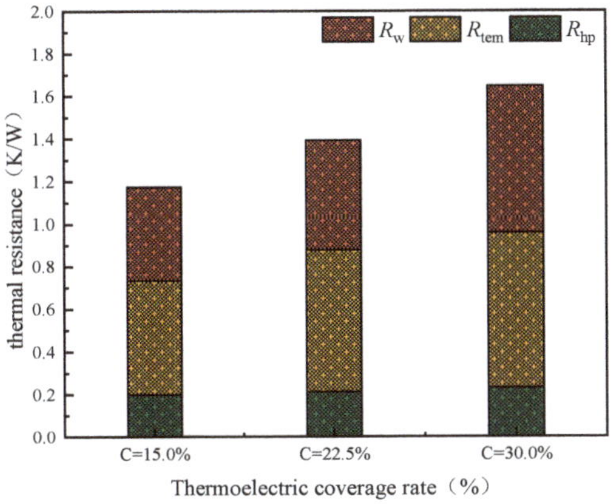

Figure 13. Heat transfer resistance distribution of coal fire waste heat conversion system.

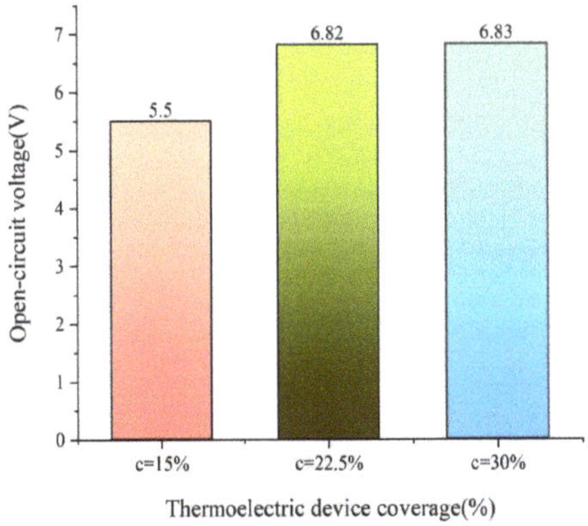

Figure 14. Open-circuit voltage of thermoelectric module.

Figure 15a,b depict that in the load resistance range of 3–20 Ω, when the load resistance increased, the output voltage increased, and the nominal power also rose. When the coverage was 15.0%, the output voltage could reach the maximum value, and the nominal power was also maximized, with the next highest values achieved at 22.5% coverage, whereas the minimum output voltage and nominal power values were observed when the coverage was 30.0%. When the load resistance was 20 Ω, the maximum output voltage and nominal power were 2.61 V and 0.34 W. However, the variation in the output voltage shows that the range of the load resistance selection was too small to achieve the optimum output performance for the system. Due to the limitation of the load coil, the system could not obtain the optimal nominal power for different coverage ranges directly by measurement. Therefore, the open-circuit voltage measured values and the thermoelectric device internal resistance estimated values were used to calculate the output voltage's theoretical values, and then they were compared with the measured values, as presented in Table 2. Equation (8) is the calculation formula for the theoretical value of the output voltage. Among them, U' represents the theoretical value of the output voltage, U_k is the open-circuit voltage value, and R_{in} and R_{out} represent the internal resistance and external resistance, respectively. The internal resistance is estimated to be 5.5 Ω. When comparing the theoretical and measured values of the output voltage under different coverage rates, the difference between the two was small, which in turn allowed us to determine the correctness of the estimated value of 5.5 Ω for the internal resistance of the thermoelectric device.

$$U' = \frac{U_k}{R_{in} + R_{out}} \quad (8)$$

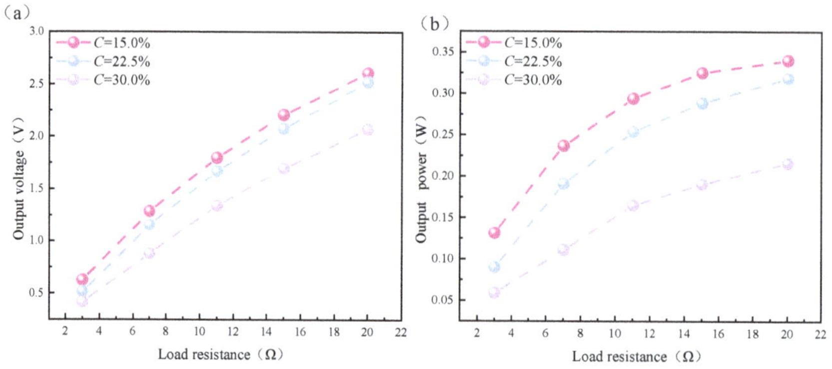

Figure 15. System's output voltage and nominal power with different coverage rates. (**a**) Output voltage, and (**b**) output power.

Table 2. Theoretical and measured values of system output voltage under different coverage rates.

Load Resistance (Ω)	C = 15.0%, Output Voltage (V)			C = 22.5%, Output Voltage (V)			C = 30.0%, Output Voltage (V)		
	Theoretical Value	Measured Value	Relative Error	Theoretical Value	Measured Value	Relative Error	Theoretical Value	Measured Value	Relative Error
3	0.66	0.63	0.048	0.56	0.52	0.077	0.43	0.42	0.024
7	1.32	1.29	0.023	1.19	1.16	0.026	0.93	0.88	0.057
11	1.83	1.80	0.017	1.7	1.67	0.018	1.36	1.34	0.015
15	2.32	2.31	0.004	2.13	2.08	0.024	1.73	1.70	0.018
20	2.61	2.61	0	2.57	2.53	0.016	2.13	2.08	0.024

Based on an internal resistance value of 5.5 Ω, the optimal matching load value, the system's maximum nominal power, and its maximum thermoelectric conversion efficiency under different coverages were determined, according to the maximum power transmission theorem. As seen in Table 3, at a 22.5% coverage, the optimal matching load value was 33 Ω, the optimal nominal power of the system was 0.35 W, and the optimal thermoelectric conversion efficiency was 0.35%. The above detailed discussion shows the impact of each factor on the thermoelectric conversion performance of the coal fire waste heat conversion system, including the cooling mode, the connection mode of the thermoelectric devices, and their coverage rate. The optimal combination of thermoelectric modules, which could ensure the optimal operation of the system, is the water-cooled mode, series connection, and a thermoelectric device coverage rate of 22.5%.

Table 3. Maximum system nominal power and thermoelectric conversion efficiency corresponding to different coverage of thermoelectric devices.

Coverage of Thermoelectric Devices (%)	Optimal Load Matching Value (Ω)	Maximum Nominal Power (W)	Thermoelectric Conversion Efficiency (%)
15.0%	22	0.34	0.34%
22.5%	33	0.35	0.35%
30.0%	44	0.26	0.26%

5. Conclusions

This work focused on the influence of the coverage of thermoelectric devices, the connection mode, and the cooling mode of the cold end on the thermoelectric generator. The main conclusions can be drawn as follows:

(1) Through changing the temperature of the condensation section and the thermoelectric module's hot end under the air-cooled mode, it was found that the addition of thermal conductive media was beneficial for increasing the heating area of the thermoelectric modules. Compared to the air-cooled mode, the water-cooled mode had a more significant impact on the system in terms of improving the thermal energy utilization rate. It can provide a larger temperature difference for thermoelectric modules, with a maximum temperature difference of 65.90 °C.

(2) When the water-cooled method is selected and the thermoelectric device coverage rate is 15.0%, the temperature difference in the level 1 thermoelectric device is significantly higher than that of level 2, and the maximum left–right temperature difference in the thermoelectric device can reach 20 °C. When the temperature difference coverage rate is 22.5%, the overall temperature difference in the thermoelectric module increases, but the maximum temperature difference decreases, and the left–right temperature difference in the thermoelectric device is concentrated within the range of 15 °C. When the coverage rate is 30%, the overall temperature difference shows a downward trend and the left–right temperature difference in the thermoelectric device is still concentrated within the range of 15 °C.

(3) The horizontal open-circuit voltage was higher in series than in parallel. When the load was added to form a loop, it could be found that the output voltage was greater in series than in parallel, the maximum horizontal open-circuit voltage value can reach 3.34 V, and the maximum output voltage is 2.61 V. The thermoelectric device's internal resistance was estimated to be 5.5 Ω.

(4) The system thermal resistance increases with the increase in the coverage range of the thermoelectric device. The thermal resistance of the thermoelectric module, the cold end of the thermoelectric module, and the heat exchanger are more affected by the coverage range. Comparing the open-circuit voltage, output voltage, and output power under three thermoelectric device coverage ranges of 15.0%, 22.5%, and 30.0%, when the coverage range is 22.5%, the output performance of the system is better, with an output power of 0.35 W and a thermoelectric conversion efficiency of 0.35%. The optimal combination of the thermoelectric modules obtained from the research results can provide ideas for the application of waste heat utilization in in situ coal fire disasters.

Author Contributions: Conceptualization, Q.B., Z.W. and Y.X.; methodology, Q.B. and X.G.; validation, B.L., W.C. and Z.W.; investigation, Q.B. and X.G.; resources, B.L. and W.C.; data curation, X.G.; writing—original draft preparation, Q.B. and X.G.; writing—review and editing, Z.W. and Y.X.; visualization, B.L.; supervision, Y.X.; project administration, Z.W. All authors have read and agreed to the published version of the manuscript.

Funding: This research received no external funding.

Data Availability Statement: The original contributions presented in the study are included in the article, further inquiries can be directed to the corresponding author.

Conflicts of Interest: Authors Qingfeng Bao, Xiuting Guo, Bo Li, Wuyi Chen were employed by the company Shanxi Lu'an Huanbao Nengyuan Kaifa Co., Ltd. The remaining authors declare that the research was conducted in the absence of any commercial or financial relationships that could be construed as a potential conflict of interest. The Shanxi Lu'an Huanbao Nengyuan Kaifa Co., Ltd. had no role in the design of the study; in the collection, analyses, or interpretation of data; in the writing of the manuscript, or in the decision to publish the results.

Nomenclature

α	Seebeck coefficient of the semiconductor material	d_i	inner diameters of the thermosyphon [m]
ΔV	potential difference in the TEG [V]	d_o	outer diameters of the thermosyphon [m]
ΔT	temperature difference in the TEG [°C]	λ	thermal conductivity coefficient of carbon steel [48 W/(m·K)]
T_c	temperatures of the cold side [°C]	$T_{w,h,o}$	wall temperatures of the evaporation section [°C]
T_h	temperatures of the hot side [°C]	$T_{w,c,o}$	wall temperatures of the condensation section [°C]
V_K	voltage generated by open circuit [V]	$T_{TEM,h}$	average temperatures of the hot ends [°C]
ΔT_{TEG}	temperature difference between the two sides of the TEG [°C]	$T_{TEM,c}$	average temperatures of the cold ends [°C]
α_{eff}	heat transfer performance of the thermosyphon	P	nominal power [W]
Q_{in}	input heat [W]	U	output voltage [V]
L_e	lengths of the evaporation section [m]	R	load resistance [Ω]
L_c	lengths of the condensation section [m]	η	thermoelectric conversion efficiency [%]
R_{hp}	thermal resistance of the thermosiphon tube [K/W]	R_{tem}	thermal resistance of thermoelectric module [K/W]
R_w	thermal resistance between the cold end of the thermoelectric module and the heat exchange [K/W]	T_{hp}	wall temperature of the thermosiphon tube [°C]
$T_{w,in}$	inlet water temperatures of the heat exchanger [°C]	$T_{w,out}$	outlet water temperatures of the heat exchanger [°C]
U'	theoretical value of the output voltage [V]	U_k	open-circuit voltage value
R_{in}	internal resistance [Ω]	R_{out}	external resistance [Ω]

References

1. Jouhara, H.; Żabnieńska-Góra, A.; Khordehgah, N.; Doraghi, Q.; Ahmad, L.; Norman, L.; Axcell, B.; Wrobel, L.; Dai, S. Thermoelectric generator (TEG) technologies and applications. *Int. J. Thermofluids* **2021**, *9*, 100063. [CrossRef]
2. Luo, D.; Yan, Y.; Wang, R.; Zhou, W. Numerical investigation on the dynamic response characteristics of a thermoelectric generator module under transient temperature excitations. *Renew. Energ.* **2021**, *170*, 811–823. [CrossRef]
3. Nozariasbmarz, A.; Poudel, B.; Li, W.; Kang, H.B.; Zhu, H.; Priya, S. Bismuth telluride thermoelectrics with 8% module efficiency for waste heat recovery application. *IScience* **2020**, *23*, 101340. [CrossRef] [PubMed]
4. Riyadi, T.W.B.; Utomo, B.R.; Effendy, M.; Wijayanta, A.T.; Al-Kayiem, H.H. Effect of thermal cycling with various heating rates on the performance of thermoelectric modules. *Int. J. Therm. Sci.* **2022**, *178*, 107601. [CrossRef]
5. Deng, J.; Qu, G.Y.; Ren, S.J.; Wang, C.P.; Xiao, Y.; Wang, J.R.; Zhao, X. Investigation on high temperature point detection of spontaneous combustion of loose coal based on optimal acoustic signal. *Process Saf. Environ.* **2024**, *185*, 423–434. [CrossRef]
6. Qu, G.Y.; Deng, J.; Ren, S.J.; Wang, C.P.; Song, Z.Y.; Wang, J.R.; Xu, Y.; Zhao, X.-Q. Research on temperature measurement model of loose coal considering humidity using acoustic wave method. *Fuel* **2024**, *373*, 132317. [CrossRef]
7. Xiao, Y.; Yin, L.; Tian, Y.; Li, S.G.; Zhai, X.W.; Shu, C.M.; Ren, S.J. Sustainable utilisation and transformation of the thermal energy from coalfield fires: A comprehensive review. *Appl. Therm. Eng.* **2023**, *233*, 121164. [CrossRef]
8. Liu, Y.; Wen, H.; Chen, C.; Guo, J.; Jin, Y.; Zheng, X.; Cheng, X.; Li, D. Research status and development trend of coal spontaneous combustion fire and prevention technology in China: A review. *ACS Omega* **2024**, *9*, 21727–21750. [CrossRef]
9. Gomaa, A.; Rady, W.A.; Youssef, A.Z.; Elsaid, A.M. Thermal performance of heat pipe at different internal groove ratios and working fluids: An experimental investigation. *Therm. Sci. Eng. Prog.* **2023**, *41*, 101827. [CrossRef]
10. Sarkar, D.; Savory, E.; DeGroot, C. Thermodynamic modeling of thermosyphons and heat pipes. *Phys. Fluids* **2023**, *35*, 123307. [CrossRef]
11. Machado, P.L.; Pereira, T.S.; Trindade, M.G.; Biglia, F.M.; Santos, P.H.; Tadano, Y.S.; Alves, T.A. Estimating thermal performance of thermosyphons by artificial neural networks. *Alex. Eng. J.* **2023**, *79*, 93–104. [CrossRef]
12. Kim, Y.; Shin, D.; Kim, J.; You, S.; Lee, J. Boiling and condensation heat transfer of inclined two- phase closed thermosyphon with various filling ratios. *Appl. Therm. Eng.* **2018**, *145*, 328–342. [CrossRef]
13. Solomon, A.B.; Danie, V.A.; Ramachandran, K.; Pillai, B.C.; Singh, R.R.; Sharifpur, M.; Meyer, J.P. Performance enhancement of a two-phase closed thermosiphon with a thin porous copper coating. *Int. Commun. Heat Mass Transf.* **2017**, *82*, 9–19. [CrossRef]
14. Alammar, A.A.; Al-Dadah, R.K.; Mahmoud, S.M. Numerical investigation of effect of fill ratio and inclination angle on a thermosiphon heat pipe thermal performance. *Appl. Therm. Eng.* **2016**, *108*, 1055–1065. [CrossRef]
15. Khazaee, I.; Hosseini, R.; Kianifar, A.; Noie, S.H. Experimental consideration and correlation of heat transfer of a two-phase closed thermosyphon due to the inclination angle, filling ratio, and aspect ratio. *J. Enhanc. Heat Transf.* **2011**, *18*, 31–40. [CrossRef]
16. Tong, Z.; Fang, C.; Zang, G.; Wu, H. Experimental study on the passive regulation of a CO2 two-phase thermosyphon loop. *Appl. Therm. Eng.* **2023**, *219*, 119457. [CrossRef]
17. Zhong, K.Q.; Xiao, Y.; Lu, X.; Deng, J.; Yin, L.; Tian, Y.; Shu, C.-M. Nanofluidic two-phase closed thermosyphon-assisted thermoelectric generator for heat recovery from coal spontaneous combustion. *Appl. Therm. Eng.* **2021**, *197*, 117397. [CrossRef]
18. He, W.; Guo, R.; Liu, S.; Zhu, K.; Wang, S. Temperature gradient characteristics and effect on optimal thermoelectric performance in exhaust power-generation systems. *Appl. Energy* **2020**, *261*, 114366. [CrossRef]

19. Chen, Y.; Hou, X.; Ma, C.; Dou, Y.; Wu, W. Review of Development Status of Bi2Te3-Based Semiconductor Thermoelectric Power Generation. *Adv. Mater. Sci. Eng.* **2018**, *2018*, 1210562. [CrossRef]
20. Luo, Y.; Zhang, L.; Li, J.; Li, C.; Xie, L.; Liu, Z.; Meng, F.; Qing, X. Study on thermal conductance allocation ratio of heat sink of thermoelectric cooler for electronic device in cold region. *Energy Procedia* **2015**, *75*, 603–607. [CrossRef]
21. Alegria, P.; Catalan, L.; Araiz, M.; Rodriguez, A.; Astrain, D. Experimental development of a novel thermoelectric generator without moving parts to harness shallow hot dry rock fields. *Appl. Therm. Eng.* **2021**, *200*, 117619. [CrossRef]
22. Ju, Y.S.; Lee, C. Evaluation of the energy efficiency of the shell coal gasification process by coal type. *Energy Conver. Manag.* **2017**, *143*, 123–136. [CrossRef]
23. Hu, X.K.; Takazawa, H.; Nagase, K.; Ohta, M.; Yamamoto, A. Three-dimensional finite element simulation for a thermoelectric generator module. *J. Electron. Mater.* **2015**, *44*, 3637–3645. [CrossRef]
24. Shi, Y.; Zhu, Z.; Deng, Y.; Zhu, W.; Chen, X.; Zhao, Y. A real-sized three-dimensional numerical model of thermoelectric generators at a given thermal input and matched load resistance. *Energy Conver. Manag.* **2015**, *101*, 713–720. [CrossRef]
25. Hans, R.; Manikandan, S.; Kaushik, S.C. Performance Optimization of Two-Stage Exoreversible Thermoelectric Converter in Electrically Series and Parallel Configuration. *J. Electron. Mater.* **2015**, *44*, 3571–3580. [CrossRef]
26. Zhao, Y.; Fan, Y.; Li, W.; Li, Y.Z.; Ge, M.H.; Xie, L.Y. Experimental investigation of heat pipe thermoelectric generator. *Energy Conver. Manag.* **2022**, *252*, 115123. [CrossRef]
27. Li, W.; Paul, M.C.; Siviter, J.; Montecucco, A.; Knox, A.R.; Sweet, T.; Min, G.; Baig, H.; Mallick, T.; Han, G.; et al. Thermal performance of two heat exchangers for thermoelectric generators. *Case Stud. Therm. Eng.* **2016**, *8*, 164–175. [CrossRef]
28. Nugraha, N.A.; Baskoro, D.C.; Riyadi, T.W.B.; Wijayanta, A.T. Thermal characteristics of thermoelectric generator at various heating rates: Experimental and numerical assessments. *Therm. Sci. Eng. Progress* **2024**, *52*, 102671. [CrossRef]
29. Liley, P.E. Thermophysical properties of fluids. In *Mechanical Engineers' Handbook, Volume 4: Energy and Power*; John Wiley & Sons: Hoboken, NJ, USA, 2015; p. 1.

processes

MDPI

Article

Research on Gas Emission Prediction Based on KPCA-ICSA-SVR

Li Liu [1,*], Linchao Dai [2,3,*], Xinyi Mao [1], Yutao Chen [2,3] and Yongheng Jing [1]

[1] College of Safety Science and Engineering, Xi'an University of Science and Technology, Xi'an 710054, China; 22220226129mxy@stu.xust.edu.cn (X.M.); jyh@stu.xust.edu.cn (Y.J.)

[2] State Key Laboratory of Coal Mine Disaster Prevention and Control, Chongqing 400037, China; cytboy71900@163.com

[3] China Coal Technology and Engineering Group, Chongqing Research Institute, Chongqing 400037, China

* Correspondence: liul@xust.edu.cn (L.L.); dailinchao@126.com (L.D.)

Abstract: In the context of deep mining, the uncertainty of gas emission levels presents significant safety challenges for mines. This study proposes a gas emission prediction model based on Kernel Principal Component Analysis (KPCA), an Improved Crow Search Algorithm (ICSA) incorporating adaptive neighborhood search, and Support Vector Regression (SVR). Initially, data preprocessing is conducted to ensure a clean and complete dataset. Subsequently, KPCA is applied to reduce dimensionality by extracting key nonlinear features from the gas emission influencing factors, thereby enhancing computational efficiency. The ICSA is then employed to optimize SVR hyperparameters, improving the model's optimization capabilities and generalization performance, leading to the development of a robust KPCA-ICSA-SVR prediction model. The results indicate that the KPCA-ICSA-SVR model achieves the best performance, with RMSE values of 0.17898 and 0.3071 for the training and testing sets, respectively, demonstrating superior robustness and generalization capability.

Keywords: gas emission prediction; data preprocessing; prediction metrics; crow search optimization algorithm

1. Introduction

Coal continues to play a pivotal role in China's energy consumption, acting as a primary energy source in critical sectors such as electricity generation, industry, and transportation, owing to its abundant reserves and relatively low cost [1–3]. However, as mining depth increases, the complexity of gas emissions also rises, posing significant safety risks in underground operations. Gas emissions not only deteriorate the mine environment and disrupt equipment function, but they also elevate underground gas concentrations, substantially increasing the risk of gas explosions and presenting a direct threat to the continuity of mining activities and miner safety. The variability and unpredictability of gas emissions, influenced by factors such as geological structures, coal seam characteristics, and mining methods, render traditional prediction approaches insufficiently responsive. Therefore, accurately forecasting gas emissions is crucial for ensuring mine safety and mitigating the occurrence of hazardous incidents [4–8].

Extensive research has been conducted both domestically and internationally on methods for predicting gas emissions in mining. Guo et al. [9] designed a gas emission prediction system using wavelet denoising combined with BP neural networks and validated its accuracy experimentally. Zhang et al. [10] applied the Bootstrap sampling method, using out-of-bag (OOB) data scores (oob_score) to tune the model and assess feature importance, ultimately determining the optimal parameters and feature variable importance proportions for constructing their predictive model. Yan et al. [11] utilized factor analysis to extract prediction indicators and developed a gas emission prediction model using a particle swarm-optimized radial basis function (RBF) neural network. However, BP neural networks are prone to noise and outliers, with training processes that can become trapped in local optima; random forest (RF) models lack interpretability, making it challenging to analyze the specific contribution of each feature to the prediction results; and RBF models face issues like the "curse of dimensionality" when handling high-dimensional data, potentially leading to decreased model performance. Furthermore, these models tend to have limited generalization abilities, low robustness to small sample sizes, and difficulties in capturing complex nonlinear relationships, thereby constraining their precision in gas emission prediction [12–14].

In this study, we address the complexity and multifactorial nature of gas emission prediction. Initially, data were preprocessed through cleaning and normalization, followed by the establishment of a comprehensive index system for gas emission prediction. To enhance predictive accuracy, Kernel Principal Component Analysis (KPCA) was introduced to perform feature extraction and dimensionality reduction, preserving critical information while reducing dimensional complexity, which improves model learning efficiency. Furthermore, an adaptive neighborhood search combined with reinforcement learning optimizes the Crow Search Algorithm (CSA) to fine-tune the SVR model's hyperparameters, enhancing convergence speed and global search capability. The resulting KPCA-ICSA-SVR model effectively captures the nonlinear characteristics of gas emissions, yielding more accurate and reliable predictions [15].

2. Formulation of Foundational Principles for Processes and Algorithms

2.1. *Process for Establishing Predictive Models*

The specific process for establishing a coal and gas outburst prediction model is as follows:

(1) Data Collection and Preprocessing: Gas emission data are collected and undergo comprehensive preprocessing. Outliers within the dataset are identified and removed using Boxplot analysis, thereby minimizing the impact of extreme values on model accuracy. Missing values are addressed through multiple imputation (MI), improving the completeness and quality of the data. Normalization is applied to remove scaling effects between features. Subsequently, Kernel Principal Component Analysis (KPCA) is employed for feature extraction and dimensionality reduction on the preprocessed data, retaining essential variable information while reducing dimensionality to optimize computational efficiency.

(2) Algorithm Optimization with Improved Crow Search (ICSA): An Improved Crow Search Algorithm (ICSA) is developed by integrating adaptive neighborhood search and reinforcement learning strategies. The adaptive neighborhood search adjusts search scopes dynamically, enhancing the algorithm's flexibility, while the reinforcement learning strategy accumulates experience to improve optimization efficiency and accuracy throughout the process. The ICSA's robust global and local search capabilities enable efficient exploration of the parameter space, fine-tuning the SVR model's hyperparameters, and yielding an optimal parameter set for gas emission prediction.

(3) Model Construction and Comparative Evaluation: After data preprocessing and hyperparameter optimization, comparative models are established to evaluate prediction effectiveness. Benchmark models such as BP, KPCA-ICSA-BP, and SVR are constructed. Evaluation metrics, including RMSE and MPE, are used to assess the performance of each model in gas emission prediction. Results indicate that the KPCA-ICSA-SVR model demonstrates a substantial advantage in prediction accuracy and generalization capability, capturing the nonlinear characteristics of gas emissions more effectively.

2.2. *Basic Principles of the Algorithm*

2.2.1. Kernel Principal Component Analysis, KPCA

Principal Component Analysis (PCA) is a classical dimensionality reduction technique that projects data into a lower-dimensional space via linear transformations, aiming to preserve as much of the original data's variance as possible. PCA is widely used in data preprocessing, feature extraction, and pattern recognition. However, the linear assumptions of

PCA limit its effectiveness in handling complex, high-dimensional nonlinear data. In many real-world applications, data often exhibit significant nonlinearity, which constrains PCA's capacity to fully capture the inherent patterns and relationships within the data [16,17].

To address this limitation, KPCA was developed [18,19]. By incorporating kernel techniques, KPCA maps data into a high-dimensional feature space where nonlinear features can be more effectively captured using kernel functions such as the Gaussian or polynomial kernel. This approach enables principal component analysis in this new feature space, retaining the dimensionality reduction advantages of PCA while significantly enhancing its ability to model nonlinear structures. KPCA has demonstrated superior performance in fields like pattern recognition and data mining, effectively improving both the accuracy and robustness of data analysis (Figure 1).

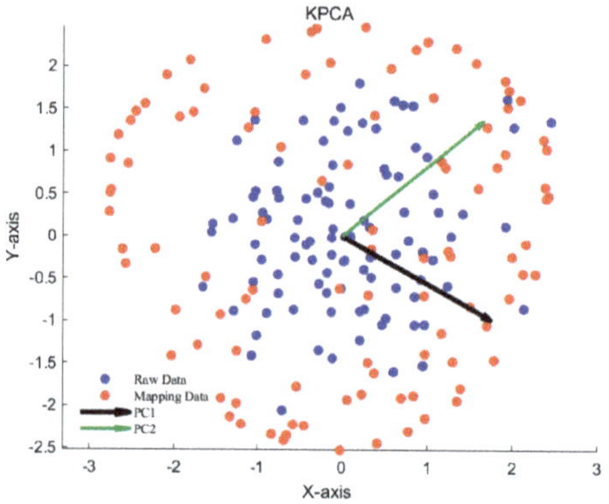

Figure 1. Schematic diagram of KPCA.

(1) Data Standardization: The initial step in KPCA involves standardizing the original dataset X = $\{x_1, x_2, \ldots, x_n\}$ to ensure that all features are on a similar scale. For each sample, $x_i \in R^m$. The standardization formula is as follows (1):

$$x_i' = \frac{x_i - \mu}{\sigma} \tag{1}$$

where μ is the mean and σ is the standard deviation.

(2) Select Kernel Function: Choose an appropriate kernel function K. In this study, the Gaussian kernel is selected, with the calculation formula given by the following (2):

$$K(x_i, x_j) = \exp\left(\frac{||x_i - x_j||^2}{2\sigma^2}\right) \tag{2}$$

(3) Construct Kernel Matrix: Calculate the kernel matrix K, where the elements are given by $K_{ij} = K(x_i, x_j)$. The dimension of the kernel matrix is $n \times n$, where n is the number of samples.

(4) Center the Kernel Matrix: Perform centering on the kernel matrix to eliminate the influence of the mean. The centering formula is given by the following (3):

$$K_{\text{centered}} = K - 1_n K - K 1_n + \frac{1}{n} 1_n K 1_n \tag{3}$$

where 1_n is an $n \times n$ matrix of ones.

(5) Compute Eigenvalues and Eigenvectors: Perform eigenvalue decomposition on the centered kernel matrix K_{centered} to obtain the eigenvalues λ_i and the corresponding eigenvectors v_i (4):

$$K_{\text{centered}} v_i = \lambda_i v_i \tag{4}$$

(6) Select Principal Components: Choose the top d eigenvectors in descending order of eigenvalues to form the feature matrix $V_d = [v_1, v_2, \ldots, v_d]$.

(7) Dimensionality Reduction Mapping: Map the original data into a lower-dimensional feature space to obtain the reduced data Y (5):

$$Y = K_{\text{centered}} V_d \tag{5}$$

Typically, the reduced representation of each sample is expressed as follows (6):

$$y_i = \sum_{j=1}^{n} K(x_i, x_j)\alpha_j \tag{6}$$

where α_j is the coefficient of the corresponding eigenvector.

(8) Result Output: Output the reduced data Y, where each column represents a principal component.

2.2.2. Improved Crow Search Algorithm, ICSA

The Crow Search Algorithm (CSA) is a population-based intelligent optimization algorithm inspired by the foraging behavior of crows [20–22]. By simulating the memory and defense strategies of crows, the CSA searches for optimal solutions within the solution space. It is characterized by its simplicity and wide applicability, having demonstrated effective results across various optimization problems. However, the CSA tends to get trapped in local optima during the search process, and its balance between exploration and exploitation is limited, leading to slower convergence speeds and insufficient precision in complex, multi-modal, high-dimensional problems [23].

To address these limitations, this study proposes an Improved Crow Search Algorithm (ICSA), which integrates adaptive neighborhood search and reinforcement learning strategies. First, adaptive neighborhood search dynamically adjusts the search radius of crows, ensuring flexibility in exploration during different search phases; this allows for extensive searching in the early stages while gradually focusing on fine-tuning in later stages. Furthermore, the incorporation of reinforcement learning allows the crows to adaptively select optimal search behaviors based on feedback from the environment, thereby enhancing the algorithm's decision-making capabilities. This improvement not only strengthens the CSA's global exploration capabilities and local search accuracy but also significantly increases convergence speed and solution precision, resulting in enhanced robustness and stability in tackling complex optimization problems. A schematic representation of the ICSA is shown in Figure 2.

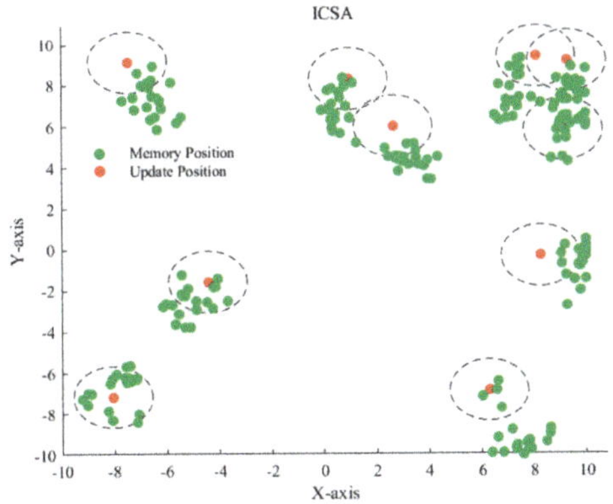

Figure 2. Schematic diagram of the ICSA.

The detailed steps and key formulas for the Improved Crow Search Algorithm (ICSA) are as follows:

(1) Initialize Population: Initialize the position x_i and memory position m_i of each crow in the population, along with their respective fitness values. Set the adaptive neighborhood radius δ_i and learning rate α_i, as well as the Q-table for reinforcement learning, which will be used to learn the optimal behaviors for different states.

(2) Adaptive Neighborhood Search: Establish the neighborhood radius δ_i and dynamically adjust the search neighborhood. Define the initial radius δ_{init} and a scaling factor γ (where $\gamma \in (0, 1)$). Adjust the neighborhood radius based on the current population density and fitness, allowing the search range to adaptively change during the optimization process (7):

$$\delta_i = \delta_{\text{init}} \times \gamma^{\frac{1}{t+1}} \tag{7}$$

where t is the current iteration, and γ controls the contraction rate.

(3) Reinforcement Learning Strategy: Apply Q-learning in reinforcement learning to establish a behavior selection strategy. Define the state S (including information like current position and fitness) and the actions A (choosing neighborhood search or extended search). Update each crow's Q-value with the learning rate η and discount factor λ. The Q-value update formula is as follows (8):

$$Q(S,A) \leftarrow Q(S,A) + \eta\left[r + \lambda \cdot \max_{A'} Q(S',A') - Q(S,A)\right] \tag{8}$$

where r is the reward obtained from the current action, and S' is the new state resulting from action A.

(4) Position Update: Based on neighborhood search and reinforcement learning decisions, choose whether to utilize the memory position (for local search) or perform global exploration (9):

$$x_i^{t+1} = \begin{cases} x_i^t + r_i \cdot \delta_i \cdot (m_j - x_i^t), & \text{if global} \\ x_i^t + r_i \cdot \alpha_i \cdot (m_i - x_i^t), & \text{if local} \end{cases} \tag{9}$$

where r_i is a random number in the range (0,1).

(5) Memory Position Update: If the new position ${X_i}^{t+1}$ is better than the memory position m_i, update m_i with ${X_i}^{t+1}$.

(6) Adaptive Adjustment of Neighborhood Radius and Learning Rate: Based on the crow's fitness and population diversity, adaptively adjust the neighborhood radius δ_i and learning rate α_i to ensure a larger exploration range in the early stages and more focused local searches later on (10):

$$\begin{aligned} \delta_i &= \delta_{\text{init}} \times \exp\left(-\beta \cdot \tfrac{t}{T}\right) \\ \alpha_i &= \alpha_{init} \times \exp\left(-\gamma \cdot \tfrac{t}{T}\right) \end{aligned} \tag{10}$$

where β and γ control the decay rates of the radius and learning rate, respectively, and T is the maximum number of iterations.

(7) Stopping Criteria: The iteration process halts if a predefined maximum number of iterations is reached or if convergence criteria are met (e.g., the fitness value shows no significant improvement), and the optimal solution is outputted.

2.2.3. Support Vector Regression, SVR

Support Vector Regression (SVR) is a regression model well suited for scenarios with small sample sizes and high-dimensional data [24–26]. In gas emission prediction, where data collection is challenging and sample sizes are typically limited, traditional methods may struggle to achieve satisfactory accuracy under small sample conditions or in high-dimensional spaces. SVR constructs an optimal regression hyperplane in the feature space, minimizing prediction error while controlling it within an ϵ-insensitive margin, providing strong generalization ability and robustness to small sample sizes. This approach effectively handles the nonlinearity and high-risk factors inherent in gas emission prediction, making it ideal for small-sample data analysis and significantly enhancing prediction accuracy and reliability. An SVR illustration is shown in Figure 3.

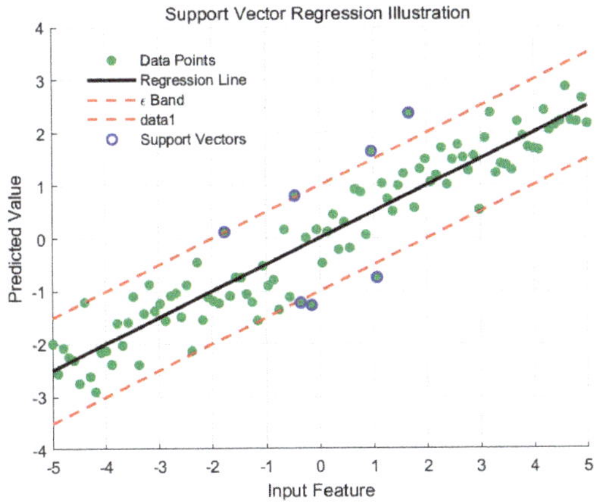

Figure 3. Schematic diagram of SVR.

The detailed steps and key formulas for the SVR are as follows:

(1) Data Preparation: A training dataset is given by $\{(x_i,y_i)\}^n{}_{i=1}$, where $x_i \in R^m$ is the feature vector of the i-th sample, and $y_i \in R$ is the corresponding continuous target value.

(2) Kernel Trick: Since the data may not be linearly separable, a kernel function is used to map the data into a higher-dimensional space, making them closer to linearly separable in the new space. In this study, the Gaussian kernel is chosen.

(3) ϵ-Insensitive Loss and Soft Margin SVR: The ϵ-insensitive loss function is introduced, allowing the model to ignore errors within the $\pm\epsilon$ range. This enables the model to tolerate some error within the ϵ margin without penalizing the optimization objective. Slack variables $\xi_i \geq 0$ and $\xi_i{}^* \geq 0$ are introduced to represent errors that exceed the ϵ margin on the upper and lower sides, respectively. The objective is to minimize the complexity of the regression model while controlling the prediction error (11):

$$\min \frac{1}{2}\|w\|^2 + C\sum_{i=1}^{n}(\xi_i + \xi_i^*) \tag{11}$$

where ξ_i and $\xi_i{}^*$ represent the errors above the upper bound and below the lower bound, respectively. C is a hyperparameter that balances the model complexity and the penalty for errors.

(4) Dual Problem: To incorporate the kernel function, the original optimization problem is transformed into its dual form. Lagrange multipliers α_i and $\alpha_i{}^*$ are introduced to represent the weights for each data point. The dual problem can be expressed as follows (12):

$$\max \sum_{i=1}^{n}(\alpha_i - \alpha_i^*)y_i - \frac{1}{2}\sum_{i=1}^{n}\sum_{i=1}^{n}(\alpha_i - \alpha_i^*)\left(\alpha_j - \alpha_j^*\right)K(x_i, x_j) \tag{12}$$

subject to $0 \leq \alpha_i, \alpha_i{}^* \leq C$ and $\sum_{i=1}^{n}\left(\alpha_i - a_i^*\right) = 0$.

(5) Support Vector Identification: After solving the dual problem, the Lagrange multipliers α_i and $\alpha_i{}^*$ are obtained. For data points x_i where $\alpha_i > 0$ or $\alpha_i{}^* > 0$, these points are considered support vectors, which play a decisive role in determining the position of the regression hyperplane.

(6) Decision Function: After identifying the support vectors, the final decision function can be written as follows (13):

$$f(\mathrm{x}) = \sum_{i=1}^{n} (\alpha_i - \alpha_i^*) K(\mathrm{x}_i, \mathrm{x}) + b \quad (13)$$

where b is the bias term, which can be calculated using the support vectors.

(7) Regression Process: Use the obtained decision function $f(\mathrm{x})$ to predict the continuous target values for new input data, evaluating the regression results based on these predicted values.

3. Data Preprocessing

3.1. Data Acquisition

This study uses data from a coal mine's comprehensive mining face, where gas over-limit situations frequently occur. To prevent major disasters such as gas explosions, this research focuses on predicting gas outburst quantities in the mining face. A total of 30 representative data samples containing 12 influencing factors were obtained from the monitoring data of the mine to serve as the original dataset, as shown in Table 1.

Table 1. Partial raw data.

	x_1	x_2	x_3	x_4	x_5	x_6	x_7	x_8	x_9	x_{10}	x_{11}	x_{12}	y
1	408	2.04	2.4	1839	0.95	163	11	2.34	1.5	20	2.4	4.6	3.54
2	415	2.46	20.3	1639	0.96	158	11	2.42	1.23	21	2.3	4.54	4.07
3	430	2.4	2.1	1770	0.96		12	2.35	1.48	23	2.1	4	4.17
4	439	2.79	2.6	2091	0.95	149	12	2.45	1.81	19	2.6	4.76	4.56
5	460	2.71	2.6	2120	0.95	165	16	2.31	1.7	109	2.6	4.59	4.3
6	541	3.6	3.1		0.95	180	12	3.05	1.6	19	3.1	4	5.07
7	545	3.25	3.3	2006	0.95	168	13	2.46	1.6	20	3.3	3.93	4.64
8	608	4.04	3.55	2586	0.91	170	11	3.18	1.72	21	3.4	4.18	6.16
9	615	4.92	6.3	3395	0.802	172	12	3.3	1.6	14	6.3	3.39	7.58
...	...	...	...	...	...	...	...	...	...	...	...	...	
30	611	4.05	6.7	3354	0.812	175	9	3.15	1.8	16	6.7	2.64	7.8

In Table 1, the 12 influencing factors are as follows: coal seam burial depth x_1 (m), coal seam gas content x_2 (m^3/t), coal seam thickness x_3 (m), mining intensity x_4 (t/d), working face recovery rate x_5 (%), working face length x_6 (m), coal seam dip angle x_7 (°), adjacent layer gas content x_8 (m^3/t), adjacent layer thickness x_9 (m), interlayer distance x_{10} (m), mining height x_{11} (m), and advance speed x_{12} (m/d); the gas outburst quantity is represented as y (m^3/t). A total of 24 samples were randomly selected for the training set, while 6 samples were reserved for the testing set.

3.2. Data Cleaning Based on Boxplot–MI

Table 1 presents the uncleaned data. MATLAB was used to clean the original dataset, beginning with the identification and handling of outliers using the Boxplot method. This involved calculating the lower and upper quartiles (Q1 and Q3) and the interquartile range (IQR) for each column of data, establishing upper and lower limits to identify which data points were outliers. These outliers were replaced with NaN to prevent them from affecting data analysis in subsequent processing. Next, multiple imputation (MI) was applied to fill in these missing values. In this study, six estimators were employed for imputing the missing values, including linear regression, logistic regression, decision trees (DTs), random forests (RFs), a support vector machine (SVM), and multilayer perceptrons (MLPs). The root mean square error (RMSE) was used to evaluate the imputation results and determine the best estimator. Figure 4 illustrates the RMSE for each estimator. Observations from Figure 4 indicate that the MLP imputation model, which achieved the lowest RMSE, demonstrates the highest accuracy. MLP neural networks excel in handling classification and regression problems, showing strong predictive performance and robustness. They are particularly well suited for complex datasets and high-dimensional data. Therefore, this study selected an MLP as the estimator for MI data imputation. The complete data are presented in Table 2 [27,28].

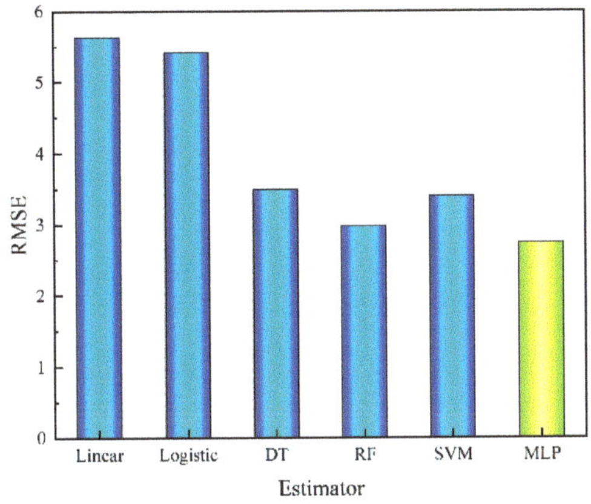

Figure 4. RMSE of different estimators.

Table 2. Partial complete data.

	x_1	x_2	x_3	x_4	x_5	x_6	x_7	x_8	x_9	x_{10}	x_{11}	x_{12}	y
1	408	2.04	2.4	1839	0.95	163	11	2.34	1.5	20	2.4	4.6	3.54
2	415	2.46	2.3	1639	0.96	158	11	2.42	1.23	21	2.3	4.54	4.07
3	430	2.4	2.1	1770	0.96	176	12	2.35	1.48	23	2.1	4	4.17
4	439	2.79	2.6	2091	0.95	149	12	2.45	1.81	19	2.6	4.76	4.56
5	460	2.71	2.6	2120	0.95	165	16	2.31	1.7	19	2.6	4.59	4.3
6	541	3.6	3.1	2250	0.95	180	12	3.05	1.6	19	3.1	4	5.07
7	545	3.25	3.3	2006	0.95	168	13	2.46	1.6	20	3.3	3.93	4.64
8	608	4.04	3.55	2586	0.91	170	11	3.18	1.72	21	3.4	4.18	6.16
9	615	4.92	6.3	3395	0.802	172	12	3.3	1.6	14	6.3	3.39	7.58
...	...	...	...	...	...	...	...	...	...	...	...	...	...
30	611	4.05	6.7	3354	0.812	175	9	3.15	1.8	16	6.7	2.64	7.8

3.3. Data Dimensionality Reduction Based on KPCA

Kernel Principal Component Analysis (KPCA) utilizes kernel functions to map data into a high-dimensional feature space, effectively extracting underlying nonlinear structures and complex features. This characteristic significantly enhances the accuracy and efficiency of data analysis when dealing with high-dimensional and complex datasets. Furthermore, KPCA not only reduces dimensionality to improve the computational performance of models but also mitigates the impact of noise, thereby enhancing the expressive power of features.

Figure 5 shows the kernel matrix heatmap, which illustrates the similarity between data samples and reflects the structure of the kernel matrix calculated using a Gaussian kernel function. This heatmap provides a clearer analysis and identification of clustering trends and potential nonlinear relationships among the data samples. Each cell in the figure represents the similarity between two samples, with the color intensity indicating the level of similarity. Darker areas indicate high similarity, while lighter areas indicate low similarity. By observing Figure 5, clustering trends and potential nonlinear relationships between the samples can be visually identified. For instance, the similarity between the data in Group 1 and the data in Groups 2 and 3 is relatively high, as shown by the darker areas in the heatmap, indicating that these three groups exhibit similar patterns or behaviors in the feature space and are likely to belong to the same or nearby categories. In contrast, the similarity between the data in Group 1 and the data in Groups 10, 14, and 15 is low, as indicated by the lighter areas, suggesting significant differences between these samples and implying that they likely belong to different categories.

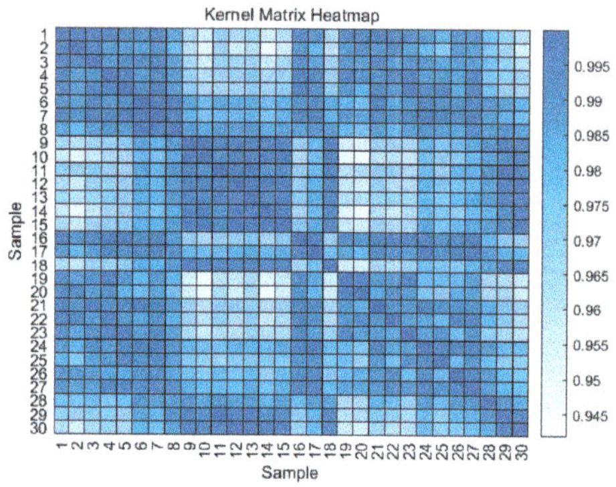

Figure 5. Kernel matrix heatmap.

The variance explanation graph is shown in Figure 6, illustrating the proportion of total variance explained by each principal component. This graph serves as a tool for evaluating the importance and effectiveness of these components. The horizontal axis represents the component number, while the vertical axis shows the proportion of variance explained by each principal component. Typically, the first few principal components account for the majority of the variance, guiding the selection of an appropriate number of components to retain as much of the original data information as possible while reducing dimensionality and lowering computational complexity. Generally, a cumulative variance explanation ratio of 85% is considered a good balance between information retention and dimensionality reduction. In Figure 6, the cumulative variance explained by the first four principal components is calculated as 0.629978 + 0.116409 + 0.080876 + 0.062901 = 0.890164, which is greater than 0.85. Therefore, the first four principal components are selected for the reduced-dimensional data, with the reduced dataset presented in Table 3.

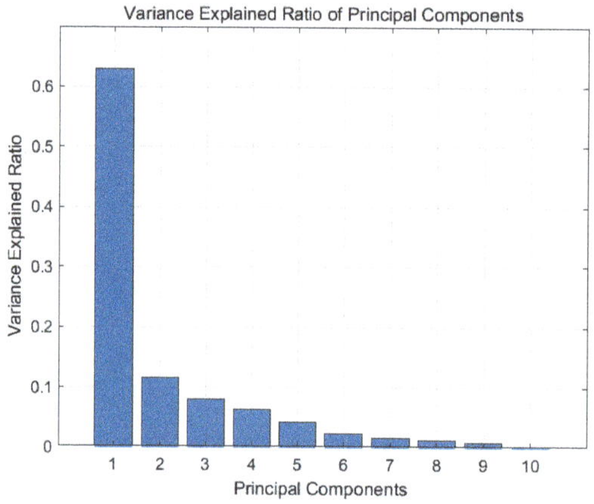

Figure 6. Variance explanation.

Table 3. Partial data after dimensionality reduction.

	P1	P2	P3	P4
1	−0.070030966	−0.007384908	0.001259093	−0.003374653
2	−0.075670036	−0.019152051	0.005231675	−0.007859444
3	−0.061941622	−0.001682007	0.003784321	−0.01279089
4	−0.057579009	0.005133138	−0.010676203	0.007044164
5	−0.0542516	0.018470685	−0.000564319	−0.002088861
6	−0.00737452	0.004864428	0.005790376	−0.010555114
7	−0.026565961	0.005300026	0.000784251	−0.004548474
8	0.009923864	0.002950055	−0.005103689	−0.00680728
9	0.080811222	−0.00030095	0.002270081	−0.001053749
...	...	...	...	...
30	0.085891336	−0.002544488	−0.005689592	0.001248051

4. Establishment of Prediction Model

4.1. Optimization of Hyperparameters Based on the ICSA

Figure 7 illustrates the optimization capabilities of the improved and original algorithms on unimodal and multimodal test functions. As shown, the fitness values of the ICSA convergence curves on all test functions improved after incorporating adaptive neighborhood search and reinforcement learning strategies. This indicates that these strategies effectively enhance the global optimization capability of the CSA.

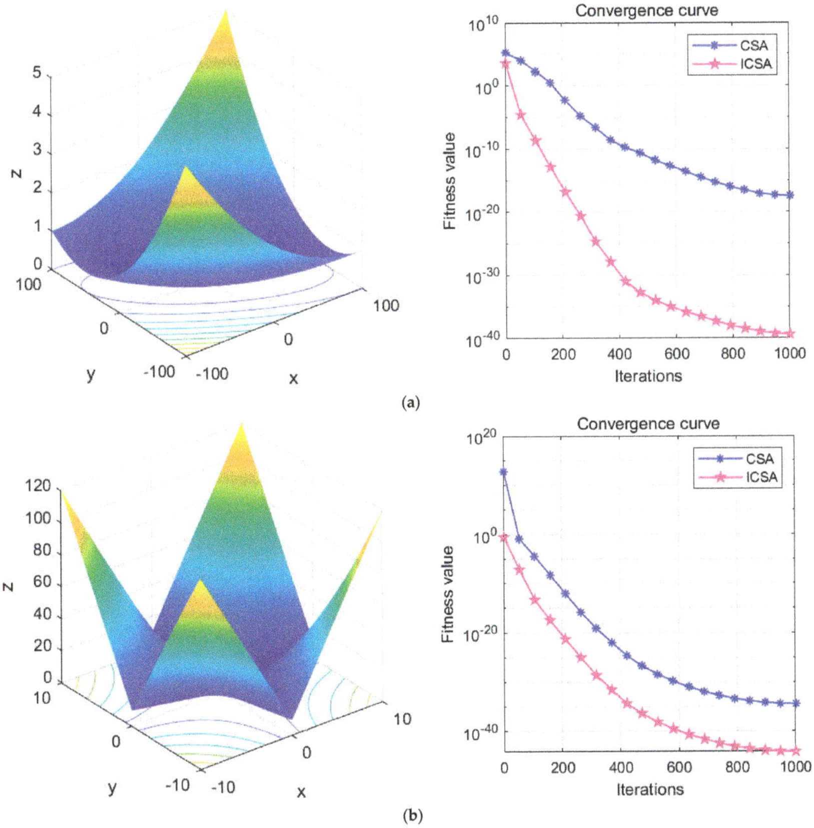

Figure 7. Optimization capabilities of the algorithm: (**a**) unimodal function; (**b**) multimodal function.

In this study, three hyperparameters are optimized: Penalty Parameter C: This parameter controls the model's tolerance for error and determines the penalty for training errors. A higher C value makes the model focus more on reducing training errors, potentially leading to overfitting. Conversely, a lower C value yields a smoother model but may result in underfitting. Gaussian Kernel Bandwidth γ: When using the RBF (Radial Basis Function) kernel, γ controls the similarity measurement between samples and influences the model's nonlinear mapping capability. A larger γ value allows the model to capture finer details, increasing flexibility but risking overfitting. A smaller γ value produces a smoother model with a higher risk of underfitting. Loss Function Parameter ϵ: This parameter in the ϵ-insensitive loss function defines the error tolerance range. A larger ϵ value allows the model to ignore minor errors, reducing sensitivity to noise. A smaller ϵ value improves model precision but may make it overly sensitive to noise.

After training, the optimal values for each parameter were obtained. The initialization range and optimal values of the ICSA parameter optimization are shown in Table 4.

Table 4. ICSA optimization parameters' initial range and optimal value.

Number	Hyperparameter	Initialization Scope	Optimal Configuration
1	C	0.1~1000	1.0587
2	γ	10^{-6}~10	0.0063
3	ϵ	0.001~1	0.1

4.2. Establishment of the KPCA-ICSA-SVR Prediction Model

Substituting the optimal hyperparameter values obtained from the ICSA optimization into the SVR model, a gas emission prediction model based on KPCA-ICSA-SVR is established. The training set prediction results are shown in Figure 8. As observed from the figure, the RMSE of the training set prediction reaches 0.17898, with an R^2 of 0.9836, MAE of 0.1651, MSE of 0.032, and MAPE of 1.67%. All performance metrics are within reasonable ranges, indicating that the KPCA-ICSA-SVR model developed in this study for gas emission prediction is both reasonable and feasible.

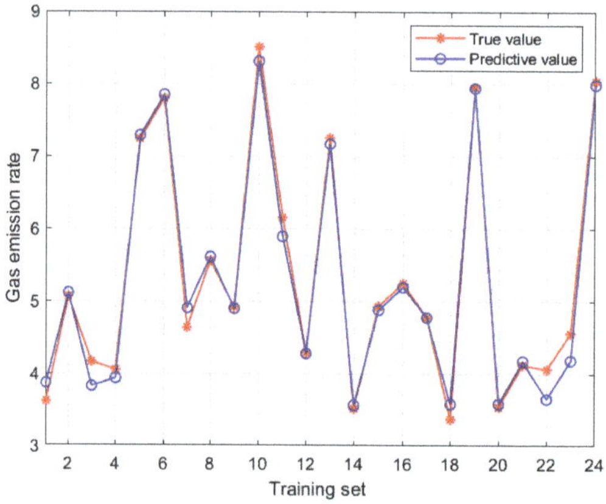

Figure 8. Training set prediction results.

The test set prediction results are shown in Figure 9. From the figure, it can be observed that the RMSE of the test set prediction reaches 0.3132, indicating that the model's accuracy is within a reasonable range. This suggests that the KPCA-ICSA-SVR model does not exhibit overfitting or underfitting, demonstrating good generalization ability and enabling accurate predictions on new data.

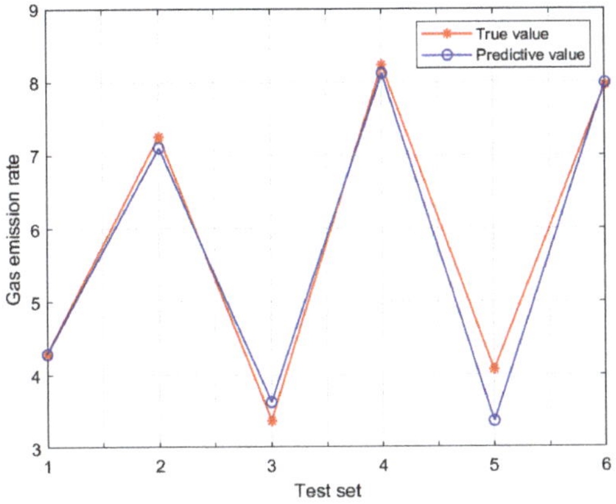

Figure 9. Test set prediction results

5. Evaluation and Comparison of Models

5.1. Test Set Performance Evaluation

To validate the superiority of the KPCA-ICSA-SVR model established in this study, three comparative models were constructed: BP neural network [29–31], KPCA-ICSA-BP, and SVR. Each model was applied to the test set for prediction, with the results presented in Table 5. Due to the small dataset size, the computation time for all models is within 15 s. As shown in the table, the KPCA-ICSA-SVR model outperforms others across all evaluation metrics. Notably, the addition of KPCA for dimensionality reduction and the ICSA to the BP model significantly enhanced its predictive accuracy, indicating that the KPCA-ICSA method can effectively optimize prediction models.

Table 5. Comparison of model prediction results.

Model	R^2	MAE	MAPE/%	MBE	RMSE	MPE/%
BP	0.8564	0.5017	6.56	0.4011	0.6041	4.51
KPCA-ICSA-BP	0.9252	0.3524	4.14	0.2519	0.4786	3.09
SVR	0.9436	0.2879	3.50	−0.1577	0.3568	−2.07
KPCA-ICSA-SVR	0.9752	0.2108	2.01	−0.1150	0.3071	−1.74

(1) BP Model

The BP model performed moderately in predicting gas emissions, with an R^2 value indicating that approximately 85% of the data variance could be explained, suggesting a degree of underfitting. Both MAE and RMSE were relatively high, reflecting the model's limitations in minimizing error, particularly in responding to larger errors (as RMSE slightly exceeded MAE). The MAPE value showed considerable variation across different data magnitudes, with a positive bias (MPE = 4.5%), indicating that BP predictions tended to overestimate actual values. This highlights the model's lower sensitivity to small samples and extreme values. An MBE value of 0.4 confirmed a notable positive bias, indicating that the BP model tends to overestimate gas emissions.

Upon incorporating KPCA for feature extraction and the ICSA to optimize the BP model's hyperparameters, the R^2 value improved significantly to 0.92, demonstrating that the model could explain 92% of data variance, with a considerable boost in fitting ability. Compared to the BP model, the KPCA-ICSA-BP model's errors were notably reduced, reflecting that feature extraction and hyperparameter optimization successfully minimized predictive bias and improved stability. The model's MAPE decreased by approximately 2.5%, enhancing prediction accuracy across various samples. A reduction in MPE indicated a better error distribution, and a lower MBE demonstrated decreased positive bias, confirming that KPCA and the ICSA effectively reduce systematic error.

(2) SVR Model

The SVR model outperformed the BP series in terms of fitting capability, achieving an R^2 value of 0.94, indicating SVR's advantage in capturing the nonlinear characteristics of gas emission data. Both MAE and RMSE were better than those of BP and KPCA-ICSA-BP, showing that SVR effectively manages larger errors. Lower overall error highlighted SVR's accuracy in predicting gas emissions, with a MAPE of 3.5%, reflecting strong stability in predictions. The negative MPE value suggested a slight tendency to underestimate gas emissions, though the error magnitude was small. The negative MBE value also confirmed a mild underestimation tendency, with minimal error.

The KPCA-ICSA-optimized SVR model achieved the highest R^2 value of 0.9752, demonstrating exceptional ability to explain the variance in gas emissions and superior fitting capability. Among the models, KPCA-ICSA-SVR had the lowest error rates, indicating robust control over both small and large errors. The MPE value was closer to zero, further affirming the model's stability and precision in gas emission prediction. The near-zero MBE value suggested a very slight underestimation, but with minimal bias, making the model highly effective for prediction.

5.2. Model Safety Adjustment

Based on the model's prediction results, a safety margin is added to the predicted gas emission values (14):

$$\hat{y}_{\text{adjusted}} = \hat{y} + \Delta_{\text{safety}} \tag{14}$$

where Δ_{safety} is a safety adjustment value set based on historical data and expert experience to ensure early warnings under high-risk conditions. The calculation formula is as follows (15):

$$\Delta_{\text{safety}} = |\text{MBE}| + \alpha \tag{15}$$

where α is a compensation coefficient determined by the historical extreme values of gas emissions and the risk tolerance level. To ensure sufficient conservativeness, this study sets α = 0.05, resulting in a Δ_{safety} value of 0.165.

After incorporating the safety margin, the Mean Bias Error (MBE) in the model evaluation metrics reaches a positive value of 0.05, effectively improving the model's safety.

5.3. Model Generalization Verification

To evaluate the applicability of the proposed model, 30 sets of field data from the reference [11] were used for validation. In [11], the test mine, located at the 12,322 main working face, has an average coal seam thickness of 6.25 m, including one to two layers of carbonaceous mudstone and mudstone interlayers. In areas affected by geological structures, the coal seam thickness varies significantly, with faults extending into the working face. These factors—variations in coal seam thickness, faulting, and interlayer disruptions—can cause periodic fluctuations in gas emission. Moreover, fault damage and stress concentration zones may lead to sudden large gas emissions, increasing the risk of unexpected outbursts. Therefore, accurate gas emission prediction is essential to ensure safe and orderly mining operations.

The sample data include 11 indicators, such as daily production of the working face, interlayer lithology, and coal seam thickness. The data samples were split into two groups; 80% were used for training and 20% for testing. The performance prediction results of different models are summarized in Table 6.

Table 6. Performance comparison.

Model		Maximum Error %	Minimum Error %	Average Error %
Reference [11]	RBF	16.29	0.77	7.28
	FA-PSO-RBF	11.64	0.25	5.65
	FA-AQPSO-RBF	9.27	0.22	2.59
This model	KPCA-ICSA-SVR	4.01	0.12	1.37

As shown in the table, the KPCA-ICSA-SVR gas emission prediction model established in this study outperforms the models referenced in the literature across all evaluation metrics. This demonstrates that the proposed model has good applicability to other coal seams.

6. Discussion

In summary, the gas emission prediction model based on KPCA-ICSA-SVR established in this study has high prediction accuracy. However, further discussions are needed in terms of data scale and other aspects.

Due to the constraints of the underground working environment, future research needs to incorporate more influencing factors for a comprehensive analysis. Gas emission is a complex, dynamic process influenced by multiple factors. Beyond the indicators considered in this study, additional potential factors, such as surrounding rock structure and the microscopic properties of coal, should be examined for a more thorough understanding and prediction of gas emission levels.

The mechanical properties and structural morphology of the surrounding rock, especially around the coal seam, significantly impact gas emissions. The strength, hardness, and fracture distribution of the surrounding rock determine the coal seam's stability and permeability. Softer or more fractured surrounding rock provides pathways for gas to seep out, leading to increased gas emissions. Additionally, structural deformations within the surrounding rock, such as faults and folds, may cause localized pressure changes in the coal seam, further influencing gas release. Stable surrounding rock can effectively restrict gas diffusion, while unstable formations exacerbate gas emissions, thereby increasing safety risks in the mine.

The microscopic properties of coal, particularly its pore structure and gas adsorption characteristics, also play a crucial role in gas emissions. The porosity, pore distribution, and particle size of coal directly affect its gas storage capacity. Coal seams with well-developed pore structures can store more gas and release it rapidly under changing external conditions. Furthermore, areas with high gas adsorption capacity tend to absorb more gas molecules, which are then released at an accelerated rate during pressure changes. Other micro-level characteristics, such as coal type, composition, coalification degree, and moisture content, also significantly influence gas release behavior.

In conclusion, while the predictive model in this study has shown certain advantages, integrating more influencing indicators will further enhance the model's accuracy and practicality, providing better decision support for gas emission prediction and disaster prevention.

7. Conclusions

This study conducts an in-depth analysis of data cleaning, feature extraction, and model optimization for gas emission prediction, proposing a high-precision prediction model based on KPCA-ICSA-SVR. Through data processing, dimensionality reduction, and algorithmic enhancements, this approach effectively captures the nonlinear relationships within complex data, significantly improving prediction accuracy and the model's generalization capability. The application of this model provides essential technical support for mine gas management.

(1) Data Cleaning and Feature Extraction: In this study, the Boxplot–MI method was employed for comprehensive data cleaning of gas emission data. After removing outliers with the Boxplot method, multiple imputation (MI) was used to handle missing data, ensuring data completeness and consistency. Next, Kernel Principal Component Analysis (KPCA) was used to extract key features and perform effective dimensionality reduction. The first four principal components, which explain a cumulative variance of 0.89016, were selected. This approach not only significantly simplified the data structure but also effectively preserved the essential information between variables.

(2) Model Optimization: An Improved Crow Search Algorithm (ICSA) was developed, integrating adaptive neighborhood search with reinforcement learning to optimize the SVR model's hyperparameters. By adaptively adjusting the neighborhood range and employing reinforcement learning strategies for dynamic exploration of the global optimum, the ICSA efficiently and stably identified the optimal parameter combination for SVR, enhancing model convergence speed and accuracy.

(3) Comparative Experiment: A series of comparative experiments, including BP, KPCA-ICSA-BP, and conventional SVR models, were conducted to assess each model's performance in gas emission prediction. Results demonstrated that the RMSE of the KPCA-ICSA-SVR model for the test set is 0.3132, which is lower than that of other models. Additionally, all other evaluation metrics show optimal performance, particularly in terms of prediction accuracy and generalization ability, surpassing the performance of other models.

Author Contributions: Conceptualization, L.L.; methodology, L.L. and X.M.; software, L.D.; validation, Y.C. and Y.J.; formal analysis, L.D. and X.M.; investigation, Y.J.; writing—original draft preparation, L.L. and L.D. All authors have read and agreed to the published version of the manuscript.

Funding: This research was funded by the Natural Science Foundation of Chongqing, grant number CSTB2022NSCQ-MSX1080; the Key Project of Science and Technology Innovation and Entrepreneurship Fund of Tiandi Technology Co., Ltd., grant number 2023-2-TD-ZD001; and the Science and Technology Project of Yulin, grant number CXY-2020-030.

Data Availability Statement: The original contributions presented in this study are included in the article. Further inquiries can be directed to the corresponding authors.

Conflicts of Interest: The authors declare that this study received funding from the Key Project of Science and Technology Innovation and Entrepreneurship Fund of Tiandi Technology Co., Ltd. The funder was not involved in the study design, collection, analysis, interpretation of data, the writing of this article or the decision to submit it for publication.

References

1. Wu, Y.; Wang, P.; Wang, D.; Bu, L.; Tian, J.; Sun, G.; Xu, S. Synergistic ductility deformation and helical design of carbon nanotube fiber composites. *Carbon* **2024**, *229*, 119441. [CrossRef]
2. Wang, L.; Sun, Y.; Zheng, S.; Shu, L.; Zhang, X. How efficient coal mine methane control can benefit carbon-neutral target: Evidence from China. *J. Clean. Prod.* **2023**, *424*, 138895. [CrossRef]
3. Harpalani, S.; Chen, G. Estimation of changes in fracture porosity of coal with gas emission. *Fuel* **1995**, *74*, 1491–1498. [CrossRef]
4. Li, J.; Xu, T.; Lu, W.; Zhao, X.; He, Q.; Li, J.; Zhuo, H. Water-gas masking effect of the primary active sites in coal and room temperature oxidation of coal after desorption. *J. China Coal Soc.* **2024**, *49*, 2298–2314.
5. Wang, K.; Li, K.; Du, F.; Zhang, X.; Wang, Y.; Sun, J. Research on prediction model of coal spontaneous combustion temperature based on SSA-CNN. *Energy* **2024**, *290*, 130158. [CrossRef]
6. Song, J.; Deng, J.; Zhao, J.; Zhang, Y.; Wang, C.; Shu, C.M. Critical particle size analysis of gas emission under high-temperature oxidation of weathered coal. *Energy* **2021**, *214*, 118995. [CrossRef]
7. Wu, Y.; Wang, P.; Wang, D.; Bu, L.; Tian, J.; Xu, S. Unveiling the microstructural evolution and interaction mechanisms for twisted structures. *Int. J. Mech. Sci.* **2024**, *279*, 109514. [CrossRef]
8. Xie, J. Research on pre-pumping gas control technology for large-diameter borehole. *J. Min. Sci. Technol.* **2024**, *9*, 529–537+652.
9. Guo, X.; Ren, Z.; Wang, Q.; Cui, C. Prediction of Gas Emission by BP Neural Network Based on Wavelet Analysis. *IOP Conf. Ser. Earth Environ. Sci.* **2019**, *252*, 052046. [CrossRef]
10. Zhang, Z.; Ma, W. Prediction of gas emission in mining face based on random forest regression algorithm. *J. Mine Autom.* **2023**, *49*, 33–39.

11. Yan, H. The Study on Prediction Method of Gas Emission Amount AQPSO-RBF in Fully Mechanized Mining Face and Its Application. Master's Thesis, Xi'an University of Science and Technology, Xi'an, China, 2020.
12. Zou, Q.; Chen, Z.; Cheng, Z.; Liang, Y.; Xu, W.; Wen, P.; Zhang, B.; Liu, H.; Kong, F. Evaluation and intelligent deployment of coal and coalbed methane coupling coordinated exploitation based on Bayesian network and cuckoo search. *Int. J. Min. Sci. Technol.* **2022**, *32*, 1315–1328. [CrossRef]
13. Zhou, J.; Lin, H.; Jin, H.; Li, S.; Yan, Z.; Huang, S. Cooperative prediction method of gas emission from mining face based on feature selection and machine learning. *Int. J. Coal Sci. Technol.* **2022**, *9*, 51. [CrossRef]
14. Zhao, W.; Dong, H.; Ren, J.; Yuan, Y.; Wang, K.; Wang, F. A software for calculating coal mine gas emission quantity based on the different-source forecast method. *Int. J. Coal Sci. Technol.* **2024**, *11*, 51. [CrossRef]
15. Zhang, B.; Wang, H.; Wang, P.; Yu, G.; Gu, S. Experimental and theoretical study on the dynamic effective stress of loaded gassy coal during gas release. *Int. J. Min. Sci. Technol.* **2023**, *33*, 339–349. [CrossRef]
16. Ma, L.; Qian, J.; Zhao, W.; Curtis, Z.; Zhang, R. Hydrogeochemical analysis of multiple aquifers in a coal mine based on nonlinear PCA and GIS. *Environ. Earth Sci.* **2016**, *75*, 716. [CrossRef]
17. Saikia, J.; Saikia, P.; Boruah, R.; Saikia, B.K. Ambient air quality and emission characteristics in and around a non-recovery type coke oven using high sulphur coal. *Sci. Total Environ.* **2015**, *530*, 304–313. [CrossRef]
18. Wang, Y.; Fu, H.; Zhang, Y. The identification model of coal and gas outburst intensity based on KPCA and CIPSO-PNN. *Chin. J. Sens. Actuators* **2015**, *28*, 271–277.
19. Liu, S.; Qu, E.; Chun, L.V.; Zhang, X. Research on the prediction of blasting fragmentation in open-pit coal mines based on KPCA-BAS-BP. *Sci. Rep.* **2024**, *14*, 16804. [CrossRef]
20. Guo, Q.; Chen, H.; Luo, J.; Wang, X.; Wang, L.; Lv, X.; Wang, L. Parameter inversion of probability integral method based on improved crow search algorithm. *Arab. J. Geosci.* **2022**, *15*, 180. [CrossRef]
21. Hichem, H.; Rafik, M.; Ouahiba, C. New Discrete Crow Search Algorithm for Class Association Rule Mining. *Int. J. Swarm Intell. Res.* **2022**, *13*, 21. [CrossRef]
22. Li, M.; Cheng, X. Identification and Prediction of Thermodynamic Disasters During Deep Coal Mining. *Int. J. Heat Technol.* **2022**, *40*, 1447. [CrossRef]
23. Ledmi, M.; Ledmi, A.; Souidi, M.; Cherif, A.; Maarouk, T.; Cherif, C. High-utility itemsets mining integrating an improved crow search algorithm and particle search optimization. *Soft Comput.* **2024**, *28*, 8471–8496. [CrossRef]
24. Li, K.; Wu, Y.; Wang, K.; Du, F.; Zhang, X.; Wang, Y.; Sun, J. Research on prediction of coal-gas compound dynamic disaster based on ICSA-CNN. *Fuel* **2024**, *359*, 130462. [CrossRef]
25. Meng, Q.; Ma, X.; Zhou, Y. Forecasting of coal seam gas content by using support vector regression based on particle swarm optimization. *J. Nat. Gas Sci. Eng.* **2014**, *21*, 71–78. [CrossRef]
26. Li, S.; Xu, K.; Xue, G.; Liu, J.; Xu, Z. Prediction of coal spontaneous combustion temperature based on improved grey wolf optimizer algorithm and support vector regression. *Fuel* **2022**, *324*, 124670. [CrossRef]
27. Boula, A.; Christine, L.M.; Gunkel-Grillon, P.; Bour, O.; Selmaoui-Folcher, N. Potential contamination of stream waters by ultramafic mining sediments: Identification of geochemical makers (New Caledonia). *J. Geochem. Explor.* **2021**, *232*, 106879. [CrossRef]
28. Farhangfar, A.; Kurgan, L.; Dy, J. Impact of imputation of missing values on classification error for discrete data. *Pattern Recognit.* **2008**, *41*, 3692–3705. [CrossRef]
29. Zhu, Z. Prediction of Coal and Gas Outburst Based on PCA-BP Neural Network. *China Saf. Sci. J.* **2013**, *23*, 45–50.
30. Zhang, F. Gas Emission Prediction Based on BP Neural Network. *Informatiz. Res.* **2016**.
31. Lan, T.; Guo, X.; Zhang, Z.; Liu, M. Prediction of microseismic events in rock burst mines based on MEA-BP neural network. *Sci. Rep.* **2023**, *13*, 9523. [CrossRef]

processes

MDPI

Article

Impact of Gas Accumulation on the Stability of Parallel Upward Ventilation in High-Temperature Sloped Shafts of Deep Wells

Xiaoping Yuan [1,2,*], Qinghua Zhang [1,2,3,*] and Zejun Wang [1,2]

[1] China Coal Technology and Engineering Group, Chongqing Research Institute, Chongqing 400037, China; wzjwzj306@163.com
[2] State Key Laboratory of Coal Mine Disaster Prevention and Control, Chongqing 400037, China
[3] School of Energy and Safety, Anhui University of Science and Technology, Huainan 232001, China
* Correspondence: 13983068516@163.com (X.Y.); zqhcqmky@163.com (Q.Z.)

Abstract: To explore the causes and influencing factors of wind flow oscillations in high-temperature inclined aisles of deep wells under parallel upward ventilation, this study conducts a comprehensive investigation using theoretical analysis and numerical simulations. Based on the kinetic analysis of gas flow, a discriminant formula for wind flow reversal in the side branch is derived. Further analysis identifies initial wind speed and branch length as key factors influencing the reversal. Both gas pressure and thermal pressure contribute to wind flow reversal in the side branch, and the opposing directions of these pressures cause high-temperature gas to periodically flow between the two branches, resulting in wind flow oscillations. A higher initial wind speed can effectively reduce the oscillation amplitude due to increased initial kinetic energy and a larger pressure difference, but it does not extend the oscillation duration. Increasing the branch length can suppress wind flow oscillations by increasing airflow frictional resistance and damping.

Keywords: gas wind pressure; heat wind pressure; airflow inverting; wind flow oscillation

1. Introduction

As shallow coal resources become depleted, coal mining in China has gradually extended to deeper levels. In some coal mines in the central and eastern regions, the mining depth has reached 1000 m [1–3]. At a mining depth of 1000 m, geothermal heat can raise the tunnel temperature to 44.7 °C, creating hot air pressure that significantly affects the ventilation system [4]. In high-temperature inclined roadways of deep wells, hot air pressure mainly arises from temperature gradients within the roadway. The temperature gradient along the tunnel's height difference reduces air density in the high-temperature region, causing gas to rise and form thermal convection. This thermal convection generates hot air pressure, leading to pressure redistribution that affects the ventilation system [5–7]. The formation of gas pressure is closely linked to gas diffusion and accumulation. Abnormal local gas outflow or insufficient ventilation in specific branches can cause an abnormal increase in gas concentration in certain roadways. The accumulation of high gas concentrations significantly alters the composition of roadway gases, creating a density difference associated with the gas concentration gradient along height variations in the roadway, which generates gas pressure that drives the gas flow [8–10]. During deep mine operations, geothermal heat causes significant hot air pressure in roadways with height variations. This places considerable strain on the ventilation system in high-temperature inclined roadways, leading to low ventilation efficiency and a higher risk of airflow stagnation or reversal in local branches [5]. Ventilation system failure can cause localized gas accumulation, generating gas pressure. Consequently, the presence of hot air pressure in high-temperature slopes in deep wells often results in gas pressure, which obstructs heat dissipation by disrupting the ventilation system, further increasing the hot air pressure [11–13]. The formation and coupling of hot air pressure and gas pressure can severely disrupt the normal operation of mine ventilation systems, increasing airflow instability. Hot air pressure alters gas distribution, while gas accumulation affects the flow characteristics of hot air. This interaction not only influences airflow patterns in the roadway but also significantly impacts ventilation safety and operational efficiency. In 2004, an abnormal localized gas buildup in the Dripping Road Coal Mine reduced airflow in the affected branch, which further elevated the localized temperature. The combined effect of hot air pressure and gas pressure caused airflow stagnation in the gas accumulation branch and a reversal of airflow in the adjacent parallel branch. As a result, gas concentration in the accumulation branch reached 100%, and the temperature rose to 48 °C. That same year, an accident occurred in Tangshan Mine at a parallel inclined tunnel, where gas accumulation caused stagnant airflow. A thorough investigation of the formation mechanisms of hot air pressure and gas pressure, as well as their interactions, is crucial for optimizing mine ventilation design and ensuring operational safety.

The failure processes of ventilation systems can be analyzed through the construction of an analogous simulation experiment [14–16]. In the numerical simulation of airflow dynamics, temperature variations, and gas concentration distribution within mines, several software packages are commonly employed, including FLUENT 2021R1, ANSYS CFX 2024R1, Ventsim 5.4, MineVent 0.3.1, and COMSOL Multiphysics 5.5. While FLUENT 2021R1 and ANSYS CFX 2024R1 are adept at simulating complex flows such as turbulence and multiphase interactions, making them suitable for high gas concentration and temperature scenarios, their computational demands are excessively high for the large-scale tunnel models utilized in this study. Conversely, Ventsim 5.4 and MineVent 0.3.1 are specialized tools for overall mine ventilation, providing rapid assessments of airflow distribution and pressure losses; however, their accuracy in modeling airflow oscillations in tunnel branches is insufficient to meet the study's requirements. In contrast, COMSOL Multiphysics 5.5 demonstrates superior flexibility in multiphysics simulations, allowing for simultaneous consideration of airflow, heat transfer, and chemical reactions. This capability is particularly beneficial for in-depth analysis of complex environments, effectively capturing the interactions between gas dynamics and airflow variations with adequate precision. Therefore, from the perspectives of simulation accuracy and computational efficiency, COMSOL Multiphysics 5.5 emerges as the most suitable choice for this research. Kai Wang experimentally investigated the changes in wind flow within the downstream ventilation system under the influence of gas pressure. He found that an increase in wind resistance in the lower side branch enhances the stability of the ventilation system and promotes the effective discharge of gas from this branch [17]. Aitao Zhou found that gas pressure can cause airflow reversal in parallel side branches, and this phenomenon is more pronounced in tunnels with lower airflow rates [18]. The impact of fire-induced pressure on the ventilation system is more pronounced. Under thermal buoyancy, the side branch experiences significant airflow reversal, and the spread of high-temperature smoke further expands the extent of the reversal [19–22]. Current research primarily focuses on the isolated effects of hot air pressure or gas pressure, leaving the coupling mechanism between them unclear. Additionally, the oscillation behavior of wind flow under their combined influence remains unexplained. To address these issues, this study investigates the wind flow reversal mechanism through theoretical analysis and elucidates the disturbance patterns of wind flow under coupled effects by developing an experimental platform for hot air and gas pressure coupling disasters, complemented by numerical simulations.

2. Theory

Gas pressure or fire pressure generates an additional upward ventilation force by affecting the density of the gases. This extra ventilation force results in a reduction in airflow volume in the branches, and may even lead to airflow reversal. When high-temperature gas flow enters the adjacent branch, it introduces further upward ventilation dynamics, which decreases the rate of airflow reversal in that branch. As the pressure differential decreases and reversal occurs due to inertial effects, the direction of airflow in the high-temperature branch will revert back to the high-temperature gas branch. This process continues in a cyclical manner, leading to the emergence of airflow oscillations.

To further quantify this process, the airflow in the tunnel can be simplified to one-dimensional flow. The interaction of gas pressure and thermal pressure on the airflow dynamics within the tunnel can be represented by Equation (1).

$$\rho L\frac{du}{dt} = -\frac{\partial p}{\partial s} + \rho g L + R|Q|Q + j_d - h_f \tag{1}$$

The schematic diagram of the ventilation network for the parallel upstream roadway is shown in Figure 1. In this diagram, branch G represents the main gas branch, while branch B represents the parallel side branch. Section CF corresponds to the main fan branch, and section FA is the atmosphere branch.

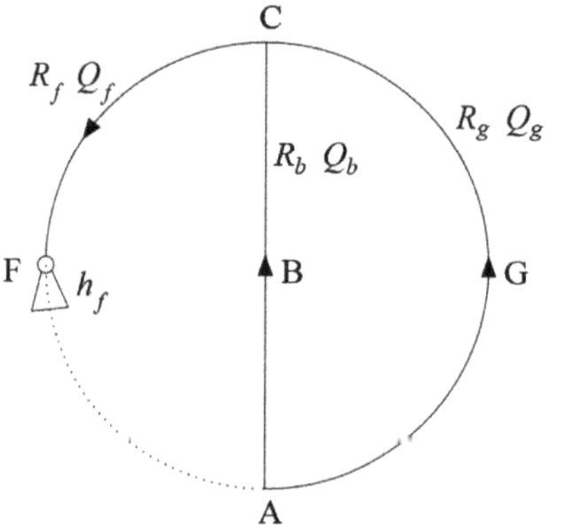

Figure 1. Schematic diagram of the parallel upstream ventilation system.

By analyzing the CF fan branch in the ventilation system shown in Figure 1, where $H_F = H_{CF} - h_f$ and $H_{CF} = \int_F^C \frac{\partial p}{\partial s} ds$, where p represents pressure and s represents length, we apply the fluid momentum equation to derive Equation (2).

$$\frac{\partial p}{\partial s} ds = \rho\frac{\partial u}{\partial t} + F + \rho\frac{\lambda u^2}{2d} \tag{2}$$

where F is the volume force and d is the equivalent diameter, defined as $d = 4S/C$; this leads to Equation (3).

$$H_{CF} = \int_F^C \rho_{FC}\frac{du}{dt}dx + \rho_{FC}g(z_F - z_C) + \int_F^C \frac{\lambda\rho_{FC}C}{8S^2}u^2dx + \rho 0 g(z_A - z_F) \tag{3}$$

By simultaneously solving Equation (3) with Equation (1), we obtain Equation (4).

$$H_F = \int_F^C \rho_{FC}\frac{du}{dt}dx + \rho_{FC}g(z_F - z_C) + \int_F^C \frac{\lambda\rho_{FC}C}{8S^2}u^2dx + \rho 0 g(z_A - z_F) - h_f \tag{4}$$

Similarly, for the side branch, the momentum equation is given by Equation (5).

$$H_{AC} = \rho_{AC}g(z_C - z_A) + \rho_{AC}L_{AC}\left(\frac{\partial u}{\partial t} + \frac{\lambda C}{8S^2}u^2\right) \tag{5}$$

When the side branch airflow has not reversed, its gas density remains constant, and therefore $\rho_{AC} = \rho_0$. The gas flow in branch G can be expressed by Equation (6).

$$H_G = \int_A^C \rho_G \frac{du}{dt} dx + \rho_G g(z_C - z_A) + \int_A^C \frac{\lambda \rho_G C}{8S^2} u^2 dx \tag{6}$$

The fan branch and side branch form one loop, while the gas branch and side branch form another, leading to Equation (7).

$$\begin{cases} H_F + H_{AC} = 0 \\ H_G - H_{AC} = 0 \end{cases} \tag{7}$$

Substituting Equations (4)–(6) into Equation (7) yields the following:

$$\begin{cases} h_f - \rho_0 L_{AC} \frac{\partial u}{\partial t} - \int_F^C \rho_{FC} \frac{du}{dt} dx - (\rho_{FC} - \rho_0) g(z_F - z_C) = \int_F^C \frac{\lambda \rho_{FC} C}{8S^2} u^2 dx + \rho_0 L_{AC} \frac{\lambda C}{8S^2} u^2 \\ \rho_0 L_{AC} \frac{\partial u}{\partial t} - \int_A^C \rho_G \frac{du}{dt} dx - (\rho_G - \rho_0) g(z_C - z_A) = \int_A^C \frac{\lambda \rho_G C}{8S^2} u^2 dx - \rho_0 L_{AC} \frac{\lambda C}{8S^2} u^2 \end{cases} \tag{8}$$

Since the wind resistance $\rho_0 L_{AC} \frac{\lambda C}{8S^2} u^2$ of the side branch is always greater than zero, Equation (8) shows that the ventilation system always satisfies Equation (9) when the side branch airflow is normal.

$$\frac{\rho_0 L_{AC} \frac{\partial u}{\partial t} - \int_A^C \rho_G \frac{du}{dt} dx - (\rho_G - \rho_0) g(z_C - z_A)}{h_f - \rho_0 L_{AC} \frac{\partial u}{\partial t} - \int_F^C \rho_{FC} \frac{du}{dt} dx - (\rho_{FC} - \rho_0) g(z_F - z_C)} < \frac{\int_A^C \frac{\lambda \rho_G C}{8S^2} u^2 dx}{\int_F^C \frac{\lambda \rho_{FC} C}{8S^2} u^2 dx} \tag{9}$$

Let $\begin{cases} a = L_{AC} \left| \frac{\partial u}{\partial t} \right| \\ b = -\int_A^C \rho_G \frac{du}{dt} dx - (\rho_G - \rho_0) g(z_C - z_A) \\ c = h_f - \int_F^C \rho_{FC} \frac{du}{dt} dx - (\rho_{FC} - \rho_0) g(z_F - z_C) \end{cases}$; This leads to Equation (8).

$$\frac{\rho_0 L_{AC} \frac{\partial u}{\partial t} - \int_A^C \rho_G \frac{du}{dt} dx - (\rho_G - \rho_0) g(z_C - z_A)}{h_f - \rho_0 L_{AC} \frac{\partial u}{\partial t} - \int_F^C \rho_{FC} \frac{du}{dt} dx - (\rho_{FC} - \rho_0) g(z_F - z_C)} = \frac{b - \rho_0 a}{c + \rho_0 a} \tag{10}$$

Equation (11) is derived from the wind pressure balance equation between the gas branch and the fan branch.

$$b + c = \int_A^C \frac{\lambda \rho_G C}{8S^2} u^2 dx + \int_F^C \frac{\lambda \rho_{FC} C}{8S^2} u^2 dx \tag{11}$$

From Equation (11), $b + c > 0$, and the right-hand side of Equation (10) is derived from α to obtain $-\frac{\rho_0(b+c)}{(c+\rho_0 a)^2}$. Thus, the left-hand side of Equation (10), which is also the left-hand side of Equation (9), is a monotonically decreasing function of $L_{AC}\left|\frac{\partial u}{\partial t}\right|$. Therefore, the greater the length of the side branch L_{AC} or the rate of change in wind speed $\left|\frac{\partial u}{\partial t}\right|$, the more likely Equation (9) will hold, meaning it becomes increasingly difficult for wind to reverse in the side branch.

3. Simulation

Based on the COMSOL Multiphysics 5.5, a geometric model was developed based on the simple ventilation network in Figure 1, and is illustrated in Figure 2. The return shaft has a height of 300 m, and the inlet shaft, with a height of 400 m, is kept at the same elevation as the exit of the return shaft. The height difference between the inlet and outlet of the gas and high-temperature branch, as well as the side branch tunnel, is 100 m. Both the return and inlet level alleys are 50 m long, and the alleys are rectangular with dimensions of 4 m in length and 3 m in height. The inlet and outlet lengths are 50 m, and the height difference between the inlet and outlet is 100 m. The tunnel airflow resistance information is presented in Table 1.

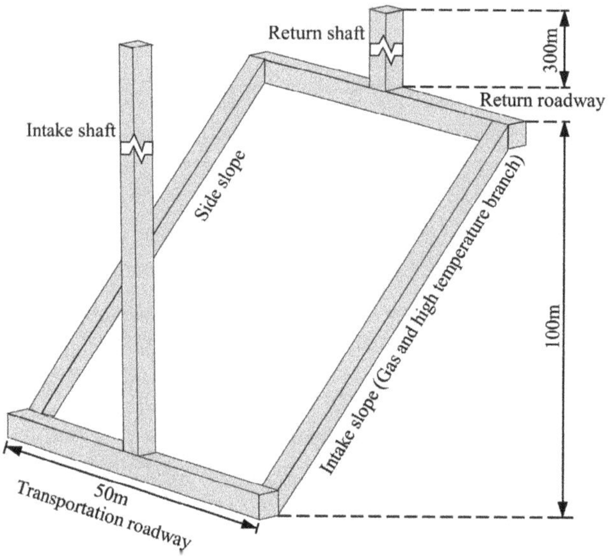

Figure 2. Geometric model used for numerical simulation.

Table 1. Resistance parameters of the roadway in numerical simulation.

Name of Roadway	Windage (Ns^2/m^8)	Length (m)
Intake shaft	0.106	400
Transportation roadway	0.102	50
Intake slope	0.540	141
Side slope	0.540	141
Return roadway	0.102	50
Return shaft	0.106	300
Intake shaft	0.106	400

At the start of the simulation, the gas and high-temperature branches are filled with 100% methane, and the initial temperature is set to 45 °C. The governing equation for gas flow in this model is provided in Equation (12).

$$\rho \frac{\partial \vec{u}}{\partial t} + \rho \nabla \vec{u} \vec{u} = -\nabla(\varepsilon p) - \mu \vec{u} + [\rho_0 g - \rho_0 \beta (T - T_0) g] \tag{12}$$

The last term of the equation represents the volumetric force generated by the high-temperature gas and includes both natural and gas pressures. The diffusion equation for gas is provided in Equation (13).

$$\frac{\partial c}{\partial t} + \nabla \left(\vec{u} c \right) = D \nabla^2 c \tag{13}$$

The time step of the numerical simulation is 1 s.

4. Results

4.1. Wind Turbulence

Because the density of gas is lower than that of air, and it further decreases at higher temperatures, gas and hot air pressures are effectively upward forces due to density differences. Under upstream ventilation conditions, these gas and hot air pressures act in the same direction as the main ventilator pressure, adding additional ventilation force. This added force increases airflow in the branch but simultaneously decreases or even reverses airflow in the parallel side branch. In the side branch, when airflow reverses in the parallel side branch, high-temperature gas also enters the side branch. Since the gas and hot air pressure caused by the high-temperature gas flow oppose the gas flow in the side branch, the airflow reversal is gradually alleviated. Some of the high-temperature gas returns to the high-temperature gas branch, and the gas flow direction in the side branch is gradually restored to its original positive direction. This process repeats in the side branch, causing a periodic change in the gas flow direction after the initial reversal. The speed and duration of the reversal decrease as the high-temperature gas is discharged. In the high-temperature gas branch, the gas flow rate increases significantly at the start, as both gas pressure and hot air pressure act in the same direction as the main ventilator. Simultaneously, a large amount of high-temperature gas enters the side branch, causing a gradual decrease in wind pressure in the high-temperature gas branch and a corresponding increase in wind pressure in the side branch, resulting in a reduction in the ventilation power of the high-temperature gas branch. As the volume of gas entering the side branch increases, the additional wind pressure generated by the high-temperature gas in the side branch surpasses that in the main branch, causing the wind pressure to oppose the local wind flow direction. The gas flow in the high-temperature gas branch continues to move toward the side branch due to inertia, further increasing the pressure difference between the branches. When this pressure difference reaches a critical level, it causes the airflow in the high-temperature gas branch to reverse direction. This process occurs cyclically, resulting in airflow oscillations in the high-temperature gas branch that are opposite to the changes observed in the side branch. Consequently, high-temperature gas airflow leads to frequent oscillations in the parallel upstream ventilation system, with periodic changes in airflow direction.

The wind flow state in the side branch during oscillation is shown in Figure 3. In the initial stage (10 s) of the oscillation, gas and hot air pressures push a large amount of gas into the side branch, causing a reversal of airflow. Due to the maximum gas pressure and thermal pressure at the initial moment of flow reversal, and under the influence of inertia effects, the gas flow velocity in the adjacent branch also reaches its peak. The gas flow velocity in the side branch reaches its maximum, with an overall airflow velocity of around 20 m/s. At this point, the wind flow in the side branch is highly unstable. As high-temperature gas gradually enters the side branch, the gas pressure and hot air pressure generated in the opposite direction of airflow cause a gradual decrease in gas flow velocity, as seen at t = 30 s. Once sufficient high-temperature gas has entered the side branch, the gas is driven to return to the positive direction by t = 50 s, with the wind flow velocity increasing accordingly. Afterward, the gas and high-temperature gas flow back to the gas and high-temperature branches, continuing the cycle that creates the wind flow state in the side branch, as shown in Figure 3.

The airflow status of the gas and high-temperature branches is shown in Figure 4. The temperature and gas concentration data were obtained using the average values from the branches. In the initial stage (t = 10 s), the direction of the main ventilator pressure aligns with the gas and hot air pressures. At this point, gas concentration and temperature remain high, resulting in the highest airflow velocity in the gas and high-temperature branches, with speeds reaching approximately 20 m/s. No wind flow reversal occurs. As the gas and high-temperature gas gradually enter the side branch, the pressure difference between the side branch and the gas and high-temperature branch decreases, along with the gas flow rate, forming a velocity distribution at t = 30 s. However, due to inertia, the gas flow direction remains unchanged, allowing excessive high-temperature gas to enter the side branch. This causes the combined gas pressure, hot air pressure, and main ventilator pressure in the side branch to exceed the pressure in the side branch, ultimately leading to a reversal of airflow at t = 50 s. This reversal causes the high-temperature gas to flow back into the local branch, gradually restoring the upward pressure, leading the airflow to cycle between reversal and recovery.

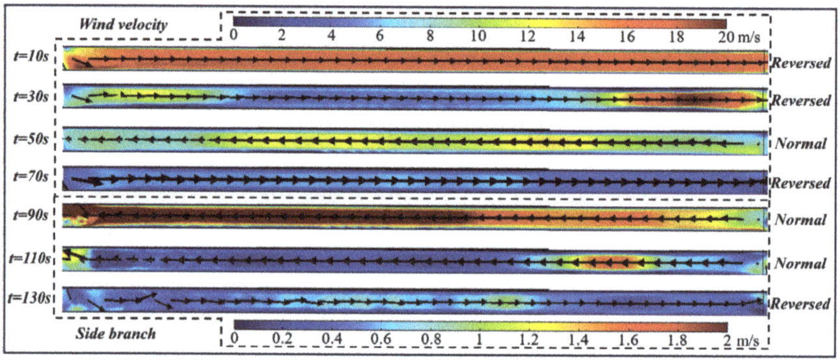

Figure 3. Wind flow state in the side branch.

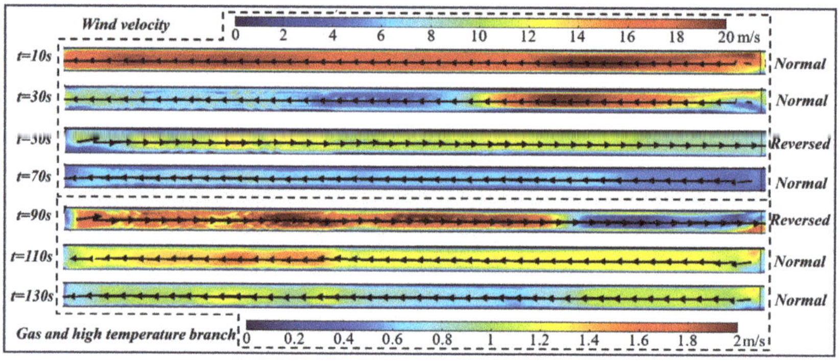

Figure 4. Wind flow state in the gas and high-temperature branches.

In the early stages of airflow oscillation, the pressure differential between branches is significant, serving as the primary driving force for airflow movement. The airflow rapidly responds to changes in this pressure differential, resulting in pronounced fluctuations. As the system undergoes several oscillations, the pressure differential gradually approaches equilibrium, diminishing the driving force and reducing airflow fluctuations. Therefore, in the initial phase, the decay of oscillation amplitude is marked, while in the later stages, the system stabilizes, the variations in pressure differentials become smaller, and the driving force for oscillation diminishes significantly, resulting in a slower decay rate.

The wind speed variation during airflow oscillation is shown in Figure 5, which illustrates that the amplitude of the first oscillation is the largest as the airflow changes direction, and the amplitude of subsequent oscillations gradually decreases. This is because, in the initial stage, the gas and hot air pressures in the gas and high-temperature branches are at their maximum values, storing the most energy. However, as the airflow oscillates, wind resistance in the roadway and losses in the roadway connections gradually dissipate the energy in the system. The damping effect from continuous energy loss causes the amplitude of each oscillation to decrease until the wind flow oscillations gradually cease and reach equilibrium. Furthermore, the redistribution of gas concentration and airflow temperature during the oscillation reduces the energy driving each subsequent oscillation, making it less than that of the first oscillation (the initial moment). Wind flow oscillations represent the system transitioning from an unbalanced state of gas concentration and temperature distribution to a steady state. During the initial oscillation, the system deviates the most, resulting in the largest amplitude. As the oscillation progresses, the system gradually returns to equilibrium, causing the amplitude to decrease and eventually stabilize. To examine the effect of gas concentration distribution on the wind flow oscillation phenomenon, we studied the variation in gas concentration in the side branch and the high-temperature gas branch, as shown in Figure 6.

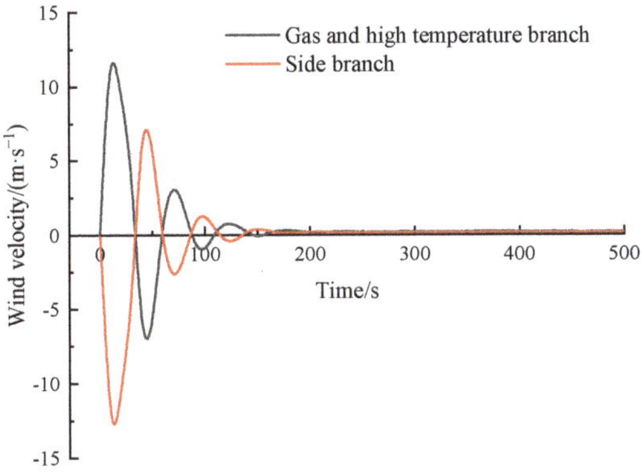

Figure 5. Wind speed variation during airflow oscillations.

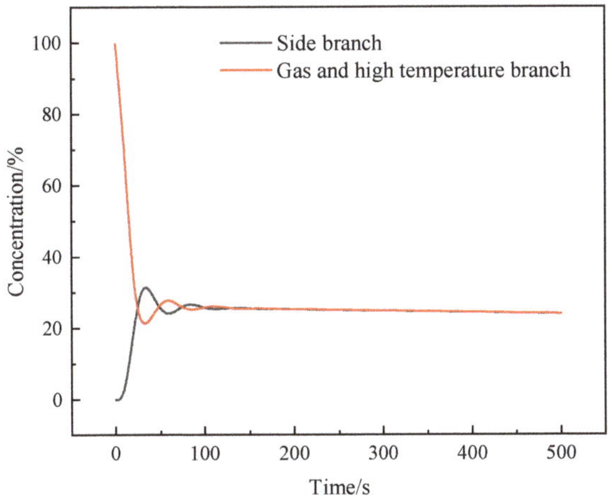

Figure 6. Gas concentration variation during airflow oscillation.

As shown in Figure 6, the first airflow oscillation causes the largest change in gas concentration in the branches. This is due to the large-scale fluctuation during the first oscillation, which results in rapid gas movement in each branch. This strong airflow exchange significantly dilutes the gas concentration in the branches. Additionally, the gas concentration decreases due to continuous gas discharge by the main ventilator during the oscillation, leading to a gradual reduction in gas concentration fluctuations afterward.

Figure 7 illustrates the temperature change during airflow oscillation. The first oscillation causes significant gas exchange, resulting in a sharp temperature change before 50 s. Additionally, the unstable airflow accelerates heat exchange within the gas. During the oscillation, fluctuations in wind pressure influence the compression and expansion of air, leading to temperature variations (compressed air temperature increases while expanded air temperature decreases). This phenomenon is particularly evident in the early stages of significant airflow disturbances. The difference in gas concentration is directly driven by pressure differentials, leading to rapid equilibrium. In contrast, the temperature difference relies more on the slower processes of thermal conduction and convective heat transfer. Consequently, the disparity in gas concentration diminishes more quickly during the initial moment of airflow turbulence.

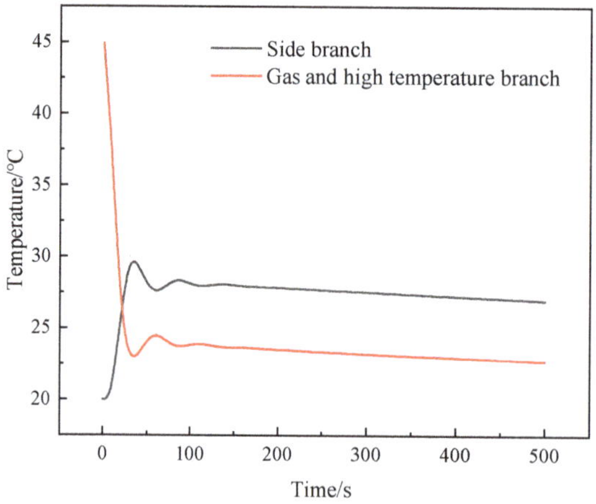

Figure 7. Temperature variation during airflow oscillations.

4.2. Impact of Initial Wind Speed

From Equations (7) and (9), it can be seen that a larger initial wind speed is more unfavorable to the wind flow reversal of the side branch. In order to verify this conclusion, the wind flow state of the parallel upstream ventilation system under different initial wind speeds is analyzed, as shown in Figure 8. It can be seen that a larger initial wind speed can indeed effectively inhibit the oscillation phenomenon of the wind flow, but this inhibition can only reduce the amplitude of the wind flow oscillation, and has no delaying effect on the time of the wind flow oscillation. A higher initial wind speed provides the airflow with greater kinetic energy, enhancing its ability to resist external disturbances and reducing the likelihood of airflow oscillations or their amplification. A higher initial wind speed indicates greater wind pressure from the main ventilator, which minimizes the impact of local gas-induced pressure fluctuations on the system's overall airflow. High air velocity in a ventilation system stabilizes the airflow, allowing fluctuations to be quickly dampened by the system's inertia, making oscillations less noticeable. Additionally, increased air velocity leads to greater ventilation resistance, enhancing the damping effect during airflow oscillations and accelerating the dissipation of perturbation energy.

4.3. Effect of Branch Length

Research indicates that the geometric parameters of flow channels are critical to the flow characteristics of gas movement influenced by gas pressure and thermal pressure, particularly due to density differentials [23,24]. In underground ventilation networks, the length of inclined tunnels affects the magnitude of natural ventilation pressure by influencing thermal pressure efficiency [25]. According to Equations (7) and (9), branch length is another factor affecting wind flow oscillation. Figure 9 illustrates wind flow oscillations at varying branch lengths. As the branch length increases, the amplitude of wind flow oscillations gradually decreases. A longer branch increases frictional resistance in the tunnel, reducing wind flow oscillations caused by external disturbances. Increased damping reduces the likelihood of significant airflow fluctuations and oscillations. Additionally, longer branches increase the time required for airflow to complete its path, progressively weakening its inertial effects. In contrast, shorter branches tend to develop stronger airflow inertia, resulting in more pronounced oscillatory behavior. Longer branches exhibit reduced airflow inertia, allowing fluctuations to be dampened more rapidly by the system. Increased branch length also extends the time required for airflow to travel between endpoints. Because initial external perturbations are instantaneous, longer branches may cause greater delays in the airflow response. As oscillations rely on timely feedback, longer delays weaken the system's feedback loop, making oscillations insignificant or unsustainable.

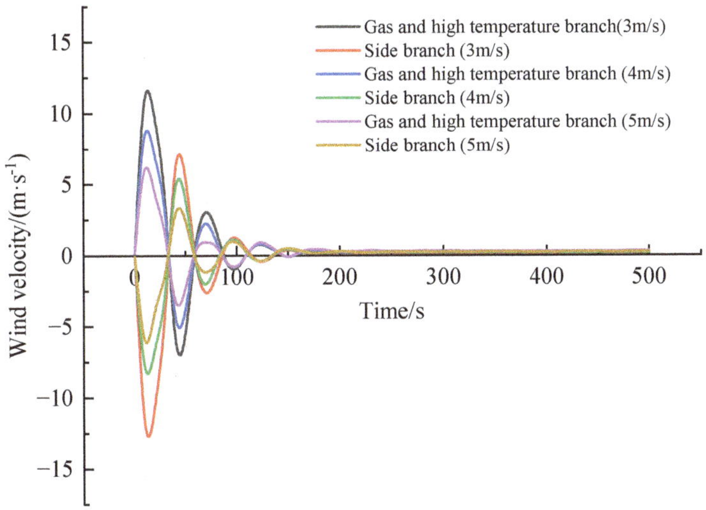

Figure 8. Wind flow oscillation phenomenon under different initial wind speeds.

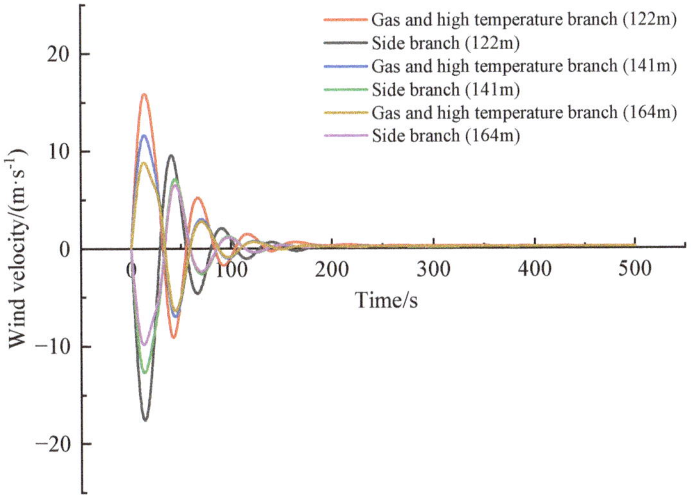

Figure 9. Wind flow oscillation phenomenon under different branch lengths.

The effects of branch length and initial wind speed on airflow oscillations are both reflected in variations in oscillation amplitude; however, their triggering mechanisms differ. Branch length primarily influences oscillations through its impact on the system's inertial effects, necessitating greater external disturbances or changes in conditions to initiate airflow oscillations. In contrast, initial wind speed affects oscillations through the kinetic energy of the airflow and turbulence effects. A higher initial wind speed renders the system more sensitive to external perturbations, making it easier to trigger airflow reversal or disruption.

5. Conclusions

This study analyzed the causes and influencing factors of wind flow oscillations under parallel upward ventilation using theoretical analysis and numerical simulations, leading to the following valuable conclusions.

(1) By simplifying the gas flow in the tunnel to one-dimensional flow and using the kinetic equation of gas flow, the theoretical analysis derives the condition for side branch airflow reversal in parallel upward ventilation, i.e., $\frac{\rho_0 L_{AC}\frac{\partial u}{\partial t} - \int_A^C \rho_G \frac{du}{dt}dx - (\rho_G - \rho_0)g(z_C - z_A)}{h_f - \rho_0 L_{AC}\frac{\partial u}{\partial t} - \int_F^C \rho_{FC}\frac{du}{dt}dx - (\rho_{FC} - \rho_0)g(z_F - z_C)} < \frac{\int_A^C \frac{\lambda \rho_G C}{8S^2}u^2 dx}{\int_F^C \frac{\lambda \rho_{FC} C}{8S^2}u^2 dx}$. It is concluded that the initial wind speed and branch length are the only factors influencing side branch airflow reversal under these conditions.

(2) In the upward parallel ventilation system, the lower density of gas and the influence of high temperatures create gas pressure and thermal pressure. These additional upward forces align with the direction of the main fan pressure, resulting in increased gas and high-temperature branch flow velocities, reaching up to 20 m/s, while a flow reversal phenomenon affecting the adjacent branch occurs within the first 30 s. As gas enters the adjacent branch, the gas and thermal pressures in the high-temperature branch gradually decrease, while the gas and thermal pressures in the adjacent branch increase. During the flow reversal state, the direction of airflow in the adjacent branch opposes the gas and thermal pressures, leading to a gradual decrease in airflow velocity, which ultimately returns to its normal direction at 50 s. This periodic change in airflow direction results in oscillations with a period of 50 s, continuing for 130 s. As gas is gradually expelled and damping effects take hold, the oscillations in airflow progressively diminish.

(3) A higher initial wind speed effectively suppresses wind flow oscillations in the parallel upward ventilation system, reducing the amplitude of the oscillations without extending their duration. Increased initial wind speed provides the airflow with greater kinetic energy, enhances its resistance to external disturbances, and inhibits the generation or amplification of oscillations. Additionally, higher wind speed increases ventilation resistance, enhancing the system's damping effect and facilitating faster dissipation of disturbed energy, thereby reducing the visibility of oscillation phenomena.

(4) Increasing branch length helps suppress wind flow oscillations, with the amplitude of these oscillations decreasing as branch length increases. Longer branches increase the frictional resistance of the flow, diminishing fluctuations caused by external perturbations, while also weakening the flow's inertial effects, making oscillations less likely to occur. Furthermore, longer branches increase the delay in airflow response to external perturbations, weakening the system's feedback loop and rendering wind flow oscillations less pronounced or unsustainable.

Author Contributions: Methodology, Writing—review & editing, X.Y.; Software, Q.Z.; Resources, Z.W. All authors have read and agreed to the published version of the manuscript.

Funding: This research was funded by the TianDi Technology Co., Ltd. Technology Innovation and Entrepreneurship Fund Special Project [2022-2-TD-ZD010].

Data Availability Statement: Data is contained within the article.

Conflicts of Interest: The authors Xiaoping Yuan, Qinghua Zhang and Zejun Wang declare that this study received funding from China Coal Technology and Engineering Group, Chongqing Research Institute. The funder was not involved in the study design, collection, analysis, interpretation of data, the writing of this article or the decision to submit it for publication.

References

1. Zhao, W.; Zhao, D.; Wang, K.; Fan, L.; Zhao, Z.; Dong, H. Will greenhouse gas emissions increase with mining depth in coal mines? An analysis of gas occurrence under varying in-situ stress conditions. *Sci. Total Environ.* **2024**, *945*, 173957. [CrossRef] [PubMed]
2. Du, F.; Ma, J.; Guo, X.; Wang, T.; Dong, X.; Li, J. Rockburst mechanism and the law of energy accumulation and release in mining roadway: A case study. *Int. J. Coal Sci. Technol.* **2022**, *9*, 96–112. [CrossRef]
3. Fan, C.; Sun, H.; Li, S.; Yang, L.; Xiao, B.; Yang, Z.; Zhou, L. Research advances in enhanced coal seam gas extraction by controllable shock wave fracturing. *Int. J. Coal Sci. Technol.* **2024**, *11*, 1–31. [CrossRef]
4. Chen, B.; Guo, A. Prediction of geotemperatures in coal-bearing strata and implications for coal bed methane accumulation in the Bide-Santang basin, western Guizhou, China. *Int. J. Min. Sci. Technol.* **2020**, *2*, 235–242.
5. Wang, E.; Li, X.; Huang, Q.; Wang, G. Research on the Influence of Natural Wind Pressure in Deep Mines on Ventilation Stability. *Adv. Civ. Eng.* **2022**, *1*, 8789955. [CrossRef]
6. El Nagdy, K.A.; Shoaib, A.M. Alternate solutions for mine ventilation network to keep a pre-assigned fixed quantity in a working place. *Int. J. Coal Sci. Technol.* **2015**, *2*, 269–278. [CrossRef]
7. Zhou, A.; Wang, K.; Wu, L.; Xiao, Y. Influence of gas ventilation pressure on the stability of airways airflow. *Int. J. Min. Sci. Technol.* **2018**, *28*, 297–301. [CrossRef]
8. Li, Z.; Yu, J.; Liu, Y.; Zhang, M.; Geng, F.; Zhang, H.; Miao, C. Numerical analysis of natural gas pressure during coal and gas outbursts. *Energy Sci. Eng.* **2021**, *9*, 1068–1079. [CrossRef]
9. Rúa, M.B.; Aragón, A.D.; Baena, P.B.; Botero, J.O. Statistical analysis to establish an ignition scenario based on extrinsic and intrinsic variables of coal seams that affect spontaneous combustion. *Int. J. Min. Sci. Technol.* **2019**, *29*, 731–737.
10. Yu, J.; Li, Z.; Wang, W. Influence of gas outburst dynamic flow on mine ventilation system. *AIP Adv.* **2021**, *11*, 075223. [CrossRef]
11. Liang, Y.; Zhang, J.; Ren, T.; Wang, Z.; Song, S. Application of ventilation simulation to spontaneous combustion control in underground coal mine: A case study from Bulianta colliery. *Int. J. Min. Sci. Technol.* **2018**, *28*, 231–242. [CrossRef]
12. Song, Z.; Huang, X.; Jiang, J.; Pan, X. A laboratory approach to CO_2 and CO emission factors from underground coal fires. *Int. J. Coal Geol.* **2020**, *219*, 103382. [CrossRef]
13. Xu, G.; Luxbacher, K.D.; Ragab, S.; Schafrik, S. Development of a remote analysis method for underground ventilation systems using tracer gas and cfd in a simplified laboratory apparatus. *Tunn. Undergr. Space Technol.* **2013**, *33*, 1–11. [CrossRef]
14. Hansen, R. Fire behaviour of multiple fires in a mine drift with longitudinal ventilation. *Int. J. Min. Sci. Technol.* **2019**, *29*, 245–254. [CrossRef]
15. Liu, Y.; Duan, Z.; Cui, Y.; Jiang, F.; Gao, F.; Wang, W. Study on Multi-Parameter Variation of Smoke Flow in Inclined Roadway Fire. *Combust. Sci. Technol.* **2024**, 1–20. [CrossRef]
16. Wang, Y.; Si, G.; Oh, J.; Belle, B. CFD modelling of longwall goaf atmosphere under vertical boreholes gas drainage. *Int. J. Coal Geol.* **2023**, *280*, 104400. [CrossRef]
17. Wang, K.; Wu, Z.; Zhou, A.; Jiang, Y.; Feng, S. Experimental study of high concentrations of coal mine methane behavior in downward ventilated tilted roadways. *J. Wind. Eng. Ind. Aerodyn.* **2016**, *158*, 69–80. [CrossRef]
18. Zhou, A.; Wang, K. Role of gas ventilation pressure on the stability of airway airflow in underground ventilatio. *J. Min. Sci.* **2018**, *54*, 111–119. [CrossRef]
19. Li, C.; Li, J.; Hu, L.; Hou, D. Visualization and simulation model of underground mine fire disaster based on Cellular Automata. *Appl. Math. Model.* **2015**, *39*, 4351–4364. [CrossRef]
20. Pach, G.; Różański, Z.; Wrona, P.; Niewiadomski, A.; Zapletal, P.; Zubíček, V. Reversal ventilation as a method of fire hazard mitigation in the mines. *Energies* **2020**, *13*, 1755. [CrossRef]
21. Fernández-Alaiz, F.; Castañón, A.M.; Gómez-Fernández, F.; Bascompta, M. Mine fire behavior under different ventilation conditions: Real-scale tests and CFD modeling. *Appl. Sci.* **2020**, *10*, 3380. [CrossRef]
22. Li, Z.; Wang, Z.; Miao, C.; Ding, J.; Wu, J. Study on the Law of Gas Transport and Influencing Factors of Coal and Gas Outburst. *Solid Fuel Chem.* **2023**, *57* (Suppl. S1), S53–S61. [CrossRef]
23. Marinosci, C.; Strachan, P.A.; Semprini, G.; Morini, G.L. Empirical validation and modelling of a naturally ventilated rainscreen facade building. *Energy Build.* **2011**, *43*, 853–863. [CrossRef]
24. Huang, Y.; Tao, Y.; Shi, L.; Liu, Q.; Wang, Y.; Tu, J.; Cao, C. Thermal and ventilation performance of a curved double-skin facade model. *Energy Build.* **2022**, *268*, 112202. [CrossRef]
25. Li, A.; Gao, X.; Ren, T. Study on thermal pressure in a sloping underground tunnel under natural ventilation. *Energy Build.* **2017**, *147*, 200–209. [CrossRef]

processes

MDPI

Article

Stress Zoning Characteristics of Goaf Side and Quantitative Evaluation of Surrounding Rock Stability

Hongkai Zhang, Xiaofei Guo * and Kunlin Yang

School of Energy and Mining Engineering, China University of Mining and Technology-Beijing, Beijing 100083, China
* Correspondence: xiaofeidr@163.com

Abstract: Aiming to maintain the stability of the mining roadway and the next working-face roadway in the goaf side of a coal mine, a systematic study was carried out, through theoretical analysis, numerical simulation, field case analysis, and other methods, of the synergistic change mechanism in the stress distribution and plastic zone development morphology of the roadway's surrounding rock in the goaf side under mining disturbance. It was revealed that, under the influence of mining, the goaf side will form a high-deviatoric-stress environment, which directly affects the shape and stability of the plastic zone of the roadway's surrounding rock. Based on the characteristics of the principal stress ratio and the morphological development law of the plastic zone of the surrounding rock, the side of the goaf is divided into four regions: unloading zone, high-deviatoric-stress zone, low-deviatoric-stress zone, and original-stress zone. And the corresponding mine pressure behavior zoning is proposed: fracture zone, butterfly failure zone, elliptical failure zone, and circular failure zone. A quantitative evaluation method for the stability of the roadway surrounding rock based on the plastic zone morphology criterion was established, and combined with the division of mine pressure, the stability of the surrounding rock on the side of the goaf was quantitatively evaluated.

Keywords: mining roadway; butterfly plastic zone; entry retained; confining pressure ratio; stability of surrounding rock

1. Introduction

With the continuous increase in coal mining intensity, many mining areas have adopted a double-lane or multilane layout to alleviate the problems of replacement tension, transportation, and the ventilation of the working face and have achieved remarkable results [1–3]. In the multilane layout, because the roadway is adjacent to the goaf from the adjacent working face, the stress environment is complex and the service cycle is long, which presents the stability control of the surrounding rock of the roadway on the side of the goaf with certain challenges. Research on the distribution laws of mine pressure and the evolution characteristics of plastic zone development in the goaf side has strong theoretical engineering significance [4–6].

Many relevant experts and scholars have made some achievements in the study of the distribution characteristics and variation rules of mine pressure on the goaf side [7–12]. Based on the stress balance theory of loose medium, Hou et al. obtained the stress distribution law of the coal body on the roadway side and the width limit in the equilibrium zone [13]. Qian et al. studied the distribution characteristics of abutment pressure in the side of the goaf, and divided the side of the goaf into decompression area, pressurization area, and pressure stabilizing area [14]; through real-time monitoring and analysis of the dynamic data of overburden movement and mine pressure appearance in the side of the goaf, Cheng et al. obtained the distribution characteristics of mine pressure and abutment pressure appearance on the goaf side [15]. Babets et al. studied the stress and strain characteristics of the surrounding rock at the intersection of the working face and the return airway in the central part of the Donetsk Basin, and obtained the variation law of the stability of the surrounding rock under different mining stages at the working face [16]. Zhao et al. used the combination of "point-surface-area" to obtain the quantitative range of the plastic zone and the stress increase zone of the coal pillar-surrounding rock under the influence of primary mining under the condition of a deep thick hard roof [17]. Paweł simulated the influence of the goaf on the stability of the surrounding rock and its stress distribution in the next working face using the FLAC3D finite difference program and cellular automaton method [18]. Xie et al. studied the stress distribution characteristics around the goaf under the close distance coal seam, and divided the goaf floor into high-stress area (area A), extremely-low-stress area (area B), stress-rebound area (area C), stress-transition area (area D), and stress-recovery area (area E), according to different stress states [19]. The above research results mainly focus on the abutment pressure, ignoring the horizontal force around the roadway. The distribution law of abutment pressure can effectively reflect the stress state of the surrounding rock, and further analyze the stability of the roadway surrounding rock, which is of great significance for guiding safe coal mine production. However, in practice, the goaf side roadway is in a three-dimensional nonuniform stress field and is affected by severe mining. It is limited to guiding the failure characteristics of the roadway surrounding rock only from the perspective of the distribution characteristics of abutment pressure.

With the in-depth study of mine pressure in recent years, related scholars [20] proposed the theory of a butterfly plastic zone to analyze the interaction between the confining pressure ratio and the plastic zone of the roadway surrounding rock. Based on M-C strength criterion, the boundary equation of the plastic zone of the roadway surrounding rock in a nonuniform stress field is derived by the approximate implicit method, and the variation law of confining pressure ratio and plastic zone development is obtained. The concept of a butterfly plastic zone is proposed for the first time and introduced into the stability study of the roadway's surrounding rock. After that, a large number of scholars studied the failure characteristics of the surrounding rock of a mining roadway under different production conditions, and achieved many research results. Ma et al. analyzed the stability of the surrounding rock of a deep mining roadway using the theory of a butterfly-shaped plastic zone, and obtained the deformation and failure characteristics of the surrounding rock of the roadway [21]; Liu et al. obtained the evolution characteristics of the plastic zone and the variation law of principal stress by studying the distribution characteristics of principal stress and the plastic zone in the double-lane layout working face [22]. Under the influence of mining, the stress of the surrounding rock is redistributed, and the principal stress deviation is large in the local area, which will lead to the formation of a high-deviatoric-stress environment around the roadway. Considering the deficiency of single bearing pressure when analyzing the stability of the surrounding rock of a roadway, based on the butterfly plastic zone theory, this paper analyzes the different characteristics of principal stress distribution on the goaf side, and explores the influence of principal stress ratio change in the goaf sides on the morphological change characteristics of the plastic zone of the surrounding rock of a roadway. A method of mine pressure zoning on the goaf side based on the principal stress ratio is proposed, and the coefficient of the plastic zone determination is introduced to quantitatively evaluate the stability of the surrounding rock on the goaf side, which provides scientific ideas for many problems such as roadway layout and stability control in engineering practices.

2. The Stress Model and Plastic Zone Morphological Characteristics of Surrounding Rock in Nonuniform Stress Field

2.1. Relationship Between Plastic Zone Morphology and Principal Stress Ratio

In reality, the stress state of a roadway is complex. In order to facilitate analysis, the elastoplastic problem of the roadway surrounding rock is often simplified to the plane strain problem of the circular hole. In order to facilitate the study, based on the mechanical model of the approximate equation of the plastic zone boundary of the surrounding rock of the circular hole in the nonuniform stress field, and combining with the research status of the stress environment and mechanical properties of the underground rock, we established the force model of the hole perimeter rock in the nonuniform stress field, as shown in Figure 1. We assume that the roadway is in an infinite underground environment, and the regional principal stress field is used to represent the complex stress environment around the hole. Here, P_1 and P_3 are the maximum principal stress and the minimum principal stress at the boundary of the hole model, θ and r represent the polar coordinates of any point in the coordinate system, a is the radius of the hole, and the shadow part is the butterfly plastic zone around the roadway.

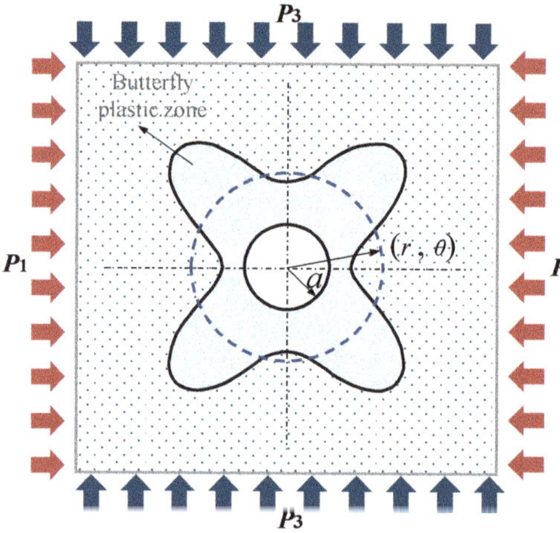

Figure 1. Stress model of surrounding rock in nonuniform stress field.

The maximum principal stress P_1 and the minimum principal stress P_3 around the roadway constitute the maximum and minimum confining pressures at the model boundary, which have a dominant effect on the stress state of the surrounding rock unit area of the perimeter rock of the borehole. It has a significant impact on the plastic zone shape, size, radius, and other parameters of the surrounding rock of the roadway. The deformation and failure of the roadway is the result of the joint action of the maximum confining pressure and the minimum confining pressure. Therefore, η is taken as the ratio of the maximum and minimum principal stresses, as in Formula (1), to further analyze the effect of the change in the ratio of the principal stresses in the nonuniform stress field on the surrounding rock of the hole, and to study the distribution of the peripheral rock stresses and the characteristics of the change in the plastic zone.

$$\eta = \frac{P_1}{P_3} \geq 1 \tag{1}$$

Based on the stress model of the surrounding rock in the nonuniform stress field combined with the implicit equation of eight times of the boundary of the plastic zone proposed in the literature [20], when η is equal to 1, the surrounding rock of the roadway is in the bidirectional isobaric, and the plastic zone morphology is circular; as η increases, the plastic zone morphology gradually evolves into elliptic and butterfly shape. Figure 2 shows the theoretical form of the plastic zone under the different confining pressure ratios obtained by Maple2021 mathematical analysis software. Combined with the mechanical properties of the conventional coal-bearing rock system in practice, the evolution characteristics of the plastic zone during the change in maximum and minimum principal stress ratios are analyzed by taking the coal rock cohesion C of 2 MPa and the internal friction angle φ as 31° [23].

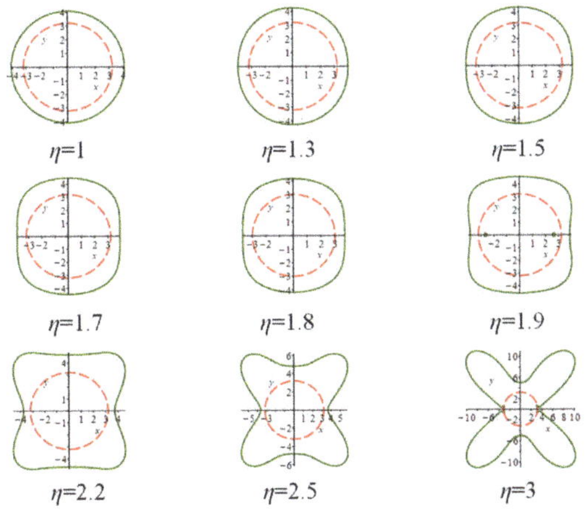

Figure 2. Evolution relationship of plastic zone morphology with principal stress ratio.

In Figure 2, the red dotted line in the diagram is the contour line of the circular roadway, the radius of the roadway is 3, and the green curve is the expansion contour of the plastic zone. It can be clearly observed that the shape of the plastic zone gradually evolves from circular to elliptical and finally shows a butterfly shape with the change in η. When η is 1.8, it is approximately the boundary between elliptical and butterfly shapes. When η is greater than 1 and less than 1.8, the length of the plastic zone in the vertical direction is greater than the length in the horizontal direction, and the overall contour shows an ellipse shape. When η is greater than 1.8, the boundary of the plastic zone gradually expands at the corner, and the expansion range of the plastic zone increases rapidly with the increase in η, showing a nonuniform distribution of the butterfly-shaped contour. Up to this point, the overall contour of the plastic zone no longer changes; only the butterfly leaves will occur to a certain degree of expansion, missing, and other phenomena [24–26].

2.2. Plastic Zone Morphology and Stability of Roadway Surrounding Rock

The formation and development of the plastic zone is a decisive factor leading to the failure of the roadway surrounding rock. As mentioned above, in the nonuniform stress field, due to the difference in the principal stress, the plastic zone will show three forms: circular, elliptical, and butterfly. The shape of the plastic zone is closely related to η. When η is 1~1.8, the plastic zone presents uniform failure with circular and elliptical shapes. When η is greater than 1.8, the edge of the plastic zone of the roadway surrounding rock shows a concave phenomenon, indicating that the nonuniform distribution of the maximum and minimum principal stresses produces a high-deviatoric-stress area in the local area of the surrounding rock of the roadway. The surrounding rock in this area reaches the strength limit, and the plastic zone of the roadway surrounding rock gradually expands in a butterfly shape, and continues to expand deeply along the direction of the butterfly leaf. The maximum radius R_{max} of the plastic zone increases significantly, and the area of the plastic zone has a significantly larger radius than that of the circular and elliptical shapes, which is unfavorable to the stability of the roadway surrounding rock.

The maximum radius R_{max} of the plastic zone can effectively reflect the stability of the roadway surrounding rock. The shape of the plastic zone of the roadway surrounding rock and the variation characteristics of the maximum radius R_{max} corresponding to the different confining pressure ratio η in the regional stress field are studied, as shown in Figure 3. During the initial change in the confining pressure ratio η in the range of 1~1.8, the maximum radius R_{max} of the plastic zone increases slowly with the increase in η. The two approximately maintain a linear relationship, and the slope fluctuation is small. The plastic zone of the roadway surrounding rock is round and oval, the plastic zone is small and evenly distributed, and the deformation and failure of the roadway surrounding rock is small. Currently, the surrounding rock of the roadway is in the "butterfly-free" stable area, and the stability of the surrounding rock is high.

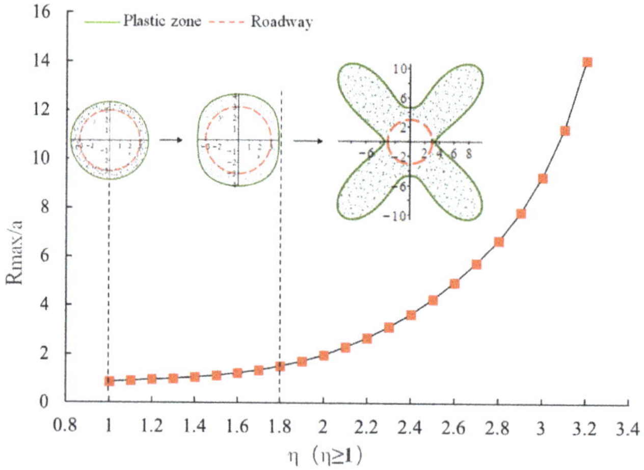

Figure 3. Relationship between the maximum radius of the plastic zone and the ratio of principal stress.

After the ratio of principal stress becomes greater than 1.8, it is obvious that the slope of the curve changes fast, and the maximum radius R_{max} of the plastic zone changes abruptly after η reaches 1.8, resulting in a sharp expansion. The relationship between the two gradually evolves from linear to exponential, and the maximum radius R_{max} of the plastic zone can reach more than ten times of the previous stage. The plastic zone is transformed from uniform distribution to nonuniform distribution of corner failure. The plastic zone of the roadway surrounding rock presents a butterfly shape, and the butterfly leaf range is large. At this time, the butterfly leaf part is subjected to high deviatoric stress. At the same time, under the influence of stress deflection angle, the roadway surrounding rock easily produces deformation damage, and roadway disasters such as rock burst, roof collapse, and floor heave can occur. This range of roadway surrounding rock is in the butterfly risk area, and the stability of the surrounding rock is very poor.

3. Mine Pressure Zoning and Development Characteristics of Plastic Zone in Goaf Side

3.1. Distribution Law and Partition of Abutment Stress in Mining Roadway

The plane layout and spatial relationship of the working face of the quarry is shown in Figure 4, and the stress distribution law around the goaf side mining roadway is analyzed. As shown in the figure, the mining roadway is located in the side of the working face, and there are also protective coal pillars, then the next working face excavation roadway, and so on. The disturbing effect produced by coal mining will cause the redistribution of the surrounding stress field. With the advancement of the working face, the load on the immediate roof strata reaches its strength limit, resulting in fracture and separation, which leads to the deformation and instability of the main roof. The energy released by the breakage of the main roof forms a large pressure on the support structure and coal pillar of the roadway in the goaf side, and produces a stress concentration phenomenon in the local area [27]. Under this action, the roadway surrounding rock easily destabilizes, and there is a severe phenomenon of mine pressure. With the transfer of stress to the deeper part of the surrounding rock, the coal pillar of the roadway in the goaf side will also be affected by the mining impact and produce obvious mechanical effects. There is an emergence of varying degrees of damage to the area, and then a gradual restoration of the original rock stress state.

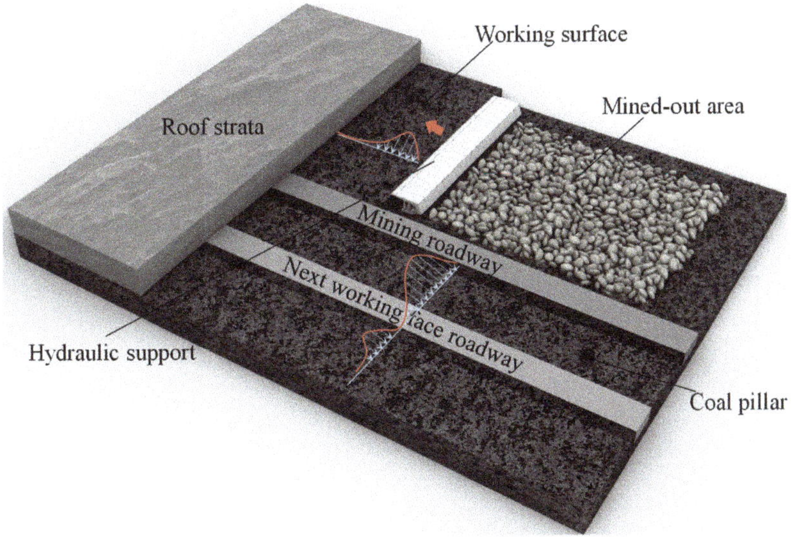

Figure 4. Layout of underground working face in coal mine.

Abutment pressure is a key factor affecting the stability of surrounding rock in stope, which can be used as an important index to measure the stress field on both sides of a mining roadway. The variation curve of abutment stress can be obtained by monitoring the distribution of stress on the goaf side, as shown in the red curve in Figure 4. On the goaf side, under the influence of small coal pillars, the abutment pressure curve shows a trend of increasing and then decreasing. The pressure has been released at the lower working face roadway, and then there is a small elevation when it reaches the solid coal side, and then it tends to flatten out.

After the mining of the working face, the stress will inevitably be redistributed, and the goaf will be gradually filled with caving gangue and gradually compacted, and produce a certain supporting force on the roof. The surface of the coal pillar on the goaf side is exposed, and the horizontal stress of the surrounding rock at the free surface is almost zero, far lower than the original rock stress state. Under the supporting pressure, the surrounding rock has been destroyed. The surrounding rock farther away is in the limit equilibrium state, and is gradually transitioning from the broken state to the plastic state; further extending to the goaf side, the surrounding rock is gradually restored to elasticity from the plastic state, but a large amount of energy is accumulated inside the rock mass. A small increase in stress or a small decrease in stress acting on the surrounding rock system will break the balance, causing the surrounding rock to reach the strength limit and become damaged. The energy in the surrounding rock system is released in large quantities, resulting in the malignant expansion of the plastic zone of the surrounding rock and causing huge damage. The surrounding rock farther away is less affected by the abutment pressure, and the stress level of the surrounding rock gradually decreases until the stress state of the original rock is restored. In this process, abutment stress on the goaf side shows the evolution process of increasing, then decreasing, tending to stabilize. Corresponding to these three evolution trends, the surrounding rock on the goaf side of is divided into three regions: decompression zone, pressurization zone, and pressure stabilization zone. At the same time, according to the stress state of the surrounding rock, it is divided into the limit equilibrium zone and elastic zone, which has significance in guiding the safe and efficient operation of the coal mine site [14].

3.2. Distribution Law and Zoning of Principal Stress Ratio in Mining Roadway

In recent years, many scholars [21,25] have conducted a large number of studies on the formation mechanism and evolution process of the butterfly-shaped plastic zone, and have studied and explored the interaction between the principal stress ratio and the development law of the plastic zone. Due to the complex geological environment of the roadway surrounding rock, the disturbance of coal mining will redistribute the stress field around the working face, resulting in a great difference between the maximum principal stress and the minimum principal stress. The lateral position of the goaf where the roadway is located is more affected by the disturbance, and the principal stress ratio increases significantly, resulting in the formation of a large area of failure zone in the roadway surrounding rock, and presenting a similar butterfly shape. It can be seen that the principal stress ratio is closely related to the development and expansion of the plastic zone of the roadway surrounding rock and its stability control. In order to better guide the stability of the roadway surrounding rock on the goaf side, the evolution law of the principal stress ratio of the stress on the goaf side is studied, and the mine pressure zoning system based on the principal stress ratio is established. The zoning method is based on the two-dimensional stress system as a criterion, and combined with the development form of the plastic zone for auxiliary analysis, which has certain guiding significance for the stability analysis of the roadway surrounding rock.

The distribution law of stress on the goaf side and the method of rock pressure zoning are shown in Figure 5. Due to the severe disturbance of the surrounding rock near the goaf, the surrounding rock in this area is broken, the rock mass is in an unloading state, and the stress of some surrounding rock is even lower than the stress level of the original rock. The surrounding rock in this area is defined as the unloading zone. The ratio of principal stress is effective in the category of elastic theory, so it has less significance to guide the surrounding rock in the unloading zone. In the elastic zone, the evolution trend of the principal stress ratio is similar to that of the vertical stress. As it moves away from the goaf, the principal stress ratio η gradually decreases from the peak and eventually approaches 1. It is known from the above that the principal stress ratio corresponds to the development relationship of the plastic zone morphology of the roadway surrounding rock, and the high-deviatoric-stress environment is a necessary condition for the formation of the butterfly plastic zone. With the change in the principal stress ratio, the shape of the surrounding rock plastic zone gradually undergoes three stages: circular–elliptical–butterfly. Among them, the butterfly plastic zone has a great influence on the stability of the roadway surrounding rock, the elliptical plastic zone has little influence, and the circular plastic zone has almost no influence. In view of this, based on the morphological difference in the plastic zone corresponding to the principal stress ratio, the surrounding rock on the goaf side is further divided into high-deviatoric-stress zone, low-deviatoric-stress zone, and original-stress zone.

(1) Unloading zone

With the advancement of the working face, the surrounding rock near the goaf gradually presents a bare state, forming a large area of free surface, the horizontal stress is significantly reduced, and the disturbance is severe. The surrounding rock in this area reaches the strength limit and presents a broken state. The rock mass is in an unloading state, and some surrounding rock stress is even lower than the original rock stress level. The surrounding rock gradually far away from the goaf gradually restores its integrity, but is still in a state of failure rather than elasticity. The surrounding rock near the goaf side in a state of failure is defined as the unloading zone, as shown in Area I in Figure 5.

(2) High-deviatoric-stress zone

As shown in Figure 5 Area II, after the stress unloading in the unloading zone, the surrounding rock gradually returns to the elastic state in the process of gradually moving away from the goaf. The maximum principal stress reaches the peak point, and then decreases gradually. The minimum principal stress is almost the original rock stress level and remains stable. The maximum and minimum principal stresses of surrounding rock in this area are very different. The ratio of principal stress is at a high level, and the value is always greater than 1.8. The surrounding rock in this area is in a high-deviatoric-stress environment, and the plastic zone presents a butterfly shape.

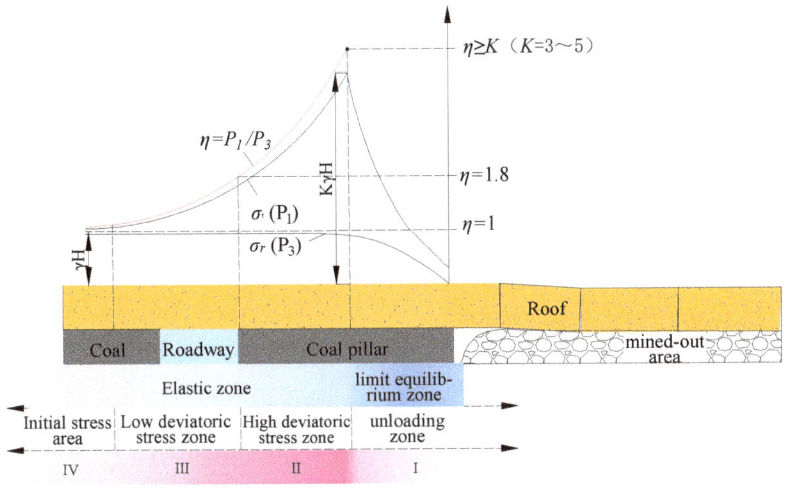

Figure 5. The distribution law of lateral stress in goaf and the zoning of mine pressure.

(3) Low-deviatoric-stress zone

As shown in Figure 5 Area III, further away from the goaf, the influence of mining on the surrounding rock is alleviated to a certain extent. The maximum principal stress is obviously reduced, the principal stress ratio continues to decrease, and the value is maintained between 1 and 1.8. The surrounding rock is no longer affected by the high-deviatoric-stress environment, the expansion degree of the plastic zone is weak, and the surrounding rock is in a relatively stable state.

(4) Original-stress zone

As shown in Figure 5 Area IV, the surrounding rock in this area is far away from the goaf, almost not affected by mining. The maximum principal stress and the minimum principal stress are close to the original rock stress state, the principal stress ratio is approximately 1, and the surrounding rock is in a stable state.

3.3. Development Characteristics and Stability Analysis of Plastic Zone in Mining Roadway

The mining activity breaks the original rock stress state of the rock mass and leads to the redistribution of stress. The difference in stress distribution makes the surrounding rock show different plastic zone forms, accompanied by different degrees of mine pressure. The above mathematical analysis model was used to demonstrate the theoretical form of the plastic zone corresponding to different principal stress ratios in the regional stress field. Here, according to the characteristics of the rock pressure zoning of the principal stress ratio on the goaf side, the numerical simulation of the plastic zone shape under the corresponding confining pressure ratio is given, as shown in Figure 6. The shape of the plastic zone of the surrounding rock corresponding to different stress zones is obviously different, which is consistent with the shape of the plastic zone obtained by theoretical calculation, showing three kinds of plastic zone shapes: circular, elliptical, and butterfly. Based on the relationship between the principal stress ratio and the plastic zone, the mine pressure behavior zoning corresponding to the mine pressure zoning on the goaf side is proposed. According to the morphological difference in the plastic zone, the two sides of the mining roadway are divided into the fracture zone, butterfly failure zone, elliptical failure zone, and circular failure zone.

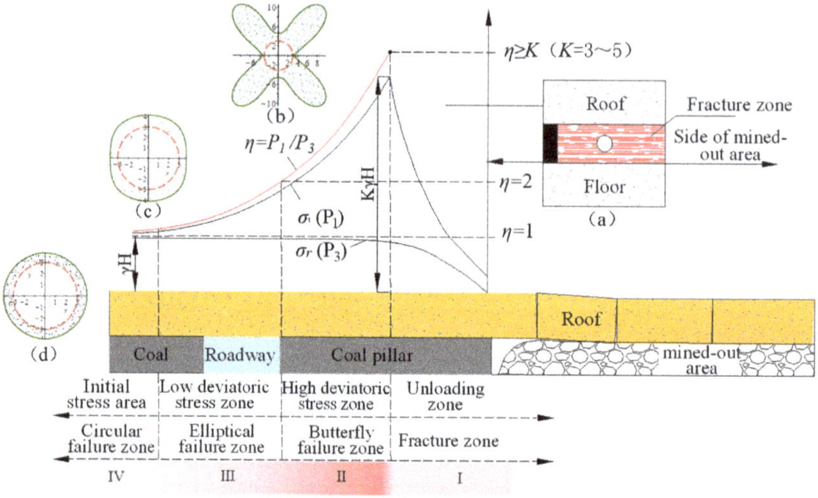

Figure 6. The law of mine pressure in different areas on the side of the goaf. (**a**) Fracture pattern, (**b**) Butterfly-shaped, (**c**) Elliptical (**d**) Circle.

(1) Fracture zone

The surrounding rock in the fracture zone is shown in Figure 6a, where the red area indicates the degree of destruction of the surrounding rock. The appearance of the free surface under the influence of mining leads to a significant decrease in the compressive strength of the local surrounding rock. Under the action of strong disturbance, the surrounding rock damage in this range is broken, with no fixed form.

(2) Butterfly failure zone

Under the influence of high-deviatoric-stress environment, the surrounding rock is in a high-stress-concentration state. If the roadway is arranged in this area, the plastic zone of the surrounding rock will show a butterfly shape, as shown in Figure 6b, in which the red dotted line is the outline of the roadway, and the green area is the range of the plastic zone. According to the shape of the plastic zone of the surrounding rock, this area is defined as a butterfly failure zone. The butterfly-shaped failure zone corresponds to the range of the high-deviatoric-stress zone. Because high stress and high confining pressure ratio are the necessary conditions for the formation of the butterfly-shaped plastic zone, the plastic zone of the surrounding rock in the butterfly-shaped failure zone is large, and the surrounding rock is in a very unstable state, which makes it prone to serious disaster accidents.

(3) Elliptical failure zone

With the increase in the distance from the goaf, the surrounding rock gradually transits to the low-deviatoric-stress zone, and the mining influence tends to ease. The principal stress ratio in this area is significantly reduced. The roadway surrounding rock is in a low-deviatoric-stress environment, and the plastic zone of the roadway surrounding rock is mainly elliptical, as shown in Figure 6c. According to the shape of the plastic zone of the surrounding rock, this area can be defined as an elliptical failure zone. In this range, the surrounding rock is relatively stable, and the damage degree of surrounding rock is low.

(4) Circular failure zone

At a considerable distance from the goaf side, the surrounding rock is not affected by mining, and is almost in an undisturbed natural stress state. The principal stress ratio is approximately 1, and the shape of the plastic zone of the surrounding rock of the roadway is approximately circular, as shown in Figure 6d. According to the shape of the plastic zone of the surrounding rock, this area can be defined as circular failure zone. The stability of the surrounding rock in the circular failure zone is good, and is not affected by mining, and the plastic zone is only formed by the influence of roadway excavation.

4. Numerical Simulation Scene Analysis

4.1. Field Engineering Overview

The Menkeqing Coal Mine 3101 working face belongs to the 3-1 coal seam, the buried depth is about 700 m, and the thickness is unstable. The thickness of the coal seam varies from 1.57 m to 6.35 m along the strike, and the average thickness of the coal seam is 4.6 m. The length of the working face is 260 m, the length of the strike is 3900 m, and the height of the coal seam is 5.0 m. The direct roof of the 3-1 coal seam is sandy mudstone, the main roof is fine sandstone, and the floor of the coal seam is mainly sandy mudstone, followed by fine sandstone. After long-term roadway surface displacement monitoring, during the mining process of the 3101 working face, due to the influence of mining, the deformation and failure characteristics of the roadway surrounding rock are significant, as shown in Figure 7. After the mining of the working face, the mining influence led to the high-deviatoric-stress environment around the roadway. There are large floor heave and rib spalling in the surrounding rock floor and the side of the auxiliary transportation roadway on the goaf side, which seriously restricts the safe and efficient production of coal mines. It is very important to deeply explore the distribution law of stress in the goaf side and grasp the stability difference in the lateral surrounding rock.

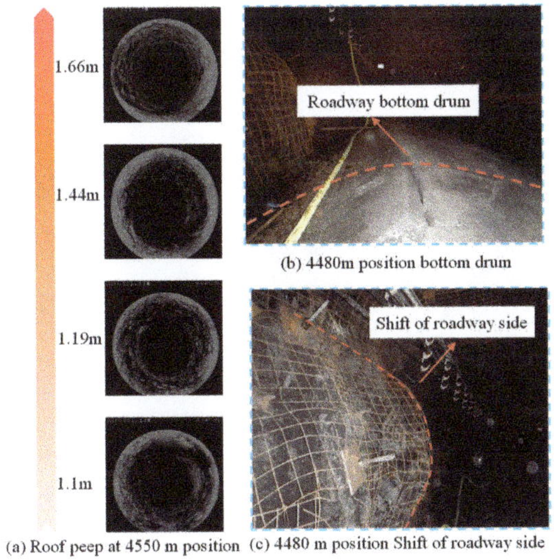

Figure 7. Deformation and instability characteristics of 3101 auxiliary transport roadway in side of the goaf.

4.2. Numerical Simulation Model

According to the actual engineering conditions of the 3101 working face in Menkeqing Coal Mine, combined with the geomechanical parameters of the rock strata, as shown in Table 1, a large three-dimensional numerical model was established by FLAC3D 7.0 numerical simulation software, as shown in Figure 8. The size of the numerical model is length × width × height (X × Y × Z) of 600 m × 600 m × 150 m. In this numerical simulation calculation, the length of the working face is 260 m, and 100 m is retained on the side to prevent the boundary effect. The X direction is the direction of the working face, and the Y direction is the direction of the working face. For the bottom of the model and the surrounding fixed boundary displacement, the upper part of the model is the free boundary. A 15 MP vertical stress is applied to simulate the overburden load on the roof of the roadway. In order to reduce the amount of calculation, the mining operation in the simulation is carried out on the working face for 200 m, the survey lines are arranged along the X direction, and the excavation simulation of multiple roadways is arranged on the side of the working face to analyze the stress distribution characteristics of the mining roadway and the development of the plastic zone.

Table 1. Physical and mechanical parameters of coal and rock strata in the 3101 working face.

Lithology	Thickness/m	Tensile Strength/MPa	Bulk Modulus/GPa	Shear Modulus/GPa	Internal Friction Angle/°	Cohesion/MPa	Density kg/m^3
2-2 medium coal	2.00	2.00	10.00	4.62	28.00	1.10	2300
Sandy mudstone	7.50	2.28	15.00	5.75	32.78	1.47	2410
Fine sandstone	18.50	0.77	20.00	8.18	38.47	2.86	2550
Sandy mudstone	4.50	2.28	15.00	5.75	32.78	1.47	2410
3-1 coal	5.00	2.00	10.00	4.62	28.00	1.10	2300
Sandy mudstone	10.50	2.28	15.00	5.75	32.78	1.47	2410
Fine sandstone	22.00	0.77	20.00	8.18	38.47	2.86	2550
Sandy mudstone	4.50	2.28	15.00	5.75	32.78	1.47	2410

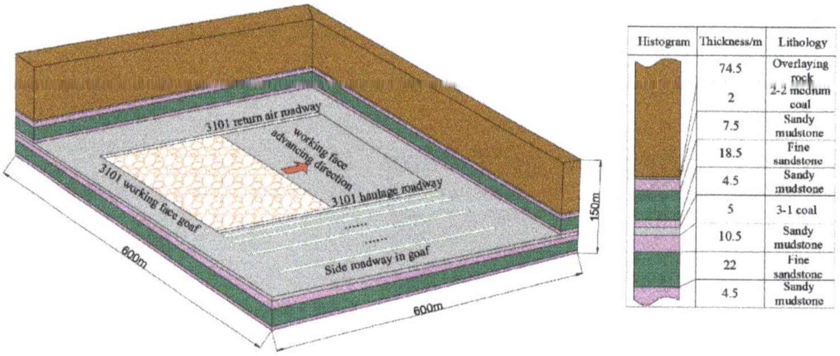

Figure 8. The 3101 working face mining numerical model.

4.3. Distribution of Lateral Principal Stress in Goaf

FLAC3D 7.0 numerical simulation software was used to simulate the excavation of the Menkeqing coal mine 3101 working face. The maximum principal stress and the minimum principal stress in the lateral range of the goaf were monitored and extracted, and the distribution curve of the principal stress in the goaf side was obtained, as shown in Figure 9.

The maximum principal stress increased sharply from below the original rock stress level to the peak point, and then began to decline slowly and tended to be stable. The minimum principal stress increased sharply to the original rock stress level at a faster rate and remained stable. Within about 5 m of the goaf side, the maximum principal stress increased rapidly and reached the peak point from below the original rock stress value, and the minimum principal stress gradually rose and recovered to the original rock stress level. At this stage, the surrounding rock was affected by mining. The local surrounding rock fragmentation caused the stress to be unloaded, so there was a situation where the minimum principal stress was lower than the original rock stress. In the range of about 5~15 m on the goaf side, the surrounding rock began to change from the failure state to the elastoplastic state. Due to the influence of mining stress disturbance, a large amount of energy was accumulated in the surrounding rock. The maximum principal stress was at a high level, and the principal stress ratio was also at a high level. In this range, the surrounding rock was greatly affected by mining, and the whole area was in a high-deviatoric-stress environment. The stability of the surrounding rock was facing a great challenge. In the range of 15~35 m on the goaf side, the maximum principal stress gradually tended to ease, the minimum principal stress was in a stable state, and the fluctuation degree of the principal stress ratio was small, slowly decreased, and tended to 1. The surrounding rock in this area was less affected by mining, the surrounding rock was in a low stress state, and the overall stability was good. The maximum and minimum principal stress changes were very small, and gradually tended to the original rock stress state and remained stable. The change in principal stress ratio was not obvious, and gradually slowed down with the increase in distance from the goaf. Finally, it was almost equal to 1. At this stage, the surrounding rock was almost not affected by mining and tended to the original rock stress state.

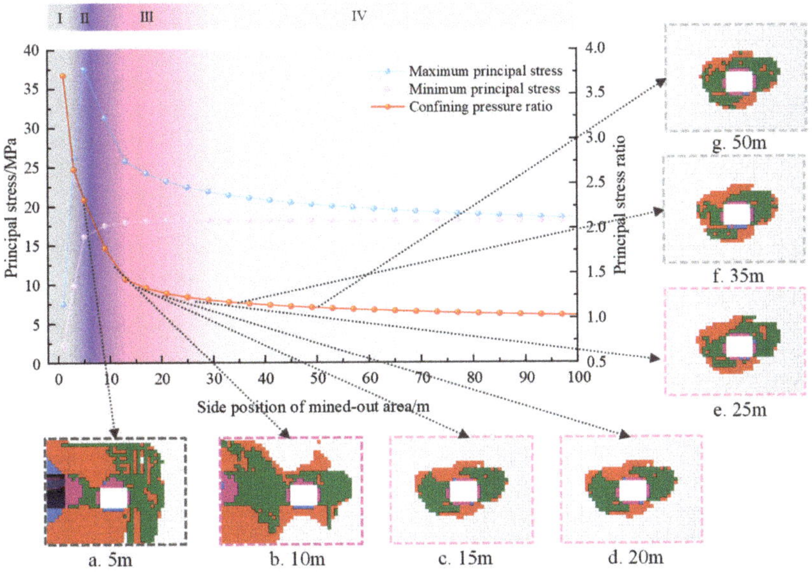

Figure 9. The development form of the plastic zone of the roadway surrounding rock at different positions of the goaf side.

The distribution characteristics of principal stress and principal stress ratio on both sides of the mining roadway obtained by the numerical simulation are similar to the stress distribution law in the above partitions. The stress around the mining roadway exists in four states: unloading state, high deviatoric stress, low deviatoric stress, and original rock stress. The mining roadway pressure zoning method based on the principal stress ratio is verified. Due to the high-deviatoric-stress environment on the goaf side, a series of mine pressure phenomena appeared in the auxiliary transportation roadway of 3101 working face in Menkeqing Mine, which induced the occurrence of floor heave and rib spalling.

4.4. The Plastic Zone Shape of Surrounding Rock of Roadway in Goaf Side

The shape of the plastic zone is closely related to the stability of the roadway surrounding rock. Several roadways were simulated and excavated at different positions on the goaf side to analyze the development characteristics and stability characteristics of the plastic zone of the surrounding rock of the roadway on both sides of the mining roadway.

This provides a scientific reference for the design of coal pillars on the goaf side and the selection of roadway layout in the next working face.

A roadway was excavated at an interval of 5 m on the side of the goaf. The FLAC3D numerical simulation software was used to slice the surrounding rock of the roadway on the goaf side, and the morphological characteristics of the plastic zone of the surrounding rock of the roadway at different positions on the goaf side could be obtained, as shown in Figure 9. It can be seen from the figure that the area with a large change in the plastic zone of the roadway affected by mining is mainly concentrated in the range of 15 m on the goaf side. The plastic zone of the roadway surrounding rock outside 15 m from the goaf is basically stable, and there is no obvious change fluctuation. The development degree of the plastic zone of the roadway about 5 m on the goaf side is more intense and irregular, which is consistent with the above partition results. The surrounding rock of this section is greatly affected by the disturbance, and the surrounding rock of the roadway reaches the strength limit and presents a broken state, which is consistent with the characteristics of the broken zone. The plastic zone of the roadway at 10 m on the goaf side has a high degree of expansion. It can be clearly observed that under the action of the nonuniform stress field, the stress concentration occurs at the corner of the roadway. The plastic zone of the surrounding rock of the roadway presents a similar "butterfly" shape, and the area of the plastic zone is larger. It is worth noting that only the right side of the roadway shows the phenomenon of butterfly leaf expansion, the left side of the roadway is missing, and there is no butterfly shape. This is mainly due to the instability of the surrounding rock induced by the influence of mining, and the surrounding rock reaches the strength limit. The goaf left side and the roadway surrounding rock are connected, which also shows that the formation of the butterfly plastic zone is not conducive to the stability of the surrounding rock. If the roadway is arranged at this position, it is prone to instability accidents. It is necessary to strengthen the monitoring of roadway surface displacement and surrounding rock stress, and make appropriate reinforcement support.

In the range of 15 m~35 m on the goaf side, the development shape of the plastic zone of the roadway surrounding rock is gradually homogenized, the overall contour is approximately elliptical, and the area of the plastic zone is significantly reduced compared with the previous stage. At this time, the surrounding rock of the coal pillar on the left side of the roadway is in an elastic state, no plastic zone is formed, and the stability of the surrounding rock is good. Further away from the goaf, the disturbance degree of the roadway surrounding rock is further reduced, even not affected by mining, and the overall surrounding rock remains stable. According to the field production conditions of the working face, the roadway surrounding rock is suitable to be arranged in the range of 15 m outside the goaf side, and the plastic zone of the roadway surrounding rock is effectively controlled.

5. Roadway Stability Quantitative Evaluation

5.1. Criterion of Plastic Zone Shape

From the above analysis, it is known that in the nonuniform stress field, the plastic zone morphology of the roadway surrounding rock will present circular, elliptical, and butterfly shapes with different principal stress ratios. The formation of the butterfly plastic zone is not conducive to the stability of the roadway surrounding rock. Mastering the formation mechanism of the butterfly plastic zone is particularly critical to ensure the stability of the roadway. On the basis of a mathematical mechanics model, Reference [28] deduced the criterion of the plastic zone shape coefficient of the circular roadway surrounding rock in the nonuniform stress field by theory, which can be seen in Formula (2), and can realize the quantitative discrimination of the plastic zone shape of the roadway surrounding rock.

$$\begin{cases} \tau = \infty - - - - - - - - \text{Circular plastic zone} \\ \tau \geq 1 \text{ or } \tau \leq 0 - - - - \text{Elliptical plastic zone} \\ 0 < \tau < 1 - - - - - - - \text{Butterfly plastic zone} \end{cases} \tag{2}$$

In Formula (2), τ is the shape coefficient of the plastic zone of the surrounding rock of the circular roadway in the nonuniform stress field, and its expression is as shown in Formula (3):

$$\tau = \frac{m_2}{2m_1} \tag{3}$$

$$\begin{aligned}
m_1 &= \left[12(1-\eta)^2 - 4(1-\eta)^2\sin^2\varphi\right]\left(\frac{-b_1+\sqrt{{b_1}^2-4a_1c_1}}{2a_1}\right)^2 - 8(1-\eta)^2\left(\frac{-b_1+\sqrt{{b_1}^2-4a_1c_1}}{2a_1}\right);\\
m_2 &= 6(1-\eta^2)\left(\frac{-b_1+\sqrt{{b_1}^2-4a_1c_1}}{2a_1}\right)^3 - 4(1-\eta^2)\left(\frac{-b_1+\sqrt{{b_1}^2-4a_1c_1}}{2a_1}\right)^2\\
&\quad + \left[2(1-\eta^2) - 4(1-\eta^2)\sin^2\varphi - \frac{4C(1-\eta)sin2\varphi}{P_3}\right]\left(\frac{-b_1+\sqrt{{b_1}^2-4a_1c_1}}{2a_1}\right);\\
a_1 &= \frac{6(\eta-1)}{1-\sin\varphi}; b_1 = (1+\eta) - \frac{(3\eta-5)(1+sin\varphi)}{(1-sin\varphi)}; c_1 = 2\eta - \frac{4C\cos\varphi}{P_3(1-sin\varphi)} - \frac{2(1+sin\varphi)}{1-\sin\varphi}.
\end{aligned}$$

Among them, m_1, m_2, a_1, b_1, and c_1 are all transition parameters, which can be expressed by the above analytical formulas of surrounding rock parameters and stress; C is the rock cohesion; and φ is the internal friction angle. Under different confining pressure conditions, the shape of the plastic zone will undergo three stages of evolution, and the development shape of the plastic zone can be judged by using the shape coefficient of the plastic zone. When the principal stress ratio is 1, that is, the roadway surrounding rock is in the equal pressure stress field, the shape coefficient of the plastic zone tends to infinity, and the shape of the plastic zone of the surrounding rock of the roadway is round. As the confining pressure gradually increases, the plastic zone shape determination coefficient gradually transitions from infinity to 1, and the plastic zone shape of the surrounding rock is elliptical. When the shape determination coefficient of the plastic zone is 1, the plastic zone of the roadway surrounding rock belongs to the critical state of ellipse and butterfly. In the process of decreasing from 1, the shape of the plastic zone gradually transitions to butterfly.

5.2. Quantitative Evaluation Method of Roadway Stability

The stability maintenance of mining roadways has always been a focus in the process of coal mining. Quantitative analysis of the stability of surrounding rock on the goaf side has a certain guiding role in ensuring safe and efficient mining at the production site. The stress field and plastic zone morphology of the roadway surrounding rock are the key factors to measure the stability of the roadway surrounding rock. The above analysis was carried out on the distribution law of stress on the goaf side, and the mine pressure zoning was carried out on the goaf side based on the principal stress ratio. At the same time, according to the morphological development characteristics of the plastic zone of the surrounding rock, the mine pressure zoning on the goaf side was obtained. The morphological characteristics of the plastic zone are closely related to the stability of the roadway surrounding rock. The plastic zone morphology of the roadway surrounding rock can be accurately determined by the plastic zone morphology criterion. Because the distribution law of stress corresponds to the morphological variation characteristics of the plastic zone of the roadway surrounding rock, the division of mine pressure on the goaf side can be realized by the criterion of the plastic zone morphology of the surrounding rock, so as to realize the quantitative division of the stability of the roadway surrounding rock and to guide the engineering practice.

As the surrounding rock of the roadway is in the complex nonuniform stress field, the homogeneity and physical and mechanical parameters of the surrounding rock are different, which has a great influence on the stability of the roadway surrounding rock. Different lithologies of surrounding rocks correspond to different confining pressure conditions when the plastic zone reaches the critical shape, but the criterion is universal for the plastic zone shape. When the plastic zone shape determination coefficient reaches 1, it is in the range of butterfly and elliptical contours. According to the above analysis, it is known that the mine pressure zoning on the goaf side and the mine pressure zoning belong to the corresponding relationship. The butterfly failure zone is in the high-deviatoric-stress zone, the maximum radius of the plastic zone is large, the degree of damage is severe, and the stability of the roadway surrounding rock is poor. The elliptical failure zone and the circular failure zone are, respectively, in the low-deviatoric-stress zone and original-stress zone, and the overall surrounding rock is relatively stable. The quantitative analysis of the stability of the roadway surrounding rock on the goaf side is equivalent to the determination of the shape of the plastic zone of the roadway. The specific quantitative analysis process is shown in Figure 10.

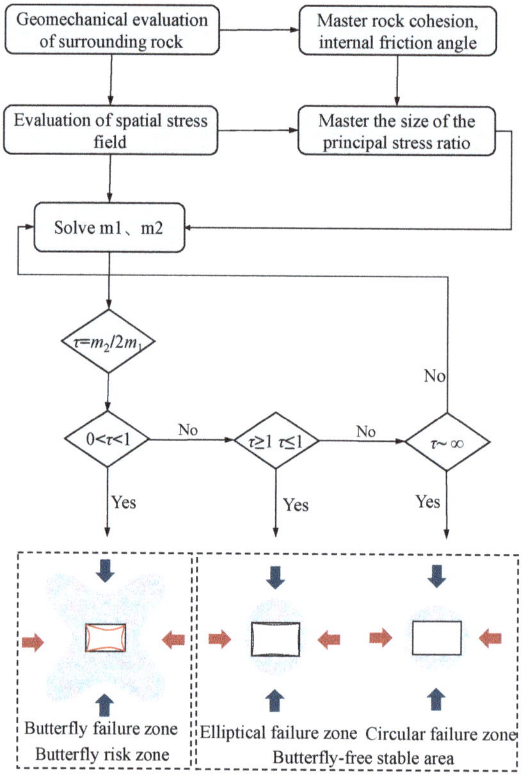

Figure 10. Quantitative evaluation of surrounding rock stability of mining roadway.

Through the preliminary evaluation of the geomechanical parameters of the surrounding rock in the area where the roadway surrounding rock is located, the basic physical and mechanical parameters of the cohesion and internal friction angle of the roadway surrounding rock are determined. The distribution characteristics of the stress field in the roadway surrounding rock are obtained, and the principal stress ratio in the roadway surrounding rock area is determined. By substituting the obtained parameters into the criterion formula of the plastic zone morphology of the surrounding rock, the transition coefficients m_1 and m_2 can be obtained, and, finally, the morphological coefficient of the plastic zone can be obtained. According to the morphological coefficient of the plastic zone, three forms of plastic zone of the roadway surrounding rock can be obtained, namely, butterfly plastic zone, elliptical plastic zone, and circular plastic zone. Based on the morphological characteristics of the plastic zone, the stability of the roadway surrounding rock can be obtained, which can be divided into butterfly risk zone and nonbutterfly stability zone. The former corresponds to the shape of the butterfly plastic zone, and the latter corresponds to the shape of the elliptical plastic zone and the circular plastic zone.

6. Discussion and Field Application Significance

With the gradual entry of coal mine production into high-intensity mining, the working conditions of the super-long working face and large mining height bring greater challenges to the stability of the roadway surrounding rock. In order to improve the mining continuity of the working face, in the mining process of this working face, the roadway of its adjacent next working face is excavated, which means that the roadway will be affected by the disturbance of the adjacent working face, which easily causes roadway instability accidents. The mine pressure zoning and the stability evaluation of the roadway surrounding rock on the goaf side can comprehensively evaluate the stability of the surrounding rock on the goaf side in combination with the general situation of the site, the characteristics of the principal stress, and the development characteristics of the plastic zone, which can provide scientific guidance for the selection of the location of the roadway on the goaf side, the support design, and the optimization of the reasonable width of the protective coal pillar.

As analyzed above, under the influence of the regional stress field, the roadway surrounding rock will appear as three kinds of plastic zone morphology: butterfly, oval, or circular. Among them, the butterfly plastic zone has a large expansion range and shows obvious nonuniform distribution characteristics. When the butterfly failure occurs in the roadway surrounding rock, the surrounding rock is seriously damaged and the stability is poor. The load on the supporting structure of the roadway surrounding rock increases, and the supporting strength and range need to be targeted. The circular plastic zone and the elliptical plastic zone will not show malignant expansion, the surrounding rock failure zone is small, and the risk of the instability of the roadway is low. Conventional support is often used to ensure the stability of the roadway surrounding rock. Based on the morphological development characteristics of the butterfly, oval, and circular plastic zones, it is clear that the circular plastic zone and the oval plastic zone are more conducive to the maintenance of the roadway. Combined with the mine pressure zoning of the goaf, it can be seen that the circular plastic zone and the elliptical plastic zone correspond to the original-rock-stress zone and the deviatoric-stress-drop zone, respectively. Therefore, when arranging the roadway, the low-deviatoric-stress zone and original-stress zone should be selected from the perspective of safety.

In this paper, the roadway surrounding rock is simplified into continuous and homogeneous surrounding rock during the analysis process. With further in-depth study, the distribution law of lateral stress in the goaf in the heterogeneous surrounding rock will be considered in the future, so as to be closer to the actual conditions of field production.

7. Conclusions

(1) Based on the principal stress ratio, the mine pressure zoning on the goaf side is proposed. The evolution law of the principal stress ratio and the plastic zone morphology is expounded. The goaf side is divided into four areas: unloading zone, high-deviatoric-stress zone, low-deviatoric-stress zone, and original-stress zone. The corresponding mine pressure behavior zoning is proposed, that is, fracture zone, butterfly failure zone, elliptical failure zone, and circular failure zone.

(2) A quantitative evaluation method for the stability of the roadway surrounding rock is proposed. The morphological coefficient of the plastic zone is introduced to quantitatively analyze the stability of the roadway surrounding rock on the goaf. Based on the morphological characteristics of the plastic zone, the roadway surrounding rock is divided into butterfly risk zone and nonbutterfly stability zone, which provides scientific guidance for the location selection of the roadway on the goaf, support design, and reasonable width optimization of the protective coal pillar.

(3) The numerical simulation combined with the field case was used to test the mining pressure zoning and the quantitative evaluation method of the surrounding rock stability on the goaf side, and the stability state of the roadway surrounding rock on the goaf side was revealed. The stress distribution characteristics and the shape of the plastic zone are consistent with the theoretical expectation. The research results

have guiding significance for the maintenance of the roadway surrounding rock in the field.

Author Contributions: Conceptualization, H.Z. and K.Y.; methodology, X.G.; writing—original draft preparation, H.Z. and K.Y.; writing—review and editing, X.G. All authors have read and agreed to the published version of the manuscript.

Funding: This research was funded by the National Natural Science Founds of China, grant number 52004289.

Data Availability Statement: The original contributions presented in this study are included in the article. Further inquiries can be directed to the corresponding author.

Conflicts of Interest: The authors declare no conflicts of interest.

References

1. Wang, W.Y.; Gao, M.Z.; Wang, M.; Xie, J.; Liu, Q.; Li, A.Q.; Cehn, H.L. Study on characteristics of deformation and stress distribution of gob-side entry retaining in the ultra-deep mine. *Chin. J. Rock Mech. Eng.* **2019**, *38*, 2955–2963.
2. Li, C.; Huo, T.H.; Wu, Z.; Lu, K.; Zhang, B. Mechanism and stability control of nonuniform and violent deformation of dynamic pressure roadway roof. *J. Cent. South Univ. (Sci. Technol.)* **2020**, *51*, 1317–1327.
3. Sun, Y.; Bi, R.; Chang, Q.; Taherdangkoo, R.; Zhang, J.; Sun, J.; Huang, J.; Li, G. Stability Analysis of Roadway Groups under Multi-Mining Disturbances. *Appl. Sci.* **2021**, *11*, 7953. [CrossRef]
4. Qiang, X.; Li, J.; Chen, C.; Dong, J.; Zheng, Y.; Chen, Z. Nonuniform Deformation Instability Mechanism of Gob-Side Entry Retained in Inclined Coal Seam and Stability Control. *Appl. Sci.* **2023**, *13*, 8727. [CrossRef]
5. Yuan, C.; Zhang, J.G.; Wang, W.J.; Lu, Y.C.; Guo, J.W.; Cao, Q.J.; Li, W.C. Controlling principles of surrounding rock in soft and broken roadway based on the distribution pattern of plastic zone. *J. Min. Saf. Eng.* **2020**, *37*, 451–460.
6. Chen, D.G.; Gao, Z.N.; Zhao, G.M.; Li, S.S.; ZHAO, C.X. Stability analysis of surrounding rock under anchorage mechanics effect. *J. China Coal Soc.* **2020**, *45*, 1009–1019.
7. Kang, H.P.; Zhang, X.; Wang, D.P.; Tian, J.Z.; Yi, Z.Y.; Jiang, W. Strata control technology and applications of non-pillar coal mining. *J. China Coal Soc.* **2022**, *47*, 16–44.
8. Hou, C.J.; Wang, X.Y.; Bai, J.B.; Meng, N.K.; Wu, W.D. Basic theory and technology study of stability control for surrounding rock in deep roadway. *J. China Univ. Min. Technol.* **2021**, *50*, 1–12.
9. Huang, B.X.; Zhang, N.; Jing, H.W.; Kan, J.G.; Meng, B.; Li, N.; Xie, W.B.; Jiao, J.B. Large deformation theory of rheology and structural instability of the surrounding rock in deep mining roadway. *J. China Coal Soc.* **2020**, *45*, 911–926.
10. Chen, D.D.; Ma, X.; Wu, Y.Y.; Xie, S.R.; Li, H.; Jiang, Z.S.; Wang, E.; Guo, F.F.; Guo, W.K.; Ye, Q.C. Failure mechanism and divisional differentiated control of surrounding rock in mining roadway under remaining coal pillar in close-distance coal seam. *J. Energy Sci. Eng.* **2023**, *11*, 1412–1435. [CrossRef]
11. Yu, Y.X.; Ke, D.; Wang, J.B.; Wang, F.Y.; Cehn, B.P. Discussion on determination method of the limit equilibrium zone width based on the deformation analysis of coal wall. *J. China Coal Soc.* **2019**, *44*, 3340–3348.
12. Wang, Q.W.; Feng, H.; Tang, P.; Peng, Y.T.; Li, C.A.; Jiang, L.S.; Mitri, H.S. Influence of Yield Pillar Width on Coal Mine Roadway Stability in Western China: A Case Study. *Processes* **2022**, *10*, 251. [CrossRef]
13. Hou, C.J.; Ma, N.J. Stress in in-seam roadway sides and limit equilibrium zone. *J. China Coal Soc.* **1989**, *04*, 21–29.
14. Qian, M.G.; Shi, P.W.; Xu, J.L. *Mining Pressure and Ground Control*; China University of Mining and Technology Press: Xuzhou, China, 2010; pp. 58–60.
15. Cheng, Y.H.; Jiang, F.X.; Pang, J.L. Research on lateral strata pressure characteristic in goaf of top coal caving in extra thick coal seam and its application. *J. China Coal Soc.* **2012**, *37*, 1088–1093.
16. Babets, D.; Sdvyzhkova, O.; Hapieiev, S.; Shashenko, O.; Prykhodchenko, V. Multifactorial analysis of a gateroad stability at goaf interface during longwall coal mining—A case study. *Min. Miner. Depos.* **2023**, *17*, 9–19. [CrossRef]
17. Zhao, S.K.; Su, Z.G.; Hou, Y.K.; Zhao, B.; Xu, Y.; Wang, H.W.; Mo, Y.L. Research on strata behavior characteristics of mining roadway and collaborative prevention and control technology of force and structure. *Coal Sci. Technol.* **2021**, *49*, 61–71.
18. Paweł, S. Simulation Possibilites of the Post-Mining Goafs Impacton the Deformations Induced by Next Underground MiningOperations with Use of the Cellular Automata Method. *Geotech. Geol. Eng.* **2021**, *39*, 1923–1935. [CrossRef]
19. Xie, S.R.; Jiang, Z.S.; Chen, D.D.; Zhai, L.W.; Yan, Z.Q. Control Study on Surrounding Rock of Gob-Side Entry Retaining below near Distance Goaf. *Processes* **2024**, *12*, 1966. [CrossRef]
20. Zhao, Z.Q. *Study on Mechanism and Control Method of Deformation and Failure of Surrounding Rock in Large Deformation Mining Roadway*; China University of Mining and Technology: Beijing, China, 2014; pp. 23–55.
21. Ma, N.J.; Zhao, X.D.; Zhao, Z.Q.; Li, J.; Guo, X.F. Stability analysis and control technology of mine roadway roof in deep mining. *J. China Coal Soc.* **2015**, *40*, 2287–2295.
22. Liu, H.T.; Wu, X.Y.; Gao, Z.; Zhao, X.D.; Guo, X.F. Evolution law and stability control of plastic zones of retained entry of working face with double roadways layout. *J. Min. Saf. Eng.* **2017**, *34*, 689–697.

23. Meng, Z.P.; Peng, S.P.; Fu, J.T. Study on control factors of rock mechanics properties of coal bearing formation. *Chin. J. Rock Mech. Eng.* **2002**, *21*, 102–106.
24. Guo, X.F.; Guo, L.F.; Ma, N.J.; Zhao, Z.Q.; Li, C. Applicability analysis of the roadway butterfly failure theory. *J. China Univ. Min. Technol.* **2020**, *49*, 646–653.
25. Zhao, Z.Q.; Ma, N.J.; Liu, H.T.; Guo, L.F. The butterfly failure theory of rock mass around roadway and its application prospect. *J. China Univ. Min. Technol.* **2018**, *47*, 969–978.
26. Gao, Z.; Sun, G.Z. Morphological characteristics and evolution law of plastic zone of mining roadway in extra-thick coal seam. *Coal Sci. Technol.* **2022**, *50*, 77–83.
27. Qian, M.G.; Xu, J.L. Behaviors of strata movement in coal mining. *J. China Coal Soc.* **2019**, *44*, 973–984.
28. Guo, X.F.; Ma, N.J.; Zhao, X.D.; Zhao, Z.Q.; Li, Y.E. General shapes and criterion for surrounding rock mass plastic zone of round roadway. *J. China Coal Soc.* **2016**, *41*, 1871–1877.

processes

MDPI

Review

Numerical Simulation of Gas Injection Displacement in Coal Seams: A Mini-Review

Xin Yang [1,2,*], Feng Du [3,*], Qingcheng Zhang [4], Yunfei Zuo [3], Feiyan Tan [2], Yiyang Zhang [3] and Yuanyuan Xu [3]

[1] Research Institute of Macro-Safety Science, University of Science and Technology Beijing, Beijing 100083, China
[2] School of Resources and Safety Engineering, University of Science and Technology Beijing, Beijing 100083, China; m202420029@xs.ustb.edu.cn
[3] School of Emergency Management and Safety Engineering, China University of Mining and Technology (Beijing), Beijing 100083, China; zuoyunfei_cumtb@163.com (Y.Z.); zhangyiyang3424@163.com (Y.Z.); 18154207168@163.com (Y.X.)
[4] Institute of Mine Safety, North China Institute of Science and Technology, Langfang 065201, China; zqc6681@ncist.edu.cn
* Correspondence: xinyangbk89@163.com (X.Y.); fengducumtb@126.com (F.D.)

Abstract

Gas injection displacement technology plays a critical role in enhancing coalbed methane (CBM) and mine gas extraction efficiency. Numerical simulation is essential for revealing multi-field coupling mechanisms and optimizing process parameters, effectively addressing challenges such as high field test costs and limited laboratory scalability. This study systematically reviews progress in modeling physical fields (e.g., flow and diffusion), focusing on multi physical field coupling mechanisms and permeability model evolution. It conducts iterative numerical model analysis—from basic flow–diffusion to fully coupled THMC models—compares simulation software (COMSOL shows greater coupling depth and compatibility than COMET3), and characterizes key mechanisms. By systematically reviewing the key advancements in the fields of numerical simulation in recent years (including important achievements such as the Buddenberg–Wilke equation and the improved Palmer–Mansoori model), a decision-making framework was proposed based on these achievements, covering "Multi-physical Field Coupling Equation Selection, Key Parameter Calibration, Permeability Equation Selection, Model Validation and Error Correction" simulation error $\leq$10% in heterogeneous coal seams. Although general-purpose tools enable high-precision multi-physics coupling, improvements are still needed in modeling flow–diffusion mechanisms, heterogeneity, and chemical field integration. This study provides a systematic methodological reference for the engineering application of gas injection displacement numerical simulation, and the framework constructed hereby can also be extended to shale hydraulic fracturing and other related fields.

Keywords: gas injection displacement technology; numerical simulation; multi-field coupling mechanism; decision-making framework; simulation error

1. Introduction

As a fundamental component of the global energy mix, coal's safe green development and synergistic greenhouse gas emission reduction are key challenges under the "dual-carbon" strategy. Traditionally, coal mine gas (methane) was managed via ventilation; now gas extraction supplements ventilation as the core method. However, complex geological conditions (varying depths, stress, gas content) hinder collection and utilization, making efficient CBM/mine gas extraction a research hotspot.

Gas injection displacement technology enhances CBM recovery and gas extraction efficiency by injecting CO_2/N_2, with the dual benefits of energy recovery and CO_2 sequestration. It applies to two scenarios: surface well CBM development (focused on CO_2 sequestration and efficient methane recovery [1,2]), and underground gas disaster prevention suppressing CBM protrusion via borehole gas injection [3]. Scholars have verified its feasibility and accumulated data through laboratory experiments [4,5] and field trials [6,7]. Numerical simulation, as a tool to reveal the gas injection multi-field coupling mechanism and optimize parameters, compensates for high field test costs and limited laboratory scales [8,9], becoming key for in-depth research. Figure 1 (from CiteSpace analysis of literature from 2015 to 2025) shows keywords' co-occurrence and evolution, indicating growing research on gas injection displacement numerical simulation, with progress in CO_2 injection–storage, simulation models, and permeability model evolution—highlighting simulation's role in unraveling displacement mechanisms.

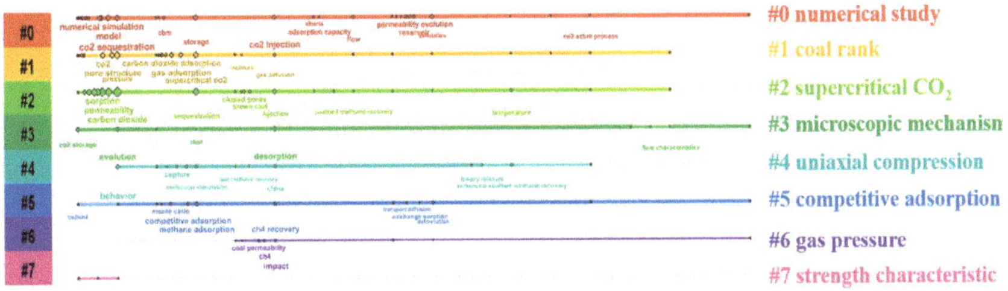

Figure 1. Gas injection displacement research hotspots (2015–2025) and keyword co-occurrence (CiteSpace 6.3.R1).

Gas injection displacement involves multi-field coupling of flow–diffusion–adsorption/desorption–stress–temperature–chemistry, constrained by complex geological conditions. Numerical simulation is pivotal for clarifying these mechanisms and optimizing applications. Recent years have seen many mathematical models characterizing physical fields' principles, coupling mechanisms, and evolution, while simulation has advanced in multi-field coupling [10–12] and permeability dynamic characterization accuracy [13,14]. Yet, with refined engineering demands, two limitations emerge: (1) a lack of systematic geological adaptation methodology—coupled equation selection, permeability characterization, and parameter acquisition rely on simplifications/empirical estimations in complex scenarios, without systematic matching logic; and (2) insufficient integration depth—previous studies [15] failed to systematically address urgent issues (precise flow–diffusion characterization, missing THMC (Thermal–Hydraulic–Mechanical–Chemical) full coupling, fracture non-homogeneity dynamic portrayal, field-laboratory parameter synergistic inversion), hindering the technology's large-scale and precise application. Thus, this review aims to systematically optimize the gas injection displacement numerical simulation framework, with the following objectives:

(1) Sort out the development of modeling theories for each physical field, analyze the coupling mechanisms of multi-physical fields, summarize the evolution and characteristics of permeability models, and provide a theoretical basis for the refinement of multi-field coupling models.

(2) Sort out numerical models at different stages, summarize their technical advantages, compare the characteristics of simulation software, and propose a decision tree of "Multi-physical Field Coupling Equation Selection, Key Parameter Calibration, Permeability Equation Selection, Model Validation and Error Correction" by integrating

model iteration, parameter integration, and error control logic, so as to provide a framework for improving simulation accuracy and engineering applicability.

(3) Analyze the bottlenecks of multi-field coupled simulation, and propose optimization paths (refined gas transport characterization, heterogeneous modeling, THMC full coupling, cross-scale parameter collaborative inversion systems) to solve the core problems in current simulation, such as flow diffusion, chemical coupling, fracture anisotropy, and key parameters relying on experience.

2. Research Methodology

This mini-review was conducted following a systematic approach guided by the Preferred Reporting Items for Systematic Reviews and Meta-Analyses (PRISMA, v4.6.1.617, Prisma Labs, Palo Alto, CA, USA) guidelines to ensure comprehensiveness and reproducibility. The literature search was performed across two major academic databases: Web of Science Core Collection and Scopus. The search strategy utilized key terms and their Boolean combinations, such as ("coal seam" or "coalbed methane") and ("gas injection" or "ECBM") and ("numerical simulation" or "modeling") and ("COMSOL" or "COMET3"). The search was confined to articles published between 2015 and 2025.

The study selection process involved three key stages, as detailed in the PRISMA flow diagram (Figure 2):

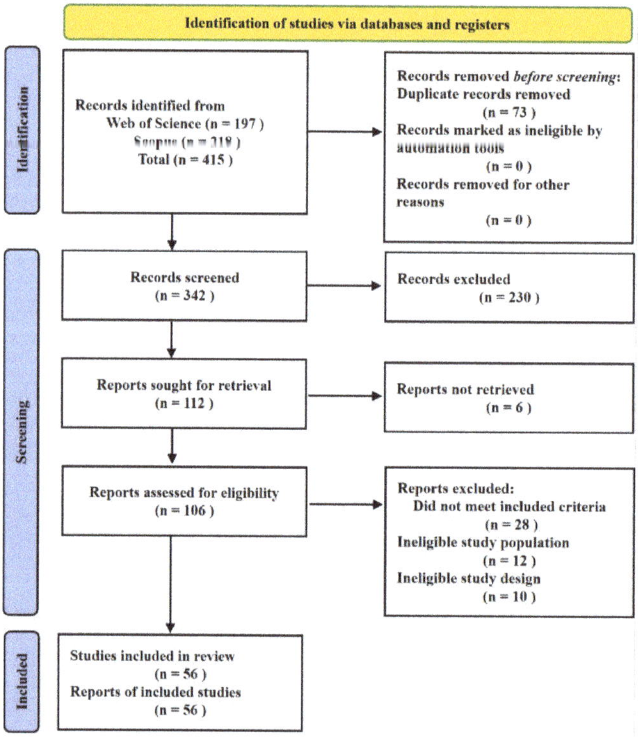

Figure 2. PRISMA flow diagram illustrating the literature identification, screening, eligibility, and inclusion process for this systematic review.

Identification: The initial database searches yielded 415 records.

Screening: After removing 73 duplicates, 342 unique records were screened based on their titles and abstracts. Of these, 230 records were excluded as they did not meet the pre-defined criteria at this stage.

Eligibility: The full texts of the remaining 112 articles were sought for retrieval, of which 106 were successfully accessed. These 106 articles were thoroughly assessed for eligibility. A further 50 articles were excluded for specific reasons: not meeting the inclusion criteria (n = 28), ineligible study population (n = 12), or ineligible study design (n = 10). Ultimately, 56 studies were identified as relevant and form the basis of this mini-review. The entire process is summarized in Figure 2 to enhance transparency.

3. Theory of Multi-Physics Field Modeling for Gas Injection Displacement

3.1. Gas Migration Theory (Flow, Adsorption/Desorption, and Diffusion)

Darcy's law was introduced in 1856 by Darcy through sand column flow tests [16]:

$$V_D = -\frac{k}{\mu}\nabla p_f \tag{1}$$

In the equation: V_D represents the flow velocity, k is the permeability, μ is the viscosity, and ∇p_f is the pressure gradient.

The law assumes linear gas flow and only applies when the Reynolds number is 1–10. Excessive flow rates or non-Newtonian fluid behavior can cause non-Darcy flow [17–19]. As shown in Figure 3: there is a minimum pressure gradient threshold in Zone A (low-speed non-Darcy flow zone) for low-speed flow, below which there is no flow, and beyond which the flow rate rises slowly with the gradient but does not follow a linear trend; the flow growth rate tends to slow down due to inertial effects deviating from linearity at high hydraulic gradients in Zone C (high-velocity non-Darcy-flow zone). For low-permeability first-mined coal seams, low-velocity gas flow exhibits the Klinkenberg effect; neglecting this leads to overestimated gas pressure [20]. At this point, the description of gas flow needs to be corrected on the basis of continuous flow through the Klinkenberg effect [21]:

$$\mathrm{k_{app}} = \mathrm{k_a}\left(1 + \frac{\mathrm{b_k}}{\mathrm{P}}\right) \tag{2}$$

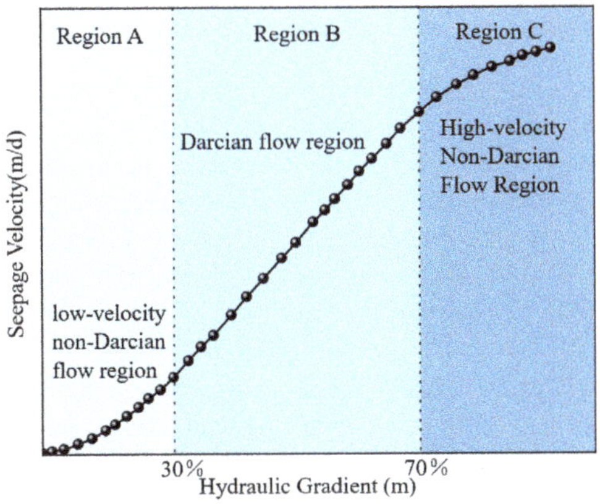

Figure 3. Hydraulic gradient–flow velocity relationship curve.

In the equation: k_{app} is the apparent permeability coefficient, and b_K represents the Klinkenberg factor.

As a porous medium, coal exhibits strong adsorption capacity for CBM. The Langmuir monolayer adsorption equation (1916), validated by numerous scholars, stands as a classic isothermal adsorption model. Other widely used alternatives include the D-A (Dubinin–Astakhov), BET (Brunauer–Emmett–Teller), and D-R (Dubinin–Radushkevich) models [22–24]. Figure 4 compares adsorption models (Langmuir, BET, D-A, D-R). The scores (1–5) in Figure 4 are based on a semi-quantitative assessment derived from a comprehensive review of the literature and the models' typical performance in practice. The evaluation criteria are as follows: computational efficiency (1 = low, 5 = high): based on model complexity and computational resource demands; parameter complexity (1 = simple, 5 = complex): based on the number and difficulty of obtaining required parameters; fitting accuracy for CH_4 (1 = poor, 5 = excellent): based on reported goodness-of-fit in studies; fitting accuracy for CO_2 (1 = poor, 5 = excellent); applicability to multi-gas (1 = single-gas, 5 = multi-gas): assesses the model's capability to handle competitive adsorption. Models may receive identical scores in a category if their performance in that specific aspect is deemed comparable based on the literature consensus.

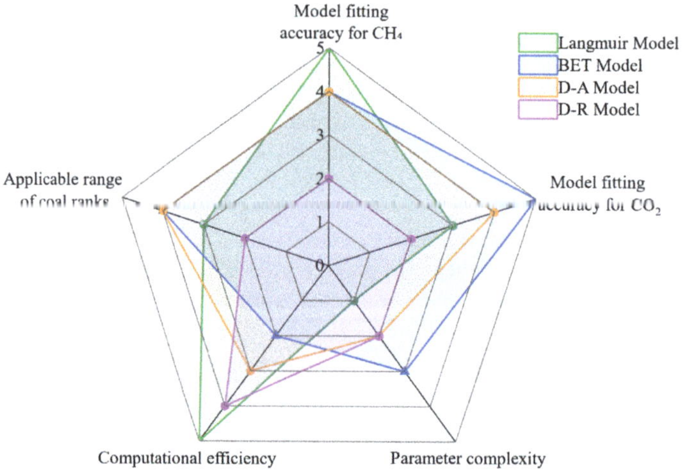

Figure 4. Radar plot comparing the multidimensional properties of coal body gas adsorption models.

Langmuir is widely used for simplicity; D-A fits CO_2 better, and all except D-R apply to CH_4/CO_2 [25,26]. With only two unknown variables, the Langmuir equation offers simplicity in solution and reasonable accuracy, making it widely employed in numerical simulations to characterize gas adsorption/desorption. The Extended Langmuir (EL) model, an advancement of the traditional Langmuir model, is designed to more precisely capture multilayer adsorption behavior [27]. It circumvents the singularity at saturated vapor pressure that plagues the BET and D-A models, preserves the simplicity of the original Langmuir model, and substantially enhances the fitting accuracy of multi-gas adsorption data. EL model achieves these improvements by extending the classic Langmuir theory to multi-component adsorption without introducing the mathematical complexities of the BET and D-A models. Unlike the BET model, which can become singular at saturated vapor pressure, and the D-A model, which requires more empirical parameters, the EL model avoids these pitfalls by utilizing a competitive adsorption formulation based on partial pressures. It preserves the simplicity and computational robustness of the original Langmuir model, requiring only the Langmuir constants for each gas, while significantly enhancing accuracy for gas mixtures by accounting for the competitive occupancy of adsorption sites. Its expression is as follows:

$$C_{pr} = \frac{\rho_{ra}\rho_c a_r b_r p_r}{1 + \sum_{i=1}^{n} b_i p_i} \tag{3}$$

where C_{pr} is the adsorption state content of gas r, ρ_{ra} is the density of gas r under the standard condition, ρ_c is the density of the coal body, a_r is the limiting adsorption capacity of gas r, b_r is the adsorption equilibrium constant of gas r, p_r is the equilibrium partial pressure of gas r, and the subscripts r = 1 and 2 denote the values of the corresponding variables of the gas and the injected gases in that order.

Coal seams have a complex multi-level pore structure. The coal matrix pores serve as both gas adsorption sites and gas transport channels. Gas first diffuses through the matrix micropores to the fracture edges, and then seeps under the action of pressure difference, forming a dynamic equilibrium of the matrix–fracture pressure gradient. Gas diffusion in coal is mainly through fine pores and can be further classified into Fickian diffusion, Knudsen diffusion, and transition diffusion based on the Knudsen number. However, in reality, these three diffusion modes are interrelated [28]. Based on the gas diffusion patterns in coal pores, scholars have developed single-pore, dual-pore, and multi-pore diffusion models. However, the single-pore model relies on the assumption of uniform isotropy, which conflicts with the heterogeneity of coal. It can only describe the diffusion in coal with a single pore structure and cannot account for the complex pore structure [29–31]. The single-pore model is overly simplified due to its assumption of a homogeneous sphere, which leads to systematic prediction bias and an inability to capture the dual-scale transport mechanism in coal. In contrast, the dual-pore model (based on Fick's law) is quantitatively superior, as it resolves two distinct diffusion coefficients for the macropores and the matrix that can differ by orders of magnitude [32]. This critical advantage is conclusively demonstrated by the direct quantitative comparison in Table 1. The data show that the dual-pore model achieves near-perfect accuracy ($R^2 > 0.998$) by utilizing diffusion coefficients differing by five orders of magnitude, thereby realistically characterizing the distinct fast (macropore) and slow (micropore) diffusion paths. Meanwhile, the single-pore model yields significantly lower fitting accuracy ($R^2 \approx 0.9666$–0.9753) with its single, inadequate diffusion coefficient, resulting in a failure to accurately describe the entire gas desorption–diffusion process, particularly the initial rapid gas release [33].

Table 1. Quantitative comparison of the single-pore and dual-pore diffusion models based on Liu et al. [33].

Model Type	Fitting Accuracy (R^2)	Key Quantitative Findings
The single-pore model	$R^2 \approx 0.9666$–0.9753	Exhibits systematic prediction bias. Fails to capture the dual-scale transport mechanism due to the use of a single diffusion coefficient.
The dual-pore model	$R^2 > 0.998$	Achieves near-perfect accuracy by introducing two diffusion coefficients differing by 5 orders of magnitude, which is essential to characterize the distinct fast (macropore) and slow (micropore) diffusion paths in coal.

As shown in Figure 5, gas flows in the coal's natural fracture network (cleat system) following Darcy's law, with its transport rate controlled by fracture openness, connectivity, and inhomogeneity. During gas injection, the fracture pressure gradient drives the injected gas to preferentially migrate along dominant flow pathways. Mathematically, this Darcy-based flow in the fracture system can be described by the following equation:

$$p_{fgi}\frac{\partial \phi_f}{\partial t} + \phi_f \frac{\partial p_{fgi}}{\partial t} = \frac{1}{\tau_i}\left(p_{mgi} - p_{fgi}\right) - \nabla \cdot \left(-\frac{k}{\mu_{gi}}\left(p_{fgi} + b_k\right)\nabla p_{fgi}\right) \quad (4)$$

where p_{mgi} is the partial pressure of the component i in coal matrix, and p_{fgi} is the partial pressure of the component i in fractures. μ_{gi} is the dynamic viscosity of component i, and b_k is the Klinkenberg factor.

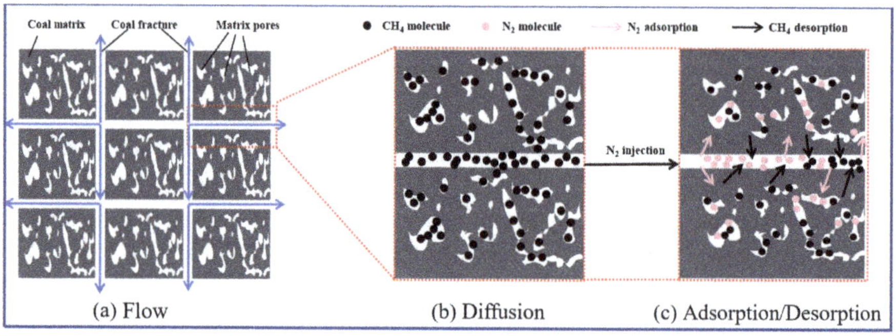

Figure 5. Schematic diagram of gas flow, diffusion, and adsorption/desorption in the coal seam.

The injected gas diffuses into coal matrix pores through the fracture–matrix interface: the matrix pore network provides a large specific surface area, and the gas diffusion path is nonlinearly meandering, with its diffusion coefficient closely related to temperature, pressure, and pore fluid properties. Gas storage on the matrix surface via physical adsorption conforms to the monomolecular layer adsorption behavior described by the EL equation. Notably, the gas injection process—along with changes in gas adsorption—causes matrix expansion/contraction, which in turn affects fracture flow and diffusion paths. The mass balance and adsorption-related processes in the matrix can be captured by the following equation:

$$\frac{\partial}{\partial t}\left(\rho_{sgi}\rho_c \frac{V_{Li} b_{Li} p_{mgi}}{1 + \sum\limits_{i=1}^{n} b_{Li} p_{mgi}} + \phi_m \frac{M_{gi}}{RT} p_{mgi}\right) = -\frac{1}{\tau_i} \cdot \frac{M_{gi}}{RT}\left(p_{mgi} - p_{fgi}\right) \quad (5)$$

where τ_i is the desorption time of the component i, M_{gi} is the molar mass of component i, R is the gas molar constant, T is the temperature in the coal seam, ρ_{sgi} is the gas density of component i in the normal state, ρ_c is the density of coal, and V_{Li} is the adsorbed gas content of component i in a unit mass of coal.

3.2. Hydraulic and Chemical Fields

The core of the hydraulic field is the migration process of gas (CH_4, CO_2/N_2) and water in the matrix–fracture dual-pore system of coal seams. After gas injection, CO_2 preferentially enters the fracture channels and comes into contact with and dissolves in the formation water in the fractures, which is the starting point of the hydraulic field triggering the chemical field effect. The dissolution of CO_2 in water will cause a series of reactions: first, CO_2 combines with water to form carbonic acid (H_2CO_3), and carbonic acid further dissociates into H^+, HCO_3^-, and even CO_3^{2-}, making the fracture water weakly acidic. This acidic water will migrate along the hydraulic gradient in the fractures and then chemically react with minerals in the coal. Common carbonate minerals in coal, such as calcite and dolomite ($CaCO_3$), will react with H^+ in the acidic water, leading to mineral dissolution; at the same time, a small number of sulfides such as pyrite in the coal will also dissolve, releasing ions such as Fe^{2+}. The direct result of these reactions is the removal of mineral blockages in the fractures, expanding the pore and fracture channels, and thereby changing the key parameters of the hydraulic field. Simulation shows that mineral dissolution can increase the porosity of the coal seam by 15–30%, and the change in permeability will in turn affect the migration efficiency of fluids in the hydraulic field [34], accelerating the diffusion of CO_2 and the desorption and migration of CH_4. If only the water–gas two-phase flow is considered and the chemical effect is ignored, the permeability change in the later stage of gas injection will be underestimated: on the one hand, the early mineral dissolution increases permeability, and if the chemical field is not coupled, the gas injection efficiency will be misjudged; on the other hand, after long-term gas injection, dissolved Ca^{2+} and Fe^{2+} may re-precipitate (secondary precipitation of $CaCO_3$), or fine particles produced by the dissolution of clay minerals may clog the fractures, leading to a decrease in permeability. This "first increase then decrease" lag effect can only be captured through the coupling of water and chemistry. In addition, acidic water will also change the hydrophilicity of the coal surface, affecting the relative permeability of water and gas, which also requires adjusting the relative permeability curve in the hydraulic field through a coupled model. The coupling of the two is the key to restoring the true process of coal seam gas injection and is also a necessary prerequisite for improving the prediction accuracy of core indicators such as permeability and CH_4 recovery rate in the model. The governing equations commonly used for simulating this hydraulic–chemical coupling are summarized and compared in Table 2.

Table 2. Comparison of hydraulic and chemical field control equations and their specificities.

Displayed Formula	Specificities
$k = k_0 \cdot \left(1 + \frac{\Delta\phi}{\phi_0}\right)^3, q_g = -\frac{kk_{rg}}{\mu_g}\left(1 + \frac{b_k}{p_{fg}}\right)\nabla p_{fg}$ [34]	The influence of temperature on the coupling process was not considered separately, focusing on the mineral dissolution kinetics under normal-temperature conditions; it includes initiative chemical reactions, involving the generation/consumption of substances, and strong two-way coupling between the chemical field and the water field.
$\rho_{fgd_i} = H_{g_i} \cdot \rho_{fg_i}$, $\frac{\partial(s_w \phi_f \rho_{fgd_i})}{\partial t} + \nabla \cdot (\rho_{fgd_i} q_w) = \frac{1}{\tau_i}\frac{M_{g_i}}{RT}(p_{mg_i} - p_{fg_i})$ [35]	Explicitly incorporate the temperature correction term to adapt to the non-isothermal gas injection environment. There is a one-way weak coupling from the water field to the chemical field. Simplify the chemical process and overlook the modification of the flow channels by mineral dissolution.

3.3. Thermo-Mechanical Response Mechanism of Coal Body Deformation

Besides flow, diffusion, adsorption/desorption, and hydraulic–chemical field effects, gas injection displacement is also influenced by temperature and stress fields [36,37]. Changes in these two fields can disrupt the coal seam's original dynamic equilibrium, forcing it to reestablish balance. During gas injection displacement, the gas adsorption/desorption reactions triggered by injected gas are accompanied by heat transfer, causing coal seam temperature changes. Studies show that as temperature rises, the Langmuir volume constant V_L changes slightly, while the adsorption equilibrium constant P_L decreases significantly—promoting desorption; alongside temperature, higher adsorption pressure and smaller coal particle size both increase gas desorption volume; coal functional group types affect the microscopic mechanism of adsorption exotherm; and elevated temperature reduces adsorption, while increased pressure stabilizes it [38,39]. The coupled equations for the temperature field during gas injection displacement are presented in Table 3.

Table 3. Comparison of temperature field control equations and their specificities.

Displayed Formula	Specificities
$\eta\nabla^2 T + qQ_t = \rho C_v \frac{\partial T}{\partial t} + \theta_T T_0 \frac{\partial e}{\partial t} + \theta_{px} T_0 \left[T_0 \ln(1+wp)\frac{\partial Q}{\partial T} + \frac{QT_0 P}{1+wp}\frac{\partial b}{\partial T} + Q\ln(1+wp)\right]\frac{\partial b}{\partial t}$ [40]	Based on the first law of thermodynamics, comprehensively consider the coupling effects of coal body deformation, gas adsorption/desorption, and heat conduction. Highlight the thermo–hydro–mechanical coupling under the natural warming of deep coal seams. This is applicable to simulating the interaction mechanism of temperature–stress–flow during deep mining. The temperature field equation needs to be solved by coupling with the flow field equation and the stress field equation.
$\frac{\partial}{\partial t}\left(\left(\rho C_p\right)_{eff} T\right) + \eta_{eff}\nabla T - \nabla \bullet (\lambda_{eff}\nabla T) + K\alpha_T T\frac{\partial \varepsilon_V}{\partial t} + \sum_{i=1}^{n} q_{sti}\frac{\rho_s \rho_{gi}}{M_{gi}}\frac{\partial V_{sgi}}{\partial t} = 0$ [41]	Based on the law of conservation of energy, it synthesizes the internal energy changes in coal skeleton and gas, heat convection/heat conduction and explicit adsorption heat terms, and is driven by adsorption and desorption of endogenous heat, which is applicable to natural warming scenarios in deep coal seams. Direct quantification of heat of adsorption modulation of permeability to support the analysis of CBM displacement mechanisms under non-isothermal conditions.

The force is the fundamental cause of the deformation and movement of the coal body; gas injection displacement of pre-drilled holes will cause plastic damage to the coal body in the mining space and around the holes, while the coal body unaffected by mining will remain in an elastic state of deformation. According to the assumption, the coal body is a linear elastic body, which will be elastically deformed by the stratum stress and fluid pressure and other effects. A tiny representative element volume (REV) is arbitrarily taken inside the coal body with side lengths dx, dy, and dz, as shown in Figure 6. The REV is subjected to both surface stresses and volume forces, which are uniformly distributed forces.

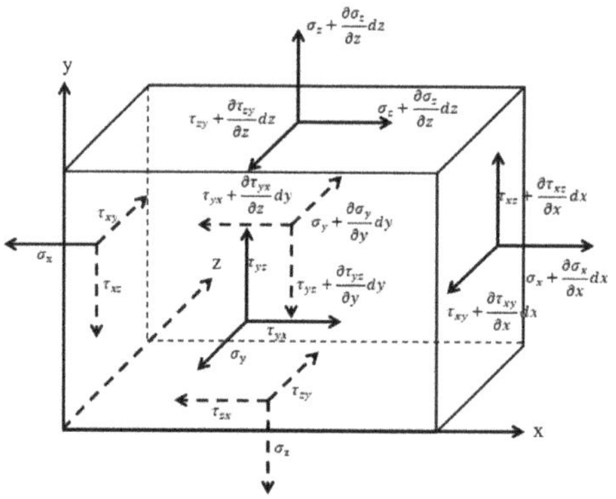

Figure 6. Schematic representation of stress balance of representative units in the coal seam.

Coal stress field mathematical models are divided into two main categories: First, the uniaxial strain assumption (Figure 7a), which ignores lateral displacement and considers only the strain caused by overlying pressure, is applicable to the analysis of a wide range of coal seams, and the classical permeability model constructed based on it couples the fracture compressibility and pore permeability relationship, while other studies derive the horizontal effective stress change and establish an exponential permeability model. Second, the triaxial compressive strain model considers three-dimensional stress–strain (Figure 7b), which is more accurate in analysis but complex in modeling, e. g., the internal expansion coefficient λ is introduced in the triaxial state, and a new permeability model is established to quantify the effect of matrix–fracture interaction and creep on permeability [42–44].

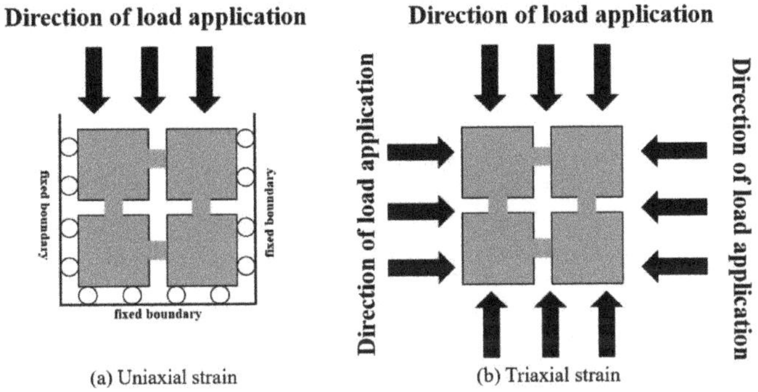

Figure 7. Schematic diagram of uniaxial strain condition and triaxial stress condition of stress field.

Coal matrix deformation (induced by adsorption/desorption behavior, stress, and temperature changes) directly affects the fracture rate and permeability of coal seams. Early studies assumed that the coal matrix was incompressible (Biot coefficient = 1), and analyzed the adsorption–flow interaction by treating the contraction/expansion of the matrix and skeleton as a unified process. Subsequent studies found that there was a "balance lag" phenomenon in the gas diffusion of the matrix: its contraction/expansion effect could not be ignored, so the Biot coefficient was revised to the range of 0 to 1 [45,46]. Currently, it is common to introduce an adsorption expansion/desorption contraction term into stress/strain equations (e.g., adsorption strain is proportional to gas content) and convert it into a pore pressure-controlled strain term. This frees the permeability model from sole reliance on effective skeleton stress, enhancing the prediction of late-stage CBM permeability rebound [47]. The strain caused by adsorbed gas in the coal seam is volumetric strain and is spatially uniform. Based on the theory of elastic mechanics of porous media and considering the effects of ground stress, gas pressure, thermal strain, and adsorption strain, the total strain of the coal body is

$$\varepsilon_{ij} = \frac{1}{2G}\sigma_{ij} - \left(\frac{1}{6G} - \frac{1}{9K}\right)\sigma_{kk}\delta_{ij} + \frac{\alpha_m P_m + \alpha_f P_f}{3K}\delta_{ij} + \frac{\alpha_T T}{3}\delta_{ij} + \frac{\varepsilon_s}{3}\delta_{ij} \tag{6}$$

where G is the shear modulus, K is the bulk modulus, α_m and α_f are the Biot coefficients of the coal matrix and fracture, P_m and P_f are the fluid pressures in the coal matrix and fracture, αT is the coefficient of thermal expansion, T is the temperature of the coal seam, ε_s is the volumetric strain of the coal matrix induced by the adsorption of the gas, δ_{ij} is the Kronecker function, and $\delta_{ij} = 1$ when i = j; $\delta_{ij} = 0$ when $i \neq j$.

All variables involved in Sections 3.2 and 3.3, including their definitions and corresponding units, are detailed in Table A1 in Appendix A.

3.4. Evolutionary Improvements in Permeability Modeling

Coal is usually treated as a dual-pore medium (matrix and fracture), so its structure needs simplification when building permeability models (Figure 8). Among dual-pore models (spherical, capillary tube, cube, etc.), the matchstick and cube models are most commonly used. For absolute permeability, the most widely applied models are the strain-based P&M model (power function form) and stress-based S&D model (exponential function form); typical absolute permeability mathematical models are shown in Table 4.

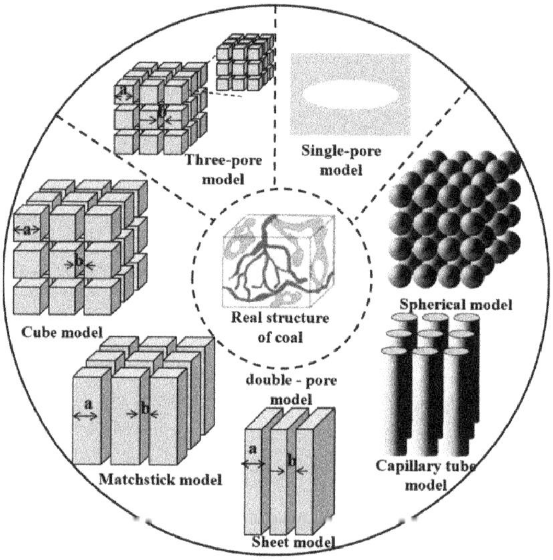

Figure 8. Schematic diagram of the pore structure composition and simplified physical model of the coal body.

Table 4. Mathematical model of absolute permeability.

Model Name	Displayed Formula	Features
Gray Model [13]	$k = k_0 e^{-3C_f(\sigma_h^e - \sigma_{h0}^e)}$	Combines pore pressure/adsorption strain; concise but limited multi-factor coupling.
P&M Model [44]	$\frac{k}{k_0} = \left(1 + (p - p_0)\frac{c_m}{\varphi_0} - (1 - \frac{K}{M})\left(\frac{bp}{1+bp} - \frac{bp_0}{1+bp_0}\right)\frac{c_1}{\varphi_0}\right)^3$	Links permeability to porosity/strain (uniaxial strain assumption; limited applicability).
S&D Model [43]	$k = k_0 e^{-3C_f(\sigma - \sigma_0)}$	"Matchstick"-based exponential model clarifies stress effect; ignores adsorption/desorption (single dimension).
C&B Model [12]	$\frac{k}{k_0} = \exp\left\{-\frac{3}{k_p}\left[(\sigma - \sigma_0) - (p - p_0)\right]\right\}$	Quantifies methane desorption strain's impact; no gas-water multiphase verification.
Liu and Rutqvist Model [14]	$\Delta\sigma = -\frac{v}{1-v}\Delta P + \frac{E(\Delta\varepsilon_s - \Delta\varepsilon_f)}{1-v}$	Integrates adsorption/desorption; uniaxial strain limits 3D scenarios.
Improved P&M Model [20]	$\frac{k_\infty}{k_{\infty 0}} = \left\{\begin{array}{l} 1 + \frac{1}{M\phi_{f0}}\left[\beta_f(p_f - p_{f0}) + \beta_m(p_m - p_{m0})\right] \\ + \frac{\varepsilon_L}{\phi_{f0}}\left(\frac{K}{M} - 1\right)\left(\frac{p_m}{p_L + p_m} - \frac{p_{m0}}{p_L + p_{m0}}\right) \end{array}\right\}^3$	Corrects dual-pore structure for gas transport; poor generalization to non-dual-pore systems (relies on P&M framework).

Notably, permeability is greatly affected by water saturation and storage morphology: hydrates forming on particle surfaces have little impact on pore size, while those forming at pore centers or throats significantly narrow pores (or even block channels), leading to a sharp drop in permeability [48]. Relative permeability (effective permeability K(Sh)) refers to the permeability of a single-phase fluid flowing in a porous medium with specific hydrate saturation. For two-phase flow, aqueous phase permeability is called water relative permeability K(rw), and gas phase permeability is gas relative permeability K(rg). As shown in Figure 9, gas–water two-phase flow is divided into Zone A (pure gas flow) and Zone B (gas–water competitive flow). Coal matrix shrinkage and the Klinkenberg effect alter pore structure to affect the permeability of both zones; low-pressure dynamic pore changes (e.g., left side of Zone B) can enhance gas-phase flow capacity. Scholars have developed various gas–water two-phase flow relative permeability models as shown in Table 5.

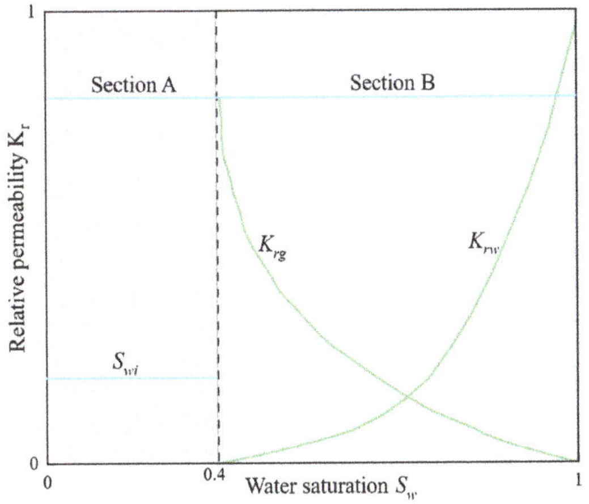

Figure 9. Typical relative permeability of gas–water two-phase flow during the displacement process.

Table 5. Mathematical model of relative permeability of gas–water two-phase flow.

Model Name	Displayed Formula	Features
Corey Model [49]	$\begin{cases} k_{rg} = k_{rg0}\left(1 - \frac{s_w - s_{wr}}{1 - s_{wr} - s_{gr}}\right)^2\left(1 - \left(\frac{s_w - s_{wr}}{1 - s_{wr}}\right)^2\right) \\ k_{rw} = k_{rw0}\left(\frac{s_w - s_{wr}}{1 - s_{wr}}\right)^4 \end{cases}$	Simple, widely used, easy parameters; ignores fracture impact.
Brooks–Corey Model [50]	$\begin{cases} k_{rw} = s_{eff}^{(2+3\lambda)/\lambda} \\ k_{rg} = (1 - s_{eff})^2\left(1 - s_{eff}^{(2+\lambda)/\lambda}\right) \end{cases}$ $s_{eff} = \frac{s_w - s_{wr}}{1 - s_{wr}}$	Includes pore size distribution; fits heterogeneous media; ignores adsorption expansion.
X-curve Model [51]	$\begin{cases} k_{rw} = a \cdot (s_w - s_{wr})^n \\ k_{rg} = b \cdot (1 - s_w - s_{gr})^m \end{cases}$	Few parameters (easy fitting, high efficiency for homogeneous media); no multiphase nonlinear capture.
V-C Model [52]	$k_{r\alpha} = k_{r\alpha 0} \cdot \left(\frac{\mu_\beta}{\mu_\alpha}\right)^c \cdot (1 - s_\beta)^n$	Considers fluid viscosity ratio (viscous-dominated multiphase flow); only for low capillary pressure/high viscosity.

3.5. Mechanism Analysis of THMC Multi-Field Coupling System in Gas Injection Displacement

During the gas injection displacement process, the gas migration field, water migration field, stress field, and temperature field of the coal seam interact through complex nonlinear mechanisms, thereby forming a multi-field coupling system of heat–fluid–solid. The core mechanism of this coupling relationship is summarized in Figure 10: in the process of gas injection displacement of CBM, the adsorption/desorption field is the starting point. The adsorption and desorption behavior of CH_4 and CO_2/N_2 in the coal matrix not only regulates the gas saturation and relative permeability, but also links the gas adsorption, diffusion, and flow field in the form of source terms, and is accompanied by volume strain and changes in thermal properties (such as specific heat capacity); the flow field plays a core role in gas transport, with gas diffusing between the matrix and fractures, and its flow characteristics are affected by gas saturation, water density, and viscosity, as well as heat conduction and convection in the temperature field; the temperature field affects the fluid in the flow field and the thermal properties of the coal matrix through heat conduction and convection, and the heat effect generated by adsorption/desorption also feeds back to the temperature field, thereby establishing a channel for the interaction between heat and fluid. In the stress field, effective stress, porosity, and permeability are regulated by the interaction of multiple fields, and the change in stress in turn affects the flow and adsorption processes, thus achieving fluid–solid coupling. These fields work together in a coordinated manner around the transfer of matter, energy, and stress.

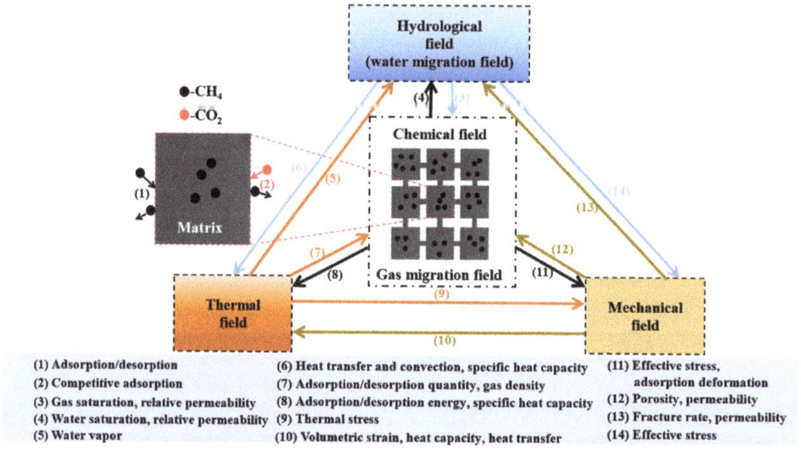

Figure 10. Multi-physical field coupling mechanism in gas injection displacement technology.

4. Evolution and Engineering Decision System of Numerical Models for Gas Injection Displacement

4.1. Model Evolution and Software Capability Comparison

Numerical simulation of gas injection displacement has evolved significantly, as shown in Table 6, progressing from single-field descriptions to multi-physics coupling (e.g., from basic flow–diffusion models [53] to comprehensive THM/THMC frameworks [34,35]), from static parameters to dynamic multi-mechanism responses, and from theoretical verification to engineering application. This evolution, summarized in Table 6, is characterized by key advancements in permeability characterization (e.g., introducing the "elastic-plastic boundary point" [14] and the λ coefficient [54]) and in the calculation of critical parameters (e.g., employing the Buddenberg–Wilke equation for real-time mixed viscosity [55] and temperature-corrected Langmuir models [35,56]).

Table 6. Comparison of different gas injection displacement numerical simulation models.

Model Source	Software	Key Progress and Highlights	Original Study Validation and Prediction Accuracy
Yang [53]	COMSOL	A "pore-fracture dual medium" coupling model was developed. This model clearly defines the mass exchange mechanism between Fickian diffusion (in pores) and Darcy flow (in fractures).	Validated by laboratory core flow tests; gas migration rate prediction deviation ≈12%.
Vishal [57]	COMET3	The Peng–Robinson equation of state was employed to calculate the gas density at different pressures and temperatures, and avoid density calculation deviations.	Compared with experimental gas density data; calculation error ≤8%.
Ren [55]	COMSOL	The Buddenberg–Wilke equation was introduced to calculate mixed viscosity in real time, addressing the flow rate deviation resulting from the assumption of constant viscosity in traditional models.	Flow rate prediction error reduced from 15% (traditional model) to 7%.
Fan [35]	COMSOL	Developed non-isothermal THMC coupling model for binary gas (CO_2/CH_4) transport, introduced the temperature-corrected Langmuir model to quantify the influence of temperature on adsorption.	Validated by high-temperature coal seam simulations; prediction error of temperature-dependent adsorption capacity ≤10%.
Shi [54]	COMSOL	The λ coefficient (internal expansion volume coefficient) was proposed, and a fully coupled permeability formula was constructed to avoid the deviation in permeability prediction.	Compared with field-measured permeability; prediction error ≤9%.
Asif [8]	COMET3	Breaking through the limitation of "empirical parameter value selection", a simulation foundation is established based on the actual measured data of coal samples, and a 3D dual-pore model is constructed using the COMET3 simulator.	Validated by field monitoring; gas migration radius prediction error ≈ 11%.
Fang [34]	COMSOL	Established fully coupled THMC models for CO_2-ECBM, overcame the deficiency of traditional models in neglecting the chemical field.	Qinshui Basin field data: methane production prediction relative error 8.64%; Baijiao Coal Mine experiment: gas volume fraction error <6%.
Liu [14]	COMSOL	The concept of "elastic-plastic deformation boundary point" was introduced for the first time, and the pressure attenuation law of N_2 was simulated through the Mises equivalent stress equation.	Agreement with experimental N_2 pressure attenuation data (R^2 = 0.92).
Yang [56]	COMSOL	The Extended Langmuir model incorporates a temperature coefficient to quantify the effects of temperature on the adsorption of various gases. Integrating Henry's law and the Kelvin–Laplace law, this model simulates the dissolution of CO_2 in water and the migration of water vapor, aiming to elucidate the mechanism underlying the change in gas injection efficiency in water-rich coal seams following initial drainage.	Multi-gas adsorption capacity: prediction error ≤6%; Water-phase processes (CO_2 dissolution, water vapor migration): simulation error ≈ 10%.

Software analysis reveals distinct strengths and application niches for COMSOL Multiphysics (Version 6.1) and COMET3 (Version 3.0). COMSOL, as a general-purpose platform, excels in flexibility, enabling the direct implementation of custom equations (e.g., the Extended Langmuir model, complex permeability laws) and the simulation of novel coupled processes like the full THMC coupling in this study. This makes it ideal for fundamental research and new model development, albeit with a steeper learning curve. In contrast, COMET3 is a dedicated, industry-proven simulator for coalbed methane. Its solvers are highly optimized for specific, large-scale production forecasting, offering superior computational efficiency for standard scenarios. However, its architecture limits the implementation of non-standard physics; it cannot directly solve fully coupled THM partial differential equations—leading to uncontrolled errors by neglecting temperature and water-phase impacts—and offers scant support for custom equations, hindering the modeling of complex processes like ternary gas competitive adsorption. Consequently, COMSOL is superior for developing and verifying innovative coupled frameworks, while COMET3 is more efficient for applied engineering and production prediction using established models.

The comparative analysis in Table 6 reveals that models achieving higher coupling depth (e.g., THM, THMC) and incorporating dynamic parameter mechanisms (e.g., real-time viscosity calculation, temperature-corrected adsorption) typically yield prediction errors below 10% for key outputs like gas migration rate and permeability. No single model is universally 'best'; however, for predicting methane production in complex, non-isothermal CO_2-ECBM scenarios, fully coupled THMC models demonstrate superior comprehensive performance by capturing a wider range of interacting physical processes. Under complex geological conditions, numerical simulation of gas injection displacement in coal seams faces challenges such as differentiated applicability of multiple models, high parameter sensitivity, and limited computational efficiency. A systematic review of the evolution of gas injection displacement mathematical models (Section 4.1) shows that no single model can universally address diverse geological characteristics (e.g., varying permeability, stress sensitivity, multiphase flow). Thus, a structured decision-making framework is needed to enhance the scientific rigor and engineering applicability of model selection.

4.2. *Engineering Decision Trees: Hierarchical Decision Trees for Numerical Simulation of Gas Injection Displacement Models*

Under complex geological conditions, numerical simulation of gas injection displacement in coal seams faces challenges such as differentiated model applicability, high parameter sensitivity, and limited computational efficiency. The systematic review in Section 4.1 confirms that no single model is universally applicable. Furthermore, analysis of high-precision models (with errors $\leq$10%) in Table 6 informed the construction of a four-dimensional decision-making framework (Figure 11) to enhance the scientific rigor and engineering applicability of model selection. This framework sequentially addresses four core questions to build confidence in simulation results, from theoretical foundation to practical reliability: (1) Governing physical equations? (Multi-physical Field Coupling Equation Selection); (2) Accurate input parameters? (Key Parameter Calibration); p(3) Evolution of the core property—permeability? (Permeability Equation Selection); and (4) Trustworthy model output? (Model Validation and Error Correction). While aspects like mesh generation are important, this framework focuses on the most critical, high-level decisions impacting model accuracy and applicability. The decision-making process is as follows (Figure 11):

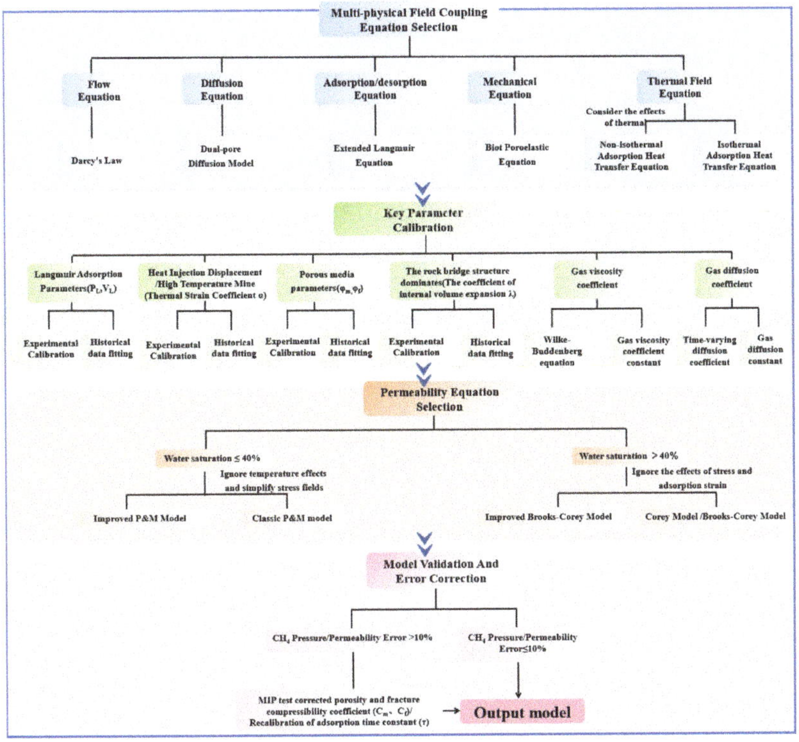

Figure 11. Engineering decision tree for model optimization of gas injection displacement technology.

(1) Multi-physics Field Coupling Equation Selection. (a) Flow equation: Darcy's law is used. (b) Diffusion equations: two-pore diffusion model, Fick's law dominant. (c) Adsorption/desorption equation: the EL equation. (d) Mechanical equation: Biot porous elasticity equation. (e) Thermal field equations: consider the influence of thermal effects using non-isothermal adsorption heat transfer equation; vice versa using isothermal adsorption heat transfer equation.

(2) Key Parameter Calibration: Acquisition of key parameters for special geological conditions and different gas injection sources. Combined with experimental acquisition or historical data fitting to obtain relevant parameters such as gas adsorption parameters (Langmuir constants V_L, P_L), thermal expansion coefficients α, rock bridge structure parameters (internal expansion coefficients λ), coal seam porosity, fracture rate (φ_m, φ_f), gas dynamic viscosity coefficients (μ), gas diffusion coefficients (D), etc., inverted based on high-precision experimental or historical data.

(3) Permeability Equation Selection:

The selection is based on the dominant flow regime, which transitions at a critical water saturation of 40%—a threshold substantiated by THMC coupled simulations [58] and aligned with the flow regime transition identified in Section 3.4 (Figure 9).

(a) Low water saturation ≤40%: In this gas–continuous-phase regime, absolute permeability, governed by stress-induced and adsorption-induced strain, is the controlling factor. The improved P&M model is selected to capture the evolution of absolute permeability under stress and sorption [20,55]. The classical P&M model is chosen if thermal effects are neglected and the stress–adsorption relationship is simplified.

(b) High water saturation >40%: In this competitive gas–water two-phase flow regime, relative permeability effects dominate. The modified Brooks–Corey model is selected to account for complex pore structure and fluid-distribution effects, consistent with approaches in simulations of high-water-saturation ECBM processes [56]. The standard Brooks–Corey model can be used if stress and temperature corrections are unnecessary. The modified Brooks–Corey model can be used for correcting the effects of stress and temperature on relative permeability, and vice versa with the Corey/Brooks–Corey model to cover gas–water two-phase flow.

(4) Model Validation and Error correction: When the error of CH_4 pressure/seam permeability was >10%, the pore/fracture compression coefficients (C_m, C_f) were corrected by piezo mercury experiments, and the adsorption time constant (τ) was re-determined. Error $\leq$10%: output the final model to ensure that the simulation accuracy meets the engineering requirements (e. g., coal mine downhole gas injection displacement, surface coalbed methane development).

Compared to conventional static models with fixed parameters, the framework's selection–calibration–validation–optimization closed-loop process provides a scientific decision-making system for optimizing both underground gas injection-enhanced drainage and surface CBM recovery designs.

4.3. Engineering Decision Tree Validation and Application Case Study

The engineering decision tree was validated using a five-well CO_2-ECBM field trial in the 3# coal seam of the Qinshui Basin [59]. A two-dimensional Cartesian coordinate model (with dimensions of 200 m by 200 m) was established, and a five-point well pattern layout (with a well spacing of 283 m) was adopted, featuring constant 10 MPa injection pressure, 0.1 MPa production drainage pressure, and adiabatic no-flow outer boundaries. The model assumed constant injection and production pressures of 10 MPa and 0.1 MPa, respectively, to simplify the initial case study and facilitate a clear comparison, acknowledging that in field operations, these parameters typically vary over time. Input parameters were derived from experimental literature data, with incorporation of the temperature-coupled governing equations (refer to Section 3.2, Equation (7) to characterize thermal–mechanical–hydraulic (THM) interactions in permeability evolution.

$$\frac{\partial(\rho(Cp)_c T)}{\partial t} + \eta\nabla T - \nabla\cdot(\lambda_c \nabla T) + \sum_{i=1}^{2} q_{sti}\frac{\rho_c \rho_{sgi}}{M_i}\frac{\partial V_{ci}}{\partial t} + K\alpha_T T\frac{\partial(\varepsilon_{s1}+\varepsilon_{s2})}{\partial t} \quad (7)$$

Due to the scarcity of time-evolving field permeability data, the simulation results from Fang et al. [59]—whose model was field-validated—served as the benchmark. For this validation case, the decision tree was applied, leading to the selection of the THM coupling equations and the improved P&M model for permeability evolution. Key parameters (e.g., Langmuir constants, initial permeability, porosity) were primarily sourced from experimental data related to the Qinshui Basin 3# coal seam [59]. Where necessary, minor calibrations were made within physically reasonable ranges to achieve a history match with the initial production data.

Figure 12 presents a comparative analysis of permeability ratio (k/k_0) evolution along the model diagonal between the present fully coupled THM model and Fang et al.'s isothermal model [59], under the condition of 10 MPa injection pressure and 5000-day simulation. The results show that the two models exhibit aligned permeability ratio trends: the relative deviation along the diagonal ranges from 5% (minimum at 9.4 m) to 8.1% (maximum at 169.8 m), and the permeability ratio calculated by our model is overall 5–8.1% higher than that of the benchmark model. Critically, the relative error reveals a distinct spatial pattern: it is minimized near the injection well (5% at 9.4 m) and increases progressively with distance, reaching a maximum in the far-field region near the production well (8.1% at 169.8 m). The prediction error was quantified using the relative error (RE) metric, defined in Equation (8):

$$RE = \left| \frac{(k/k_0)_{m,i} - (k/k_0)_{F,i}}{(k/k_0)_{F,i}} \right| \times 100\% \quad (8)$$

where $(k/k_0)_{m,i}$ and $(k/k_0)_{F,i}$ are the permeability ratios from our model and the benchmark model at the i-th location, respectively.

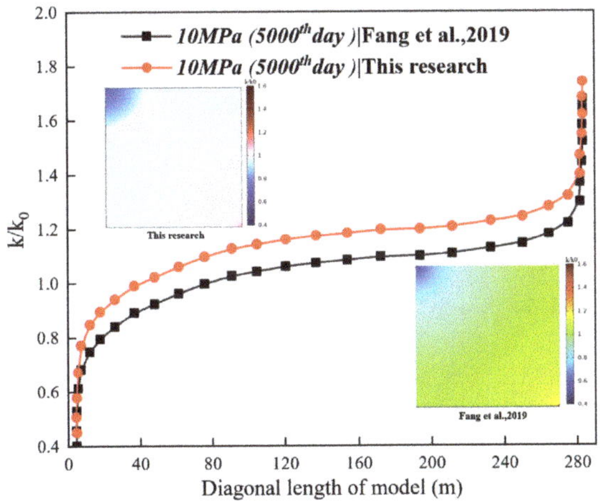

Figure 12. Comparison of permeability ratio (k/k_0) evolution between the fully coupled THM model and the isothermal model under 10 MPa injection pressure at 5000 days [59].

This divergence primarily stems from two mechanisms: (1) after 3000 days, the mutual cancelation of CO_2 adsorption swelling and CH_4 desorption shrinkage minimized effective stress impacts; and (2) the synergistic effects of late-stage production well permeability rebound and stabilized injection elevated whole-seam permeability above initial values. This mechanism of permeability influence has been verified [59]. The framework demonstrated $\leq$10% prediction errors, effectively addressing systematic deviations in prior thermal-neglected models and providing reliable guidance for CO_2-ECBM parameter optimization. The discrepancies observed in Figure 12, particularly the minimal deviation at 9.4 m (closer to injection well) and the maximum deviation at 169.8 m (further into the reservoir), can be attributed to the coupled THM mechanisms. Near the injection well (e.g., 9.4 m), strong forcing (high CO_2 pressure) dominates the permeability evolution, leading to better agreement between models. At farther distances (e.g., 169.8 m), the interplay between CO_2 adsorption swelling, CH_4 desorption shrinkage, and the resulting effective stress changes becomes more complex and spatially heterogeneous. Our fully coupled THM model captures the delayed permeability rebound in these regions more accurately than the isothermal model, which neglects temperature-mediated strain effects and their spatial variation, leading to the larger discrepancy.

It should be noted that the use of a two-dimensional model represents a simplification of the inherently three-dimensional gas injection process, adopted here for computational efficiency and conceptual clarity. While it serves the purpose of demonstrating the framework's utility under controlled conditions, future applications should aim for 3D simulations to more accurately capture spatial heterogeneities and complex flow paths. In addition, although the constant-pressure assumption is adopted in this study to simplify the initial case, dynamic pressure changes in field operations significantly affect the coalbed methane displacement process: on the one hand, fluctuations in injection/production pressures directly disturb the adsorption–desorption equilibrium of CH_4 and CO_2, altering the displacement efficiency of the two gases; on the other hand, the dynamic evolution of pressure not only affects the fluid seepage pattern (e.g., triggering non-Darcy seepage) by changing the pressure gradient in the seepage field, but also causes dynamic changes in coal permeability through the effective stress effect, and the permeability changes in turn act back on the seepage process. In future work, we will conduct multi-physics simulations under dynamic pressure coupling to improve the model's adaptability to field conditions.

5. Urgent Problems and Future Prospects

5.1. Precision Modeling of Flow–Diffusion Mechanisms

During the gas injection displacement process, gas transportation is controlled by both flow and diffusion fields. Current characterization often suffers from idealized distortions: the flow field is predominantly described by Darcy's law, while the diffusion field is homogenized based on Fick's law. In reality, flow behavior may exhibit complex nonlinear relationships, and diffusion involves mixed mechanisms of Knudsen and transition regime diffusion in nanopores, further modulated by pore surface roughness. To address these limitations, precise modeling of both fields is urgently needed:

(a) For the flow field, it should be treated as a composite flow with nonlinear terms. The dominant flow equation should be selected based on flow velocity and pore size dynamics, choosing from the Darcy, Forchheimer, or Swartzendruber equations. As compared in Table 7, the Swartzendruber equation offers a more flexible framework for capturing non-Darcy behaviors that deviate from the quadratic velocity dependence of the Forchheimer model, making it potentially more accurate for complex fluid–coal interactions.

(b) For the diffusion field, a multi-mechanism quantification framework is required, delineating the dominant diffusion regime (Fick, Knudsen, or transition) based on pore size distribution from BET test data. It is crucial to note that the BET method has limitations in ultra-micropores, potentially introducing uncertainty. Thus, a surface roughness correction factor (obtained from experiment or molecular simulation) should be introduced to dynamically calibrate the diffusion coefficients.

Table 7. Comparison of different flow control equations for coal seam gas injection.

Equation Name	Mathematical Form	Applicability and Characteristics
Darcy's Law	$-\nabla p = (\mu/k)v$	Applies to linear, low-velocity flow. Simple but inaccurate for high flow rates or non-Newtonian fluids.
Forchheimer Equation	$-\nabla p = \frac{\mu}{k}v + \beta\rho v^2$	Accounts for inertial effects at high velocities. Standard for non-Darcy flow in many porous media.
Swartzendruber Equation	$-\nabla p = \frac{\mu}{k}v + \beta\rho v^m$ (m < 2)	Describes nonlinear flow where the power-law index m deviates from 2. Potentially more accurate for certain non-Newtonian behaviors or specific pore structures in coal.

5.2. Characterization of Anisotropic Fracture Networks

Fracture networks in deep coal reservoirs demonstrate intricate topological configurations characterized by fractal dimensions, with natural fracture apertures conforming to a Weibull distribution. Traditional dual-porosity models, however, approximate these structures using isotropic permeability tensors, resulting in significant deviations in the representation of key flow parameters. Although anisotropic permeability models based on tensor decomposition have been proposed by some scholars, and fractal geometry theory (e.g., Mandelbrot fractal dimension) and discrete fracture network models have been validated in the field of rock mechanics to describe inhomogeneous fractures [60], these models have not yet been integrated into the mainstream simulation framework of gas injection displacement. However, these models have not been integrated into the mainstream gas injection displacement simulation framework and lack field engineering validation. Future efforts should focus on the engineering adaptation of anisotropic models. The first priority is to develop an anisotropic permeability interface in numerical simulation software, integrating existing anisotropic permeability models into gas injection simulation platforms. This will enable the software to receive and compute anisotropic permeability parameters, thereby overcoming the traditional isotropic default setting. Secondly, the rationality and accuracy of anisotropic models should be validated by comparing simulation results before and after their application, based on field gas injection data and laboratory experimental results. Finally, based on the validation outcomes, the anisotropic model should undergo iterative optimization—refining its input parameter system to progressively enhance the accurate characterization of anisotropic features in coal seams.

5.3. THMC Full-Coupling Deficit Mitigation

The existing THMC coupling models have not comprehensively resolved the issue of chemical mineralogical changes during the interaction of CO_2, water, and coal. For example, they do not meticulously calculate the dissolution and precipitation rates of minerals such as calcite. Moreover, they do not take into account the influence of the altered water absorption of the coal surface after mineral changes on gas flow, which results in inaccurate permeability predictions. Although some studies have attempted to incorporate chemical processes into the models [60], they either fail to thoroughly explore the dynamic processes of chemical reactions, or do not integrate chemical changes with permeability and mechanical deformation in real time. Additionally, they do not capture the hysteresis phenomenon where permeability increases due to short-term mineral dissolution and then decreases due to long-term clay particle clogging. There are also combined simulations using COMSOL and PHREEQC (a computer program for speciation, batch-reaction, and transport calculations), yet most of them are focused on sandstones and shales and not adapted to the specific structure of coal reservoirs. To bridge this gap, a COMSOL-PHREEQC coupled simulation framework can be constructed. First, employ the geochemical database of PHREEQC to track the dissolution–precipitation changes in coal minerals in real time [61]. Second, correlate the pore changes resulting from mineral reactions with the permeability model to dynamically update the permeability. Third, integrate experimental measurements of the water absorption changes in the coal surface to adjust the relevant parameters of gas–water flow. This approach enables the accurate capture of the hysteresis characteristics of permeability (i.e., increasing first and then decreasing), thereby making the THMC model more consistent with practical scenarios.

5.4. Joint Inversion of Key Parameters in the Field and Laboratory

Accurate acquisition of key parameters (e.g., gas adsorption constants (V_L, P_L)), diffusion coefficients (D), fracture compression coefficients (C_f), and other key parameters is the key to the reliability of the gas injection displacement model. There are significant discrepancies between current laboratory measurement conditions and actual geological environments. During gas injection, temperature and pressure variations induce dynamic parameter evolution, yet existing models treat these parameters as constants. Laboratory data are predominantly derived from small-scale core samples, whereas field parameter inversion relies on kilometer-scale inter-well monitoring data. This substantial spatial-scale mismatch (spanning multiple orders of magnitude) introduces experimental parameter errors that ultimately compromise model accuracy. The construction of the "laboratory calibration—field inversion—dynamic update" trinity framework is the key to breakthroughs; constructing a database containing coal rock microstructure and fundamental parameters by simulating the initial distribution characteristics of parameters under real stratigraphic conditions in the laboratory. Machine learning algorithms [62] are implemented to construct multiscale correlation models bridging microscale laboratory measurements with macroscale field datasets. The machine learning prediction results are used as the initial values, coupled with a mathematical model of gas injection displacement, and the key parameters are inverted by a history-fitting optimization algorithm (e.g., genetic algorithm) to construct a dual-drive inversion model. In the future, integrating geophysical data (e.g., seismic exploration) to constrain the inversion of geomechanical properties will be an important direction for improving the spatial characterization accuracy of model input parameters.

5.5. Extensibility of the Framework to Other Technologies and Reservoirs

The proposed four-dimensional decision-making framework, while developed for gas injection in coal seams, possesses inherent extensibility to other subsurface energy technologies, notably hydraulic fracturing in shale and tight sandstone reservoirs. The core logic—selecting governing equations, calibrating key parameters, choosing constitutive models, and validating against field data—remains universally applicable. Taking the study by [63] as a concrete example, which established a seepage–damage coupling model for hydraulic fracturing considering material heterogeneity and solved it using COMSOL, the workflow aligns closely with our framework. Adapting our framework for hydraulic fracturing simulations can be specified as follows:

(a) Governing Equations Selection: The core at this stage is the coupling between geomechanics and fluid flow. For instance, an elastoplastic or elastic-damage constitutive model describing rock failure (e.g., the Mohr–Coulomb and maximum tensile stress criteria used in the reference) should be adopted and coupled with Darcy's law or more complex non-Darcy flow equations.

(b) Key Parameter Calibration: The key parameters to be calibrated become rock mechanical properties (e.g., elastic modulus, Poisson's ratio, compressive/tensile strength) and the in situ stress field. These can be acquired through laboratory core tests or field measurements (e.g., hydraulic fracturing tests). The reference's use of Weibull and log-normal distributions to assign heterogeneous mechanical and hydraulic parameters exemplifies the implementation of the "Key Parameter Calibration" step in addressing reservoir heterogeneity.

(c) Constitutive Model Selection: At this stage, the focus shifts to selecting a fracture propagation model. Depending on the simulation focus, one might choose continuum methods (e.g., damage mechanics-based models like the one used in the reference) or discrete fracture network models.

(d) Model Validation and Error Correction: Finally, the model must be validated and corrected against field data, such as microseismic monitoring events, fracturing treatment pressure curves, or production history data. The reference validated its model by comparing the simulated fracture initiation pressure against theoretical solutions from classical H-F and H-W equations and by aligning fracture patterns with engineering expectations.

6. Conclusions

(1) This study systematically reviews the theoretical development and engineering application of numerical simulation for coal seam gas injection displacement, focusing on its core role in enhancing coal mine gas drainage and safety. It clarifies the modeling principles of key physical fields (flow–adsorption–diffusion, thermo–mechanical, hydraulic–chemical) and their coupling mechanisms: the flow field relies on Darcy's law corrected by the Klinkenberg effect to characterize fracture flow; the adsorption/desorption field adopts the Extended Langmuir model to quantify multi-gas competitive adsorption; the thermo-mechanical field integrates Biot porous elasticity theory with adsorption-induced strain; the hydraulic–chemical field couples CO_2 dissolution–mineral dissolution kinetics to reveal permeability evolution. Additionally, it summarizes the evolution of permeability models—from static constant assumptions to dynamic coupling of stress, adsorption, and chemical effects, laying a theoretical foundation for refined simulation of gas control.

(2) This study traces the evolution of numerical models from basic flow–diffusion models to THMC full-coupling models, and compares the performance of simulation software: COMSOL demonstrates greater coupling depth and flexibility for complex scenarios, supporting customized equations and THMC nonlinear partial differential equation solving, whereas COMET3 is optimized for standard production forecasts. Based on this, a four-dimensional decision-making framework ("Multi-physical Field Coupling Equation Selection, Key Parameter Calibration, Permeability Equation Selection, Model Validation and Error Correction") is proposed. Validation in the 3# coal seam of Qinshui Basin shows the framework controls simulation error $\leq$10%, effectively resolving the problems of empirical parameter selection and low model adaptability in traditional simulation. It provides a standardized technical path for optimizing underground gas injection drainage parameters and improving gas control efficiency.

(3) This study identifies four key bottlenecks restricting the application of gas injection displacement simulation: imprecise flow–diffusion characterization (over-reliance on Darcy's law, neglect of Knudsen diffusion in nanopores); insufficient fracture network anisotropy portrayal (isotropic permeability tensors failing to reflect fractal features); incomplete THMC coupling (inadequate integration of mineral dissolution–precipitation with permeability–mechanical deformation); and cross-scale parameter mismatch. The corresponding future research directions are proposed to address these issues: (1) refine gas transport modeling by classifying flow regimes (Darcy/Forchheimer/Swartzendruber) based on flow rate and pore size, and calibrate diffusion coefficients with surface roughness correction; (2) develop anisotropic permeability interfaces in simulation software, integrate discrete fracture network models, and validate with field data; (3) construct a COMSOL-PHREEQC coupled framework to track coal mineral reactions and capture permeability hysteresis; and (4) establish a "laboratory calibration-field inversion-dynamic update" parameter system, using machine learning to bridge micro–macro scale gaps. These efforts aim to develop high-precision simulation tools for coal mine gas control, promoting large-scale application of gas injection displacement technology in complex geological conditions.

Author Contributions: X.Y.: Methodology, Investigation, Writing—original draft; F.D.: Formal analysis, Writing—review and editing; Q.Z.: Conceptualization, Formal analysis, Writing—review and editing; Y.Z. (Yunfei Zuo): Writing—review and editing; F.T.: Investigation; Y.Z. (Yiyang Zhang): Data curation; Y.X.: Investigation. All authors have read and agreed to the published version of the manuscript.

Funding: This project was supported by the National Natural Science Foundation of China (52204196, 52374249).

Data Availability Statement: No new data were created or analyzed in this study.

Conflicts of Interest: The authors declare that they have no known competing financial interests or personal relationships that could have appeared to influence the work reported in this paper.

Appendix A

Table A1. Variables corresponding to different chapters.

Variables from Section 3.2		**Variables from Section 3.3**	
η	thermal conductivity of the coal body, J/(m·s·K)	$(\sigma - \sigma_0)$	effective stress variation, PA
Q_t	gas content, m^3/t	K_p	bulk modulus of pores, MPa
C_v	constant volume specific heat of a coal body, J/(kg·K)	$\Delta\varepsilon_f$	coal fracture adsorption volume strain, dimensionless
θ_T	thermal strain coefficient, K^{-1}	β_m, β_f	Biot coefficient of coal matrix, dimensionless Biot coefficient of coal fracture, dimensionless
$\frac{\partial e}{\partial t}$	volumetric strain rate, s^{-1}	p_L	Langmuir pressure, MPa
θ_{px}	phase change pressure coupling coefficient, dimensionless	C_{pc}	perimeter pressure sensitivity factor, MPa^{-1}
w	pressure correction factor, Pa^{-1}	k_{rg}	relative gas-phase permeability, dimensionless
h	gas adsorption phase fraction, dimensionless	k_{rog}	relative permeability of gas-phase endpoints, dimensionless
$(\rho C_p)_{eff}$, η_{eff}, λ_{eff}	effective specific heat capacity, J/(m^3·K); effective thermal convection coefficient, J/(m^2·s); effective thermal conductivity, W/(m·K)	k_{rw}	relative permeability of aqueous phase, dimensionless
q_{sti}	isothermal heat of adsorption/desorption of gases, kJ/mol	K_{rw0}	relative permeability of aqueous phase endpoints, dimensionless
C_s, C_{gi} and C_w	specific heat capacity of coal, gas mixtures and water, J/(kg K)	S_w	water saturation, %
Variables from Section 3.3		S_{wr}	bound water saturation, %
K_0	initial penetration rate, mD	S_{gr}	residual gas saturation, %
C_f	coal fracture compression factor, MPa^{-1}	S_{eff}	effective saturation, %
$(\sigma^e_h - \sigma^e_{h0})$	adsorption volume strain, %	λ	pore size distribution parameters, dimensionless
φ_0, φ	initial porosity, porosity, dimensionless	a, b	empirical fitting parameters, dimensionless
c_m, c_p	coal matrix compression factor, pore compression coefficient, MPa^{-1}	$k_{r\alpha}$	relative permeability of a phase, dimensionless
M	bulk modulus, MPa	$k_{r\alpha0}$	endpoint relative permeability of a phase
E	modulus of elasticity, MPa	μ_α, μ_β	dynamic viscosity of α- and β- phase fluids, P·s
b	gas adsorption constant, MPa^{-1}	C, n	empirical constant, dimensionless
C_1	contribution weight of adsorption strain to porosity, dimensionless	s_β	saturation of the β-phase, dimensionless

References

1. Ding, S.; Li, Y.; Zhang, M.; Xu, C.; Wang, S.; Gao, Y.; Yu, H.; Du, Y.; Ma, J. Holistic Review on CO_2 Geological Storage Potential Evaluation. *Energy Fuels* **2024**, *38*, 19946–19965. [CrossRef]
2. Zhou, L.; Feng, Q.; Chen, Z.; Liu, J. Modeling and Upscaling of Binary Gas Coal Interactions in CO_2 Enhanced Coalbed Methane Recovery. *Procedia Environ. Sci.* **2012**, *12*, 926–939. [CrossRef]
3. Gong, H.; Wang, K.; Wang, G.; Yang, X.; Du, F. Underground coal seam gas displacement by injecting nitrogen: Field test and effect prediction. *Fuel* **2021**, *306*, 121646. [CrossRef]
4. Yang, H.; Zhou, K.; Chen, L.; Lu, X.; Kang, N. Differential Analysis of Various Moisture Phases on N_2 Displacement of CH_4 in Coal. *ACS Omega* **2024**, *9*, 15633–15640. [CrossRef]
5. Zhou, K.; Yang, H.; Guan, J.; Si, Z. Study on the Differences between Various Gas Injection Sources in the Process of Coal Seam Methane Replacement by Gas Injection. *ACS Omega* **2022**, *7*, 37572–37580. [CrossRef]
6. Yang, X.; Wang, G.; Du, F.; Jin, L.; Gong, H. N_2 injection to enhance coal seam gas drainage (N_2-ECGD): Insights from underground field trial investigation. *Energy* **2022**, *239*, 122247. [CrossRef]
7. Yang, X.; Wang, G.; Ni, M.; Shu, L.; Gong, H.; Wang, Z. Investigation on Key Parameters of N_2 Injection to Enhance Coal Seam Gas Drainage (N_2-ECGD). *Energies* **2022**, *15*, 5064. [CrossRef]
8. Asif, M.; Wang, L.; Panigrahi, D.C.; Ojha, K.; Hazlett, R. Integrated assessment of CO_2-ECBM potential in Jharia Coalfield, India. *Sci. Rep.* **2022**, *12*, 7533. [CrossRef] [PubMed]
9. Sarkar, S.; Singh, I.V.; Mishra, B.K. A simple and efficient implementation of localizing gradient damage method in COMSOL for fracture simulation. *Eng. Fract. Mech.* **2022**, *269*, 108552. [CrossRef]
10. Fan, C.; Yang, L.; Wang, G.; Huang, Q.; Fu, X.; Wen, H. Investigation on Coal Skeleton Deformation in CO_2 Injection Enhanced CH_4 Drainage from Underground Coal Seam. *Front. Earth Sci.* **2021**, *9*, 766011. [CrossRef]
11. Fang, H.-H.; Zheng, C.-S.; Qi, N.; Xu, H.-J.; Liu, H.-H.; Huang, Y.-H.; Song, S.-L. Coupling mechanism of THM fields and SLG phases during the gas extraction process and its application in numerical analysis of gas occurrence regularity and effective extraction radius. *Pet. Sci.* **2022**, *19*, 990–1006. [CrossRef]
12. Cui, X.; Bustin, R.M. Volumetric strain associated with methane desorption and its impact on coalbed gas production from deep coal seams. *AAPG Bull.* **2005**, *89*, 1181–1202. [CrossRef]
13. Gray, I. Reservoir Engineering in Coal Seams: Part 1—The Physical Process of Gas Storage and Movement in Coal Seams. *SPE Reserv. Eng.* **1987**, *2*, 28–34. [CrossRef]
14. Liu, H.-H.; Rutqvist, J. A New Coal-Permeability Model: Internal Swelling Stress and Fracture–Matrix Interaction. *Transp. Porous Media* **2010**, *82*, 157–171. [CrossRef]
15. Ma, X.; Zhou, A.; Cheng, X.; Cheng, C.; Zhao, W. Multi-Field Coupling Models of Coal and Gas and Their Engineering Applications to CBM in Deep Seams: A Review. *Energies* **2024**, *17*, 6221. [CrossRef]
16. Darcy, H. *Recherches Expérimentales Relatives au Mouvement de L'eau dans les Tuyaux*; Mallet—Bachelier: Paris, France, 1857.
17. Javadi, M.; Sharifzadeh, M.; Shahriar, K.; Mitani, Y. Critical Reynolds number for nonlinear flow through rough-walled fractures: The role of shear processes. *Water Resour. Res.* **2014**, *50*, 1789–1804. [CrossRef]
18. Wang, X.; Thauvin, F.; Mohanty, K.K. Non-Darcy flow through anisotropic porous media. *Chem. Eng. Sci.* **1999**, *54*, 1859–1869. [CrossRef]
19. Wang, X.; Sheng, J. Gas Sorption and Non-Darcy Flow in Shale Reservoirs. *Pet. Sci.* **2017**, *14*, 746–754. [CrossRef]
20. Liu, Q.; Cheng, Y.; Zhou, H.; Guo, P.; An, F.; Chen, H. A Mathematical Model of Coupled Gas Flow and Coal Deformation with Gas Diffusion and Klinkenberg Effects. *Rock Mech. Rock Eng.* **2015**, *48*, 1163–1180. [CrossRef]
21. Klinkenberg, L.J. The Permeability of Porous Media to Liquids and Gases. In *Drilling and Production Practice*; American Petroleum Institute: Washington, DC, USA, 1941.
22. Brunauer, S.; Emmett, P.H.; Teller, E. Adsorption of Gases in Multimolecular Layers. *J. Am. Chem. Soc.* **1938**, *60*, 309–319. [CrossRef]
23. Dubinin, M.M. The Potential Theory of Adsorption of Gases and Vapors for Adsorbents with Energetically Nonuniform Surfaces. *Chem. Rev.* **1960**, *60*, 235–241. [CrossRef]
24. Dubinin, M.M. Physical Adsorption of Gases and Vapors in Micropores. In *Progress in Surface and Membrane Science*; Cadenhead, D.A., Danielli, J.F., Rosenberg, M.D., Eds.; Elsevier: Amsterdam, The Netherlands, 1975; Volume 9, pp. 1–70.
25. Dutta, P.; Harpalani, S.; Prusty, B. Modeling of CO_2 sorption on coal. *Fuel* **2008**, *87*, 2023–2036. [CrossRef]
26. Clarkson, C.R. Adsorption potential theories to coal methane adsorption isotherms at elevated temperature and pressure. *Carbon* **1997**, *35*, 1689–1705. [CrossRef]
27. Zhang, P.; Wang, L. Extended Langmuir equation for correlating multilayer adsorption equilibrium data. *Sep. Purif. Technol.* **2010**, *70*, 367–371. [CrossRef]
28. Wang, J.; Cheng, Y.; Wang, L.; Hu, B.; Wang, C.; Yi, M. Diffusion Process of Methane in Coal Particles and Its Relationship with the Pore Scale. *Energy Fuels* **2022**, *36*, 14734–14747. [CrossRef]

29. Nandi, S.P.; Walker, P.L. Activated diffusion of methane in coal. *Fuel* **1970**, *49*, 309–323. [CrossRef]
30. Nandi, S.P.; Walker, P.L. Activated diffusion of methane from coals at elevated pressures. *Fuel* **1975**, *54*, 81–86. [CrossRef]
31. Clarkson, C.R.; Bustin, R.M. The effect of pore structure and gas pressure upon the transport properties of coal: A laboratory and modeling study. 1. Isotherms and pore volume distributions. *Fuel* **1999**, *78*, 1333–1344. [CrossRef]
32. Shi, G.; Wei, F.; Gao, Z.; Yan, J.; Zhang, H.; Hao, T. Gas desorption-diffusion behavior from coal particles with consideration of quasi-steady and unsteady crossflow mechanisms based on dual media concept model: Experiments and numerical modelling. *Fuel* **2021**, *298*, 120729. [CrossRef]
33. Liu, Q.; Wang, J.; Zhao, W.; Huang, W.Y.; Ling, H.; Liang, W.; Cheng, Y.P. A novel bidisperse diffusion model for coal particles considering micropores as the major sites of CH_4 adsorption. *Fuel* **2023**, *333*, 126505. [CrossRef]
34. Fang, H.; Yu, S.; Zhang, S.; Sang, S.; Guo, J.; Wang, Z.; Wang, X. Coupling Mechanism of THMC Fields in Crushed Soft Coal with Low Permeability after CO_2 Injection and Its Application in CO_2-ECBM Technology. *Energy Fuels* **2024**, *38*, 6891–6911. [CrossRef]
35. Fan, C.; Elsworth, D.; Li, S.; Zhou, L.; Yang, Z.; Song, Y. Thermo-hydro-mechanical-chemical couplings controlling CH_4 production and CO_2 sequestration in enhanced coalbed methane recovery. *Energy* **2019**, *173*, 1054–1077. [CrossRef]
36. Guo, W.; Xiong, W.; Gao, S.; Hu, Z.; Liu, H.; Yu, R. Impact of temperature on the isothermal adsorption/desorption of shale gas. *Pet. Explor. Dev.* **2013**, *40*, 514–519. [CrossRef]
37. Wang, X.; Lee, B.; Chua, H.T. Methane desorption and adsorption measurements on activated carbon in 281–343 K and pressures to 1.2 MPa. *J. Therm. Anal. Calorim.* **2012**, *110*, 1475–1485. [CrossRef]
38. Ye, Q.; Li, C.; Yang, T.; Wang, Y.; Li, Z.; Yin, Y. Relationship between desorption amount and temperature variation in the process of coal gas desorption. *Fuel* **2023**, *332*, 126146. [CrossRef]
39. Zhang, Q.-L. Adsorption mechanism of different coal ranks under variable temperature and pressure conditions. *J. China Univ. Min. Technol.* **2008**, *18*, 395–400. [CrossRef]
40. Zhou, H.W.; Zhang, L.; Wang, X.Y.; Rong, T.L.; Wang, L.J. Effects of matrix-fracture interaction and creep deformation on permeability evolution of deep coal. *Int. J. Rock Mech. Min. Sci.* **2020**, *127*, 104236. [CrossRef]
41. Tao, Y.; Xu, J.; Liu, D.; Liang, Y. Development and validation of THM coupling model of methane-containing coal. *Int. J. Min. Sci. Technol.* **2012**, *22*, 879–883. [CrossRef]
42. Fan, C.; Li, S.; Luo, M.; Du, W.; Yang, Z. Coal and gas outburst dynamic system. *Int. J. Min. Sci. Technol.* **2017**, *27*, 49–55. [CrossRef]
43. Connell, L.D. A new interpretation of the response of coal permeability to changes in pore pressure, stress and matrix shrinkage *Int. J. Coal Geol.* **2016**, *162*, 169–182. [CrossRef]
44. Palmer, I.; Mansoori, J. How Permeability Depends Stress Pore Press. Coalbeds: A New Model. *SPE Reserv. Eval. Eng.* **1996**, *1*, 539–544. [CrossRef]
45. Biot, M.A. Theory of Propagation of Elastic Waves in a Fluid-Saturated Porous Solid. I. Low-Frequency Range. *J. Acoust. Soc. Am.* **2005**, *28*, 179–191. [CrossRef]
46. Levine, J.R. Model study of the influence of matrix shrinkage on absolute permeability of coal bed reservoirs. *Geol. Soc. Publ.* **1996**, *109*, 197–212. [CrossRef]
47. Miao, T.; Chen, A.; Li, Z. Stress-Dependent models for permeability and pososity of fracture of fractured rock based on fractal. *Fractals* **2023**, *31*, 2350093. [CrossRef]
48. Lenhard, R.J.; Oostrom, M.; White, M.D. Modeling fluid flow and transport in variably saturated porous media with the STOMP simulator. 2. Verification and validation exercises. *Adv. Water Resour.* **1995**, *18*, 365–373. [CrossRef]
49. Corey, A.T. The Interrelation Between Gas and Oil Relative Permeabilities. *Prod. Mon.* **1954**, *19*, 38–41.
50. Brooks, R.H.; Corey, A.T. Properties of Porous Media Affecting Fluid Flow. *J. Irrig. Drain. Div.* **1964**, *92*, 61–88. [CrossRef]
51. Pan, X.; Wong, R.C.; Maini, B.B. Steady State Two-phase in a Smooth Parallel Fracture. In Proceedings of the Annual Technical Meeting, Calgary, AB, Canada, 9–11 June 1996. [CrossRef]
52. Ehrlich, R. Viscous coupling in two-phase flow in porous media and its effect on relative permeabilities. *Transp. Porous Media* **1993**, *11*, 201–218. [CrossRef]
53. Yang, T. Numerical simulation of coal-bed methane displacement by underground gas injection based on multi-physics coupling. *J. China Coal Soc.* **2010**, *35*, 1479–1484.
54. Shi, Y.; Lin, B.; Liu, T.; Zhao, Y.; Hao, Z. Synergistic ECBM extraction technology and engineering application based on hydraulic flushing combing gas injection displacement in low permeability coal seams. *Fuel* **2022**, *318*, 123688. [CrossRef]
55. Ren, T.; Wang, G.; Cheng, Y.; Qi, Q. Model development and simulation study of the feasibility of enhancing gas drainage efficiency through nitrogen injection. *Fuel* **2017**, *194*, 406–422. [CrossRef]
56. Yang, L.; Fan, C.-J.; Luo, M.-K.; Wen, H.-O.; Sun, H.; Zhou, L.-J.; Cheng, Z.-H.; Zhang, Z.-P. Geological parameters and gas mixture composition on enhanced coalbed methane recovery: A THM modeling approach. *Pet. Sci.* **2025**, *22*, 2078–2095. [CrossRef]
57. Vishal, V.; Singh, L.; Pradhan, S.P.; Singh, T.N.; Ranjith, P.G. Numerical modeling of Gondwana coal seams in India as coalbed methane reservoirs substituted for carbon dioxide sequestration. *Energy* **2013**, *49*, 384–394. [CrossRef]

58. Fang, H.; Du, G.; Gu, W.; Weng, J.; Yu, S. Displacement Effect of CO_2/N_2-ECBM Process with Different Water Saturations Based on THMC Coupling. *Energy Fuels* **2025**, *39*, 8086–8106. [CrossRef]
59. Fang, H.; Sang, S.; Liu, S. Establishment of dynamic permeability model of coal reservoir and its numerical simulation during the CO_2-ECBM process. *J. Pet. Sci. Eng.* **2019**, *179*, 885–898. [CrossRef]
60. Liang, X.; Hou, P.; Liu, G.; Xue, Y.; Liu, J.; Gao, F.; Zhang, Z. A clustered fractal discrete fracture network model for fractured coal. *Fractals Interdiscip. J. Complex Geom. Nat.* **2024**, *2*, 32. [CrossRef]
61. Parkhurst, D.; Appelo, C.A.J. User's Guide to PHREEQC (Version 2): A Computer Program for Speciation, Batch-Reaction, One-Dimensional Transport, and Inverse Geochemical Calculations (99-4259). 1999. Available online: https://pubs.usgs.gov/publication/wri994259 (accessed on 10 October 2025).
62. Meng, M.; Qiu, Z.; Zhong, R.; Liu, Z.; Liu, Y.; Chen, P. Adsorption characteristics of supercritical CO_2/CH_4 on different types of coal and a machine learning approach. *Chem. Eng. J.* **2019**, *368*, 847–864. [CrossRef]
63. Yuan, Z.G.; Shao, Y.H. Numerical Modeling on Hydraulic Fracturing in Coal-Rock Mass for Enhancing Gas Drainage. *Adv. Civ. Eng.* **2018**, *2018*, 1485672. [CrossRef]

processes

MDPI

Review

Recent Advances in Zero Discharge Treatment Technologies for Desulfurization Wastewater in Coal-Fired Power Plants: A Mini-Review

Binsheng Liao [1], Xianyang Zeng [2,*], Zhongqian Ling [2], Sanmei Zhao [3], Bin Li [3] and Xinlu Han [4,*]

1 Guoneng Jiujiang Power Generation Co., Ltd., Jiujiang 332000, China
2 College of Energy Environment and Safety Engineering, China Jiliang University, Hangzhou 310018, China
3 Nanjing Yitao Environmental Protection Technology Co., Ltd., Nanjing 210000, China
4 College of New Energy, China University of Petroleum (East China), Qingdao 266580, China
* Correspondence: 06a0204057@cjlu.edu.cn (X.Z.); hanxinlu@upc.edu.cn (X.H.)

Abstract: Zero Liquid Discharge (ZLD) is a wastewater management strategy that eliminates liquid waste while maximizing water use efficiency. This article reviews the primary ZLD technologies used for desulfurization wastewater (DWW) treatment in coal-fired power plants. These technologies include the thermal process and the membrane process. The thermal process includes "concentrated crystallization" technology and "gas evaporation and drying" technology. The paper also highlights recent advances in membrane technology for power plant wastewater treatment. The advantages and limitations of each technique are discussed. Membrane technology is considered a promising solution for wastewater recycling, while thermal technology offers easy operation and maintenance without the need for pretreatment. Finally, the paper outlines possible future directions for the treatment of DWW.

Keywords: desulfurization wastewater; zero liquid discharge; thermal process; membrane process

1. Introduction

Coal is a significant source of energy in industrial processes and is often used in power plant boilers. However, burning coal releases large amounts of carbon dioxide and harmful gases. Among them, sulfur compounds (SO_2) contribute significantly to environmental problems such as haze and acid rain [1]. Flue gas desulfurization is therefore essential to decrease pollutant emissions. Currently, the primary flue gas desulfurization technology is the wet desulfurization process based on the limestone–gypsum process [2,3]. However, this plant produces large amounts of desulfurization wastewater (DWW) during operation. DWW is characterized by low acidity, suspended solids, high heavy metal ion content, high hardness, and high chemical oxygen demand (COD) [4,5]. The detailed basic characteristics of DWW are shown in Table 1 [6–8]. Due to its complex composition, treatment is difficult.

Recently, the concept of ZLD (Zero Liquid Discharge) has gained considerable attention in the treatment of DWW [9,10]. This is due to the increasing scarcity of water resources and the tightening of pollutant discharge regulations worldwide. The ZLD concept originated from the industrial high-salt wastewater treatment in the 1970s. Its core goal is to recover >95% of the water resources through a multi-stage process (pretreatment, concentration, crystallization) and convert dissolved solids into reusable salts (e.g., NaCl, $CaSO_4$) or stabilizing residues, completely eliminating liquid waste [11,12]. ZLD technology uses a closed water cycle, minimizing wastewater disposal and environmental risks.

 https://doi.org/10.3390/pr13040982

It also reduces pressure on ecosystems by reducing the need for freshwater extraction. ZLD has emerged as a crucial strategy for sustainable wastewater management. It has been successfully used in industries such as pharmaceuticals, oil refining, and textiles and has achieved promising results. It is also strongly promoted by the Clean Water Act in the United States and the Industrial Discharge Directive in the European Union [13]. However, despite the rapid development of ZLD technology, there are limited reports on its application in coal-fired power plants (CFPPs). It is therefore necessary to examine its applications in a timely manner. Moreover, the economic viability of the ZLD system remains a bottleneck for its promotion, as its high initial investment and operational costs need to be balanced through technological innovations (such as waste heat recovery and the development of anti-pollution membrane materials) and policy subsidies [14–16].

Table 1. The basic characteristics of DWW.

Parameters	Description	Typical Range
pH value	Low acidity, mainly caused by the accumulation of acidic substances (e.g., sulfate) in the desulfurization process	2.0–7.85
Suspended solids	Contain unreacted limestone, gypsum particles, and dust	100–1000 mg/L
Heavy metal ions	Contain mercury (Hg)	0.1–50 mg/L
Total hardness	Mainly caused by calcium (Ca^{2+}) and magnesium (Mg^{2+}) ions, which can easily lead to equipment scaling	500–5000 mg/L
Chemical oxygen demand (COD)	Reflects the content of organic matter and reducing inorganic matter in wastewater, and its value is affected by desulfurizer and coal composition	200–2000 mg/L
Chloride ions (Cl^-)	High concentrations of chloride ions are highly corrosive and may be the result of the combustion of coal or the desulfurization processes	500–10,000 mg/L
Sulfates (SO_4^{2-})	Mainly come from the absorption reaction of SO_2 in flue gas	1000–20,000 mg/L

This article reviews the latest developments in ZLD technology for desulfurization of CFPPs' wastewater, including the thermal membrane process like concentrated crystallization, flue gas evaporation, as well as membrane separation. The literature reviewed is listed in the References section. The advantages and limitations of new ZLD technologies are discussed. Finally, the paper outlines possible future directions for the treatment of DWW.

2. ZLD Technology for the Recovery and Reuse of DWW

A typical ZLD system consists of three main components: pretreatment, concentration, and crystallization [17–19] (see Figure 1). Pretreatment uses traditional physical and chemical methods to adjust the inflow for the next processing step. This stage includes a reactor or clarifier that removes metals, silica, toxins, and large suspended solids by adding chemical reagents to facilitate precipitation and solidification. In some cases, the pH value is adjusted to support the solidification process. The wastewater then flows into a flocculation system, where solidified particles are changed into precipitable particles. Then, the wastewater enters a settling chamber in which the flocculated particles are taken away. Depending on the specific properties of the wastewater, additional treatment units may be required to remove toxins and metals.

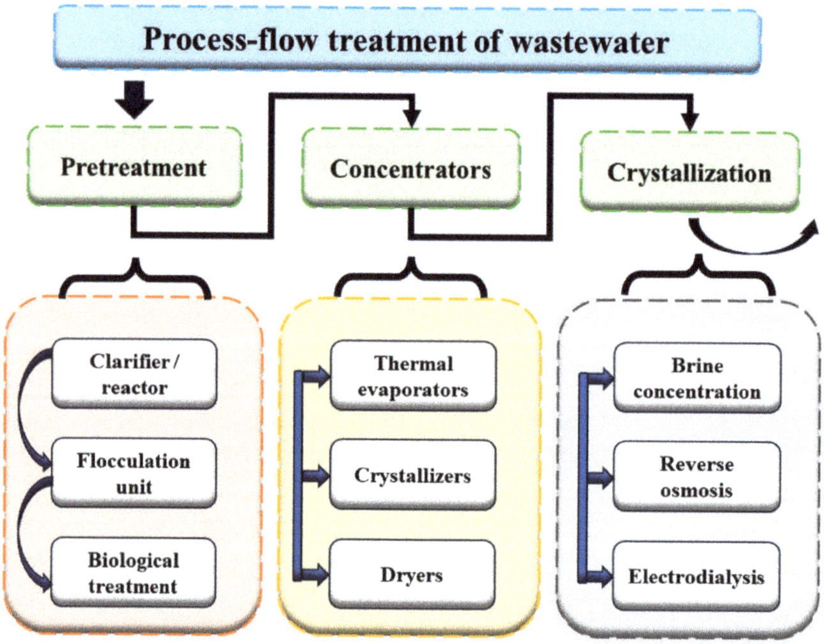

Figure 1. Wastewater treatment process.

The next step is the process of concentration. This process concentrates the wastewater to a high salt content while recycling a large portion of the wastewater for reuse. The dissolved salts are taken away from the wastewater through a combination of electrodialysis, brine concentration, and reverse osmosis techniques [20,21].

The final stage of the ZLD process is evaporation and crystallization. In this phase, highly mineralized wastewater undergoes solid phase conversion, salt recovery, and final treatment. This process is achieved through heat evaporators, crystallizers, and dryers, all of which consume energy. To decrease the energy consumption, a combination of solar-powered systems, as well as vacuum crystallizers and vacuum evaporator technologies, have proven successful [22].

ZLD systems are mainly realized through thermal engineering methods and membrane-based technologies.

2.1. Thermal Process Method

The traditional ZLD system is based on thermal processes, with the majority of energy consumption occurring during the brine concentration and evaporation–crystallization stages. During the treatment process, the distillate is collected and recovered as clean water. The simultaneous production of solid waste is considered a valuable by-product because it contains usable minerals and salts. A thermal ZLD system typically contains a pretreatment unit, a brine concentrator, a crystallizer, a solids recovery unit, and an evaporation tank [22–24] (see Figure 2). The system essentially includes two key technologies: "thermal concentration–crystallization" and "flue gas evaporation–drying".

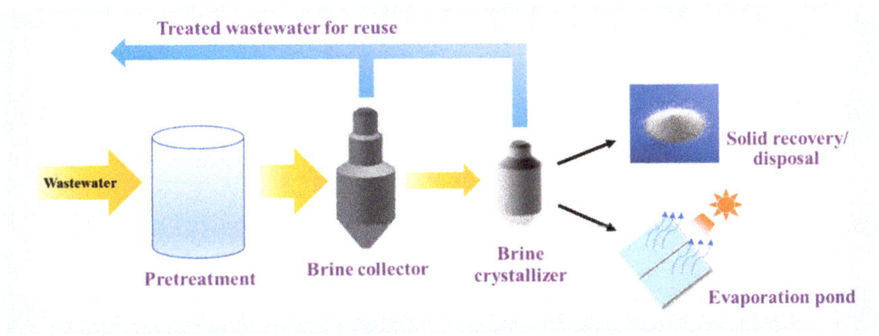

Figure 2. Schematic diagram of the thermal ZLD system [22].

2.1.1. Thermal Concentration–Crystallization Technology

Thermal concentration–crystallization technology plays a crucial role in the emission-free treatment of DWW. This process involves heating the wastewater using an external heat source to initiate crystallization once the concentration exceeds saturation. Based on the rule that the solubility of substances in solution changes with temperature, when a solute's concentration exceeds its saturation point, the excess solute crystallizes to form solid particles. This process effectively separates pollutants from wastewater and facilitates recovery [25]. Pollutants in DWW will be removed through heating, evaporation, concentration, and subsequent crystallization upon cooling.

2.1.2. Flue Gas Evaporation–Drying (FGED) Technology

The technology of FGED works under humid conditions and high temperature [26,27]. DWW is mixed with compressed air and introduced into the flue gas as liquid droplets. The water in the wastewater evaporates quickly and diffuses into the boiler flue, effectively separating pollutants. The dried, solidified material resulting from evaporation typically ends up in the dust collector, where it is collected along with the dust. This process achieves "zero discharge" of DWW. A diagram illustrating this technique is displayed in Figure 3 [25].

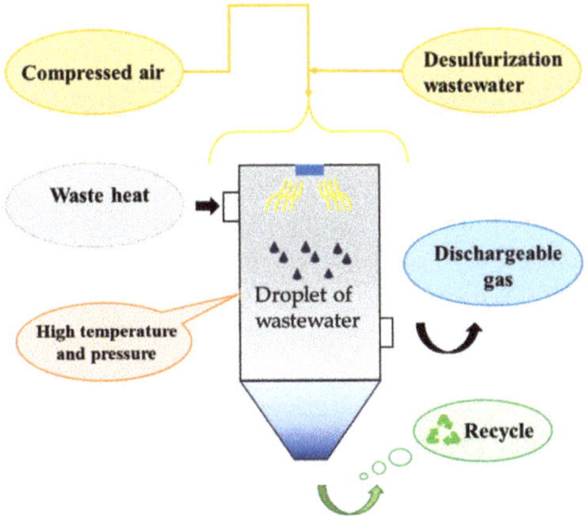

Figure 3. Flue gas evaporation and drying (FGED) [25].

2.2. Membrane Method

Membrane technology is a proven method for seawater desalination. It is often used to concentrate wastewater in brine enrichment plants prior to heat treatment [22] (see Figure 4). Unlike thermal processes, the membrane process does not cause phase changes in wastewater, which reduces treatment costs (by 48–67%) and energy consumption (by 58–75%) when replacing the brine concentration and evaporation stages in thermal systems [28].

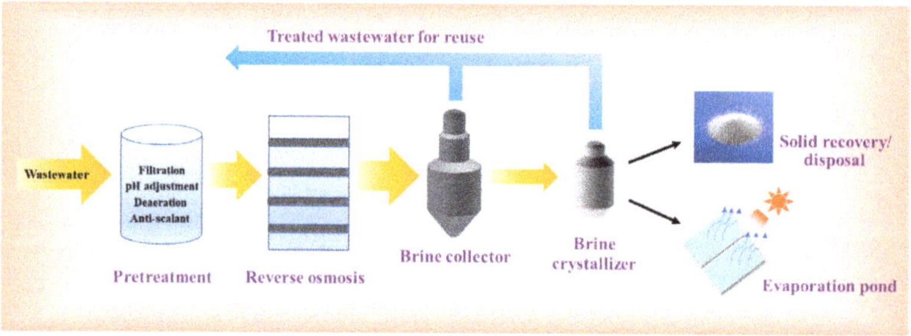

Figure 4. Schematic diagram of the membrane method.

Although the membrane method has lower energy requirements, it is limited by the upper salinity threshold of the wastewater (<7000 mg/L). This method is also susceptible to membrane contamination and has a relatively low efficiency (approx. 50–60%) [29,30]. Membrane contamination often results from the existence of particles in wastewater. Pretreatment methods can reduce contaminants, but they can increase operating costs. In recent years, advanced membrane technologies, e.g., high efficiency reverse osmosis (HERO), forward osmosis (FO), membrane distillation (MD), as well as electrodialysis (ED), have been developed for ZLD applications.

2.2.1. High Efficiency Reverse Osmosis (HERO)

HERO is an advanced wastewater treatment technology that is widely applied in fields such as wastewater treatment in CFPPs. This technology relies on the collaborative effect of ultrafiltration pretreatment as well as pH adjustment to achieve a higher clean water recovery rate and reduce membrane fouling. Ultrafiltration pretreatment can remove impurities such as macromolecular organic matter, colloids, bacteria, and suspended particles from wastewater, reducing the risk of subsequent reverse osmosis membrane fouling. Precise pH adjustment optimizes the charge distribution on the membrane surface, further alleviating membrane fouling [31]. Currently, HERO has been successfully applied in field-scale wastewater treatment projects, demonstrating excellent performance in treating high-salinity wastewater. For example, in CFPPs, it can effectively treat DWW, recycle clean water, reduce the consumption of fresh water resources, and reduce environmental pollution [32]. HERO is successfully used worldwide in on-site wastewater treatment. It effectively reduces membrane contamination and treats wastewater with high salinity (close to or more than 100,000 mg/L) [12]. However, the HERO technology also faces some challenges, such as its secondary deposition of inorganic substances and its complex cleaning requirements.

2.2.2. Forward Osmosis (FO)

Forward osmosis works based on osmotic pressure differences across a semi-permeable membrane. This allows freshwater molecules to pass through while blocking salts [33]. Compared with HERO, the FO method is a cost-effective alternative for treating wastewater with high salinity (~200,000 mg/L). A limitation of this method is that water purification molecules are separated from the extract. However, FO membranes are less susceptible to contamination than reverse osmosis membranes, which reduces maintenance costs. In addition, FO is more energy efficient and economical compared with thermal desalination technologies.

2.2.3. Membrane Distillation (MD)

MD combines thermal separation and membrane filtration. It concentrates feed water on a hydrophobic microporous membrane and creates water vapor based on partial vapor pressure gradients. MD operates at low temperatures (60–90 °C) and low pressures and has a reduced tendency for membrane fouling [34,35]. However, the presence of volatile compounds or surfactants in the feed water can affect the qualities of the water. Most MD research has been conducted in laboratory settings. Further studies are needed to assess the potential for large-scale applications.

2.2.4. Electrodialysis (ED)

ED uses an electrical potential to eliminate dissolved ions (DI) from wastewater. The membrane selectively allows counterions to pass through while inhibiting other ions. This selective ion movement reduces salinity and creates concentrated brine. ED is suitable for pretreatment and brine concentration prior to evaporation and crystallization in a ZLD system. Compared with reverse osmosis, electrodialysis can concentrate the feed water to a larger salt content (>100,000 mg/L) [36].

Table 2 summarizes the advantages and disadvantages of different technologies in ZLD systems.

Table 2. Technologies involved in ZLD.

ZLD Technology	Main Features	Advantages	Disadvantages	References
Thermal	Evaporation of wastewater feed in brine concentrators	Solid waste recycling	Energy intensive	Yaqub et al. [30] Amanda et al. [15]
HERO	For steel industry wastewater 3.96 kWhe/m^3 of wastewater feed	Minimizes solid waste disposal Efficient water recovery (91%)	Secondary deposition of inorganics Complicated cleaning method	Huaqiang Chu et al. [37]
FO	Based on osmotic pressure differences 21 KWH$_e$/m^3 of wastewater feed	Cost-efficient compared with RO High salinity upper limit Economical Less likely to clog than RO	Limited performance in the field Lower water flux Concentration polarization and the regeneration of draw solutions	Shaffer et al. [38] Patel at al. [39]
MD	High salinity limitation (>200,000 ppm) 22–60 KWH$_e$/m^3	Lower menbrane clogging Resistance to fouling and high recovery rates	Ineffective for volatiles or surfactants Lower water flux	Curcio et al. [40] Chan He et al. [41]
ED	Selective motion on ion exchange membranes 7–15 KWH$_e$/m^3 of wastewater feed	Relatively stable Less fouling for silica-rich feedwater	High energy requirement Cost investment Not effective against neutral pollutants	Bo Sun et al. [42] Juan Wang et al. [43] McGovern et al. [44]

3. ZLD in DWW Recovery Applications

3.1. Thermal Examples

3.1.1. Thermal Concentration–Crystallization Technology

Two new gas-FPPs in Texas have implemented GE's ZLD system to deal with circulating cooling water in accordance with the Environmental Protection Agency standards in the U.S. The system primarily uses thermal concentration and crystallization processes of the brine and achieves a 98% reuse rate [45]. Li et al. [46] conducted experimental studies on DWW from CFPPs, focusing on the combination of wastewater pretreatment and concentrated crystallization (see Figure 5a). In the pretreatment phase, the $NaSO_4$ is added to the wastewater, which is then pumped into the crystallization unit. The results show that the ZLD system is effective for treating DWW and enables the recovery of high-purity chlorination crystals. The purity of the sodium chloride crystals is over 99.10.

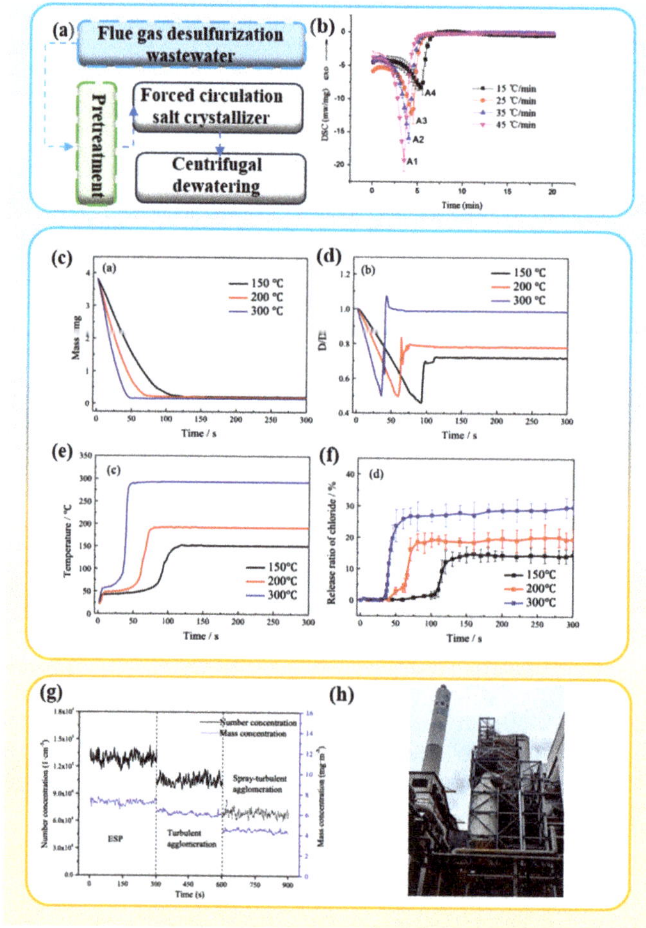

Figure 5. (**a**) Diagram of an experimental system [46]. (**b**) DSC curves for increasing evaporation time and temperature rates [47]. Evaporation characteristic curve and chloride release at different evaporation temperatures [48]: (**c**) temperature; (**d**) quality; (**e**) diameter; (**f**) chloride release ratio. (**g**) Quantity and mass concentration after ESP spraying [49]. (**h**) Liquid zero discharge program carried out by Guoneng Jiujiang Power Generation Co., Ltd. [25].

3.1.2. Flue Gas Evaporation–Drying Technology

Liang et al. [47] investigated the evaporation and crystallization properties of droplets from DWW in CFPPs under different operating conditions. Their results show that in the 15–45 °C/min range, a higher heating rate leads to a faster evaporative crystallization rate. During crystallization, a faster SO_4^{2-} concentration accelerates evaporation because an increased SO_4^{2-} concentration increases the vapor pressure. The study also found that Mg^{2+} concentration had little influence on the evaporation rate, but higher Mg^{2+} levels increased the crystallization rate (see Figure 5b).

Chen et al. [48] investigated the release of chloride during evaporation using the single drop drying approach. Based on the chloride release rate and wastewater droplet properties such as mass, temperature, and diameter, they identified three distinct chloride release periods. Almost no chloride was released during the evaporation of free water. In the instable shell phase, 19.8% of the chloride was released, primarily because of the elevated temperature of the droplet. The chloride release rate rises from 14.56% to 29.8% when the temperature is increased from 150 °C to 300 °C. This occurs due to the destruction of the shell and the increasing hydrolysis at higher temperatures (see Figure 5c–f). Therefore, it is crucial in technical applications to balance the release of chloride ions with the evaporation efficiency.

Conventional evaporation processes involve high investment and energy costs, low flue gas and droplet contact efficiency, and a lack of cost control. To meet the demands for lower investment and energy consumption, more efficient and energy-saving wastewater evaporation and crystallization technologies need to be developed.

Furthermore, a new technique for treating DWW via spray evaporation, as well as turbulent agglomeration, was proposed by Sun et al. [49] They experimentally examined the removal effect of spray DWW on fine particles. The spray turbulent agglomeration system can effectively promote the agglomeration and remove fine particles. This was achieved by the adhesion of precipitated crystals and capillary force in the wastewater droplets. Treatment with an electrostatic precipitator (ESP) increased the efficiency of fine particle agglomeration to 39.3%, which is 10.9% higher than that of single turbulent agglomeration (see Figure 5g).

However, flue gas evaporation technology has a limited agglomeration efficiency and a small evaporation rate in removing fine particles from DWW. This limits its application for this purpose. To address this problem, Guoneng Jiujiang Power Generation Co., Ltd. (Jiujiang, China) introduced a "flash-concentrated high-temperature flue gas bypass evaporation technology" for fluid-free delivery of DWW (Figure 5h). A combination of waste gas heating, as well as flash evaporation, is used to warm the wastewater [25]. This approach not only utilizes waste energy, but also lowers the temperature of the flue gas that is fed into the desulfurization system, thereby reducing water consumption in the systems. The reported recovery rate of DWW after evaporation treatment is 97%. The quality of the treated water is suitable for reuse as boiler water.

This method has several advantages: ease of operation, low investment in equipment, low energy consumption, and high-quality wastewater. In addition, supporting products can be tailored to meet the specific requirements. However, there are some problems with flue gas evaporation technology. If the DWW does not completely evaporate, or if the ESP does not fully capture the condensate, the highly corrosive chlorine-containing material will damage the stack and downstream systems. Furthermore, the complex composition of DWW, which contains heavy metals and other pollutants, may hinder the extensive use of fly ash. Furthermore, there is a lack of theoretical research on the mechanisms of product transformation and migration during evaporation. The heat transfer process between flue gas and wastewater could be better understood through numerical simulation and high-fit dynamic modeling, which would serve as a guide for practical engineering applications.

3.2. Membrane Method Example

3.2.1. FO Method

Lee et al. [50] has developed a hybrid FO and MD technology (FO–MD) to treat DWW under high pollution conditions. This innovation successfully addressed the challenges of treating DWW using membrane distillation (Figure 6a). Anderson et al. [51] employed a permeable FO–MD system to evaluate its effectiveness on waste salt removal, water recovery, as well as its thermal requirements for DWW (Figure 6c). The FO–MD system achieved a water recovery rate of up to 89%, removed 99.7% of the ions, and produced low conductivity distilled water, making it suitable for field use or discharge. By capitalizing on on-site exhaust heat, the FO–MD method could be an affordable and sustainable alternative for treating DWW. Ma et al. [52] used FO technology combined with soda ash (SA) or scale inhibitor (FOSI) to treat DWW. The total dissolved solids (TDS) of the influent were concentrated from about 820 mg/L to more than 120,000 mg/L (concentrated nearly 8 times), and the removal rates of Cl^{-}, Mg^{2+}, and Ca^{2+} were more than 99%. The system recovery rate was 73.02–85.38% and the operating cost was 21.77–25 yuan/ton, which verified the high efficiency and economic feasibility of the technology. Despite these successes, scalability and long-term economic viability remain critical hurdles, particularly due to membrane fouling and dependency on policy subsidies.

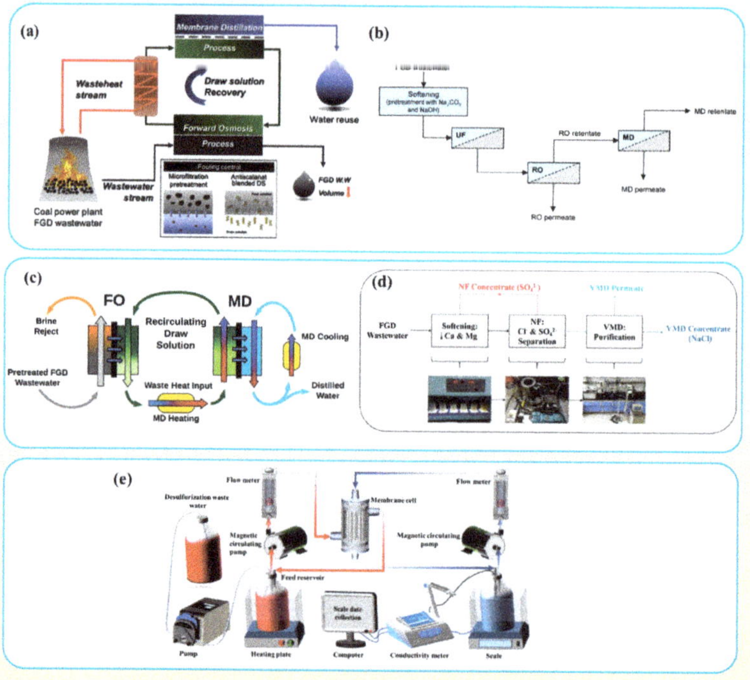

Figure 6. (**a**) Industrial wastewater treatment by the FO–MD hybrid systems [50]. (**b**) Integrated RO–MD desulfurization process for wastewater [53]. (**c**) FO–MD DWW treatment system [51]. (**d**) Flow diagram of the integrated FGD wastewater treatment system without discharge [54]. (**e**) Schematic representation of a direct contact MD system [55].

3.2.2. RO Method

Conidi et al. [53] also used an integrated membrane treatment technology that combines RO and MD to treat DWW (Figure 6b). The process includes ultrafiltration (UF) and chemical softening as pretreatment steps to remove Mg^{2+} ions, Ca^{2+} ions, and organic chemicals. The properties of the RO and MD membranes were assessed based on desalination rate, osmotic flow, water recovery rate, and pollution index. The combined system achieved an overall recovery rate of approximately 94%, reducing the amount of FGD wastewater. The conductivity of the MD permeate was 80 μS/cm, making it suitable for reuse in power plants and reducing the need for fresh water. Chen et al. [56] treated DWW through the integration of reverse osmosis (RO) membranes with low-frequency, high-intensity ultrasonic cleaning technology. This approach effectively inhibited membrane fouling, restoring permeation flux by 49% within 10 min and doubling membrane lifespan. However, broader implementation of this method requires cost-effective fouling mitigation strategies, reduced energy demands, and validation under variable wastewater compositions and long-term operational conditions.

3.2.3. MD Method

Jia et al. [54] studied a near-zero discharge process employing a combined nanofiltration MD method (Figure 6d). The desalination rate of the system exceeded 99.99% and the water recycling rate exceeded 92%. The salt content of the MD permeate was very low ($SO_4^{2-} \leq 1$ mg/L, $Cl^- \leq 3$ mg/L), making it suitable for recovery from thermal power plants. After MD concentration, highly pure brine (≥97%, 10% NaCl) could be produced. Chemical softening and nanofiltration pretreatment significantly reduced the contamination of the MD membrane. Li et al. [55] used the direct contact MD method with softening pretreatment for DWW treatment (Figure 6e), and they successfully recycled the wastewater. The results showed that impurities caused by Mg–Si complexes, $CaSO_4$, organic crystals, and Al colloids reduced the generated water flow. Qian et al. [57] used mixed denitrification (MD) technology to treat wet DWW and freshwater wastewater through the MD–SANI process: 80% COD removal rate was achieved in the sulfate–sulfite reduction reactor, and complete denitrification was completed in the anoxic reactor through the synergistic effect of autotrophic and heterotrophic denitrification. Microbial analysis showed that the up-flow sludge bed was dominated by sulfate–sulfite reducing bacteria. The process reduces sludge production by 90% and reduces energy consumption by 35%, which verifies the high efficiency and feasibility of the sulfur cycle coupled with mixed denitrification technology.

To sum up, membrane separation technologies for DWW treatment provide high efficiency in heat transfer, excellent sulfide removal, a small footprint, easy operation, minimal waste generation, and permeate recycling. However, these technologies come with high costs, stringent water quality requirements, and possible membrane contamination. Therefore, both economic and technical feasibility should be carefully considered when selecting an appropriate treatment method.

3.3. Technical Comparison

3.3.1. Comparison with Traditional Techniques

ZLD technologies have demonstrated significant advantages in achieving full wastewater recovery and resource utilization. However, their economic viability remains a core challenge that restricts their large-scale application. Compared with traditional wastewater treatment technologies, ZLD systems generally have higher initial investment and operating costs. Nevertheless, through the closed-loop water cycle, salt recovery, and implicit benefits brought by environmental compliance, their long-term economic feasibility is gradually emerging. Taking the DWW of CFPPs as an example, although the traditional chemical precipitation method has a relatively low cost per ton of wastewater treatment (about $4–6), it cannot avoid the continuous consumption of fresh water and the risk of sewage discharge. In contrast, the cost per ton of water for an integrated membrane-based ZLD system (such as the combined reverse osmosis–membrane distillation technology) is $10–15. However, it can offset more than 30% of the operating costs by recovering high-purity salts (such as sodium chloride and calcium sulfate), and at the same time avoid high-value environmental protection fines. Research shows that in water-scarce areas, the water-saving benefits of ZLD systems can reduce the cost of freshwater procurement by 40%, and that the Levelized Cost of Water (LCOW) is gradually approaching that of traditional technologies.

3.3.2. Comparison of Thermal and Membrane Techniques

In zero-discharge wastewater treatment, reducing the concentration effectively decreases the treatment load on the evaporation and curing sections, thereby lowering overall operating costs. Figure 7 shows a comparison of the process routes between the thermal and membrane methods. No crystalline salts are formed during thermal treatment. Instead, the wastewater is dried and converted into fly ash. By using waste heat, energy efficiency is improved and the wastewater quality is excellent. In addition, this method offers low investment and operating costs. However, it requires multiple pieces of equipment, consumes a lot of energy, takes up a lot of space, and requires high operational precision. In addition, the inorganic salts resulting from the treatment are not recycled.

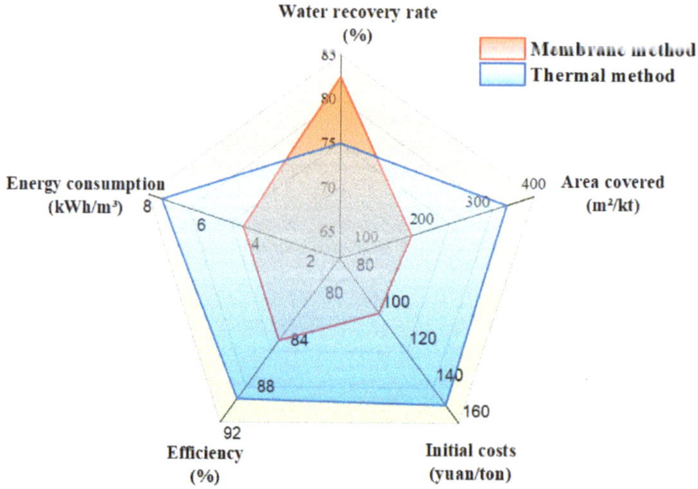

Figure 7. Comparison of the process routes between the thermal and membrane methods.

In contrast, the membrane process requires less equipment, requires lower capital and operating costs, and takes up less space. Waste heat recovery reduces energy consumption to 15–30 kWh/m^3 in thermal systems, but these systems incur higher initial costs (100,000–200,000 yuan/ton) compared with membrane-based forward osmosis–membrane distillation (FO–MD) systems (80,000–120,000 yuan/ton) [58,59]. The footprint of thermal systems is approximately 300–400 m^2 per thousand tons of water, whereas membrane methods occupy only 100–200 m^2 for the same capacity [60–62]. Energy consumption also differs significantly: thermal methods consume 7–12 kWh/m^3, while membrane methods require 3–6 kWh/m^3 [63,64]. Similarly, thermal systems achieve efficiencies of 85–95%, slightly higher than the 75–90% efficiency of membrane systems [65,66]. Membrane technologies demonstrate superior water recovery rates (70–90%) under low-salinity conditions ($Cl^- < 7000$ mg/L) [67–69]. However, the pretreatment requirements of nanofiltration and reverse osmosis membranes are high, which makes the total investment and operating costs of the zero-discharge membrane process relatively high. Furthermore, the potential for recycling inorganic salts is still limited and membrane fouling remains a significant problem. Another challenge is that the economic viability of both thermal and membrane methods needs to be improved.

3.4. Challenges and Future Directions

Technical comparisons and literature analyses indicate that conventional thermal treatment methods, such as evaporation–crystallization, face significant limitations. Thermally driven processes require substantial heat energy, often relying on external sources (e.g., coal or electricity), resulting in poor economic viability. Additionally, high-concentration salts (e.g., $CaSO_4$, $CaCO_3$) in wastewater tend to crystallize and form scale during evaporation, which reduces heat transfer efficiency and increases maintenance costs. Furthermore, the recovered crystalline salts often exhibit low purity due to contamination by heavy metals and impurities, necessitating costly post-treatment modifications for resource utilization. In membrane-based treatment, operational challenges persist. Suspended solids, organic matter, and polyvalent ions (e.g., Ca^{2+}, $SO_4{}^{2-}$) in wastewater cause rapid membrane fouling, demanding frequent cleaning or replacement of membrane modules and elevating operational expenses. Both thermal and membrane methods also suffer from high initial capital investments, underscoring the need for technological optimization to reduce energy consumption and material costs.

Future breakthroughs in zero-discharge technologies for desalination of wastewater will focus on interdisciplinary technologies and integration of materials science. In thermal treatment, research should focus on enhancing energy efficiency and waste heat recovery. For instance, low-energy evaporation–crystallization systems driven by power plant waste heat or renewable energy (e.g., solar thermal) could reduce operational costs [70,71]. Concurrently, scaling inhibitors must be developed to minimize fouling during crystallization while enabling the recovery of high-purity salts (e.g., NaCl, $CaSO_4$) [72,73]. Further improvements in wastewater atomization efficiency and flue gas mixing uniformity are also critical to mitigate adverse impacts on boiler performance [74,75]. These challenges may be addressed through multi-phase flow simulations and optimized nozzle designs. For membrane-based treatment, the development of advanced anti-fouling materials is essential. Novel composite membranes with high selectivity and chemical resistance—such as graphene oxide or metal–organic frameworks (MOFs)—could significantly reduce membrane contamination [76–78]. Integrated multi-stage systems combining high-pressure reverse osmosis (HPRO), electrodialysis (ED), as well as membrane distillation (MD) represent a promising direction for high-salinity wastewater treatment [79–81]. Techno-economic and life cycle analyses of ZLD recycling, reuse, and regeneration methods need to be revised to promote the practical application of ZLD technology [22,82,83]. Additionally, intelligent monitoring systems leveraging machine learning could predict scaling trends in thermal processes or membrane fouling, enabling dynamic adjustments to operational parameters (e.g., pressure, flow rate) and optimizing overall process efficiency [84–86].

4. Conclusions and Prospect

ZLD is a wastewater management solution designed to eliminate liquid waste from desulfurization processes in CFPPs while maximizing water efficiency. This article reviews the main ZLD technologies for desulfurization of wastewater, including concentrated crystallization in thermal processes, gas evaporation and drying, as well as high efficiency reverse osmosis (HERO), membrane distillation (MD), forward osmosis (FO), and electrodialysis (ED) technologies, which are used in power plants. Membrane-based technologies are considered a promising strategy for future wastewater recycling. Thermal processes, on the other hand, require no pretreatment and are easier to operate and maintain. Integrating these technologies into ZLD systems can help achieve complete reuse, recycling, and regeneration of wastewater. However, both thermal and membrane methods suffer from high initial capital investments, underscoring the need for technological optimization to reduce energy consumption and material costs.

Future efforts should include the introduction of numerical simulation methods and additional research to reduce the capital and operating costs of these technologies and make them more cost-effective. It is also important to explore sustainable models that use waste heat from facilities and renewable energy sources to power these treatment processes. In addition, techno-economic and life cycle analyses of ZLD recycling, reuse, and regeneration methods need to be revised to promote the practical application of ZLD technology. Additionally, intelligent monitoring systems leveraging machine learning could predict scaling trends in thermal processes or membrane fouling, enabling dynamic adjustments to operational parameters and optimizing overall process efficiency.

Author Contributions: Writing—original draft preparation, B.L. (Binsheng Liao); writing—review and editing, X.Z.; investigation, Z.L.; project administration, S.Z.; conceptualization, B.L. (Bin Li); resources, X.H. All authors have read and agreed to the published version of the manuscript.

Funding: This research was funded by the "Pioneer" and "Leading Goose" R&D Program of Zhejiang grant number (2023C01124)

Data Availability Statement: The data that support the findings of this study are available from the corresponding author upon reasonable request.

Conflicts of Interest: Author Binsheng Liao was employed by Guoneng Jiujiang Power Generation Co., Ltd. Author Sanmei Zhao and Author Bin Li were employed by Nanjing Yitao Environmental Protection Technology Co., Ltd. The remaining authors declare that the research was conducted in the absence of any commercial or financial relationships that could be construed as a potential conflict of interest. Guoneng Jiujiang Power Generation Co., Ltd. and Nanjing Yitao Environmental Protection Technology Co., Ltd. had no role in the design of the study; in the collection, analyses, or interpretation of data; in the writing of the manuscript, or in the decision to publish the results.

References

1. Sivalingam, S.; Vishal, G.; Anush, B. Chapter 4—Environmental and health effects of acid rain. In *Health and Environmental Effects of Ambient Air Pollution*; Dehghani, M.H., Karri, R.R., Vera, T., Hassan, S.K.M., Eds.; Academic Press: Cambridge, MA, USA, 2024; pp. 91–107.
2. Li, W.; Wu, H.; Tong, H.; Du, Z.; Wang, H.; Zhou, C.; Zhang, Z.; Yang, H. Formation and migration of soluble ions in condensable particulate matter in limestone-gypsum wet flue gas desulfurization system. *Fuel* **2024**, *357*, 129807.
3. Han, Y.; Zhu, Y. SO_3 removal rate and emission test of the limestone-gypsum wet desulfurization system for coal-fired power plants. *J. Phys. Conf. Ser.* **2023**, *2598*, 012014. [CrossRef]
4. Yin, T.; Zhang, Y.; Dong, D.; Wang, T.; Wang, J. Highly efficient capacitive removal of Cd^{2+} over MoS_2-Carbon framework composite material in desulphurisation wastewater from coal-fired power plants. *J. Cleaner Prod.* **2022**, *355*, 131814.
5. Sun, Z.; Zhao, N.; Feng, Y.; Liu, F.; Cai, C.; Che, G.; Zhang, Y.; Wu, H.; Yang, L. Experimental study on the treatment of desulfurization wastewater from coal-fired power plant by spray evaporation. *Environ. Sci. Pollut. Res.* **2022**, *29*, 90791–90802. [CrossRef] [PubMed]
6. Li, Z.; Chen, H.; Zhan, L.; Zheng, S.; Wu, H.; Yang, L. Review of Migration, Transformation, and Control of Volatile Components during Desulfurization Wastewater Evaporation: Advances and Perspectives. *Energy Fuels* **2023**, *37*, 15248–15266.
7. Zhang, B.; Guo, X.; Sun, H.; Mao, Z.; Wang, X.; Wang, X. Zero discharge advanced treatment control system for wet desulfurization wastewater in power plant adapted to water quality fluctuation. *Desalin. Water Treat.* **2025**, *321*, 100985.

8. Cattaneo, C.R.; Muñoz, R.; Korshin, G.V.; Naddeo, V.; Belgiorno, V.; Zarra, T. Biological desulfurization of biogas: A comprehensive review on sulfide microbial metabolism and treatment biotechnologies. *Sci. Total Environ.* **2023**, *893*, 164689. [CrossRef]
9. Pundir, A.; Thakur, M.S.; Radha; Goel, B.; Prakash, S.; Kumari, N.; Sharma, N.; Parameswari, E.; Senapathy, M.; Kumar, S.; et al. Innovations in textile wastewater management: A review of zero liquid discharge technology. *Environ. Sci. Pollut. Res.* **2024**, *31*, 12597–12616.
10. Zhang, H.; Zhao, H.; Feng, B.; Wang, X.; Liu, X.; Dong, Y. Solvent extraction desalination applied to desulphurization wastewater towards zero liquid discharge: Parameters analysis and energy evaluation. *J. Cleaner Prod.* **2024**, *441*, 140961.
11. Buyukada-Kesici, E.; Topuz, E.; Pala, B.; Koseoglu-Imer, D.Y.; Aydiner, C. 10—Implementation of zero liquid discharge policy in industrial water management. In *Resource Recovery in Industrial Waste Waters;* Sillanpää, M., Khadir, A., Gurung, K., Eds.; Elsevier: Amsterdam, The Netherlands, 2023; pp. 199–228.
12. Abdelfattah, I.; El-Shamy, A.M. Review on the escalating imperative of zero liquid discharge (ZLD) technology for sustainable water management and environmental resilience. *J. Environ. Manag.* **2024**, *351*, 119614.
13. Florides, F.; Giannakoudi, M.; Ioannou, G.; Lazaridou, D.; Lamprinidou, E.; Loukoutos, N.; Spyridou, M.; Tosounidis, E.; Xanthopoulou, M.; Katsoyiannis, I.A. Water Reuse: A Comprehensive Review. *Environments* **2024**, *11*, 81. [CrossRef]
14. Yang, J.; Hou, Z.; Meng, F.; Qi, J.; Zhu, Z.; Wang, Y.; Gao, J.; Cui, P. Sustainability Analysis for the Wastewater Treatment Technical Route for Coal-to-Synthetic Natural Gas Industry through Zero Liquid Discharge Versus Standard Liquid Discharge. *ACS Sustain. Chem. Eng.* **2020**, *8*, 8425–8435. [CrossRef]
15. Prado de Nicolás, A.; Molina-García, Á.; García-Bermejo, J.T.; Vera-García, F. Desalination, minimal and zero liquid discharge powered by renewable energy sources: Current status and future perspectives. *Renew. Sustain. Energy Rev.* **2023**, *187*, 113733. [CrossRef]
16. Jensen, F.; Whitfield, L. Leveraging participation in apparel global supply chains through green industrialization strategies: Implications for low-income countries. *Ecol. Econ.* **2022**, *194*, 107331. [CrossRef]
17. Xiong, R.; Wei, C. Current status and technology trends of zero liquid discharge at coal chemical industry in China. *J. Water Process Eng.* **2017**, *19*, 346–351. [CrossRef]
18. Yadav, A.; Labhasetwar, P.K.; Shahi, V.K. Membrane distillation crystallization technology for zero liquid discharge and resource recovery: Opportunities, challenges and futuristic perspectives. *Sci. Total Environ.* **2022**, *806*, 150692. [CrossRef] [PubMed]
19. Moltedo, J.J.; Schwarz, A.; Gonzalez-Vogel, A. Evaluation of percrystallization coupled with electrodialysis for zero liquid discharge in the pulping industry. *J. Environ. Manag.* **2022**, *303*, 114104. [CrossRef]
20. Arola, K.; Van der Bruggen, B.; Mänttäri, M.; Kallioinen, M. Treatment options for nanofiltration and reverse osmosis concentrates from municipal wastewater treatment: A review. *Crit. Rev. Environ. Sci. Technol.* **2019**, *49*, 2049–2116. [CrossRef]
21. Azerrad, S.P.; Isaacs, M.; Dosoretz, C.G. Integrated treatment of reverse osmosis brines coupling electrocoagulation with advanced oxidation processes. *Chem. Eng. J.* **2019**, *356*, 771–780. [CrossRef]
22. Date, M.; Patyal, V.; Jaspal, D.; Malviya, A.; Khare, K. Zero liquid discharge technology for recovery, reuse, and reclamation of wastewater: A critical review. *J. Water Process Eng.* **2022**, *49*, 103129. [CrossRef]
23. Panagopoulos, A. Techno-economic assessment of zero liquid discharge (ZLD) systems for sustainable treatment, minimization and valorization of seawater brine. *J. Environ. Manag.* **2022**, *306*, 114488. [PubMed]
24. Shetty Kodialbail, V.; Sophia, S. Chapter 1—Concept of zero liquid dischare—Present scenario and new opportunities for economically viable solution. In *Concept of Zero Liquid Discharge;* Hussain, C.M., Kodialbail, V.S., Eds.; Elsevier: Amsterdam, The Netherlands, 2023; pp. 3–31.
25. Xu, F.; Zhao, S.; Li, B.; Li, H.; Ling, Z.; Zhang, G.; Liu, M. Current Status of Zero Liquid Discharge Technology for Desulfurization Wastewater. *Water* **2024**, *16*, 900. [CrossRef]
26. Liu, X.; Yang, B.; Li, R. Flue Gas Treatment of Desulphurization Wastewater. *IOP Conf. Ser. Earth Environ. Sci* **2021**, *634*, 012022.
27. Shuangchen, M.; Jin, C.; Gongda, C.; Weijing, Y.; Sijie, Z. Research on desulfurization wastewater evaporation: Present and future perspectives. *Renew. Sustain. Energy Rev.* **2016**, *58*, 1143–1151.
28. Elimelech, M.; Phillip, W.A. The Future of Seawater Desalination: Energy, Technology, and the Environment. *Science* **2011**, *333*, 712–717.
29. Ahmed, S.F.; Mehejabin, F.; Momtahin, A.; Tasannum, N.; Faria, N.T.; Mofijur, M.; Hoang, A.T.; Vo, D.-V.N.; Mahlia, T.M.I. Strategies to improve membrane performance in wastewater treatment. *Chemosphere* **2022**, *306*, 135527. [PubMed]
30. Muhammad, Y.; Lee, W. Zero-liquid discharge (ZLD) technology for resource recovery from wastewater: A review. *Sci. Total Environ.* **2019**, *681*, 551–563.
31. Nazmkhah, A.; Oghyanous, F.A.; Etemadi, H.; Yegani, R. Optimizing dose of coagulant and pH values for membrane fouling control in a submerged membrane bioreactor. *J. Chem. Technol. Biotechnol.* **2022**, *97*, 2794–2804.
32. Subramani, A.; Jacangelo, J.G. Treatment technologies for reverse osmosis concentrate volume minimization: A review. *Sep. Purif. Technol.* **2014**, *122*, 472–489.

33. Valladares Linares, R.; Li, Z.; Sarp, S.; Bucs, S.S.; Amy, G.; Vrouwenvelder, J.S. Forward osmosis niches in seawater desalination and wastewater reuse. *Water Res.* **2014**, *66*, 122–139.
34. Kiss, A.A.; Kattan Readi, O.M. An industrial perspective on membrane distillation processes. *J. Chem. Technol. Biotechnol.* **2018**, *93*, 2047–2055. [CrossRef]
35. Abdel-Karim, A.; Leaper, S.; Skuse, C.; Zaragoza, G.; Gryta, M.; Gorgojo, P. Membrane cleaning and pretreatments in membrane distillation—A review. *Chem. Eng. J.* **2021**, *422*, 129696. [CrossRef]
36. Loganathan, K.; Chelme-Ayala, P.; Gamal El-Din, M. Treatment of basal water using a hybrid electrodialysis reversal–reverse osmosis system combined with a low-temperature crystallizer for near-zero liquid discharge. *Desalination* **2015**, *363*, 92–98. [CrossRef]
37. Liu, X.; Ma, J.; Li, E.; Zhu, J.; Chu, H.; Zhou, X.; Zhang, Y. Multistage membrane-integrated zero liquid discharge system for ultra-efficient resource recovery from steel industrial brine: Pilot-scale investigation and spatial membrane fouling. *J. Membr. Sci.* **2024**, *699*, 122655. [CrossRef]
38. Shaffer, D.L.; Werber, J.R.; Jaramillo, H.; Lin, S.; Elimelech, M. Forward osmosis: Where are we now? *Desalination* **2015**, *356*, 271–284. [CrossRef]
39. Patel, D.; Mudgal, A.; Patel, V.; Patel, J.; Park, K.; Davies, P.; Alegre, R.R. Energy, exergy, economic and environment analysis of standalone forward osmosis (FO) system for domestic wastewater treatment. *Desalination* **2023**, *567*, 116995. [CrossRef]
40. Curcio, E.; Drioli, E. Membrane Distillation and Related Operations—A Review. *Sep. Purif. Rev.* **2005**, *34*, 35–86. [CrossRef]
41. He, C.; Xu, D.; Luo, C.; Liu, Z. Review on wastewater treatment technology in coal-fired power plants towards zero liquid discharge in China. *Desalination* **2025**, *601*, 118520. [CrossRef]
42. Sun, B.; Zhang, M.; Huang, S.; Cao, Z.; Lu, L.; Zhang, X. Study on mass transfer performance and membrane resistance in concentration of high salinity solutions by electrodialysis. *Sep. Purif. Technol.* **2022**, *281*, 119907. [CrossRef]
43. Wang, J.; Liu, M.; Feng, Z.; Liu, J.; Li, X.; Yu, Y. Effects of di-ions structure of the core layer ionomer on anion exchange membrane characteristics of anti-protein fouling and electrodialysis desalination. *Desalination* **2024**, *576*, 117334.
44. Nayar, K.G.; Fernandes, J.; McGovern, R.K.; Al-Anzi, B.S.; Lienhard, J.H. Cost and energy needs of RO-ED-crystallizer systems for zero brine discharge seawater desalination. *Desalination* **2019**, *457*, 115–132.
45. EPA, U.S. *Steam Electric Power Generating Point Source Category: Final Detailed Study Report*; Environmental Protection Agency: Washington, DC, USA, 2009.
46. Ruozheng, L.; Chong, Z.; Wanqiang, Y.; Wenming, M.; Zhilong, J.; Can, W.; Xuan, C.; Haibin, J. Experimental Study of Flue Gas Desulfurization Wastewater Zero Discharge from Coal-fired Power Plant. In Proceedings of the 2016 International Forum on Energy, Environment and Sustainable Development, Shenzhen, China, 16–17 April 2016.
47. Liang, Z.; Zhang, L.; Yang, Z.; Qiang, T.; Pu, G.; Ran, J. Evaporation and crystallization of a droplet of desulfurization wastewater from a coal-fired power plant. *Appl. Therm. Eng.* **2017**, *119*, 52–62.
48. Chen, H.; Zhan, L.; Gu, L.; Feng, Q.; Zhao, N.; Feng, Y.; Wu, H.; Yang, L. Chloride release characteristics of desulfurization wastewater droplet during evaporation process using the single droplet drying method. *Fuel* **2021**, *305*, 121551. [CrossRef]
49. Sun, Z.; Yang, L.; Chen, S.; Bai, L.; Wu, X. Promoting the removal of fine particles and zero discharge of desulfurization wastewater by spray-turbulent agglomeration. *Fuel* **2020**, *270*, 117461.
50. Lee, S.; Kim, Y.; Hong, S. Treatment of industrial wastewater produced by desulfurization process in a coal-fired power plant via FO-MD hybrid process. *Chemosphere* **2018**, *210*, 44–51.
51. Anderson, W.V.; Cheng, C.-M.; Butalia, T.S.; Weavers, L.K. Forward Osmosis–Membrane Distillation Process for Zero Liquid Discharge of Flue Gas Desulfurization Wastewater. *Energy Fuels* **2021**, *35*, 5130–5140.
52. Ma, C.; Li, Q.; Dai, C.; Wang, L.; Zhao, B.; Zhang, Z.; Xue, S.; Tian, W. Pilot-scale study of forward osmosis for treating desulfurization wastewater. *Water Sci. Technol.* **2020**, *82*, 2857–2863. [CrossRef]
53. Conidi, C.; Macedonio, F.; Ali, A.; Cassano, A.; Criscuoli, A.; Argurio, P.; Drioli, E. Treatment of Flue Gas Desulfurization Wastewater by an Integrated Membrane-Based Process for Approaching Zero Liquid Discharge. *Membranes* **2018**, *8*, 117. [CrossRef]
54. Jia, F.; Wang, J. Treatment of flue gas desulfurization wastewater with near-zero liquid discharge by nanofiltration-membrane distillation process. *Sep. Sci. Technol.* **2018**, *53*, 146–153.
55. Li, B.; Yun, Y.; Liu, G.; Li, C.; Li, X.; Hilal, M.; Yang, W.; Wang, M. Direct contact membrane distillation with softening Pre-treatment for effective reclaiming flue gas desulfurization wastewater. *Sep. Purif. Technol.* **2021**, *277*, 119637.
56. Chen, X.; Li, T.; Dou, X.; Meng, L.; Xu, S. Reverse Osmosis Membrane Combined with Ultrasonic Cleaning for Flue Gas Desulfurization Wastewater Treatment. *Water* **2022**, *14*, 875. [CrossRef]
57. Qian, J.; Liu, R.; Wei, L.; Lu, H.; Chen, G.-H. System evaluation and microbial analysis of a sulfur cycle-based wastewater treatment process for Co-treatment of simple wet flue gas desulfurization wastes with freshwater sewage. *Water Res.* **2015**, *80*, 189–199.

58. Panagopoulos, A.; Giannika, V. Techno-economic analysis (TEA) of zero liquid discharge (ZLD) systems for treatment and utilization of brine via resource recovery. *Chem. Eng. Process. Process Intensif.* **2024**, *200*, 109773.
59. Tong, T.; Elimelech, M. The Global Rise of Zero Liquid Discharge for Wastewater Management: Drivers, Technologies, and Future Directions. *Environ. Sci. Technol.* **2016**, *50*, 6846–6855. [PubMed]
60. Bernardo, P.; Drioli, E. Membrane gas separation progresses for process intensification strategy in the petrochemical industry. *Pet. Chem.* **2010**, *50*, 271–282.
61. Al-Sahali, M.; Ettouney, H. Developments in thermal desalination processes: Design, energy, and costing aspects. *Desalination* **2007**, *214*, 227–240.
62. Roy, S.; Ragunath, S. Emerging Membrane Technologies for Water and Energy Sustainability: Future Prospects, Constraints and Challenges. *Energies* **2018**, *11*, 2997. [CrossRef]
63. Suwaileh, W.; Johnson, D.; Hilal, N. Membrane desalination and water re-use for agriculture: State of the art and future outlook. *Desalination* **2020**, *491*, 114559.
64. Ahmed, F.E.; Lalia, B.S.; Hashaikeh, R.; Hilal, N. Alternative heating techniques in membrane distillation: A review. *Desalination* **2020**, *496*, 114713.
65. Castel, C.; Favre, E. Membrane separations and energy efficiency. *J. Membr. Sci.* **2018**, *548*, 345–357.
66. Radha; Sharma, N.; Prakash, S.; Kumari, N.; Sharma, D.; Laller, R.; Pundir, A.; Puri, S. From Challenges to Opportunities: Exploring Minimum Liquid Discharge and Zero Liquid Discharge Strategies for Wastewater Management and Resource Recovery. In *Role of Science and Technology for Sustainable Future: Volume 2—Applied Sciences and Technologies*; Sobti, R.C., Ed.; Springer Nature Singapore: Singapore, 2024; pp. 371–394.
67. Turek, M.; Mitko, K.; Piotrowski, K.; Dydo, P.; Laskowska, E.; Jakóbik-Kolon, A. Prospects for high water recovery membrane desalination. *Desalination* **2017**, *401*, 180–189.
68. Venzke, C.D.; Giacobbo, A.; Ferreira, J.Z.; Bernardes, A.M.; Rodrigues, M.A.S. Increasing water recovery rate of membrane hybrid process on the petrochemical wastewater treatment. *Process Saf. Environ. Prot.* **2018**, *117*, 152–158.
69. Zhao, S.; Yan, S.; Wang, D.K.; Wei, Y.; Qi, H.; Wu, T.; Feron, P.H.M. Simultaneous heat and water recovery from flue gas by membrane condensation: Experimental investigation. *Appl. Therm. Eng.* **2017**, *113*, 843–850.
70. Wu, Z.-W.; Yang, H.-C. Solar energy technologies for desalination and utilization of hypersaline brines. *Sustain. Energy Fuels* **2025**, *9*, 673–692.
71. Lu, H.; Wang, J.; Wang, T.; Wang, N.; Bao, Y.; Hao, H. Crystallization techniques in wastewater treatment: An overview of applications. *Chemosphere* **2017**, *173*, 474–484.
72. Nakhodazadeh, M.R.; Hashemifard, S.A.; Matsuura, T.; Abbasi, M.; Khosravi, A. Challenges and potentials of hybrid Membrane-crystallization processes in sustainable zero liquid discharge process and energy cost estimation. *Sep. Purif. Technol.* **2025**, *354*, 128644.
73. Ting, W.H.T.; Tan, I.A.W.; Salleh, S.F.; Abdul Wahab, N.; Atan, M.F.; Abdul Raman, A.A.; Kong, S.L.; Lam, L.S. Sustainable saline wastewater treatment using eutectic freeze crystallization: Recent advances, challenges and future prospects. *J. Environ. Chem. Eng.* **2024**, *12*, 112919.
74. Ye, X.; An, X.; Zhang, H.; Wang, S.; Guo, B.; Yu, A. Process simulation on atomization and evaporation of desulfurization wastewater and its application. *Powder Technol.* **2021**, *389*, 178–188.
75. Wang, Y.; Zhan, L.; Luo, Q.; Chen, H.; Mao, J.; Wan, J.; Liu, C.; Chen, H.; Zheng, S.; Chen, Z.; et al. Investigation on the rotary atomization evaporation of high-salinity desulfurization wastewater: Performance and products insights. *J. Environ. Manag.* **2024**, *371*, 123044.
76. Jafarian, H.; Dadashi Firouzjaei, M.; Aghapour Aktij, S.; Aghaei, A.; Pilevar Khomami, M.; Elliott, M.; Wujcik, E.K.; Sadrzadeh, M.; Rahimpour, A. Synthesis of heterogeneous metal organic Framework-Graphene oxide nanocomposite membranes for water treatment. *Chem. Eng. J.* **2023**, *455*, 140851.
77. Rao, Z.; Feng, K.; Tang, B.; Wu, P. Surface Decoration of Amino-Functionalized Metal–Organic Framework/Graphene Oxide Composite onto Polydopamine-Coated Membrane Substrate for Highly Efficient Heavy Metal Removal. *ACS Appl. Mater. Interfaces* **2017**, *9*, 2594–2605. [CrossRef] [PubMed]
78. Kamran, U.; Rhee, K.Y.; Lee, S.-Y.; Park, S.-J. Innovative progress in graphene derivative-based composite hybrid membranes for the removal of contaminants in wastewater: A review. *Chemosphere* **2022**, *306*, 135590. [CrossRef] [PubMed]
79. Pasaoglu, M.E.; Kaya, R.; Koyuncu, I. Novel Membrane Technologies in the Treatment and Recovery of Wastewaters. In *Wastewater Management and Technologies*; Debik, E., Bahadir, M., Haarstrick, A., Eds.; Springer Nature: Cham, Switzerland, 2023; pp. 87–106.
80. Rastgar, M.; Moradi, K.; Burroughs, C.; Hemmati, A.; Hoek, E.; Sadrzadeh, M. Harvesting Blue Energy Based on Salinity and Temperature Gradient: Challenges, Solutions, and Opportunities. *Chem. Rev.* **2023**, *123*, 10156–10205. [CrossRef]
81. Osman, A.I.; Chen, Z.; Elgarahy, A.M.; Farghali, M.; Mohamed, I.M.A.; Priya, A.K.; Hawash, H.B.; Yap, P.-S. Membrane Technology for Energy Saving: Principles, Techniques, Applications, Challenges, and Prospects. *Adv. Energy Sustain. Res.* **2024**, *5*, 2400011. [CrossRef]

82. Panagopoulos, A. Assessing the Energy Footprint of Desalination Technologies and Minimal/Zero Liquid Discharge (MLD/ZLD) Systems for Sustainable Water Protection via Renewable Energy Integration. *Energies* **2025**, *18*, 962. [CrossRef]
83. Panagopoulos, A.; Giannika, V. Decarbonized and circular brine management/valorization for water & valuable resource recovery via minimal/zero liquid discharge (MLD/ZLD) strategies. *J. Environ. Manag.* **2022**, *324*, 116239. [PubMed]
84. Krishnan, A.; Sundaram, T.; Nagappan, B.; Devarajan, Y.; Bhumika. Integrating artificial intelligence in nanomembrane systems for advanced water desalination. *Results Eng.* **2024**, *24*, 103321. [CrossRef]
85. Osman, A.I.; Nasr, M.; Farghali, M.; Bakr, S.S.; Eltaweil, A.S.; Rashwan, A.K.; Abd El-Monaem, E.M. Machine learning for membrane design in energy production, gas separation, and water treatment: A review. *Environ. Chem. Lett.* **2024**, *22*, 505–560. [CrossRef]
86. Ignacz, G.; Bader, L.; Beke, A.K.; Ghunaim, Y.; Shastry, T.; Vovusha, H.; Carbone, M.R.; Ghanem, B.; Szekely, G. Machine learning for the advancement of membrane science and technology: A critical review. *J. Membr. Sci.* **2025**, *713*, 123256.

MDPI AG
Grosspeteranlage 5
4052 Basel
Switzerland
Tel.: +41 61 683 77 34

Processes Editorial Office
E-mail: processes@mdpi.com
www.mdpi.com/journal/processes

www.ingramcontent.com/pod-product-compliance
Lightning Source LLC
LaVergne TN
LVHW070502061126
851858LV00011B/279